Wang

SEMICONDUCTORS AND SEMIMETALS

VOLUME 22

Lightwave Communications Technology

Part D

Photodetectors

Semiconductors and Semimetals

A Treatise

Edited by R. K. Willardson
CRYSCON TECHNOLOGIES, INC.
PHOENIX, ARIZONA

Albert C. Beer
BATTELLE COLUMBUS LABORATORIES
COLUMBUS, OHIO

SEMICONDUCTORS AND SEMIMETALS

VOLUME 22

Lightwave Communications Technology

Volume Editor

W. T. TSANG

AT&T BELL LABORATORIES
HOLMDEL, NEW JERSEY

Part D

Photodetectors

1985

ACADEMIC PRESS, INC.
(Harcourt Brace Jovanovich, Publishers)

Orlando San Diego New York London
Toronto Montreal Sydney Tokyo

ACADEMIC PRESS, INC.
Orlando, Florida 32887

United Kingdom Edition published by
ACADEMIC PRESS INC. (LONDON) LTD.
24–28 Oval Road, London NW1 7DX

LIBRARY OF CONGRESS CATALOG CARD NUMBER: 65-26048

ISBN 0–12–752153–4

PRINTED IN THE UNITED STATES OF AMERICA

85 86 87 88 9 8 7 6 5 4 3 2 1

Contents

Chapter 4 **Sensitivity of Avalanche Photodetector Receivers for High-Bit-Rate Long-Wavelength Optical Communication Systems**

S. R. Forrest

Chapter 5 **Phototransistors for Lightwave Communications**

J. C. Campbell

List of Contributors

Numbers in parentheses indicate the pages on which the authors' contributions begin.

J. C. CAMPBELL, *AT&T Bell Laboratories, Crawford Hill Laboratory, Holmdel, New Jersey 07733* (389)

FEDERICO CAPASSO, *AT&T Bell Laboratories, Murray Hill, New Jersey 07974* (1)

S. R. FORREST, *AT&T Bell Laboratories, Murray Hill, New Jersey 07974* (329)

TAKAO KANEDA, *Optical Semiconductor Devices Laboratory, Fujitsu Laboratories Ltd., Atsugi, Kanagawa 243-01, Japan* (247)

T. P. PEARSALL, *AT&T Bell Laboratories, Murray Hill, New Jersey 07974* (173)

M. A. POLLACK, *AT&T Bell Laboratories, Crawford Hill Laboratory, Holmdel, New Jersey 07733* (173)

Treatise Foreword

This treatise continues the format established in the books of Volume 21, in which a subject of outstanding interest and one possessing ever-increasing practical applications is treated in a multivolume work organized by a guest editor of international repute. The present series, which consists of five volumes (designated as Volume 22, Parts A through E) deals with an area that is experiencing a technological revolution and is destined to have a far-reaching impact in the near future—not only in the communications and data-processing fields, but also in numerous ancillary areas involving, for example, control systems, interconnects that maintain individual system isolation, and freedom from noise emanating from stray electromagnetic fields.

That the excitement engendered by the rapid pace of developments in lightwave communications technology is universal is borne out by the large number of contributions to this series by authors from abroad. It is indeed fortunate that W. T. Tsang, who is most highly knowledgeable in this field and has made so many personal contributions, has been able to take the time to put together a work of the extent and excellence of the present series. The treatise editors are also greatly indebted to Dr. Patel and the other colleagues of Dr. Tsang at AT&T Bell Laboratories, without whose understanding and encouragement this group of books would not have been possible.

R. K. Willardson
Albert C. Beer

Foreword

Lightwave technology is breaking down barriers in communications in a manner similar to the way barriers in computing came down thanks to semiconductor integrated circuit technology. Increased packing densities of components on integrated circuit chips made possible a phenomenal amount of information processing capacity at continually decreasing cost. The impact of lightwave technology on communications is quite similar. We are reaching a point where an exponentially increasing transmission capacity is resulting in our capability to provide vast amounts of information to the most distant reaches of the world at a nominal cost. This revolution in information transmission capacity is engendered by the rapid developments in lightwave communications.

Along with the very large transmission capacity predicted in the late fifties when the laser was invented have come a number of additional advantages. Of these advantages, I single out those arising from the nonmetallic nature of the transmission medium. These fall under the broad category of what may be called an immunity from unanticipated electromagnetic coupling. The following rank as very important benefits: freedom from electromagnetic interference, absence of ground loops, relative freedom from eavesdropping (i.e., secure links), and potential for resistance to the electromagnetic pulse problems that plague many conventional information transmission systems utilizing metallic conductors as well as satellite and radio technology. Each of these benefits arises naturally from the medium through which the light is propagated and is, therefore, paced by the progress in optical fibers.

However, what we take for granted today was not so obvious for many decades following the first practicable use of light for communications by Alexander Graham Bell in 1880. The use of heliographs in ancient Greece, Egypt, and elsewhere and the smoke signaling by various American Indian tribes notwithstanding, Bell's experiments on the use of sunlight for transmitting spoken sounds over a distance of a few hundred meters was undoubtedly the first step toward practical optical communications, since it represents a quantum jump in the increase in the bandwidth used for information transmission. The excitement he felt is keenly expressed in his words:

I have heard articulate speech produced by sunlight. I have heard a ray of sun laugh and cough and sing. I have been able to hear a shadow, and I have even perceived by ear the passing of a cloud across the sun's disk.

The results of his experiments were presented at a meeting of the American Association of Scientific Persons in Boston, Massachusetts. But the generally favorable reaction to Bell's photophone in the popular press was tempered with some skepticism. The following paragraph is taken from an article that appeared on the editorial pages of the August 30, 1880, issue of the *New York Times,* which reported on Bell's results.

> What the telephone accomplishes with the help of a wire the photophone accomplishes with the aid of a sunbeam. Professor Bell described his invention with so much clearness that every member of the American Association must have understood it. The ordinary man, however, may find a little difficulty in comprehending how sunbeams are to be used. Does Professor Bell intend to connect Boston and Cambridge, for example, with a line of sunbeams hung on telegraph posts, and, if so, of what diameter are the sunbeams to be, and how is he to obtain them of the required size? . . .

Bell reported optical communication through free atmosphere, but the reporter unintentionally seemed to have foreseen the time when optical-fiber cables would be strung from pole to pole or buried underground.

A unique set of circumstances and a host of advances resulting from extensive interdisciplinary efforts have fueled the revolution in lightwave communications and the acceptance of this new technology. The tremendous progress in lightwave communications is a result of necessity as well as of the response of the scientists and engineers to the formidable challenges. The large bandwidth possible with lightwave communications is a direct result of the very high carrier frequency of electromagnetic radiation in the optical region. This advantage was recognized at least as early as the late fifties and early sixties. Yet almost fifteen years elapsed before lightwave communications technology became economically viable. Two primary components of the communications technology paced this development: the light source and the transmission medium. A third component, the receiver, is also important but was not the pacing one in the early years of development of lightwave systems.

The laser was invented in 1958, and within a very few years laser action was demonstrated in a variety of solids, liquids, and gases. The semiconductor injection laser, the workhorse of contemporary optical communications, was invented in 1962, but its evolution to a practical transmitter in a lightwave system took another eight years. In 1970 Hayashi and Panish (and, independently, Alferov in the Soviet Union) demonstrated the first continuous wave (cw) room-temperature-operated semiconductor laser. The potentials of small size, high reliability, low cost, long life, and ability to modulate the light output of the semiconductor laser at very high rates by merely modulating the drive current were recognized early in the game. With the demonstration of the cw room-temperature operation the race was on to exploit all these advantages.

Again, while laser light propagation through the atmosphere was considered in the mid-sixties, everyone recognized the limitations due to unpredictable and adverse weather conditions. To avoid these limitations, propagation in large hollow pipes was also studied, but again practical difficulties arose. It was the development of optical fiber technology to reduce transmission losses to acceptable levels that has led to the practical implementation of lightwave communications. While light transmission through very small-diameter fibers was demonstrated in the early fifties, it was a combination of theoretical advances by Kao and inventive experimentation by Maurer in the late sixties that resulted in the realization of 20-dB/km fiber. Additional fuel was thus provided to speed up the revolution.

Today, new records are continually being set for the longest and the highest-capacity lightwave communications system. Yet these records are thousands of times below the fundamental bandwidth limits set by the carrier frequency of optical radiation on the rate of information transmission. Furthermore, from very fundamental considerations of light-transmitting materials, there is no reason why the currently achieved lowest losses for optical fibers, in the region of 0.1 dB/km at 1.55 μm, will not be considered too high in the future. It is not inconceivable that fiber losses as low as 10^{-4} dB/km may someday be achieved. It does not take a great deal of imagination to realize the impact of such development.

This is where we are. What future developments will pace the exploitation of lightwave communications? The five-volume minitreatise on lightwave communications technology aims both to recapitulate the existing developments and to highlight new science that will form the underpinnings of the next generation of technology. We know a lot about how to transmit information using optical means, but we know less than enough about how to switch, manipulate, and process information in the optical domain. To take full advantage of all the promise of lightwave communications, we have to be able to push the optical bits through the entire communications system with the electronic-to-optical and optical-to-electronic interfaces only at the two ends of the lightwave communications system. To achieve this, we will need practical and efficient ways of switching, storing, and processing optical information. This is a must before lightwave communications is able to touch every single subscriber of the present telephone and other forms of communications technology.

We have come a long way since Bell's experiments of 1880, but there is a lot more distance ahead. That is what the field of lightwave communications is all about — more challenges, more excitement, more fun for those who are the actors, and a greater opportunity for society to derive maximum benefit from the almost exponentially increasing information capacity of lightwave systems.

AT&T Bell Laboratories C. K. N. PATEL
October 9, 1984

Preface

When American Indians transmitted messages by means of smoke signals they were exploiting concepts at the heart of modern optical communications. The intermittent puffs of smoke they released from a mountaintop were a digital signal; indeed, the signal was binary, since it encoded information in the form of the presence or absence of puffs of smoke. Light was the information carrier; air was the transmission medium; the human eye was the photodetector. The duplication of the signal at a second mountaintop for the transmission to a third served as signal reamplification, as in today's electronic repeater. Man had devised and used optical communications even long before the historic event involving the "photophone" used over a hundred years ago (1880) by Alexander Graham Bell to transmit a telephone signal over a distance of two hundred meters by using a beam of sunlight as the carrier. It was not until 1977, however, that the first commercial optical communications system was installed. Involved in the perfection of this new technology are the invention and development of a reliable and compact near-infrared optical source that can be modulated by the information-bearing signal, a low-loss transmission medium that is capable of guiding the optical energy along it, and a sensitive photodetector that can recover the modulation error free to re-treat the information transmitted.

The invention and experimental demonstration of a laser in 1958 immediately brought about new interest and extensive research in optical communications. However, the prospect of practical optical communications brightened only when three major technologies matured. The first technology involved the demonstration of laser operation by injecting current through a semiconductor device in 1962 and the achievement of continuous operation for over one million hours in 1977. The second technology involved the attainment of a 20-dB/km doped silica fiber in 1970, the realization that pure silica has the lowest optical loss of any likely medium, the discovery in 1973 that suitably heat-treated, boron-doped silica could have a refractive index less than that of pure silica, and the recent achievement of an ultralow loss of 0.157 dB/km with Ge-doped silica-based fibers. The third technology is the development of low-noise photodetectors in the 1970s, which made possible ultrahigh-sensitivity photoreceivers. It is the simultaneous achievement of reliable semiconductor current-injection lasers, low loss in optical fibers, and low-noise photodetectors that thrusts lightwave communications technology into reality and overtakes the conventional transmission systems employing electrical means.

Since optical-fiber communications encompasses simultaneously several other technologies, which include the systems area of telecommunications and glass and semiconductor optoelectronics technologies, a tremendous amount of research has been conducted during the past two decades. We shall attempt to summarize the accumulated knowledge in the present series of volumes of "Semiconductors and Semimetals" subtitled "Lightwave Communications Technology." The series consists of seven volumes. Because of the subject matter, the first five volumes concern semiconductor optoelectronics technology and, therefore, will be covered in "Semiconductors and Semimetals." The last two volumes, one on optical-fiber technology and the other on transmission systems, will be covered in the treatise "Optical Fiber Communications," edited by Tingye Li and W. T. Tsang.

Volume 22, Part A, devoted entirely to semiconductor growth technology, deals in detail with the various epitaxial growth techniques and materials defect characterization of III–V compound semiconductors. These include liquid-phase epitaxy, molecular beam epitaxy, atmospheric-pressure and low-pressure metallo-organic chemical vapor deposition, and halide and chloride transport vapor-phase deposition. Each technique is covered in a separate chapter. A chapter is also devoted to the treatment of materials defects in semiconductors.

In Volume 22, Parts B and C, the preparation, characterization, properties, and applications of semiconductor current-injection lasers and light-emitting diodes covering the spectral range of 0.7 to 1.6 μm and above 2 μm are reviewed. Specifically, Volume 22, Part B, contains chapters on dynamic properties and subpicosecond-pulse mode locking, high-speed current modulation, and spectral properties of semiconductor lasers as well as dynamic single-frequency distributed feedback lasers and cleaved-coupled-cavity semiconductor lasers. Volume 22, Part C, consists of chapters on semiconductor lasers and light-emitting diodes. The chapters on semiconductor lasers consist of a review of laser structures and a comparison of their performances, schemes of transverse mode stabilization, functional reliability of semiconductor lasers as optical transmitters, and semiconductor lasers with wavelengths above 2 μm. The treatment of light-emitting diodes is covered in three separate chapters on light-emitting diode device design, its reliability, and its use as an optical source in lightwave transmission systems. Volume 22, Parts B and C, should be considered as an integral treatment of semiconductor lasers and light-emitting diodes rather than as two separate volumes.

Volume 22, Part D, is devoted exclusively to photodetector technology. It includes detailed treatments of the physics of avalanche photodiodes; avalanche photodiodes based on silicon, germanium, and III–V compound semiconductors; and phototransistors. A separate chapter discusses the sensitivity of avalanche photodetector receivers for high-bit-rate long-wavelength optical communications systems.

Volume 22, Part E, is devoted to the area of integrated optoelectronics and other emerging applications of semiconductor devices. Detailed treatments of the principles and characteristics of integrable active and passive optical devices and the performance of integrated electronic and photonic devices are given. A chapter on the application of semiconductor lasers as optical amplifiers in lightwave transmission systems is also included as an example of the important new applications of semiconductor lasers.

Because of the subject matter (although important to the overall treatment of the entire lightwave communications technology), the last two volumes will appear in a different treatise. The volume on optical-fiber technology contains chapters on the design and fabrication, optical characterization, and nonlinear optics in optical fibers. The final volume is on lightwave transmission systems. This includes chapters on lightwave systems fundamentals, optical transmitter and receiver design theories, and frequency and phase modulation of semiconductor lasers in coherent optical transmission systems.

Thus, the series of seven volumes treats the entire technology in depth. Every author is from an organization that is engaged in the research and development of lightwave communications technology and systems.

As a guest editor, I am indebted to R. K. Willardson and A. C. Beer for having given me this valuable opportunity to put such an important and exploding technology in "Semiconductors and Semimetals." I am also indebted to all the contributors and their employers who have made this series possible. I wish to express my appreciation to AT&T Bell Laboratories for providing the facilities and environment necessary for such an endeavor and to C. K. N. Patel for preparing the Foreword.

CHAPTER 1

Physics of Avalanche Photodiodes

Federico Capasso

AT&T BELL LABORATORIES
MURRAY HILL, NEW JERSEY

ISBN 0-12-752153-4

I. Introduction

In recent years there has been a renewed interest and widespread research effort in avalanche photodiodes (APDs). The driving force behind this work has been the development of new lightwave communication systems exploiting the low-loss and low-dispersion windows at 1.3 and 1.55 μm of silica optical fibers.

For wavelengths below 1.06 μm, the silicon APD represents the ideal detector choice in a fiber-optic communication link. For longer wavelengths, a lot of work has concentrated on the development of suitable low-noise APDs using combinations of binary (e.g., InP, GaSb) and ternary/quaternary (e.g., InGaAs, InGaAsP, AlGaAsSb) III–V semiconductors.

It has long been realized that to attain such high-performance low-noise APDs, the ratio of the ionization coefficients of electrons and holes should be very different from unity. Many measurements of the ionization rates have been reported in recent years; these experiments have shed light on possible material options for long-wavelength APDs.

This work, coupled with a deeper understanding of the energy bands of semiconductors, has also uncovered an important link between ionization coefficients and the band structure of III–V semiconductors. In the impact ionization regime, carriers typically gain energies of the order of the band gap; thus the APD not only is an important device for practical applications, but it also represents an exciting tool for probing the band structure at energies inaccessible by other methods and for studying the physics of very hot carriers.

An understanding of the physics APDs is also important for applications, since it may guide in the choice of materials for APDs. An excellent example of this is the discovery of an important band-structure-related phenomenon, the resonant enhancement of impact ionization in certain III–V and II–VI alloys.

These investigations have also indicated that most III–V materials, with one or two exceptions, have comparable ionization rates for electrons and holes, so that they are unsuitable for APDs with noise performance comparable to that of silicon. This has led to the exploration of ways of artificially altering the α/β ratio by using novel structures such as superlattices, graded-gap materials, and periodic doping profiles. These new materials allow one to modify in an almost arbitrary fashion the conventional energy-band diagram of a p–n-junction detector and thus to tailor the α/β ratio. [For a review of band-gap engineering, see Capasso (1983a,b)].

From this band-gap engineering approach a powerful new detector concept, the staircase APD, has emerged. This device represents the solid-state analog of a photomultiplier and has the potential of achieving essentially noise-free avalanche multiplication. This could lead to unprecedented receiver sensitivity improvements in fiber-optic communication systems. This chapter represents a review of these important developments.

In Part II, the microscopic foundations of the physics of APDs are discussed. The important role of ionization threshold energies is analyzed, along with developments in the theory of impact ionization. In Part III, the macroscopic description of impact ionization, relating ionization rates to the measured avalanche gains, is reviewed. Advances in measurement techniques are discussed, and a complete list of ionization-rate data for Si, Ge, and III–V materials is given. New band-structure phenomena are also discussed.

Finally, Part IV is devoted to work on new APD structures with enhanced ionization-rate ratios. A detailed analysis of the staircase solid-state photomultiplier is also presented.

II. Theory of Impact Ionization

1. Ionization Threshold Energies

In the impact-ionization effect in semiconductors, free carriers (electrons and holes) are accelerated by a high electric field until they gain sufficient energy to promote an electron from the valence into the conduction band. The electric fields required to observe impact ionization depend on the band gap of the material and may range at room temperature from $\simeq 10^4$ V cm^{-1} in low-gap semiconductors, such as InAs ($E_g = 0.33$ eV), to values well in excess of 10^5 V cm^{-1} in wide-gap materials, such as GaP ($E_g = 2.24$ eV).

The minimum energy needed for impact ionization is the ionization threshold energy E_i. Threshold energies are important because they strongly influence the ionization rates (or coefficients) for electrons and holes. These quantities, denoted respectively by α and β, are defined as the reciprocal of the average distance, measured along the direction of the electric field, traveled by an electron or a hole to create an electron–hole pair. Many consecutive impact ionization events produce avalanche multiplication. The values of α and β, along with the length of the field region and the carrier-injection conditions, determine the magnitude of the avalanche gain.

Impact ionization is a three-body collision process, and the final carriers are left with finite kinetic energy and momentum. This means that in general the energy of the impact-ionizing carrier is greater than the band gap.

The threshold energies depend on the band structure of the semiconductor. For parabolic energy bands with equal effective masses for electrons and holes, it is a simple matter to calculate E_i from simple energy- and momentum-conservation considerations (Wolff, 1954). However, extensive experimental and theoretical work in the past two decades has shown that in general energy bands have a complex structure in k space, characterized by nonparabolicities, satellite valleys, and strong anisotropies. Nonlocal pseudopotential calculations can accurately determine the band structure of semiconductors (Chelikowsky and Cohen, 1976). Thus, a general accurate algorithm for calculating the threshold energies is needed.

The conditions for an impact-ionization process to occur at threshold were first established by Keldysh (1960) and are based on minimization of the total energy of the final carriers. Anderson and Crowell (1972), not aware of this work, rederived these conditions and developed them into an algorithm applicable to a general type of band structure. This method has since been extensively employed to calculate threshold energies of many semiconductors, especially III–V semiconductors, using realistic band structures (Anderson and Crowell, 1972; Pearsall *et al.*, 1977a; Pearsall, 1979). An alternative and probably more accurate procedure for calculating thresholds from realistic band structures has been developed by Balliger *et al.* (1973).

In this review, we shall discuss only the first method, which has the advantage of being more physically intuitive, and we shall illustrate applications to the band structure of several III–V semiconductors. [For a comprehensive review of the mathematical aspects of the theory of impact-ionization thresholds, including the relationship with the reverse process (Auger effect), the reader is referred to Robbins (1980a–c).]

Consider a typical impact-ionization event as shown in Fig. 1. An initial electron in conduction band i impact-ionizes by promoting electron 3 from

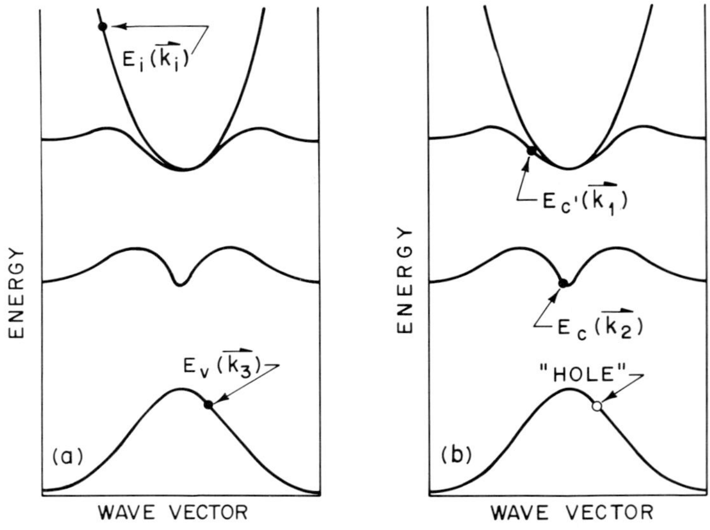

FIG. 1. Hypothetical impact-ionization process: (a) primary particles before ionization; (b) resultant particles immediately after ionization. [From Anderson and Crowell (1972).]

the valence band into conduction band c′. In many practical cases, c and c′ are the same band. We shall consider this case only.

The total energy E_f and the total momentum of the final carriers $\mathbf{K}_f$ are given by

$$E_f = E_c(\mathbf{k}_1) + E_c(\mathbf{k}_2) - E_v(\mathbf{k}_3) \tag{1}$$

and

$$\mathbf{K}_f = \mathbf{k}_1 + \mathbf{k}_2 - \mathbf{k}_3, \tag{2}$$

where $E_\zeta(\mathbf{k})$ is the energy in band ζ of a charge carrier with vector $\mathbf{k}$.

To find the ionization threshold energy one must

1. minimize, for a given momentum $\mathbf{K}_f$, the total energy E_f, with respect to small changes of wave vector of any of the final carriers; and
2. conserve energy and momentum.

Step 1 is expressible mathematically by the equations

$$d\mathbf{K}_f = d\mathbf{k}_1 + d\mathbf{k}_2 - d\mathbf{k}_3 = 0 \tag{3}$$

and

$$dE_f = d\mathbf{k}_1 \cdot \nabla_{\mathbf{k}} E_v(\mathbf{k}_1) + d\mathbf{k}_2 \cdot \nabla_{\mathbf{k}} E_c(\mathbf{k}_2) - d\mathbf{k}_3 \cdot \nabla_{\mathbf{k}} E_c(\mathbf{k}_3) = 0. \tag{4}$$

Substituting Eq. (3) into Eq. (4) and recognizing that $\mathbf{v}_g = \nabla_{\mathbf{k}} E(\mathbf{k})/\hbar$, where $\mathbf{v}_g$ is the group velocity of a carrier of momentum $\mathbf{k}$ and energy E, one obtains

$$d\mathbf{k}_2 \cdot (\mathbf{v}_1 - \mathbf{v}_3) + d\mathbf{k}_3 \cdot (\mathbf{v}_2 - \mathbf{v}_3) = 0. \tag{5}$$

Since $d\mathbf{k}_2$ and $d\mathbf{k}_3$ are linearly independent when $\mathbf{K}_f$ is constant, Eq. (5) is satisfied if and only if

$$\mathbf{v}_1 = \mathbf{v}_2 = \mathbf{v}_3; \tag{6}$$

i.e., a necessary condition for an impact-ionizing carrier to have minimum energy consistent with pair production is that the final carriers have equal group velocities (Keldysh, 1960; Anderson and Crowell, 1972).

Step 2 (energy and momentum conservation) requires that the minimum total energy of the final carriers $E_{fm}(\mathbf{K}_f)$ be equal to the energy $E_i(\mathbf{K}_f)$ of a primary electron with wave vector $\mathbf{K}_f$ in some band i. There is no guarantee that this occurs at any $\mathbf{K}_f$; in other terms, Eq. (6) is not a sufficient condition for ionization.

To better understand this point let us consider two hypothetical sets of curves $E_i(\mathbf{K})$ and $E_{fm}(\mathbf{K})$ (Fig. 2). For $\mathbf{K} < \mathbf{K}_2$, $E_i(\mathbf{K}) < E_{fm}(\mathbf{K})$, so that the carrier has not enough energy to ionize. At wave vector $\mathbf{K}_2$, $E_i(\mathbf{K}_2) = E_{fm}(\mathbf{K}_2)$, so that a threshold exists at $\mathbf{K}_2$. For $\mathbf{K} > \mathbf{K}_2$, $E_i(\mathbf{K}) > E_{fm}(\mathbf{K})$, so that the initiating carrier has more energy than required for ionization.

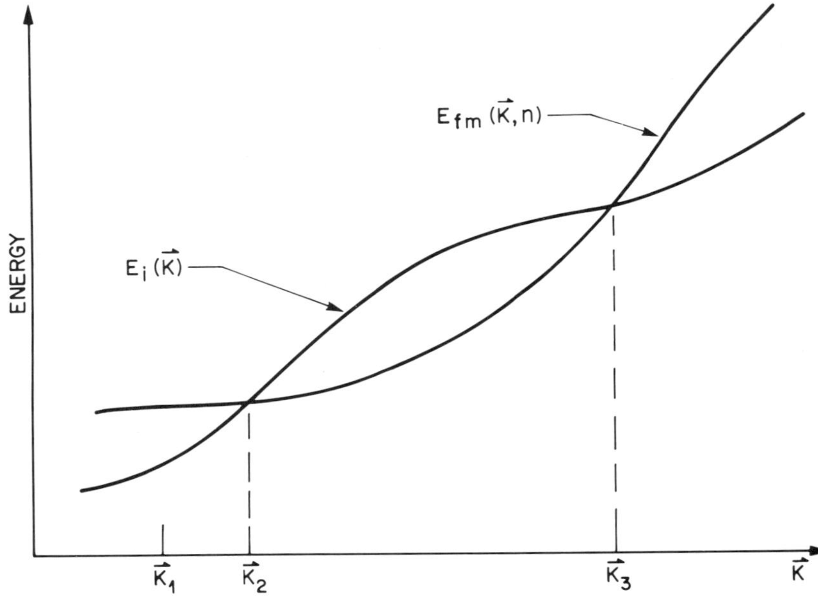

FIG. 2. Hypothetical set of curves of E_i and E_{fm} as a function of **K** for a process that exhibits a threshold at $\mathbf{K}_2$ and an antithreshold at $\mathbf{K}_3$. [From Anderson and Crowell (1972).]

Beyond $\mathbf{K}_3$, $E_i < E_{fm}$, making impact-ionization impossible. Note that the behavior of E_i and E_{fm} in the neighborhood of $\mathbf{K}_3$ differs that of E_i and E_{fm} near $\mathbf{K}_2$. For this reason, $E_i(\mathbf{K}_3)$ is called an antithreshold. Thus, $\mathbf{K}_1$ and $\mathbf{K}_3$ define a "pair-production window." This interesting effect may occur for the process in which a light hole gives two heavy holes in a parabolic band and an electron in a parabolic conduction band (Camphausen and Hearn, 1972). The finite pair-production window may reduce the cross section for the impact-ionization process if the threshold and antithreshold are very close in energy.

Another important point is the possible existence of several thresholds. For example, for the situation depicted in Fig. 2, $E_i(\mathbf{K})$ may again cross over $E_{fm}(\mathbf{k})$ at some $\mathbf{k} > \mathbf{k}_3$; this would be a second threshold. A multiplicity of thresholds has been found by Anderson and Crowell (1972) in materials such as GaAs, Ge, Si, GaP, and InSb. This indicates that the actual effective ionization threshold can be field dependent. In simple terms, the high-energy tail of the distribution of carriers determines impact ionization; as the field is increased, the tail moves to higher energy and can thus reach several thresholds.

It is also worth mentioning that impact-ionization processes at threshold may be indirect, involving either a reciprocal lattice vector (umklapp process) or the absorption of a phonon. In the latter case, the ionization energy may be substantially reduced and become comparable to the band gap, since the phonon wave vector facilitates momentum conservation (Shockley, 1961). However, the criterion that governs the final states [similar to Eq. (6)] in this case becomes extremely restrictive (Anderson and Crowell, 1972).

In the following we shall be exclusively interested in direct processes and consider only the lowest threshold, henceforth referred to as ionization threshold energy E_{i}. To better illustrate threshold energies, it is useful to consider first several simple models of energy bands.

a. Two Parabolic Bands

We consider one conduction band with effective mass m_{e} and one valence band effective mass m_{h}. Using the threshold condition Eq. (6), one finds the following relationship among the wave vectors of the final particles:

$$\mathbf{k}_1 = \mathbf{k}_2 = \mathbf{k}_3(m_{\mathrm{e}}/m_{\mathrm{h}}). \tag{7}$$

The total final wave vector for an electron-initiated ionization is

$$\mathbf{k}_f = 2\mathbf{k}_1 + \mathbf{k}_3 = \mathbf{k}_1(2 + m_{\mathrm{h}}/m_{\mathrm{e}}), \tag{8}$$

and the minimum total energy of the final particles is

$$E_{f\mathrm{m}} = (\hbar^2 k_1^2/2m_{\mathrm{e}})(2 + m_{\mathrm{h}}/m_{\mathrm{e}}) + E_{\mathrm{g}} = E_{\mathrm{i,e}}. \tag{9}$$

To conserve energy and momentum, the conduction band must supply an initiating electron with wave vector $\mathbf{k}_i$ and energy $E_{\mathrm{c}}(\mathbf{k}_i)$ such that

$$E_{\mathrm{c}}(\mathbf{k}_{\mathrm{i}}) = (\hbar^2 k_1^2/2m_{\mathrm{e}})(2 + m_{\mathrm{h}}/m_{\mathrm{e}})^2 = E_{\mathrm{i,e}}. \tag{10}$$

Eliminating k_1 from Eqs. (9) and (10) leads to the ionization threshold energy for electrons

$$E_{\mathrm{i,e}} = E_{\mathrm{g}}(1 + m_{\mathrm{e}}/m_{\mathrm{h}}). \tag{11}$$

For the hole-ionization threshold, one has similarly

$$E_{\mathrm{i,h}} = E_{\mathrm{g}}(1 + m_{\mathrm{h}}/m_{\mathrm{e}}). \tag{12}$$

Note that when $m_{\mathrm{h}} = m_{\mathrm{e}}$, then

$$E_{\mathrm{i,e}} = E_{\mathrm{i,h}} = \tfrac{3}{2}E_{\mathrm{g}}, \tag{13}$$

which is the well-known 3/2-band-gap rule (Wolff, 1954). Note that this model assigns the lowest threshold to the carrier with the smallest effective mass and that one always has

$$E_{\mathrm{i,e}} + E_{\mathrm{i,h}} = 3E_{\mathrm{g}}, \tag{14}$$

independent of the effective mass ratio.

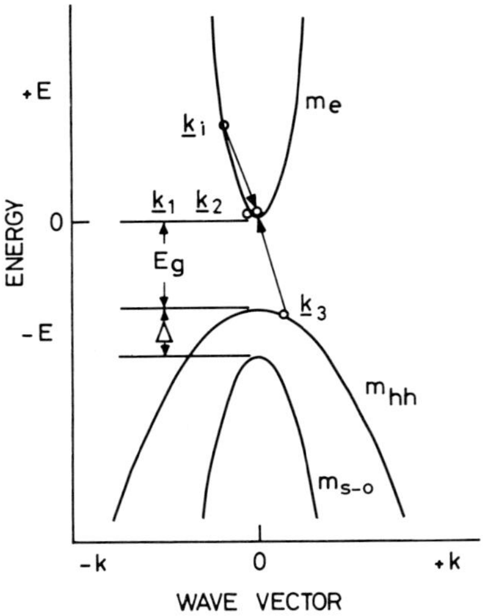

FIG. 3. Hypothetical band-structure diagram showing an electron-initiated impact-ionization process at threshold: initial state has energy E_i and wave vector $\mathbf{k}_i$; final states are labeled 1, 2, and 3. [From Pearsall *et al.* (1977a).]

b. Three Parabolic Bands

We can make the model more realistic by considering two valance bands, one for heavy holes with mass m_{hh} and the other separated toward lower energy by an amount Δ, with mass m_{s-o} due to the spin–orbit coupling (as shown in Fig. 3). This band structure exhibits some of the relevant features of a zinc blende semiconductor such as GaAs. The expressions for the threshold energies E_i, the wave vector of the initiating carrier at threshold k_i and the wave vectors of the final carriers then become (Pearsall *et al.*, 1977a, Pearsall, 1979),

$$E_{i,e} = E_g\left(1 + \frac{m_e}{m_{hh} + m_e}\right), \tag{15}$$

$$k_{i,e} = \left[E_g\left(\frac{2m_e}{\hbar^2}\right)\left(1 + \frac{m_e}{m_{hh} + m_e}\right)\right]^{1/2}, \tag{16}$$

$$k_1 = k_2 = k_{i,e}\left(\frac{m_e}{2m_{hh} + m_e}\right), \tag{17}$$

$$k_3 = k_{i,e}\left(\frac{m_{hh}}{2m_{hh} + m_e}\right), \tag{18}$$

for electron-initiated ionization, and

$$E_{i,h} = E_g\left[1 + \frac{m_{s-o}(1 - \Delta/E_g)}{(2m_{hh} - m_{s-o} + m_e)}\right], \tag{19}$$

$$k_{\mathrm{i,h}} = \left[(E_{\mathrm{g}} - \Delta) \left(\frac{2m_{\mathrm{s-o}}}{\hbar^2} \right) \left(\frac{2m_{\mathrm{hh}} + m_{\mathrm{e}}}{2m_{\mathrm{hh}} - m_{\mathrm{s-o}} + m_{\mathrm{e}}} \right) \right]^{1/2}, \tag{20}$$

$$k_1 = k_2 = k_{\mathrm{i,h}} \left(\frac{m_{\mathrm{hh}}}{2m_{\mathrm{hh}} + m_{\mathrm{e}}} \right), \tag{21}$$

$$k_3 = k_{\mathrm{i,h}} \left(\frac{m_{\mathrm{e}}}{2m_{\mathrm{hh}} + m_{\mathrm{e}}} \right), \tag{22}$$

for hole-initiated ionization. Figure 3 illustrates the electron-initiated impact-ionization processes at threshold. Equations (15)–(22) give a good qualitative idea of the band-structure dependence of the ionization thresholds. They represent a reasonable approximation for materials in which bands do not depart substantially from parabolas, for energies up to ionization thresholds (e.g., GaSb, $\mathrm{Al}_x\mathrm{Ga}_{1-x}\mathrm{Sb}$).

Equations (15)–(22) show that spin–orbit splitting can have a pronounced effect on the hole threshold by reducing it, leaving the electron threshold unaltered.

For a material in which $E_{\mathrm{g}} = \Delta$, the hole-ionization energy equals the band gap, and $k_{\mathrm{i}} = k_1 = k_2 = k_3 = 0$. In other words, the zero momentum transferred in this vertical transition reduces the threshold energy. This effect is responsible for the strong resonant enhancement of the hole-ionization rate experimentally observed in the $\mathrm{Al}_x\mathrm{Ga}_{1-x}\mathrm{Sb}$ alloys near $x = 0.065$ (Hildebrand *et al.*, 1980) and is discussed in Part III.

c. Nonparabolic Conduction Band and One Parabolic Valence Band

The conduction-band central valley in several III–V materials is well represented by a band structure of the type

$$\hbar^2 k^2 / 2m_{\mathrm{e}} = E(1 + \alpha E), \tag{23}$$

where the nonparabolicity α is a measure of the deviation of the band $E(k)$ from the parabolic shape. In this case, the threshold energy for electrons becomes (Ridley, 1977)

$$E_{\mathrm{i,e}} = E_{\mathrm{g}} \left(\frac{1 + 2\mu}{1 + \mu} \right) \left[1 + \frac{\mu(1 + 2\mu)}{(1 + \mu)^2} \alpha E_{\mathrm{g}} \right], \tag{24}$$

where $\mu = m_{\mathrm{e}}/m_{\mathrm{h}}$, and m_{e} and m_{h} are the electron and hole band-edge masses, respectively.

d. Realistic Band Structures of III–V *Materials*

For an accurate and meaningful calculation of the threshold energies, it is important to use the real band structure, especially for materials such as GaAs and InP with large anisotropies and nonparabolicities. To this end,

Anderson and Crowell (1972) have developed a simple graphical technique based on Eq. (6) for the case of initiating carriers traveling along the principal crystal axes ⟨111⟩, ⟨110⟩, and ⟨100⟩ of several semiconductors. They used the pseudopotential band structure of Cohen and Bergstresser (1966) and presented results for Si, Ge, GaAs, GaP, and InSb. Threshold energy calculations for GaAs, GaSb, and InP, based on the same method but on a more accurate nonlocal pseudopotential band structure (Chelikowsky and Cohen, 1976), have also been reported (Pearsall *et al.*, 1977a, 1978; Pearsall, 1979). We shall illustrate here these latest results in view of their importance in the interpretation of several experiments.

Figure 4 represents the band structure of GaAs as given by Chelikowsky and Cohen (1976); the bands are computed along the symmetry directions Λ, Δ, Σ (corresponding respectively to the crystal axes ⟨111⟩, ⟨100⟩, and ⟨110⟩) and the line $U(k)$. Most of the structure in the optical data appears to

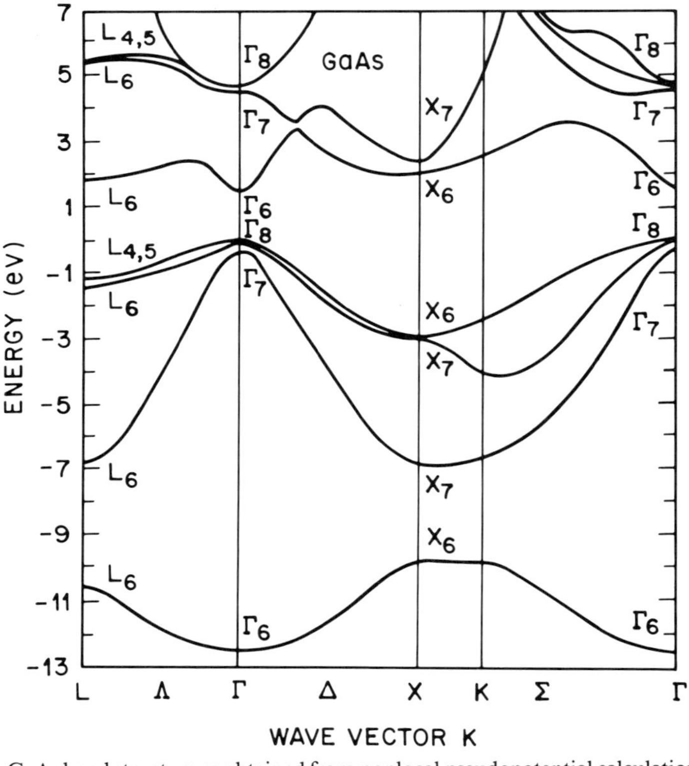

FIG. 4. GaAs band structure as obtained from nonlocal pseudopotential calculations. [From Chelikowsky and Cohen (1976).]

rise from states near these symmetry lines. Strong deviations from a parabolic or a simply isotropic band model are evident for conduction-band energies starting well below the value of the band gap.

A simple inspection of Fig. 4 shows that along the Λ direction, the width of the conduction band is smaller than the band gap. Thus, an electron traveling along one of the corresponding crystal axes ($\langle 111 \rangle$) cannot gain sufficient energy for impact ionization. The next conduction band lies more than 2 eV higher in energy so that there is no ionization threshold state for an electron moving in the $\langle 111 \rangle$ direction in an electric field of the magnitude typically reached in GaAs $p-n$ junctions ($\leq 10^6$ V cm^{-1}).

The electron threshold states show a considerable variation with orientation. For electrons moving along an electric field oriented in the $\langle 110 \rangle$ direction, the ionization threshold state occurs near the top of the conduction band at $k/k_{max} = 0.325$ with $E_i = 2.01$ eV (Fig. 5a). On the other hand, the threshold in the $\langle 100 \rangle$ direction does not occur in the principal conduction band but is separated from it by a "pseudogap" of ≈ 0.2 eV (Fig. 5b). For sufficiently high fields, electrons can be expected to tunnel across this gap to threshold. In fact, a simple calculation based on the uncertainty principle shows that the uncertainty in the electron energy in k space near the pseudogap at an electric field of 5×10^5 V cm^{-1} is greater than the width of the pseudogap (Pearsall *et al.*, 1978).

Note that since along the $\langle 110 \rangle$ direction the threshold state lies near the very top of the first conduction band, the cross section for this ionization process may be very small. In addition, more recent calculations by Crowell (1983, private communication) suggest that this threshold is actually an antithreshold. Thus, in GaAs, electrons most likely must transfer to the second conduction band in order to impact ionize, regardless of the electric field orientation.

Considering holes instead, one can see that the conditions for hole-initiated impact ionization are rather similar for all three orientations, a reflection of the nearly isotropic nature of the split-off band in GaAs (Fig. 6). Second, although a hot hole can always reach threshold in the direction of the electric field, holes must scatter at least once to enter the split-off band. This scattering tends to randomize the distribution of hot carriers. The combination of the scattering and the similar threshold conditions leads one to predict that orientation effects in the hole ionization rate should be small. Table I gives a list of the threshold states for electrons and holes in $\langle 111 \rangle$-, $\langle 110 \rangle$- and $\langle 100 \rangle$-oriented GaAs as calculated by Pearsall *et al.* (1978).

Next we discuss InP, a III–V material with a room-temperature band gap $E_g = 1.35$ eV. The results of these calculations are given in Table II and in Figs. 7 and 8. Again, as for GaAs, an electron moving parallel to an electric field oriented in the $\langle 111 \rangle$ direction cannot ionize because the width of the

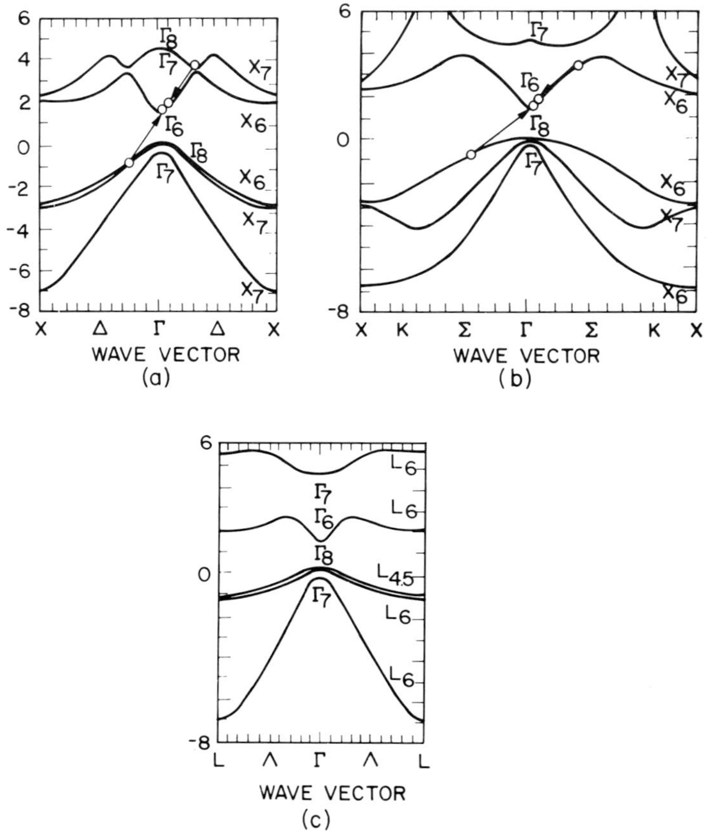

FIG. 5. Electron-initiated impact-ionization processes in GaAs at threshold along three symmetry axes: (a) ⟨100⟩, (b) ⟨110⟩, and (c) ⟨111⟩. [From Pearsall *et al.* (1978).]

conduction band in this direction is relatively narrow (≈0.8 eV). In the other two major symmetry directions, ⟨110⟩ and ⟨100⟩, the threshold state exists in the main conduction band. It is principally the smaller energy band gap of InP, compared to that of GaAs, which permits impact ionization at room temperature in both the ⟨100⟩ and the ⟨110⟩ directions (Pearsall, 1979).

Impact-ionization events at threshold for electron and holes moving along the ⟨100⟩ crystal orientation of GaSb are illustrated in Fig. 9. The results of these calculations are summarized in Table III (Pearsall *et al.*, 1977a).

Gallium antimonide approximates fairly well the simple three parabolic-band model. In fact, using Eqs. (15) and (19), one calculates $E_{ie} = 0.9$ eV for

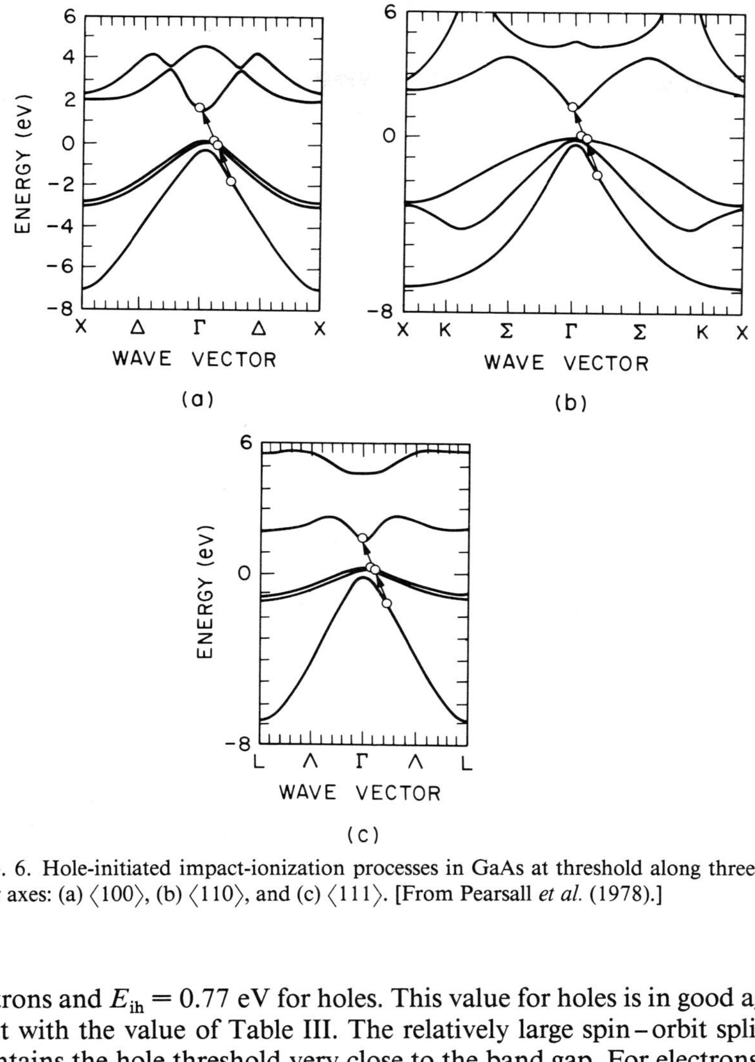

FIG. 6. Hole-initiated impact-ionization processes in GaAs at threshold along three symmetry axes: (a) ⟨100⟩, (b) ⟨110⟩, and (c) ⟨111⟩. [From Pearsall *et al.* (1978).]

electrons and $E_{ih} = 0.77$ eV for holes. This value for holes is in good agreement with the value of Table III. The relatively large spin–orbit splitting maintains the hole threshold very close to the band gap. For electrons, the simple calculation does not give as good agreement because the ionization event is more dependent on the real conduction band of GaSb. Note instead that in the case of GaAs and InP, the simple parabolic model fails completely. It predicts a threshold energy for electrons lower than that of holes (since $m_h > m_e$), whereas in these two compounds it is the hole that has the lower ionization threshold. This is a result of the relatively large central valley conduction-band nonparabolicities.

TABLE I

THRESHOLD ENERGIES FOR IMPACT IONIZATION BY ELECTRONS AND HOLES IN GaAs AT 300 K

Crystal direction	Threshold energy[a] (eV)	Wave vector[b] Relative units (k/k_{max}) k_i	k_1	k_2	k_3	Absolute units ($2\pi/a_0$) k_i	k_1	k_2	k_3
⟨100⟩									
Electrons	2.05	0.3	0.04	0.04	−0.22	0.300	0.04	0.04	−0.22
Holes	1.81	0.240	0.11	0.11	−0.02	0.240	0.11	0.01	−0.02
⟨110⟩									
Electrons	2.01	0.325	0.02	0.02	−0.285	0.459	0.028	0.028	−0.403
Holes	1.58	0.146	0.07	0.07	−0.006	0.206	0.099	0.099	−0.0084
⟨111⟩									
Electrons	—	—	—	—	—	—	—	—	—
Holes	1.58	0.215	0.1	0.1	−0.015	0.186	0.086	0.086	−0.013

[a] Threshold energies for electrons are measured from the bottom of the conduction band, whereas threshold energies for holes are measured from the top of the valance band.

[b] Wave vectors are given both as fractions of the zone-boundary wave vector and in absolute units of $2\pi/a_0$ ($a_0 = 5.65$ Å).

TABLE II

THRESHOLD ENERGIES AND PARTICLE-STATE WAVE VECTORS FOR IMPACT IONIZATION BY ELECTRONS AND HOLES IN InP AT 300 K

Crystal direction	Threshold energy[a] (eV)	Wave vector[b] Relative units (k/k_{max}) k_i	k_1	k_2	k_3	Absolute units ($2\pi/a_0$) k_i	k_1	k_2	k_3
⟨100⟩									
Electrons	1.84	0.300	0.030	0.030	−0.240	0.300	0.030	0.030	−0.240
Holes	1.65	0.284	0.135	0.135	−0.014	0.284	0.135	0.135	−0.014
⟨110⟩									
Electrons	1.47	0.190	0.0075	0.0075	−0.175	0.269	0.011	0.011	−0.247
Holes	1.38	0.164	0.081	0.081	−0.002	0.233	0.115	0.015	−0.003
⟨111⟩									
Electrons	—	—	—	—	—	—	—	—	—
Holes	1.42	0.258	0.125	0.125	−0.008	0.223	0.108	0.108	−0.007

[a] Threshold energies for electrons are measured from the bottom of the conduction band, whereas threshold energies for holes are measured from the top of the valence band.

[b] Particle wave vectors are given both as functions of the zone-boundary wave vector and in absolute units of $2\pi/a_0$ ($a_0 = 5.86$ Å).

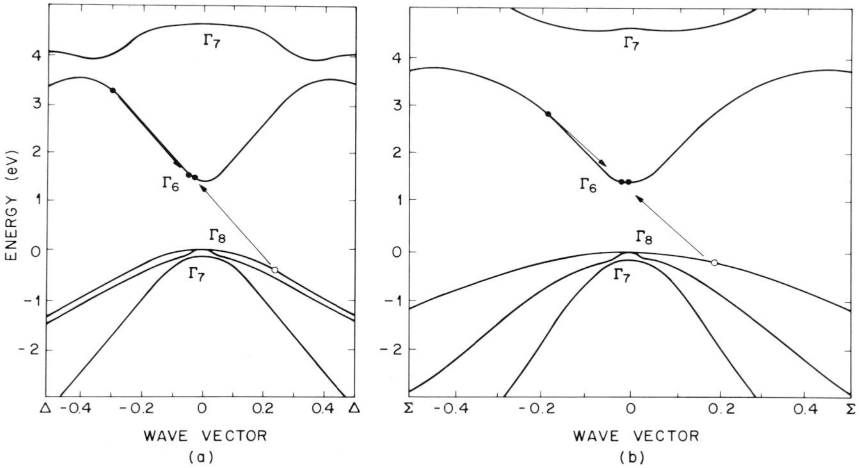

FIG. 7. Electron-initiated impact-ionization processes at threshold in InP along the (a) ⟨100⟩ and (b) ⟨110⟩ symmetry axes; no threshold state exists along the ⟨111⟩ axis. [From Pearsall (1979).]

2. Impact-Ionization Cross Section and Tunnel-Impact Ionization

For the impact-ionization effect to occur, the carrier energy must exceed the ionization threshold. Right at E_i, the ionization cross section is zero. Because there are three final state particles, density-of-states considerations lead one to expect that the cross section will be a strongly rising function of electron energy above threshold.

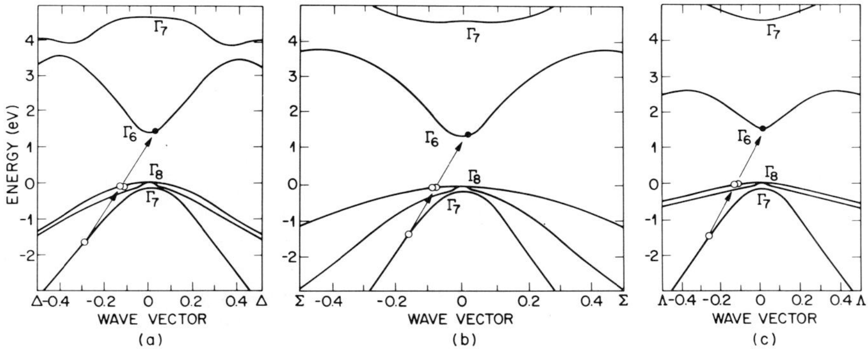

FIG. 8. Hole-initiated impact-ionization processes at threshold in InP along the (a) ⟨100⟩, (b) ⟨110⟩, and (c) ⟨111⟩ directions. [From Pearsall (1979).]

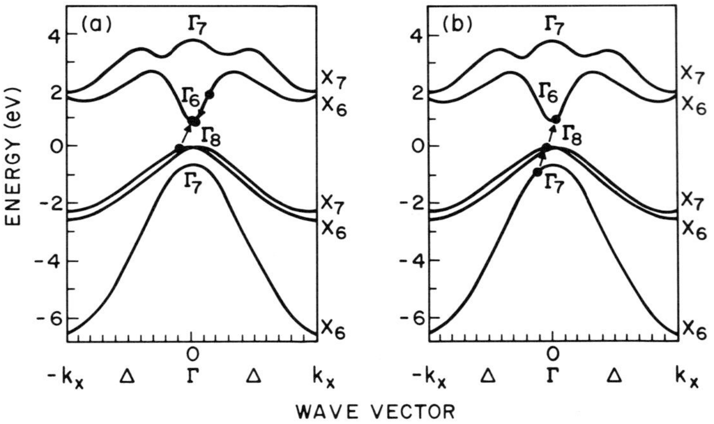

FIG. 9. Impact-ionization transitions at threshold in GaSb in the ⟨100⟩ direction: (a) electron initiated; (b) hole initiated. [From Pearsall *et al.* (1977a).]

The probability per unit time of electron-initiated impact ionization, which is proportional to the cross section, may be expressed in the Born approximation as (Keldysh, 1960)

$$\frac{1}{\tau_i(E)} = \frac{2\pi}{\hbar} V^3 \int |M(\mathbf{p}_1\mathbf{p}_1\mathbf{p}_2)|^2\, \delta[E_e(\mathbf{p}) - E_e(\mathbf{p}_1) - E_e(\mathbf{p}_2) + E_h(\mathbf{p}_3)] \times \delta\left[\frac{(\mathbf{p} - \mathbf{p}_1 - \mathbf{p}_2 + \mathbf{p}_3)}{\hbar}\right] \frac{d^3\mathbf{p}_1\, d^3\mathbf{p}_2\, d^3\mathbf{p}_3}{(2\pi\hbar)^9}, \tag{25}$$

TABLE III

THRESHOLD ENERGIES[a] AND WAVE VECTORS[b] FOR IMPACT IONIZATION IN GaSb IN THE ⟨100⟩ DIRECTION AT 300 K

	E_i (eV)	k_i	k_1	k_2	k_3
Electrons	0.80	0.12	0.015	0.015	−0.09
Holes	0.79	0.08	0.035	0.035	−0.01

[a] The ionization threshold energies E_i for electrons and holes are measured from the bottom of the conduction band and the top of the valence band, respectively.

[b] Wave vectors are measured in fractions of the zone-boundary wave vector.

where $E_e(\mathbf{p})$ is the initial energy of the impact-ionizing electron; $E_e(\mathbf{p}_1)$, $E_e(\mathbf{p}_2)$, and $E_h(\mathbf{p}_3)$ the energies of the two final electrons and the hole; $\mathbf{p}_1$, $\mathbf{p}_2$, and $\mathbf{p}_3$ the corresponding momenta; V the volume of the unit cell when the reduced Brillouin-zone scheme is considered; and M the matrix element of the transition.

Keldysh (1960) showed that for semiconductors with large dielectric constants, Eq. (25) can be approximated near threshold by

$$\frac{1}{\tau_i(E)} = \frac{1}{\tau_{ph}(E_i)} P \left(\frac{E - E_i}{E_i} \right)^2, \tag{26}$$

where $1/\tau_{ph}(E_i)$ is the phonon-scattering rate of the electron at threshold, and P is a dimensionless constant often assumed much greater than one. If this is the case, one can then see that above threshold the ionization probability can increase extremely rapidly with energy and quickly exceed the phonon-scattering rate. This produces a dramatic drop of the magnitude of the distribution function above threshold.

In general, the effective ionization energy will depend on the maximum in the product of a rapidly falling (exponentially) distribution function and a rising (with energy) cross section (Kane, 1967). If the ionization cross section increases rapidly with energy, this leads to the prediction that the effective ionization energy lies very close to threshold (≈ 0.1 eV or so).

This is the reason for the assumption routinely made in the theory of ionization rates that carriers ionize as soon as they reach threshold and are therefore left with a negligible residual energy after impact ionization. We shall return to this point later. However, the reader is cautioned that there may be several important exceptions such as silicon. Kane (1967) has calculated the ionization probability per unit time for electrons in silicon using a realistic band structure. This probability expressed in energy units $[\hbar/\tau_i(E)]$, is plotted as a function of electron energy (measured from the conduction-band maximum) in Fig. 10. The cross section appears to go to zero at a threshold $\simeq 1.1$ eV, in agreement with the value calculated by Anderson and Crowell (1972) using the previously discussed threshold criterion. Note, however, that the rise of the cross section is very gradual, giving, for example, at 0.8 eV above threshold an ionization scattering rate of only 10^{12} sec^{-1}, much smaller than the corresponding phonon-collision rate. Only at 4.2 eV above threshold, do the two scattering rates become equal ($\simeq 10^{14}$ sec^{-1}). The slow rise of the cross section in silicon suggests that the effective ionization energy may be much higher than the 1.1 eV threshold. Thornber (1981) points out that if one were to use this low value for the ionization energy, there would be so few electrons with energies above E_i that it would be "very hard to understand how enough electrons could reach the energies needed to account for the observed oxide charging near the Si/SiO_2 interface where a

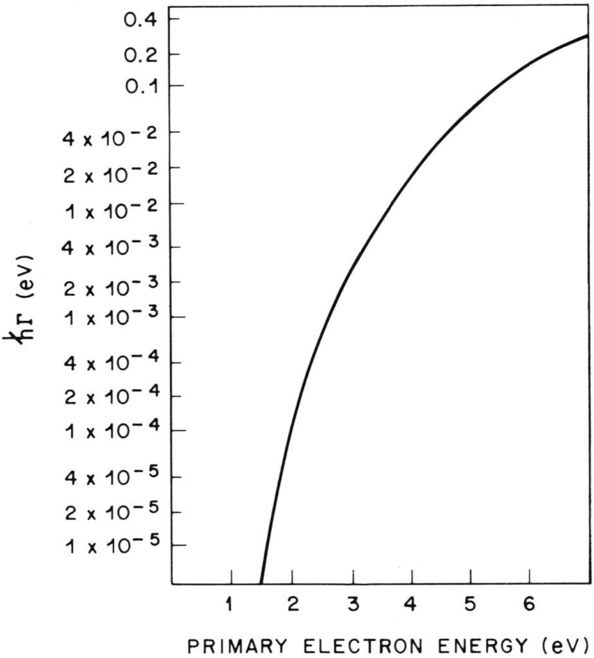

FIG. 10. Scattering rate for pair production by a primary electron in Si, expressed in energy units; Γ is scattering rate; zero electron energy is the conduction-band minimum. [From Kane (1967).]

3.1-eV barrier for electrons must be exceeded." In fact, Monte Carlo simulations of electron-ionization rates in Si and of hot-electron emission from Si into SiO_2 give best fit to the experimental data when $P = 0.01$ is used in Eq. (26) (Tang and Hess, 1983a,b). This corresponds to a slowly rising cross section, leading to a effective ionization energy significantly higher than the threshold.

Similar findings on the slow rise of the ionization cross section have been reported for InSb by Devreese *et al.* (1982). In addition, Monte Carlo calculations of the electron-ionization rate in GaAs show that an excellent fit of the experimental data is only obtained when a soft threshold [$P = 0.5$ in Eq. (26)] is used (Bulman *et al.,* 1983).

It is important to point out that in high electric fields, electron–hole pairs may be produced at energies below the threshold for band-to-band impact ionization by a mechanism called "tunnel-impact ionization."

This effect, although predicted many years ago by Keldysh (1958), has apparently gone unnoticed. This phenomenon is a combination of impact ionization and Zener tunneling and is conceptually related to the Franz–Keldysh effect. In the Franz–Keldysh effect, briefly sketched in Fig. 11a, an incoming photon with sub-band-gap energy promotes a valence electron to a virtual state in the forbidden gap from where it can tunnel into the conduction band (photon-assisted tunneling). This leads to a lowering and broadening of the absorption threshold below the band gap.

Tunnel-impact ionization is very similar, with the difference that here the energy required for this field-assisted transition is provided not by a photon but by a conduction-band electron with an energy below the ionization threshold (Fig. 11b). Subthreshold production of electron–hole pairs may be important in relatively low-gap materials with low effective masses. Kyuregyan (1976) has calculated the transition rate for this process and its effect on the ionization coefficient assuming parabolic bands. He finds that the electrons that tunnel-impact-ionize typically have energies slightly below threshold.

It is desirable that experimental investigations of this effect too long neglected be undertaken. A good candidate for such studies seems to be the ternary alloy $In_{0.53}Ga_{0.43}As$ ($E_g = 0.73$ eV at 300 K), which can be grown

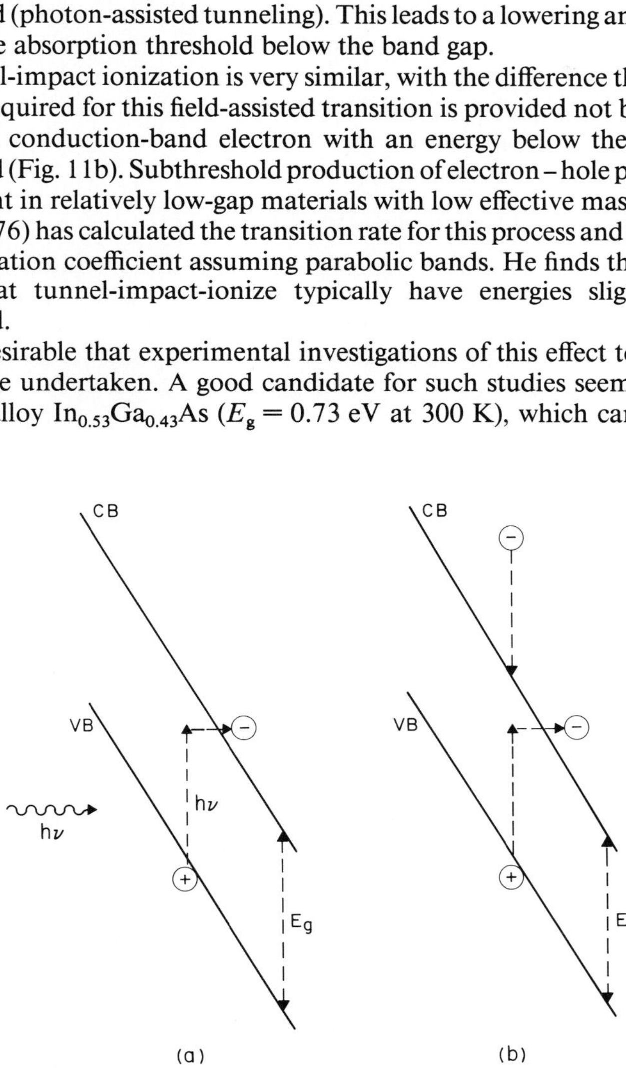

FIG. 11. Real-space energy-band diagram illustrating (a) Franz–Keldysh effect (photon-assisted tunneling); (b) tunnel-impact ionization. In the latter effect, a valence-band (VB) electron acquires energy by collision with an energetic electron and tunnels into the conduction band (CB). This process effectively reduces the ionization threshold energy.

lattice matched to InP. This alloy has a low conduction-band effective mass, and experiments have shown that Zener tunneling in this material occurs at fields slightly lower than those of band-to-band impact ionization (Forrest *et al.,* 1980).

Important information on the ionization cross sections can be obtained experimentally by measuring the wavelength dependence of the quantum efficiency of the so-called internal photoelectric effect (Antončik, 1967; Shockley, 1961). This phenomenon consists of the creation of an electron–hole pair by impact ionization using photons rather than the electric field to give carriers the ionization energy.

3. Phonon Scattering in High Electric Fields

Impact-ionization rates are not only affected by the ionization energy but also by the phonon scattering rates. Collisions with phonons control the energy and momentum losses of the carriers and thus influence the average distance required to create an electron–hole pair by impact ionization. In the absence of phonon collisions, this distance would be E_i/eF where F is the electric field. Phonon scattering greatly increases this distance. Scattering against the direction of the field is very effective in this respect, since after suffering these collisions, carriers are slowed down by the electric field and lose a considerable portion of their energy. This increases considerably the distance required to gain the ionization energy.

For semiconductors with band gaps in excess of 0.5 eV, the scattering process that affects impact ionization most is the deformation potential interaction, such as scattering by longitudinal optical and acoustic intervalley phonons with relatively large k, characterized by an energy E_p of the order of the center-zone optical phonon energy (typically, several tens of millivolts in most semiconductors). Polar optical photon scattering is of limited importance except in small-gap semiconductors, such InSb and InAs.

Scattering by ionized impurities ought in principle be considered given that avalanche multiplication in most instances occurs in the space-charge region of reverse-biased $p-n$ junctions. However, the scattering rate for this process decreases with increasing energy, due to its Coulomb nature, and becomes negligible for hot carriers typical of this transport regime.

Neutral impurity scattering, an energy-independent process, is also neglected in theories of impact ionization. Nevertheless, another effect, impact ionization of impurities by hot carriers, may become important in practical circumstances, depending on the depth of the associated energy levels, the temperature, and their concentration. This ionization process has seldom been studied in the literature. Clearly from a fundamental point of view, it can be neglected if we consider ultrapure semiconductor materials, such as presently available germanium.

Another collision mechanism to be considered is alloy scattering in ternary and quaternary alloys associated with the irregular distribution of column III and/or column V atoms in their respective lattice sites. This is also of limited importance in the impact-ionization regime due to its decreasing scattering rate at high carrier energies. However, another type of alloy disorder, clustering can be very important in altering the impact-ionization-rate ratio. It is considered in, Part IV of this chapter because of its interesting relationship with the problem of impact ionization in a superlattice.

Carrier–carrier scattering should also be considered. It is a good approximation to neglect this mechanism in reverse-biased APDs operating in the prebreakdown region; here the carrier density typically does not exceed 10^{10} cm^{-3}. However, it is not a valid approach in IMPATT diodes, where the density can easily exceed 10^{16} cm^{-3}.

The purpose of this section is to give a brief description of the most important scattering mechanisms in the avalanche regime and to emphasize their relationship to the band structure of the semiconductor. A short discussion of quantum effects in high field scattering is also given. The interested reader is referred to the appropriate treatises for detailed accounts of this topic (Conwell, 1967; Ferry *et al.,* 1980).

A carrier in a state with wave vector $\mathbf{k}$ can make a transition to a state $\mathbf{k}'$ by absorbing or emitting one or more phonons. If we consider only single-phonon processes, conservation of momentum and energy requires that

$$\mathbf{k}' = \mathbf{k} \pm \mathbf{q}, \tag{27}$$

$$E(\mathbf{k}') = E(\mathbf{k}) \pm \hbar\omega_{\mathbf{q}}, \tag{28}$$

where $\mathbf{q}$ and $\hbar\omega_{\mathbf{q}}$ are the wave vector and the energy of the phonon, respectively, and $E(\mathbf{k})$ and $E(\mathbf{k}')$ the initial and final energies of the carrier.

Usually, two key assumptions, too often forgotten or overlooked, are made in calculating the total scattering rate $1/\tau(k)$ out of state $\mathbf{k}$:

1. The transition from state $\mathbf{k}$ to one of the many final states $\mathbf{k}'$ occurs instantaneously (zero collision duration).
2. The collision frequency $1/\tau$ is sufficiently small so that the final and initial states are well defined in momentum and energy. This requires that the collision broadening of the levels $E(\mathbf{k}')$ and $E(\mathbf{k})$, $\Delta E \approx \hbar/\tau$, imposed by the uncertainty principle, be much smaller than the energy $\hbar\omega$ of the transition.

Under these two assumptions, the probability per unit time of a carrier being scattered out of state $\mathbf{k}$ (total scattering rate) can be calculated using the

Fermi golden rule of first order time-dependent perturbation theory to give

$$\frac{1}{\tau(\mathbf{k})} = \frac{2\pi}{\hbar} \sum_{\mathbf{q}} |H'_{\mathbf{k}+\mathbf{q},\mathbf{k}'}|^2 [(1 + N_{\omega_\mathbf{q}})\, \delta(E(\mathbf{k}) - E(\mathbf{k}+\mathbf{q}) + \hbar\omega_\mathbf{q})$$
$$+ N_{\omega_\mathbf{q}}\, \delta(E(\mathbf{k}) - E(\mathbf{k}+\mathbf{q}) - \hbar\omega_\mathbf{q})], \tag{29}$$

where $H'_{\mathbf{k}+\mathbf{q},\mathbf{k}}$ is the matrix element of the interaction between the electron and the phonon with wave vector $\mathbf{q}$ and $N_{\omega_\mathbf{q}}$ (from now on written as N) the equilibrium number of such phonons,

$$N = [\exp(\hbar\omega_\mathbf{q}/kT) - 1]^{-1}. \tag{30}$$

In Eq. (29), the term in $N + 1$ represents spontaneous and stimulated phonon emission, whereas that in N takes into account phonon absorption. The summation in Eq. (29) can be approximated by the relation

$$\sum_{\mathbf{q}} F(\mathbf{q}) = \frac{V}{(2\pi)^3} \int F(\mathbf{q})\, d\mathbf{q}, \tag{31}$$

which, using spherical coordinates leads to

$$\frac{V}{(2\pi)^3} \int F(\mathbf{q})\, d\mathbf{q} = \int_0^q \int_0^\pi \int_0^{2\pi} \frac{V}{(2\pi)^3} F(\mathbf{q}) |\mathbf{q}|^2 \sin\theta\, d\theta\, d\phi\, dq, \tag{32}$$

where V is the volume of the crystal and $F(\mathbf{q})$ is a function of the phonon wave vector $\mathbf{q}$. Obviously, Eq. (29) can also be expressed as an integral over all allowed final states $\mathbf{k}'$, satisfying Eq. (27). Using the dispersion relation for the carrier energy $E = E(\mathbf{k})$ and Eqs. (29)–(32), one obtains the total scattering rate as a function of the carrier energy.

We shall only consider here the dominant phonon-scattering mechanisms relevant to very high field transport, i.e., polar phonon scattering and intervalley scattering. The first is important in the impact-ionization regime in low-gap III–V materials, such as InAs and InSb, whereas the second strongly affects avalanche multiplication in silicon and in most other important III–V materials, such as GaAs, InP, and GaSb, including ternary and quaternary alloys with band gaps in the long-wavelength 1.3–1.7-μm region.

a. Polar Mode Scattering

In polar semiconductors, lattice vibrations of the longitudinal optical branch cause an electrical polarization of the crystal, that scatters carriers. In III–V materials, such as InSb and InAs, this scattering mechanism is the dominant one for electron energies up to ionization thresholds: E_{ie}(InAs) = 0.42 eV, E_{ie}(InSb) = 0.25 eV at 77 K (Mikhailova *et al.*, 1976; Van Welzenis, 1981).

Intervalley scattering does not play a role in the avalanche regime in these two materials because the separation between the satellite valleys and the Γ minimum is greater than the electron-ionization energy.

It can be shown (Conwell, 1967) that the polar phonon-scattering rate $W_{\mathbf{k},\mathbf{k}'}$ from state **k** to **k**′ is proportional to $1/|\mathbf{k}' - \mathbf{k}|^2$. The $|\mathbf{k}' - \mathbf{k}|^2$ term causes small-angle scattering to be strongly favored over large-angle scattering. This is a manifestation of the Coulomb component of the interaction between the electron and the electrical polarization of lattice vibrations. As a consequence, momentum is not randomized, which leads to a focusing of the electron velocities in the direction of the electric field **F**. In addition, at high values of F, the electric field itself has a tendency to focus the electron distribution function in k space along the direction of the field, since there is a substantial momentum and energy gain between small-angle collisions. This is clearly shown in Fig. 12 in which an electron initially away from the direction of the field gradually approaches this direction (Dumke, 1968).

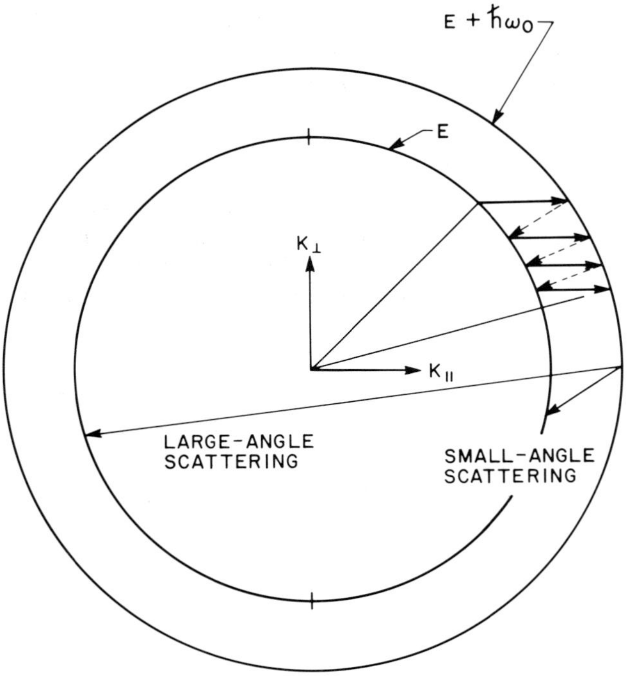

FIG. 12. Schematic representation of focusing effect of the electric field. An electron is accelerated in k space in the direction of the field F but scatters between energy surfaces with a probability that is symmetric about the radius to the initial state. For the purpose of illustration, we have assumed that the scattering takes place when the electron reaches the upper surface and that the scattering occurs at the center of the distribution of scattering probability. [From Dumke (1968)].

The anisotropy of the electron distribution is stronger at higher fields and also persists in the avalanche regime in InAs and InSb. The distribution function and ionization rate (per unit time) for these two semiconductors have been investigated in great detail by Dumke (1968).

The total scattering rate as a function of energy can be calculated by integration of $W_{\mathbf{k},\mathbf{k}'}$ over the final states (Conwell, 1967). Figure 13 illustrates this probability for the case of GaAs at 300 K (Fawcett *et al.*, 1970). Note that for energies greater than the optical phonon energy (>35 meV), the scattering rate rapidly increases due to polar phonon emission and then gradually decreases with increasing energy. This latter feature implies that at high fields, the decrease in the collision rate causes the rate of energy input from the field to increase, leading to a rapid rise in the electron temperature and in low-gap materials (InSb, InAs), to the so-called polar breakdown (Conwell, 1967). In practice, this does not occur because energy losses by impact ionization take over when polar phonon scattering is less effective. The loss in effectiveness of polar scattering at high energy and the small-angle nature of these collisions are a clear manifestation of the Coulomb nature of the interaction between the electron and the electrical polarization of longitudinal lattice vibrations.

b. Intervalley and Interband Scattering

In materials such as GaAs and InP, polar mode scattering is not the dominant collision mechanism at the fields typical of avalanche multiplication ($>10^5$ V cm^{-1}). At fields well below breakdown (10^3-10^4 V cm^{-1}), electrons have gained sufficient energy in the central valley that they can

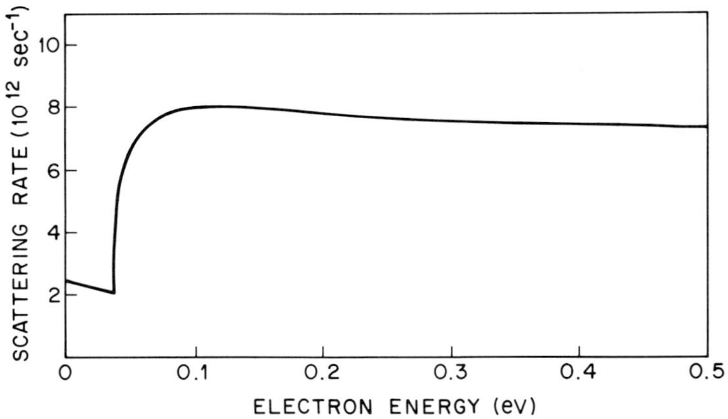

FIG. 13. Polar phonon scattering rate in GaAs at 300 K versus electron energy. Zero energy is the conduction-band minimum. [From Fawcett *et al.* (1970).]

transfer by intervalley phonon collisions to the satellite valleys or to higher conduction bands.

The reader is referred to the band structure of GaAs for the following discussion (Fig. 4). There are two sets of valleys into which the carrier can scatter: three X valleys along the equivalent $\langle 100 \rangle$ orientations and four L valleys along the $\langle 111 \rangle$ directions. Note that because of the position of the satellite minima, this scattering mechanism involves large momentum transfers in the collisions. Thus, intervalley transfer, unlike polar scattering, is very effective in randomizing the momentum gained from the electric field. Intervalley transitions are due to deformation potential scattering, i.e., scattering by the strain field of the acoustic waves. The phonons of interest in these transitions are predominantly relatively large-k LA and LO phonons. In most theories of impact ionization, only one phonon energy, the center-zone optical phonon energy, is assumed. Since in reality different types of phonons are involved, a better approximation is to take the average of the LO and LA zone-edge phonon energies (Ridley, 1983b).

It can be shown that for phonon wave vectors not too different from the Brillouin zone radius, the matrix element for intervalley scattering is independent of the initial and final states of the electron, thus making the collision probability strictly isotropic (Conwell, 1967). This assumption of total isotropy may therefore be questioned at the high energies typical of the impact-ionization regime in III–V materials such as InP and GaAs, although it is routinely made in theories of impact ionization.

In addition to intervalley scattering among nonequivalent valleys, there can also be scattering among, e.g., the four equivalent $\langle 111 \rangle$ valleys or the $\langle 100 \rangle$ valleys (equivalent intervalley scattering). For nonequivalent intervalley scattering between valley i and a set j of equivalent valleys with constant-energy-surface ellipsoids of revolution, the scattering rate at energy E can be easily calculated from Eq. (29) to give (Conwell, 1967; Capasso *et al.*, 1979a)

$$\frac{1}{\tau_{ij}} = \frac{D_{ij}^2 (m_{\mathrm{de}})^{3/2}}{2^{1/2}\pi\hbar^3 \rho\omega_{ij}} [N(E - \Delta E_j + \hbar\omega_{ij})^{1/2} + (N+1)(E - \Delta E_j - \hbar\omega_{ij})^{1/2}], \tag{33}$$

where m_{de} is the density-of-states mass given by

$$(m_{\mathrm{de}})^{3/2} = n m_{\mathrm{t}}^* m_{\ell}^{*1/2}, \tag{34}$$

where n is the number of equivalent j valleys and m_{t}^* and m_{ℓ}^* are their transverse and longitudinal masses, respectively; ΔE_j is the energy separation between the i and j valleys, ω_{ij} the angular frequency of the phonon, ρ the semiconductor density, and D_{ij} the coupling constant between the i and j valleys. This constant is proportional to the matrix element of the transition

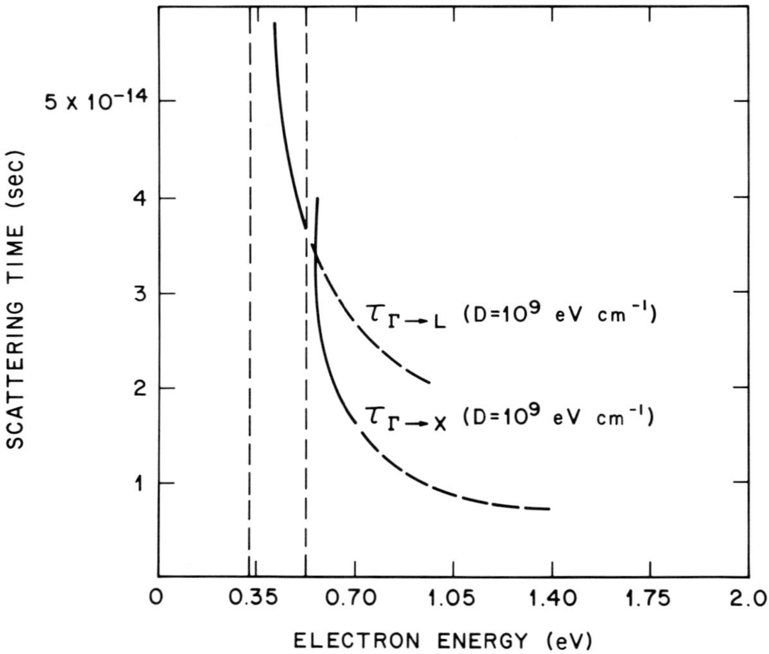

FIG. 14. Phonon scattering time for Γ–L and Γ–X intervalley transitions in GaAs at 300 K versus electron energy. Parabolic X and L valleys have been assumed. Zero energy is the conduction-band minimum. [From Capasso *et al.* (1979a).]

and is often called the deformation potential, assumed here to be independent of the initial and final states of the electron.

The dominant role played by intervalley scattering in GaAs and in InP in the avalanche regime has only recently been recognized (Capasso *et al.*, 1979a; Capasso and Bachelet, 1980; Shichijo and Hess, 1981). Figure 14 shows the $\tau_{\Gamma L}$, $\tau_{\Gamma X}$ intervalley scattering times versus electron energies in GaAs as calculated by Capasso *et al.* (1979a). Parabolic X and L valleys were assumed with effective density-of-state masses $m_{deX} = 0.85$, $m_{deL} = 0.55 m_e$ and energy minima $\Delta E_{\Gamma L} = 0.33$ eV, $\Delta E_{\Gamma X} = 0.55$ eV (Aspnes, 1976). A value of 28 meV was taken for the phonon energy $\hbar\omega_{ij}$, corresponding to the zone-edge LO phonon and $\rho = 5.36$ g cm^{-3}. The quantity most subject to uncertainty is the deformation potential. A value $D = 10^9$ eV cm^{-1} for both ΓL and ΓX coupling was chosen that gives best fit of Monte Carlo simulations of transport properties to high-field drift velocities (Littlejohn *et al.*, 1977). Kash and Wolff (1983) have determined experimentally the value of the coupling constant $D_{\Gamma L}$ to be 7×10^8 eV cm^{-1} via an accurate measurement of the intervalley relaxation times using nonlinear optical techniques.

The use of this experimental value would double the scattering time $\tau_{\Gamma L}$, shown in Fig. 14. It is seen that the total scattering rate $\tau_{\Gamma L}^{-1} + \tau_{\Gamma X}^{-1}$ is $\simeq 10^{14}$ sec^{-1} at high energies, one order of magnitude higher than that for polar scattering. Note, however, that this calculation underestimates τ^{-1}, since the satellite valleys have been assumed parabolic.

In the general case, the scattering rate by deformation potential at high energies ($E \gg \hbar\omega$) can be written as (Ridley, 1983a)

$$1/\tau(E) = (\pi D^2/\rho\omega)(2N+1)\rho_F(E), \tag{35}$$

where $\rho_F(E)$ is the density of final states. Equation (35) reduces to Eq. (33) in the case of nonequivalent intervalley scattering between a valley and a set of equivalent parabolic valleys.

Shichijo *et al.* (1981) have calculated the total scattering rate in GaAs for an electron in the central valley using realistic pseudopotential band structures (Cohen and Bergstresser, 1966). Their results are illustrated in Fig. 15. For these calculations, the same set of parameters as those used by Littlejohn *et al.* (1977) in a simulation of high field-drift velocities was employed. We shall focus mainly on the high-energy part of the curve between 0.7 and 2.0 eV, which is the important region in the impact-ionization regime. Note the extremely high scattering rates with a maximum of 4.5×10^{14} sec^{-1} at 1.5 eV, corresponding to a time between collisions of $\simeq 2.5 \times 10^{-15}$ sec.

These scattering rates in the high-energy region are dominated by intervalley phonons and can be obtained from Eq. (35). This point is very clear by noting that in the high-energy region, the scattering rate of Fig. 15 is proportional to $D^2\rho_F$. This result, which is essentially Eq. (35), can be easily obtained when the coupling constant is independent of the initial and final states of the electron by integrating the Fermi golden rule equation over the final energies of the electron [Eq. (29)]. Thus, the scattering rates of Shichijo and Hess are derived from perturbation theory. The application of the golden rule in this high-scattering-rate region is questionable, as pointed out by Capasso *et al.* (1981a) and extensively discussed by Barker (1973, 1980). The reason for this is that the collision broadening of the initial and final states of the electron, implied by such high scattering rates is such that the assumptions behind the golden rule are no longer valid, and collisions can no longer be treated perturbatively. The collision broadening or full collision width may be expressed as (Barker, 1973)

$$\Delta E = \hbar/\tau \tag{36}$$

This can be seen as a consequence of the uncertainty principle. The collision broadening resulting from the scattering rates of Shichijo and Hess can be read off the scale on the right of Fig. 15. It is clear that for energies greater than 0.7 eV; the broadening is a significant fraction of the electron energy, so

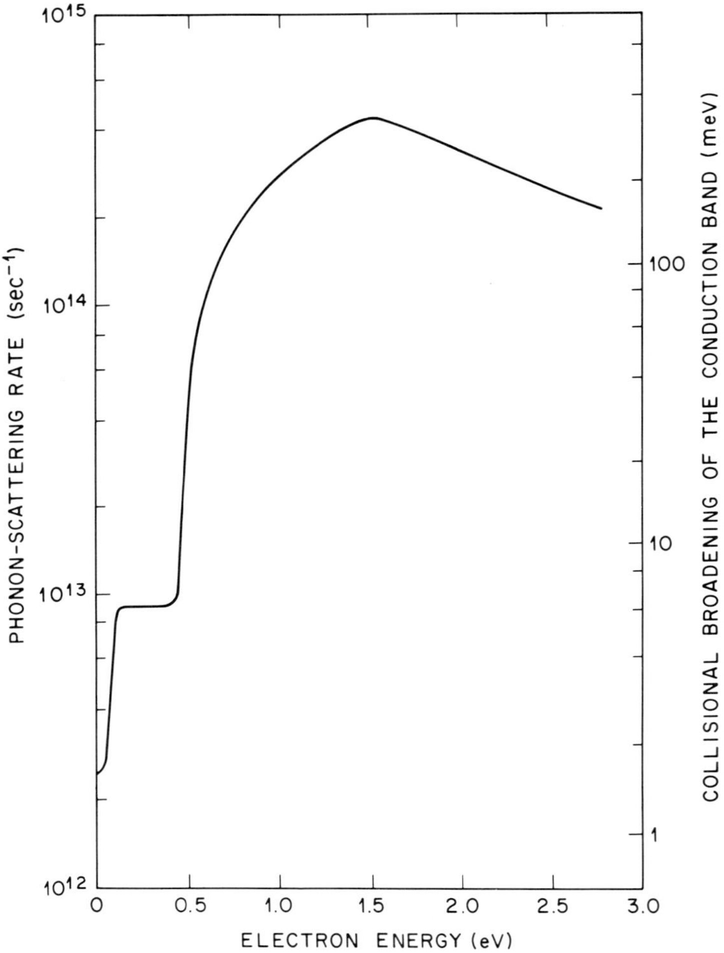

FIG. 15. Phonon scattering rate $1/\tau$ in GaAs at 300 K as a function of electron energy. Zero energy is the conduction-band minimum, (after Shichijo and Hess, 1981). The vertical scale on the right gives the amount of collision broadening ($\Delta E = \hbar/\tau$) associated with the scattering. [From Capasso *et al.* (1981a). © 1981 IEEE.]

that scattering rate calculations using the Fermi golden rule may no longer be trusted, and higher order quantum corrections should be included. Basically in this regime, as a result of the broadening, collisions can no longer be treated as transitions between well-defined momentum states.

Capasso *et al.* (1981a) were the first to propose that proper inclusion of quantum effects, such as collisional broadening and the intracollisional field effect, could be important in the impact-ionization regime in semiconductors, in addition to modifying the phonon scattering rates.

The intracollisional field effect is a phenomenon associated with the finite collision duration. As collisions become very frequent, the energy gained during a collision duration τ_c and the associated broadening of the electronic states (given that τ_c is a statistical variable) may not be negligible (Barker, 1973; Thornber, 1978; Arora, 1983).

Chang *et al.* (1983) undertook a calculation of the scattering rates in the high-energy region in GaAs that goes beyond perturbation theory and includes many body corrections. They consider only spontaneous intervalley phonon emission, which is the dominant scattering mechanism at such high energies. When the time between collisions becomes smaller than the reciprocal of the phonon angular frequency, the electron can no longer be treated as a semiclassical particle propagating among independent phonon collisions. The electron must then be pictured as a particle "dressed" or followed by a cloud of virtual phonons that it constantly emits and reabsorbs (quasiparticle). The effect of this phonon cloud (polaron effect) is to broaden and shift the states of the otherwise quasi-free electron (self-energy corrections).

Chang *et al.* (1983) find that the main effect of including these higher order quantum corrections is to reduce by $\simeq 20\%$ the peak in the total scattering rate of Fig. 15 and to broaden the energy range of final states by $\simeq 100$ meV near $\simeq 1.7$ eV. The energies of the quasi-particles are shifted by 70 meV. This reduction in the scattering rate is considerable and confirms that the golden rule should be used with caution in this region. However, the reduced collision rate is still substantially higher than the value $\tau^{-1} = 10^{14}\ \text{sec}^{-1}$ proposed by Capasso *et al.* (1981a).

Another material in which intervalley scattering plays an important role at very high field is silicon. Figure 16 shows the band structure of silicon (Chelikowsky and Cohen, 1976). It is seen that due to the large separation between the X and the L valleys ($\simeq 1$ eV), X–L intervalley scattering becomes important only at very high fields (100 kV cm^{-1}). Tang and Hess (1983) also established the importance of the second conduction band, whose minimum lies only $\simeq 0.1$ eV above the main conduction-band minimum. Electrons can scatter into this band even at relatively low fields. For electron energies below the L minimum, the main scattering mechanism is equivalent intervalley scattering among the three X valleys. Nonpolar phonon scattering within the X valley is forbidden by symmetry considerations except at the highest energies.

Interband scattering between the main and the second conduction bands may also be important in GaAs for electron energies greater than the separation between the Γ and the X_7 points, as can be seen from Fig. 4, especially because electrons must transfer to the second conduction band to impact

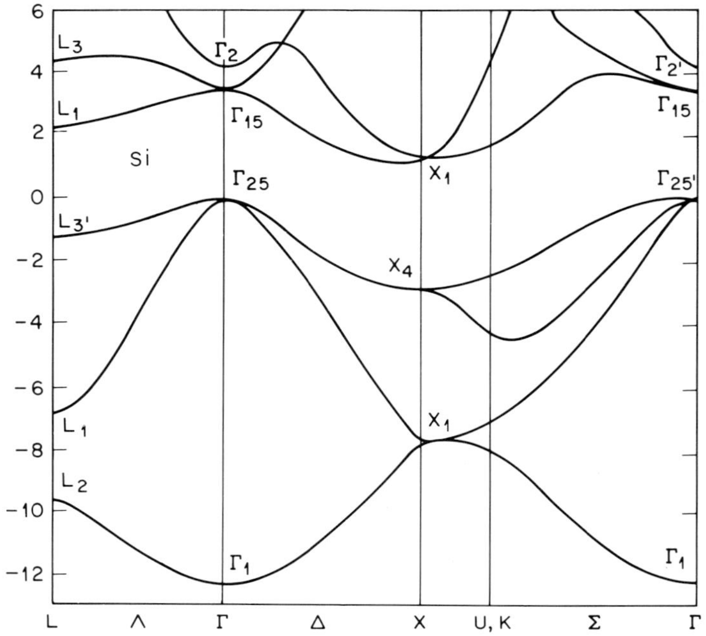

FIG. 16. Silicon band structure as obtained from nonlocal pseudopotential calculations. [From Chelikowsky and Cohen (1976).]

ionize. Intervalley scattering between the L and the X valleys is also dominant in the avalanche regime in germanium (Meyer and Jorgensen, 1965).

4. Distribution Functions and Ionization Rates for Electrons and Holes

The theory of impact ionization in semiconductors has been reviewed by Chynoweth (1968) and Stillman and Wolfe (1977) in previous volumes of this series and by Mönch (1969). Since then there has been considerable theoretical progress in this field. In this section, we shall summarize the main features of each theory and particularly focus on important developments such as the inclusion of realistic band structures in the calculations; physical interpretation, rather than mathematical aspects, is emphasized.

Theories of impact ionization attack the problem of calculating the carrier-distribution function at very high fields and the ionization rate. Most of these theories give the ionization rates in the form of analytical expressions or numerical plots expressed in terms of basic parameters, such as mean free paths and threshold energies. These can be then determined by fitting the experimental data. These theories usually assume parabolic bands or simple nonparabolic valleys.

More recently there have been several attempts to introduce in the theory of impact ionization realistic band structures, with special regard to III–V materials. In these cases, the ionization rates are conveniently calculated using the Monte Carlo method.

Microscopic theories of impact ionization make some basic and important assumptions concerning the ionization process and the geometry of the avalanche region. Because of their apparent obviousness or simplicity they go often unmentioned, and the various theories are acritically used to fit or to interpret experimental ionization coefficients measured in devices that do not satisfy the following assumptions:

1. spatially uniform and time-independent electric fields;
2. very long avalanche regions; and
3. carriers rapidly reaching a spatial steady state under the competing influences of the field and the phonon collisions.

Under these conditions, the ionization coefficient, defined as the probability per unit distance along the direction of the field that a carrier impact ionizes, is independent of position and only dependent on the electric field strength. This ionization coefficient is then identified and calculated as the reciprocal of the average distance $\langle x \rangle$ along the direction of the field between ionizing collisions or, equivalently, as the ratio of the average probability per unit time of impact ionization to the average drift velocity. It is very important to stress that such identification is correct only if

$$\langle x \rangle \gg E_i/eF, \tag{37}$$

where the right-hand side is the minimum distance required by a carrier to create an electron–hole pair in a uniform field F (Okuto and Crowell, 1972).

At very high fields, however, this distance becomes comparable with $\langle x \rangle$ so that $\langle x \rangle^{-1}$ can no longer be identified as the probability per unit distance of creating a pair. In this case, the ionization coefficient becomes a nonlocal position-dependent quantity (Okuto and Crowell, 1972).

Experimental devices quite often do not satisfy the previous assumptions, since either the electric field varies greatly over a short distance or the avalanche region is very thin ($\ll 1\ \mu$m). If, however the field does not vary too much over the ionization distance $1/\alpha$, i.e.,

$$\frac{1}{\alpha(F_m)} \frac{1}{F_m} \frac{dF}{dx} < 1, \tag{38}$$

where $\alpha(F_m)$ is the calculated ionization coefficient at the maximum field F_m, and the width of the avalanche region is much larger than the minimum ionization distance E_1/eF_m, one can still define a localized ionization coefficient $\alpha(F(x))$ and calculate it as outlined before.

The ionization rate is determined mainly by the probability of a carrier gaining in the field an energy equal to the ionization threshold energy E_i, provided that the ionization cross section rises sufficiently fast above E_i that we can assume that most carriers ionize very close to the threshold (Keldysh, 1965). If we denote this probability by $P(E_i)$, the electron-ionization rate can be expressed in a fairly general way as (Thornber, 1981)

$$\alpha = (eF/E_i)P(E_i), \tag{39}$$

where $P(E_i)$ is a strongly increasing function of the electric field F. An identical expression of course holds for the hole-ionization rate. The factor multiplying $P(E_i)$ can be understood as a limitation due to energy conservation. At extremely high fields, the probability of reaching the threshold tends to unity, since virtually every carrier escapes collisions with phonons. The average distance between ionizing collisions α^{-1} must then tend to E_i/eF, which is the minimum distance required to create an electron–hole pair. This energy-conservation limitation was first clearly stated by Wolff (1954). It is also an important consideration in the analysis of experimental data. Values of the ionization coefficients exceeding the asymptotic limit eF/E_i should be discarded as unphysical (Okuto and Crowell, 1972).

Thornber (1981) points out that Eq. (39) is also valid for the case when the carrier, on the average, ionizes at energies significantly greater than the threshold. In this case, E_i should be substituted by the effective ionization energy $\langle E_i \rangle$; i.e.,

$$\alpha(F) = (eF/\langle E_i \rangle)P(\langle E_i \rangle), \tag{40}$$

where $\langle E_i \rangle$ in general depends on the electric field but much less strongly than $P(\langle E_i \rangle)$. The effective ionization energy $\langle E_i \rangle$ is discussed in Section 8.

The probability $P(E_i)$ is proportional to the energy-distribution function $f(E_i)$, which in turn can be obtained by solving the Boltzmann equation for the momentum distribution function $f(\mathbf{p})$. The Boltzmann equation, assuming time independence, a uniform field, and absence of diffusion effects (Chynoweth, 1968), becomes

$$e\mathbf{F} \cdot \nabla_{\mathbf{p}} f(\mathbf{p}) = \int d\mathbf{p}'[f(\mathbf{p}')W(\mathbf{p}', \mathbf{p}) - f(\mathbf{p})W(\mathbf{p}, \mathbf{p}')], \tag{41}$$

which simply states that at steady state, the rate of change of the distribution due to acceleration by the field must balance the rate of change due to collisions; $W(\mathbf{p}, \mathbf{p}')$ is the probability per unit time of a carrier with momentum $\mathbf{p}$ being scattered into the element of volume $d\mathbf{p}'$ at $\mathbf{p}'$. The distribution function can then be represented as a series of Legendre polynomials:

$$f(\mathbf{p}) = \sum_{n=0}^{\infty} f_n(E)P_n(\cos\theta), \tag{42}$$

where $P_n(\cos\theta)$ is the Legendre polynomial of nth order and θ the angle between the momentum $\mathbf{p}$ and the direction of the field.

Usually, only a few terms of Eq. (42) are retained, and the expansion is substituted in the Boltzmann equation. This converts Eq. (41) into a set of coupled differential equations for the coefficients $f_n(E)$. We are interested in $f_0(E)$, which is proportional to the carrier density and represents the energy-distribution function.

This scheme of solving the Boltzmann equation has been used in many theories of impact ionization. More recently, the problem has been attacked by the Monte Carlo iterative technique, which is equivalent to the Boltzmann equation.

5. Shockley, Wolff, and Baraff Theories

The starting underlying assumptions of these theories are as follows:

(1) Energy bands are parabolic. This is not a crucial assumption, in spite of general opinion in the sense that the principal physical conclusions of these theories are also valid for nonparabolic bands, as is illustrated later. In addition, the use of several adjustable parameters can absorb band-structure effects even in the case of more complex band structures. This may explain the surprising success of the Baraff theory when applied to silicon and germanium, which certainly do not satisfy this assumption.

(2) Isotropic optical phonon scattering characterized by an energy-independent mean free path. Only optical phonon emission is considered. In the rest of this chapter, the phonon energy $\hbar\omega_{ij}$ is denoted by E_{p}, following the notation of the theory of impact ionization. However, absorption processes can easily be included, as is later illustrated.

For materials where the dominant scattering mechanism at high energy is deformation potential scattering (e.g., GaAs, Si, Ge), the assumption of isotropic scattering is reasonable. The assumption of an energy-independent mean free path is valid only for parabolic bands. This can easily be seen by noting that the phonon scattering mean free path for a carrier in a state characterized by an energy E is

$$\lambda(E) = v(E)\tau(E), \tag{43}$$

where v is the carrier-group velocity and $\tau(E)$ the total scattering time in that state.

For a parabolic band

$$v = \sqrt{2E/m^*} \tag{44}$$

the scattering rate $[\tau(E)]^{-1}$ for a carrier of energy E in a parabolic band is proportional to $\sqrt{E}$ if E is much greater than the phonon energy by simple density of states considerations [see, e.g., Eq. (33)]. Thus, λ is energy independent.

(3) The rise of the ionization probability above threshold is described in terms of an abrupt steplike increase of the ionization cross section from zero to a constant value σ_i, greater than the phonon-scattering cross section σ_{0p}.

The important parameter that determines the shape of the high-energy distribution function is the ratio of the energy gained between phonon collisions $eF\lambda$ to the energy lost per collision E_p (Baraff, 1962). Let us assume first that

$$eF\lambda \leq E_p. \tag{45}$$

In this situation, carriers will not be able to gain the high energies typical of impact ionization unless they escape phonon collisions. These are the "lucky" or ballistic carriers first considered by Shockley (1961). The probability that an electron, starting at zero energy, ballistically gains the threshold for ionization is equal to the probability that it travels for the ionization distance E_i/eF without undergoing phonon collision; i.e.,

$$P(E_i) = \exp(-E_i/eF\lambda). \tag{46}$$

Thus, the ionization rate will be proportional to $P(E_i)$; i.e.,

$$\alpha = (eF/E_i)\exp(-E_i/eF\lambda). \tag{47}$$

At the other extreme, let us consider situations in which the energy gained between collisions is much greater than the energy lost per collision; i.e.,

$$eF\lambda \gg E_p. \tag{48}$$

This situation is well described by a theory first developed by Wolff (1954). He writes the distribution function as the first two terms of the expansion given in Eq. (42); i.e.,

$$f(p, \cos\theta) = f_0(E) + f_1(E)\cos\theta. \tag{49}$$

This is essentially a diffusion theory approximation in which electrons undergo many collisions while gaining energy; the energy lost per collision is so much smaller than the energy gained that the collisions serve to keep the distribution nearly isotropic. The result is an energy distribution $f_0(E)$, which, below threshold, is well approximated by a Maxwellian characterized by an effective temperature T_e. This temperature can be easily found by the following heuristic rate equation argument. The energy rate equation is, at steady state,

$$dE/dt = eFv_d - E_p v_\theta/\lambda = 0, \tag{50}$$

where v_d is the drift velocity and v_θ the random Maxwellian thermal velocity.

Note that λ/v_θ is the time between collisions. The momentum rate equation is

$$dp/dt = eF - (m^* v_d)v_\theta/\lambda = 0. \tag{51}$$

In addition, we recall that for a Maxwellian distribution with an effective temperature T_e, the following equation holds:

$$\tfrac{3}{2}kT_e = \tfrac{1}{2}m^* v_\theta^2. \tag{52}$$

From Eqs. (50) and (51), we find that

$$v_d = \sqrt{E_p/m^*}, \tag{53}$$

which is the well-known expression for the saturated drift velocity. Using Eq. (53) in Eq. (51), we solve for v_θ and substitute this expression in Eq. (52) to give

$$kT_e = (eF\lambda)^2/3E_p. \tag{54}$$

This is exactly the formula found by Wolff (1954) via the Boltzmann equation. The probability that a carrier has the ionization energy is then proportional to

$$\exp(-E_i/kT_e). \tag{55}$$

Thus, the ionization rate in Wolff's theory is given approximately by

$$\alpha = \frac{eF}{E_i}\exp\left(-\frac{3E_pE_i}{(eF\lambda)^2}\right). \tag{56}$$

As pointed out by Baraff (1962), the electric fields typical of avalanche multiplication are such that neither Eq. (47) nor Eq. (56) can be used. This means that the distribution function is neither a spike in the direction of the field (Shockley's ballistic electrons) nor the nearly isotropic distribution of Wolff; the two aspects are complementary.

Baraff assumed a velocity distribution of the form

$$f(v, \cos\theta) = A(v) + B(v)\,\delta(1 - \cos\theta), \tag{57}$$

which is a mixture of a spherical part and a "spike" in the direction of the electric field. By inserting this expression in the Boltzmann equation [Eq. (41)], one can find the coefficients $A(v)$ and $B(v)$ that best represent the true distribution function. The result for the energy distribution function is

$$f_0(E) = E^{-[2/3+eF\lambda/3E_p]}\exp\left[-\frac{E_i}{(eF\lambda)^2/3E_p + \frac{2}{3}eF\lambda}\right]. \tag{58}$$

The ionization rate is proportional to $f_0(E_i)$. In the limit $eF\lambda = E_p$, Eq. (58) reduces to the energy distribution proposed by Shockley, whereas for $eF\lambda \gg E_p$, it goes to Wolff's Maxwellian.

Baraff also gave a rigorous solution of the Boltzmann equation, arriving at a more accurate distribution function than Eq. (58). The ionization rate was calculated as the reciprocal of the average distance between ionizing collisions defined as

$$\frac{1}{\langle x \rangle} = \frac{eF}{E_i + \langle n \rangle E_p}, \tag{59}$$

where $\langle n \rangle$ is the average number of phonons emitted between ionizing collisions. Equation (59) can easily be obtained from energy conservation; i.e.,

$$eF\langle x \rangle - \langle n \rangle E_p = E_i. \tag{60}$$

Baraff's results can be expressed in terms of the product $\alpha\lambda$ (or $\beta\lambda$) plotted against $E_i/eF\lambda$, with the ratio of the optical phonon energy to the ionization threshold energy as a parameter. These plots were modified by Crowell and Sze (1966) who discussed the temperature dependence of the mean free paths by also considering phonon absorption processes.

The mean free path is then written as

$$\lambda = \lambda_0/(2N+1) = \lambda_0 \tanh(E_p/2kT), \tag{61}$$

where N is the Bose–Einstein factor and T the absolute temperature. Equation (61) can be easily understood by noting that for energies $\gg E_p$, the scattering rate is proportional to $2N + 1$ [Eq. (35)] and by recalling the definition of λ; λ_0 is the mean free path at 0 K, which is only due to spontaneous emission of phonons. At higher temperatures, scattering is also due to stimulated emission and absorption processes whose combined rate is proportional to $2N$.

The phonon energy must also be modified, and one introduces the effective phonon energy (Crowell and Sze, 1966):

$$\langle E_p \rangle = E_p \tanh(E_p/2kT). \tag{62}$$

Figures 17 and 18 show typical modified Baraff plots for germanium and silicon, along with the experimental data (Logan and Sze, 1966; Crowell and Sze, 1966). For a given set of ionization measurements, the values of the ionization rates α and β, their field dependence, and the ionization energy E_i (assumed to be equal to $\frac{3}{2}E_g$ in Figs. 17 and 18) are fixed so that one can obtain λ by fitting the data to the Baraff plot. For Ge at 300 K, the value of $\langle E_p \rangle / E_i$ is 0.022. We find 64 Å for the room-temperature electron mean free path and 69 Å for the hole mean free path. From Eq. (61), one can then

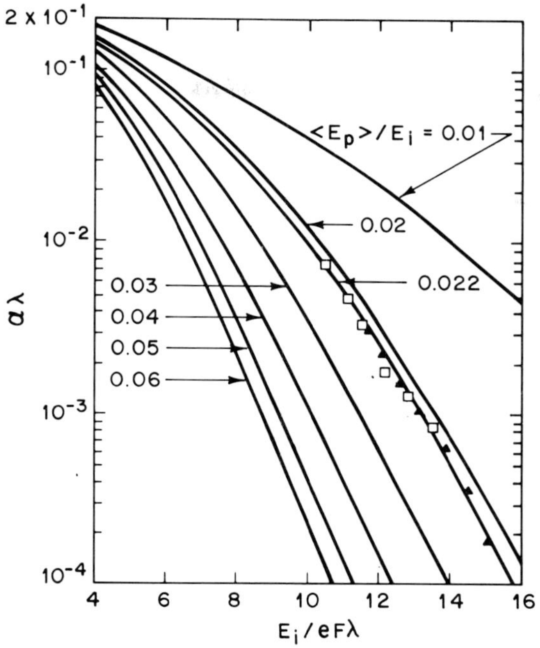

FIG. 17. Baraff's plot of the product of ionization rate and phonon mean free path ($\alpha\lambda$ versus $E_i/eF\lambda$) for Ge at 300 K, where E_i is the ionization energy and F the electric field. The running parameter is the ratio $\langle E_p \rangle/E_i$, where $\langle E_p \rangle$ is the average energy lost per phonon collision. The solid curves are theoretical results (after Baraff, 1962). The experimental data are from Ge $p-n$ junctions (▲, electrons, $\lambda = 64$ Å; □, holes, $\lambda = 69$ Å) (after Logan and Sze, 1966). The electron and hole phonon mean free paths are obtained by fitting the ionization rate data to the appropriate Baraff plot (after Sze, 1969).

obtain λ_0 and calculate λ at various temperatures; from the temperature dependence of $\langle E_p \rangle$, one can then choose the correct Baraff plot corresponding to the appropriate $\langle E_p \rangle/E_i$ and thus predict α and β at different temperatures.

The theoretically predicted electron-ionization rates in silicon as obtained from the previous approach are shown in Fig. 18, along with experimental data at three different temperatures (Crowell and Sze, 1966). The agreement is satisfactory.

Thus, Baraff's theory is fairly successful in describing the ionization rates in Si and Ge. Certainly, the theory cannot well describe ionization in materials where strong band-structure effects involving more than one band have been observed, such as the resonant enhancement of hole-impact ionization in $Al_xGa_{1-x}Sb$ (Hildebrand *et al.*, 1980).

One important finding of the Baraff theory is also that the ionization rate

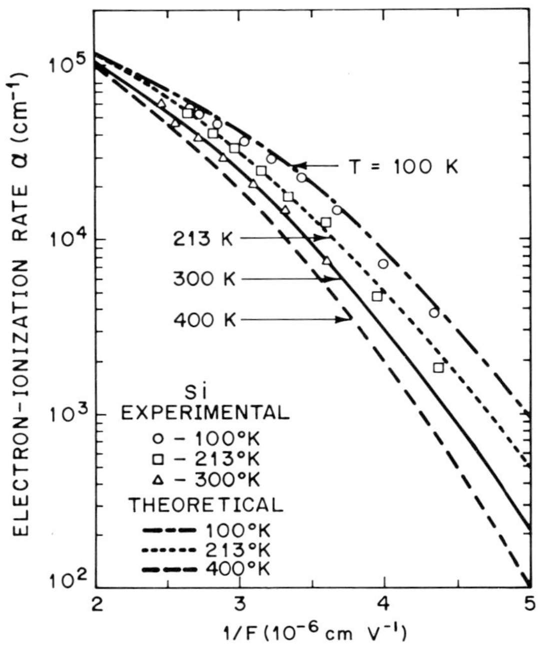

FIG. 18. Electron ionization rate versus reciprocal electric field for Si. The theoretical curves are obtained from Baraff's theory. [From Crowell and Sze (1966).]

depends weakly on the ionization cross section above threshold, provided that it is greater than approximately one-fourth of the phonon cross section. This is equivalent to assuming a large P in Keldysh's formula [Eq. (26)] for the ionization probability. This assumption, however, appears to be questionable in material such as silicon, as discussed in Section 1 on threshold energies.

6. Physical Interpretation of Baraff's Theory and the Lucky-Drift Model

Although Baraff's theory clarifies the nature and shape of the distribution function at very high electric fields, it leaves somewhat obscure the detailed physical mechanism by which carriers reach the ionization energy. This point has been clarified by Shichijo and Hess (1981) who performed a Monte Carlo simulation of the impact-ionization phenomenon in GaAs. The results of this calculation are discussed in great detail later. They are consistent with Baraff's theory, and in addition they provide a very clear physical picture of the ionization process, which can be easily explained as follows.

Carriers spend most of their time around an average energy under the heating influence of the electric field and the cooling influence of phonon collisions. These electrons are in a thermalized, quasi-steady-state distribution and correspond to those in the nearly symmetric part of Baraff's distribution. Occasionally, however, one of these carriers will escape all but a few phonon collisions and reach the ionization energy. These "quasi-ballistic" electrons are those in the "spike" of Baraff's distribution. The main difference from Shockley's lucky electrons is that they start from the average energy rather than from zero energy. In addition, they still undergo a few collisions before impact ionizing.

More recently, Ridley (1983a) has developed a simple analytical theory of impact ionization (the "lucky drift model"), which is based on a physical picture similar to that of Shichijo and Hess (1981) and gives results that are in excellent agreement with Baraff's theory.

This approach is based on the distinction between the rates of momentum and energy relaxation τ_m and τ_E. If $\tau_m \ll \tau_E$, which is true for many semiconductors at high energies, "it is possible for carriers to drift in an electric field with a drift velocity determined for a time by momentum-relaxing collisions without there being at the same time significant energy relaxation. This state is called lucky drift, and is is intermediate between Shockley's ballistic state and Wolff's equilibrium state" (Ridley, 1983a).

In Ridley's model, the dominant component of the ionization rate is due to electrons in the lucky-drift mode that start from the average energy; these carriers correspond to those in the simulation of Shichijo and Hess that set out from the average energy and ionize after one or two phonon collisions.

The lucky-drift model starts by considering the rate equations for energy and momentum, which read as follows:

$$d(m^*(E)v_{\mathrm{d}}(E))/dt = eF - m^*(E)v_{\mathrm{d}}(E)/\tau_m(E), \tag{63}$$

$$dE/dt = eFv_{\mathrm{d}}(E) - E/\tau_E(E). \tag{64}$$

In the lucky drift mode, the electron drifts for a time $t(<\tau_E)$ without significant energy relaxation if $\tau_m \ll \tau_E$. Thus, the momentum is randomized but not the energy; we can therefore equate to zero the right-hand side of Eq. (63) and neglect the relaxation term in Eq. (64). This yields

$$v_{\mathrm{d}}(E) = e\tau_m(E)F/m^*(E), \tag{65}$$

$$dE/dt = e^2F^2\tau_m(E)/m^*(E). \tag{66}$$

Once a steady state has been reached (for times $t > \tau_E$), the energy has relaxed also, and the average carrier energy and drift velocities are

$$E_{\mathrm{av}} = e^2F^2\tau_m(E_{\mathrm{av}})\tau_E(E_{\mathrm{av}})/m^*(E_{\mathrm{av}}), \tag{67}$$

$$v_{\mathrm{d}}(E_{\mathrm{av}}) = e\tau_m(E_{\mathrm{av}})F/m^*(E_{\mathrm{av}}). \tag{68}$$

The next step is to calculate the probability that a carrier gains the ionization energy E_i. The first way it can do that is by the lucky ballistic mode of Shockley, which has a probability

$$P_1(F, E_i) = \exp\left(-\int_0^{E_i} \frac{dE}{eF\lambda(E)}\right), \tag{69}$$

which is a simple generalization of Eq. (46) to the case of an energy-dependent mean free path.

Another way the electron can gain the energy E_i is by lucky drift. The probability that an electron initially at zero energy travels for a time t without suffering appreciable energy relaxation is

$$P_2(F, t) = \exp\left(-\int_0^{t} \frac{dt}{\tau_E(E)}\right). \tag{70}$$

Converting to energy through Eq. (66) gives

$$P_2(F, 0, E_i) = \exp\left(-\int_0^{E_i} \frac{m^*(E)\, dE}{e^2F^2\tau_m(E)\tau_E(E)}\right). \tag{71}$$

However, there are other possibilities also. One is a combination of purely ballistic motion up to an energy E, followed by a momentum-randomizing collision and a subsequent lucky drift up to the ionization threshold. It is assumed that one collision is sufficient to randomize momentum and transfer the electron from the ballistic to the lucky drift mode. This is a reasonable assumption, since once the electron is deflected by a collision, its path to threshold becomes much longer, making other collisions likely.

The probability that an electron gains the energy E_i after making a collision is therefore

$$P_2(F, E_i) = \int_0^{E_i} P_1(F, E)P_2(F, E, E_i)\frac{dE}{eF\lambda(E)}, \tag{72}$$

where $P_1(F, E)[dE/eF\lambda(E)]$ is the probability that an electron gains ballistically an energy E and then suffers a collision in the energy range dE; $P_2(F, E, E_i)$ is the probability of lucky drift from energy E to E_i, which is obtained by a straightforward generalization of Eq. (71); i.e.,

$$P_2(F, E, E_i) = \exp\left(-\int_E^{E_i} \frac{m^*(E)\, dE}{e^2F^2\tau_m(E)\tau_E(E)}\right). \tag{73}$$

We can call $P_2(F, E_i)$ the total probability of lucky drift from zero energy to E_i.

Finally, one must consider those electrons that start from the average energy and reach threshold either by a ballistic flight or by a ballistic flight followed by a lucky drift. This total probability for this process can be written as

$$P_T(P_1^T(F, E_i) + P_2^T(F, E_i)), \tag{74}$$

where $P_1^T(F, E_i)$ and $P_2^T(F, E_i)$ are obtained from Eqs. (69) and (72) by changing the lower limit of integration from zero to the average energy E_T; P_T is the probability of thermalization, i.e., that a carrier gains the energy E_T. Note that in general E_T is smaller than the steady-state average energy E_{av}, since at high electric fields the ionization distance E_i/eF is not long enough to allow sufficient time for the electrons to reach steady state. The average energy has been simply and somewhat arbitrarily estimated by Ridley (1983a) by solving the energy rate equation [Eq. (64)], converting to the space coordinate x and setting x equal to $E_i/eF - 3\lambda_{av}$. Note that $3\lambda_{av}$ (where λ_{av} is some average momentum relaxation mean free path) is a characteristic distance required for hot-electron thermalization to start. The factor of 3 is suggested by the fact that after a distance $l = 3\lambda_{av}$, $\exp(-l/\lambda_{av}) \simeq 0.1$, thus making it likely that the momentum is randomized and therefore that thermalization has started. In other terms, this cutoff procedure implies that no hot-electron thermalization occurs for electric fields so high that

$$E_i/eF < 3\lambda_{av}. \tag{75}$$

The average energy therefore reads

$$E_T = E_{av}\left[1 - \exp\left(-\frac{E_i/eF - 3\lambda_{av}}{v_d \tau_E}\right)\right]. \tag{76}$$

The quantity in parentheses can be interpreted as the probability of hot-electron thermalization.

The next step is to add all the calculated probabilities of reaching threshold. Using the general definition of the ionization coefficient [Eq. (39)], one then has

$$\alpha = eF/E_i[P_1(F, E_i) + P_2(F, E_i) + P_T(P_1^T(F, E_i) + P_2^T(F, E_i)]. \tag{77}$$

The lucky drift mechanism is possible only if the energy relaxation time is much longer than the momentum relaxation time. At the high energies typical of the impact-ionization regime, deformation potential scattering (which is isotropic and proportional to the density of final states) is the dominant electron–phonon interaction. In this case, the ratio of the two relaxation times takes the simple form (Ridley, 1983a)

$$\tau_E(E)/\tau_m(E) = E/rE_i, \tag{78}$$

where r is the ratio of the effective phonon energy [Eq. (62)] to the ionization energy. Lucky drift is therefore a valid process if $E \gg rE_i$, which is true in semiconductors with a band gap of 0.5 eV or more. For materials where polar-mode scattering is the dominant interaction, the lucky drift model is obviously not valid.

To obtain an explicit expression for α from Eq. (77), one must know the energy dependence of the scattering rates. In the case of simple parabolic bands, the mean free path λ is independent of energy, and using Eq. (78) one has

$$\frac{m^*(E)}{\tau_m(E)\tau_E(E)} = 2rE_i/\lambda^2. \tag{79}$$

Substituting Eq. (79) for the relevant terms in Eq. (77) and performing the integration, one then has

$$\alpha\lambda = \left\{ e^{-x} + \left[\frac{e^{-2rx^2} - e^{-x}}{1 - 2rx}\right] + P_T \left[e^{-x(1-\xi)} + \frac{(e^{-2rx^2(1-\xi)} - e^{-x(1-\xi)})}{1 - 2rx} \right]\right\}, \tag{80}$$

where

$$x = E_i/eF\lambda, \qquad \xi = E_T/E_i = P_T/2rx^2, \qquad P_T = 1 - e^{-2rx(x-3)}|_{x\geq 3}. \tag{81}$$

Note that the parameters appearing in Eq. (81) are the same as those of Baraff's theory; i.e., E_i, λ, and r. In Fig. 19 a comparison between Baraff's theory (dashed curves) and the lucky drift theory (solid curves) is shown for various values of r. Good agreement is generally observed.

Note that Eq. (80) can be simplified by maintaining only the terms representing lucky drift from the average energy to give

$$\alpha\lambda = [x(1 - 2rx)]^{-1}[e^{-2rx^2} - e^{-x} + P_T(e^{-2rx^2(1-\xi)} - e^{-x(1-\xi)})]. \tag{82}$$

This confirms Baraff's remarks that Shockley's lucky electrons do not play a major rule in the ionization process.

Ridley (1983a) has also shown that Eq. (81) is a reasonably valid expression for the case of a real band structure if λ is interpreted now as an average mean free path in the range of energies $0 \leq E \leq E_i$. This average mean free path is roughly equal to 1.5 times the mean free path at E_i. Ridley notes that at high energies, the scattering rate is predominantly determined by density of states [Eq. (35)]. In addition, at these elevated energies the density-of-states function looks roughly the same for all semiconductors, and the same is true for holes. The density of states increases with energy significantly faster than for a parabolic band; this increase is roughly linear with energy.

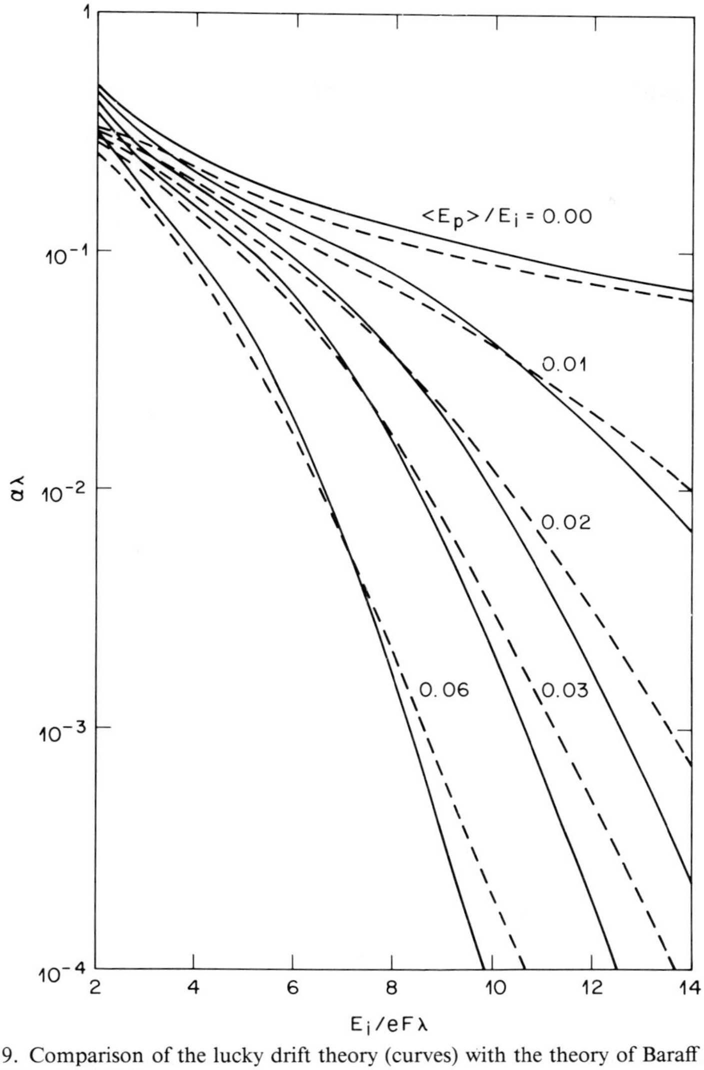

FIG. 19. Comparison of the lucky drift theory (curves) with the theory of Baraff (curves). [From Ridley (1983a).]

Thus, the major effect of a real band structure is to cause the mean free path to decrease with increasing energy with a dependence of the type $\lambda \simeq E^{-1/2}$, which must be then averaged over the energy range, as discussed earlier.

By fitting the lucky drift model to the experimental ionization rates data in many materials, Ridley (1983b) was able to derive an approximate universal expression for the average mean free path at 0 K valid for electron and holes:

$$\lambda(0) \simeq 600(\rho E_p / E_i^{1/2}) \quad \text{Å}, \tag{83}$$

where E_p is taken to be the mean of the LO and LA zone-edge phonons. Both E_p and E_i are expressed in electron volts, the density ρ is given in grams per cubic centimeter. From Eq. (83) one can then obtain λ at any temperature by using Eq. (61).

The previously defined average mean free path is independent of the electric field. In reality, if one does a rigorous averaging procedure using the energy distribution function, one finds a field-dependent mean free path. This has been done via a Monte Carlo simulation in GaAs by Shichijo and Hess (1981). They find a relatively weak dependence of the high-field average mean free path λ on the field.

The previous discussion indicates that at least for certain materials, the constant mean free path approximation is not too bad even in the presence of strong nonparabolicities. This also explains the surprising success of Baraff's theory in spite of the many simplifying assumptions.

7. Other Analytical Expressions for the Ionization Rates

It is convenient to express Baraff's plots in an analytical way. This is done very accurately by Sutherland's (1980) cubic fit. The cubic polynomial is expressed as

$$\alpha = \lambda^{-1} \exp[C_0(r) + C_1(r)x + C_2(r)x^2 + C_3(r)x^3], \tag{84}$$

where

$$C_0(r) = -0.07238 - 51.5r + 239.6r^2 + 3357r^3, \tag{85}$$

$$C_1(r) = -0.4844 + 12.45r + 363r^2 - 5836r^3, \tag{86}$$

$$C_2(r) = 0.02982 - 0.07571r - 148.1r^2 + 1627r^3, \tag{87}$$

$$C_3(r) = -1.841 \times 10^{-5} - 0.1851r + 10.41r^2 - 95.65r^3, \tag{88}$$

$$r = \langle E_p \rangle / E_i, \tag{89}$$

$$x = E_i / F\lambda. \tag{90}$$

The units of α are number of ionizations per centimeter, assuming λ is in centimeters; F is in volts per centimeter, and E_i in electron volts. Identical equations hold for the hole ionization rate β.

An alternative approach to the theory of impact ionization, physically equivalent to the theories of Baraff and Ridley, was chosen by Keldysh (1965). Rather than numerically integrating the Boltzmann equation for a series of chosen parameters, he solved for the distribution function in analytic form for arbitrary values of temperatures and electric fields. For the case of parabolic bands, isotropic masses, and energy-independent phonon scattering mean free paths, he finds for the energy distribution function,

$$f_0(E) = \text{const}\ E^{\nu} \exp - [(E/eF\lambda)s_0(F, T)], \tag{91}$$

where s_0 is the positive root of the transcendental equation

$$\frac{[1 + (\lambda_{op}/\lambda_{ac}] \cosh(E_p/2kT)}{(\lambda_{op}/\lambda_{ac}) \cosh(E_p/2kT) + \cosh((E_p/2kT) - (s_0 E_p/eF\lambda))} + \frac{1}{2s_0} \ln\left(\frac{1 - s_0}{1 + s_0}\right) = 0, \tag{92}$$

and the exponent ν is a function of s_0,

$$\begin{aligned} \nu = -\frac{1}{2}\Bigg\{1 - \frac{E_p}{eF\lambda} \tanh\left(\frac{E_p}{2kT}\right) \frac{s_0}{1 - s_0^2} \\ \times \Bigg[\frac{1 + \lambda_{op}/\lambda_{ac}}{1 - s_0^2} + \frac{1 + \lambda_{op}/\lambda_{ac}}{2s_0} \ln\left(\frac{1 - s_0}{1 + s_0}\right) \\ - \frac{s_0 E_p}{eF\lambda}\left(\frac{1}{2s_0} \ln\left(\frac{1 - s_0}{1 + s_0}\right)\right)^2 \times \frac{\sinh[(E_p/2kT) - s_0(E_p/eF\lambda)]}{\cosh(E_p/2kT)}\Bigg]^{-1}\Bigg\}. \end{aligned} \tag{93}$$

Note that

$$1/\lambda = 1/\lambda_{op} + 1/\lambda_{ac} \tag{94}$$

where λ_{op} and λ_{ac} are the mean free paths for optical and acoustic phonon scattering, respectively.

For $eF\lambda \gg E_p$,

$$s_0 \cong \frac{3}{1 + \lambda_{op}/\lambda_{ac}} \frac{E_p}{eF\lambda} \tanh\left(\frac{E_p}{2kT}\right). \tag{95}$$

If we substitute Eq. (95) in Eq. (91) and neglect acoustic phonons ($\lambda_{ac} = \infty$), then the energy-distribution function is essentially that of Wolff [Eq. (56)], with the phonon energy substituted by the effective energy lost per collision [Eq. (62)]. On the other hand, in the opposite limit of $eF\lambda \ll E_p$, $f_0(E)$ is proportional to $\exp(-E/eF\lambda)$, thus describing carriers that accelerate to an energy E without collisions (Shockley's lucky electrons).

Keldysh was also able to generalize his theory to the case of nonparabolic bands. In this case, the mean free path is energy dependent, and the energy distribution function is given by

$$f_0(E) = \text{const} \exp\left[-\int_0^E \frac{dE'}{eF\lambda(E')} s_0(E')\right], \tag{96}$$

where $s_0(E')$ is given by Eq. (92), with the mean free paths substituted by energy-dependent quantities. The impact-ionization coefficient in Keldysh's theory is calculated analogously to a scattering cross section as the number of ionizations per unit time per unit volume normalized to the incident flux; i.e.,

$$\alpha(F) = \frac{1}{nv_{\mathrm{d}}} \int_0^\infty \tau_{\mathrm{i}}^{-1}(E) f_0(E) \rho(E)\, dE, \tag{97}$$

where $[\tau_{\mathrm{i}}(E)]^{-1}$ is the impact-ionization probability per unit time as given by Eq. (26), $\rho(E)$ the density of states, n the carrier density, and v_{d} the drift velocity. This definition is equivalent, under steady-state conditions, to the reciprocal of the average distance between ionizing collisions. However, it has the advantage of being more general in the sense that it represents the best definition for the nonlocal case; i.e., when α is an explicit function of position, as demonstrated by Thornber (1981). For the case of parabolic bands with isotropic effective masses and a quadratic increase of the ionization cross section above threshold, Eq. (97) yields

$$\begin{aligned} \alpha(F) = \frac{eF}{E_{\mathrm{i}}} & \left[\frac{E_{\mathrm{i}} s_0}{eF\lambda + E_{\mathrm{p}} \exp[(s_0 - 1)/s_0]} \right]^{\nu+2} \\ & \times \phi_2 \left[\frac{E_{\mathrm{i}} s_0}{eF\lambda} \left(\frac{s_0^2}{12P} \right)^{1/2} \right] \exp\left(-\frac{E_{\mathrm{i}}}{eF\lambda} s_0 \right), \end{aligned} \tag{98}$$

where the special function ϕ_2, calculated at $z = (E_{\mathrm{i}} s_0/eF\lambda)(s_0^2/12P)^{1/2}$, is defined as

$$\phi_2(z) = z \int_0^\infty \exp\left\{ -\frac{z}{2} \int_0^x [1 + (1 + y^2)^{1/2}]\, dy \right\} x^2\, dx, \tag{99}$$

and P is the dimensionless constant appearing in the Keldysh formula for the ionization probability [Eq. (26)]. The energy-conserving factor in Eq. (98) before the exponential is equivalent to $eF/\langle E_{\mathrm{i}} \rangle$ of Eq. (40).

Chuenkov (1967) used an approach similar to that of Keldysh but considered the more general case of anisotropic scattering with an energy-dependent mean free path. This applies to materials such as InSb and InAs. He finds a field dependence of the ionization coefficient that is intermediate between that of Shockley and Wolff and given by

$$\alpha \simeq \exp[-\mathrm{const}\; F^{-\gamma(F)}], \tag{100}$$

where $\gamma(F)$ is a weak function of the electric field and is in the range $0 < \gamma \leq 2$.

Dumke (1968) calculated the impact-ionization rate per unit time g for InAs and InSb and pointed out that the distribution function is strongly peaked in the direction of the field. The reason for this was discussed in detail in Section 3 on phonon scattering. Unfortunately, it is not easy to obtain from Dumke's results the ionization rate α, since a knowledge of the drift velocity v_d in the avalanche regime is needed (note that $\alpha = g/v_d$). Dumke's theory has been improved by van Welzenis and de Zeew (1983) and found to give good agreement with their time-dependent pair-generation-rate measurements in InSb at 77 K.

Okuto and Crowell (1972) developed a simple empirical expression for the ionization coefficients that fits experimental data for various materials (Ge, Si, GaAs, and GaP) and requires only one adjustable parameter (the phonon mean free path). This expression is

$$\alpha = (eF/E_i) \exp\{0.217(E_i/E_p)^{1.14} - [(0.217(E_i/E_p)^{1.14})^2 + (E_i/eF\lambda)^2]^{1/2}\}. \tag{101}$$

Equation (101) reduces to Shockley's and Wolff's expressions in the low- and high-field limits, respectively, and is consistent with Baraff's plots at intermediate fields.

More recently, Chwang *et al.* (1979) presented a theory of impact ionization valid for nonpolar semiconductors, such as Si and Ge, based on a Markov chain approach. Although this formulation does not contain much more physical information than Baraff's theory, these authors were able to express results in a relatively simple analytical formula, which, unlike Eq. (101), contains the effect of a finite cross section above threshold via the effective ionization energy $\langle E_i \rangle$. This expression is

$$\alpha = (eF/\langle E_i \rangle) \exp[A(r, x)/r - [A^2(r, x)/r^2 + x^2(1 - r)^2]^{1/2}], \tag{102}$$

where

$$A(r, x) = \sum_{i=0}^{3} (r - 0.06)^i \frac{a_{i0} + a_{i1}x + a_{i2}x^2}{1 + a_{i3}x + a_{i4}x^2}, \tag{103}$$

$$r = \langle E_p \rangle / \langle E_i \rangle, \tag{104}$$

$$x = \langle E_i \rangle / qE\lambda, \tag{105}$$

where the adjustable parameters in Eq. (103) can be obtained by least-squares fitting the experimental data, as discussed in detail by Chwang *et al.* (1979).

Thornber (1981) presented a formula that he claimed to be the first simple, physical, analytical expression for the ionization coefficients valid for all fields and for arbitrary band structures. This formula is

$$\alpha(F) = \frac{eF}{\langle E_i \rangle} \exp\left[-\frac{F_i}{F(1 + F/F_p) + F_{kT}}\right], \tag{106}$$

where F_{kT}, F_p, and F_i are the threshold fields for carriers to overcome the decelerating effects of thermal (acoustic) phonon, optical phonon, and ionization scattering respectively. Note that no mean free paths enter this expression. This is a desirable feature, since as noted previously, phonon scattering cannot be characterized in general by a constant mean free path. In Eq. (106) one can assume that $F_i/F_{kT} = E_i/kT$; this is required by Boltzmann statistics, since at very low fields ($F \to 0$) such that $F < F_{kT}$, the exponential term in Eq. (106) must tend to $\exp(-E_i/kT)$.

Thornber's theory has been particularly successful in fitting experimental data for silicon. The results of these fits for silicon are summarized in Table IV. Note that the data of Woods *et al.* (1973) and Grant (1973) are among the most reliable available to date in Si. It is interesting to note that every set of data yields the same effective ionization energy 3.6 eV for electrons, which is substantially higher than the threshold energy (1.1 eV) and is consistent with the slow increase of the ionization cross section, as calculated by Kane (1967), and with the observed charging of the Si–SiO_2 interface, as previously discussed in this section. It thus appears that previous fits of data in silicon that assume low ionization energies have limited physical significance. The author of this chapter has found a strong physical justification of Thornber's formula.

TABLE IV

EXPERIMENTAL IONIZATION RATE PARAMETER FOR Si

Ref.	Parameter[a]	Electrons	Holes
Van Overstraeten and DeMan (1970)			
	E_i	3.6 eV	6.2 eV
	F_i	1.404×10^6 V cm^{-1}	5.933×10^6 V cm^{-1}
	F_r	2.229×10^6 V cm^{-1}	5.910×10^4 V cm^{-1}
	F_{kT}	9.747×10^3 V cm^{-1}	2.392×10^4 V cm^{-1}
Woods *et al.* (1973)			
	E_i	3.6 eV	6.2 eV
	F_i	1.659×10^6 V cm^{-1}	1.411×10^6 V cm^{-1}
	F_r	2.384×10^5 V cm^{-1}	1.574×10^6 V cm^{-1}
	F_{kT}	1.152×10^4 V cm^{-1}	5.689×10^3 V cm^{-1}
Grant (1973)			
	E_i	3.6 eV	5.0 eV
	F_i	1.954×10^6 V cm^{-1}	3.091×10^6 V cm^{-1}
	F_r	1.069×10^5 V cm^{-1}	1.110×10^5 V cm^{-1}
	F_{kT}	1.357×10^4 V cm^{-1}	1.545×10^4 V cm^{-1}

[a] Values of the parameters are obtained by fitting the experimental ionization rates with Thornber's law Eq. (106); $F_i/F_{kT} \equiv E_i/kT$ has been assumed.

Let us consider the case where optical phonon scattering can be described by an energy-independent mean free path (parabolic bands) and where acoustic phonon scattering is negligible ($F_{kT} = 0$). The threshold fields F_p and F_i can then be expressed by (Thornber, 1981)

$$F_i = \langle E_i \rangle / e\lambda, \tag{107}$$

$$F_p = 3E_p / e\lambda. \tag{108}$$

Substitution of these expressions in Eq. (106) gives

$$\alpha(F) = \frac{eF}{\langle E_i \rangle} \exp\left[-\frac{\langle E_i \rangle}{[(eF\lambda)^2/3E_p] + eF\lambda}\right]. \tag{109}$$

It is very interesting to note that the exponential term in Eq. (109) is essentially proportional to the energy-distribution function calculated at the ionization energy obtained heuristically by Baraff (1962) by assuming a momentum distribution function consisting of a "spike" and a spherically symmetric part [Eq. (58)]. Thus, Thornber's theory contains the same physical information as Baraff's theory. Note also that Eq. (109) reduces to Shockley's and Wolff's [Eqs. (47) and (56)] in the low- and high-field limits, respectively.

The "tunnel-impact ionization" effect discussed in Part II has been incorporated in the expression of the ionization coefficient by Kyuregyan (1976) for the case of parabolic bands. The expression for α is

$$\alpha(F) = \exp(\delta)\alpha_0(F), \tag{110}$$

where $\alpha_0(F)$ is the ionization rate Eq. (98) calculated by Keldysh (1965). The factor $\exp(\delta)$ represents the contribution to ionization of those electrons with energies below threshold. These carriers, upon colliding with an atom of the lattice, excite a valence electron to a virtual state in the gap from whence it tunnels into the conduction band. For $eF\lambda \gg E_p$, $\delta \propto \text{const}/F^4\lambda^6$, for $kT \leq eF\lambda \leq E_p$, $\delta \propto \text{const}/F\lambda^3$, whereas $\delta \propto \text{const} \times E^2$ when $eF\lambda \ll kT$. Consequently, δ and the contribution of electrons below the ionization threshold (α/α_0) at first increase with field ($\delta \propto F^2$), reach a maximum when $eF\lambda \simeq kT$, and then decrease as the field is increased further, first according to $\delta \propto F^{-1}$, then as $\delta \propto F^{-4}$ when $eF\lambda \gg E_p$. This apparent paradoxical effect occurs because as the field is increased, at first the probability of tunnel-impact ionization increases more rapidly than the probability of direct ionization, whereas for $eF\lambda \geq kT$, the ionization probability increases with field faster than the tunnel-ionization rate.

Next, we shall briefly discuss the effect of high carrier densities on the impact-ionization rates. Most theories of impact ionization neglect carrier–carrier scattering. Although this is a good approximation in avalanche photodiodes, it is a poor one in IMPATT diodes where carrier densities typically exceed 10^{16} cm^{-3}. Very little work has been done to understand the effect of intercarrier scattering on α and β.

Kafka and Hess (1981) have used a carrier temperature model to derive an expression for α in the presence of large current densities. They assume that high-energy carriers are in a Maxwell–Boltzmann distribution characterized by an effective temperature T_e. This is a valid approximation in view of the expected high electron–electron scattering rate. The final expression for α is proportional to $\exp(-E_i/kT_e)$. Note, however, that this electron temperature is different from that given in Eq. (54) (Wolff's approximation), since it is dependent on the carrier density.

Reuter and Hübner (1971) have studied theoretically electron-impact ionization in crossed electric and magnetic fields. The ionization rate was found to increase with the quotient of the effective electric field $(F_a^2 + F_H^2)^{1/2}$ and the magnetic field B; F_a is the bias field and F_H the Hall field.

8. Effective Ionization Energy

It was previously noted that hot carriers impact ionize at energies higher than the ionization threshold energy E_i as a result of the finite ionization cross section. It is thus important to calculate the average or effective ionization energy $\langle E_i \rangle$ that appears in the general expression of the ionization coefficient [Eq. (40)]; $\langle E_i \rangle$ is in general going to be determined by two competing effects; the drop and decrease of the distribution function above threshold and the increase of ionization cross section (Kane, 1967).

Chwang and Crowell (1976) have calculated $\langle E_i \rangle$ assuming that carriers that reach the ionization energy are streaming ballistically along the direction of the field. For this case, the ionization density function (probability per unit energy interval of creating an electron–hole pair) reads

$$\rho(E')\,dE' = \frac{1}{eF\lambda_i}\exp\left[-\int_0^{E'}\left(\frac{1}{\lambda_{op}} + \frac{1}{\lambda_i}\right)\frac{dE''}{eF}\right]dE', \qquad (111)$$

where the energy E' is measured from the threshold energy E_i, and the ionization mean free path λ_i is obtained from Keldysh's expression [Eq. (26)]; i.e.,

$$1/\lambda_i = (P/\lambda_{op})(E'/E_i)^2, \qquad (112)$$

where λ_{op} is the mean free path for scattering by optical phonons. Note that $\rho(E')$ is proportional to the number of carriers having an energy greater than the threshold E_i by an amount E', times the ionization probability $1/\lambda_i$.

Making the following substitutions,

$$x = E'/E_i, \tag{113}$$

$$a = E_i/eF\lambda_{op}, \tag{114}$$

$$b = P(E_i/eF\lambda_{op}), \tag{115}$$

and performing the integration yields

$$\rho(E')\,dE' = bx^2 \exp\{-[ax + (b/3)x^3]\}\,dx. \tag{116}$$

The average ionization energy is then calculated as

$$\langle E_i \rangle - E_i = \frac{\int_0^\infty bE'^3 \exp\{-[aE' + (b/3)E'^3]\}\,dE'}{\int_0^\infty bE'^2 \exp\{-[aE' + (b/3)E'^3]\}\,dE'}. \tag{117}$$

The values of these integrals can be expressed analytically in terms of gamma functions and are given by Chwang and Crowell (1976). For $P \simeq 10^2$, they find that on the average the impact-ionization event takes place at 0.6 of $eF\lambda_{op}$ above the ionization threshold. For $\lambda_{op} = 50$ Å and $F = 4 \times 10^5$ V cm^{-1}, this corresponds to an energy of 0.12 eV.

A quick evaluation of the effective ionization threshold can be obtained using Kane's observation (1967) that $\langle E_i \rangle$ may be estimated from the maximum of the product of the ionization cross section and the distribution function above threshold. This product is simply proportional to Eq. (111).

Chwang *et al.* (1979) have generalized the method previously outlined to a more realistic distribution function of the Baraff type obtained in a Markov chain approach and presented an empirical equation to calculate $\langle E_i \rangle$. This equation reads

$$\langle E_i \rangle = \langle E_{is} \rangle [1 - 0.173P^{-0.4} \exp(-P^{1/2}/50) - 0.0395E_i/eF\lambda_{op}P], \tag{118}$$

where

$$\langle E_{is} \rangle = E_i \left\{ 1 + \left[\frac{3}{(PE_i/eF\lambda_{op})} \right]^{1/3} \Gamma\left(\frac{4}{3}\right) \right\}, \tag{119}$$

and $\langle E_{is} \rangle$ is an effective ionization threshold calculated using Eq. (117), valid for a ballistic distribution but neglecting phonon scattering above threshold. Figure 20 is a plot of $\langle E_{is} \rangle / E_i$ versus $E_i/eF\lambda_p$ for various values of P. Let us be reminded that the dimensionless constant P that appears in Keldysh's formula [Eq. (26)] can be viewed physically as a parameter that specifies the energy at which the phonon and impact-ionization scattering

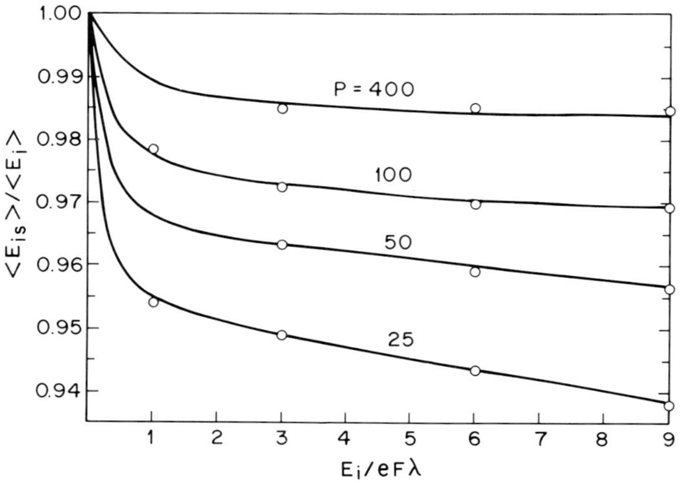

FIG. 20. The ratio of the calculated exact effective ionization threshold $\langle E_i \rangle$ to that deduced assuming a ballistic energy distribution function $\langle E_{is} \rangle$ versus $E_i/eF\lambda$ for various values of P. [From Chwang *et al.* (1979).]

rates are equal. It is seen that $\langle E_i \rangle$ differs from $\langle E_{is} \rangle$ by not more than 10% for P and $E_i/eF\lambda_{op}$ in the ranges 25–400 and 1–10, respectively, so that Eq. (119) can be used as a reliable estimate of the effective ionization energy in many cases.

9. Band-Structure-Dependent Theories

The theories of impact ionization previously considered assume either parabolic or simple nonparabolic bands or incorporate the band structure in appropriate parameters. Several attempts have been made to incorporate more realistic energy bands in the theory of impact ionization.

Capasso and Bachelet (1980) generalized Shockley's lucky electron approach to the case of strongly anisotropic conduction bands (GaAs and InP). This work was motivated by the observation of the crystal orientation dependence of the ionization coefficients in GaAs (Pearsall *et al.*, 1977b, 1978). It was then suggested that the observed anisotropy of the electron ionization rate could be due to the orientation dependence of ballistic motion. This interpretation is controversial, however, and should be reconsidered or at least modified in the light of more recent calculations and experimental results.

Capasso *et al.* (1979b) gave the criterion necessary for ballistic electrons to give a significant contribution to impact ionization. The time required by a carrier to gain the threshold energy can be calculated from the equation of motion

$$e\mathbf{F} = \hbar(d\mathbf{k}/dt), \tag{120}$$

which upon integration yields

$$\mathbf{k} = (e\mathbf{F}/\hbar)t + \mathbf{k}_0. \tag{121}$$

Setting $k = k_i$ (wave vector corresponding to the threshold energy E_i) and assuming that the carrier starts from zero energy ($k_0 = 0$), one has

$$t_i = \hbar k_i / eF. \tag{122}$$

This time should be compared with the phonon scattering time, which in general is energy dependent. For a conservative estimate, one can take a scattering time corresponding to the highest density of states in the energy range from 0 to E_i. This will be the minimum value of the scattering time τ_{min} and should be compared with t_i.

If $t_i \leq \tau_{min}$, then a significant fraction of electrons will be able to reach the threshold energy ballistically in the direction of the field if indeed a threshold exists along that orientation. The probability of ballistically reaching E_i is (Capasso and Bachelet, 1980)

$$P(E_i) = \exp\left[\int_0^{E_i} - \frac{dt'}{\tau(E(t'))}\right], \tag{123}$$

where $\tau(E(t))$ is the scattering time or average time between collisions with phonons at an electron energy E; E depends (through the electron wave vector k) on the time t. This is the time required to gain ballistically the energy E in the direction of the field and is obtained from Eq. (121) by setting $k_0 = 0$, i.e.,

$$t' = \hbar k / eF. \tag{124}$$

Using Eq. (124), Eq. (123) can be rewritten as

$$P(E_i) = \exp\left[-\frac{\hbar}{eF}\int_0^{E_i} \left(\frac{dE}{dk}\right)^{-1} \frac{dE}{\tau(E(k))}\right]. \tag{125}$$

This expression clearly shows that $P(E_i)$ depends on the orientation of the field via the band-structure slope along its direction. Note that the group velocity along this direction is given by

$$v_g = \frac{1}{\hbar}\frac{dE(k)}{dk} \tag{126}$$

Substituting Eq. (121) in Eq. (126), one can obtain the time dependence of the velocity of a ballistic carrier moving along a given crystal orientation.

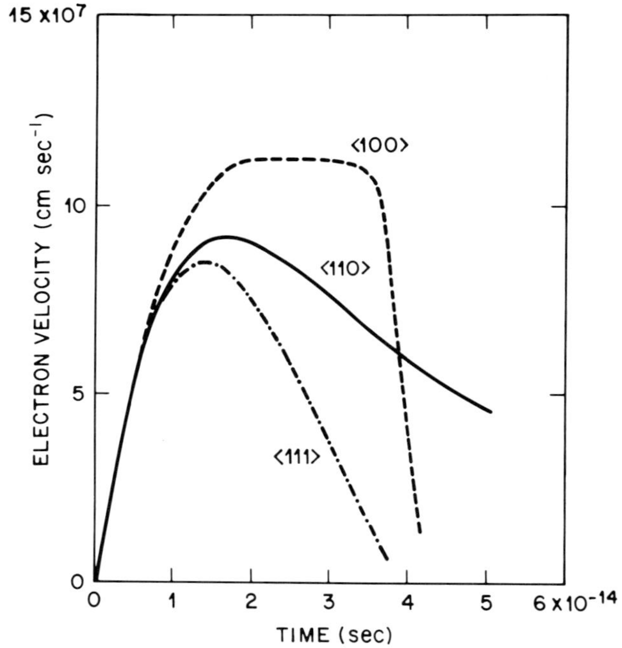

FIG. 21. Variation of ballistic electron velocity with time in three crystallographic directions of GaAs ($F = 500$ kV cm^{-1}). [From Shichijo *et al.* (1980).]

This has been done by Shichijo *et al.* (1980) for GaAs using the band structure of Cohen and Bergstresser (1966) and is illustrated in Fig. 21. It is seen that ballistic electrons have the highest velocity along the $\langle 100 \rangle$ orientation. Thus, this orientation is the most favorable for ballistic motion in GaAs.

For the $\langle 100 \rangle$ orientation actually, Eq. (125) should be multiplied by the tunneling probability through the pseudogap between Γ_6 and Γ_7–X_7, since to reach threshold ballistically along this orientation the electron must tunnel into the upper conduction band, as illustrated in Fig. 5b. Assuming a triangular barrier of height ΔE, T is given by

$$T = \exp\left[-\frac{\pi m^{*\,1/2}(\Delta E)^{3/2}}{eF\sqrt{2}\hbar}\right], \tag{127}$$

where m^* is the electron effective mass at the top of the first conduction band along the $\langle 100 \rangle$ orientation.

The ionization rate can then be calculated using the general definition of α [Eq. (39)] (Capasso and Bachelet, 1980). Capasso and Bachelet (1980) calculated $P(E_i)$ for $\langle 100 \rangle$ and $\langle 110 \rangle$ GaAs using the band structure of Cohen and

Chelikowsky for the Γ valley and assuming parabolic bands for the satellite valleys. For the intervalley scattering time, they used Eq. (33). From Fig. 14 it is seen that $\tau \simeq 10^{-14}$ sec for energies ≥ 1 eV.

At a field of 4×10^5 V cm^{-1}, the ballistic time required to reach the threshold momentum $\hbar k_i$ along the ⟨100⟩ direction is given by

$$t_i = 2\hbar\pi/3eFa \simeq 4 \times 10^{-14} \quad \text{sec} \tag{128}$$

where a is the GaAs lattice constant. Thus, $t_i \geq \tau$, and one must rely on a calculation of $P(E_i)$ to assess the importance of ballistic electrons.

In Fig. 22 we have plotted the ballistic probability of reaching the threshold momentum along the ⟨100⟩ and ⟨110⟩ orientations, respectively. We see that the effect of the anisotropy of the group velocity is indeed large and

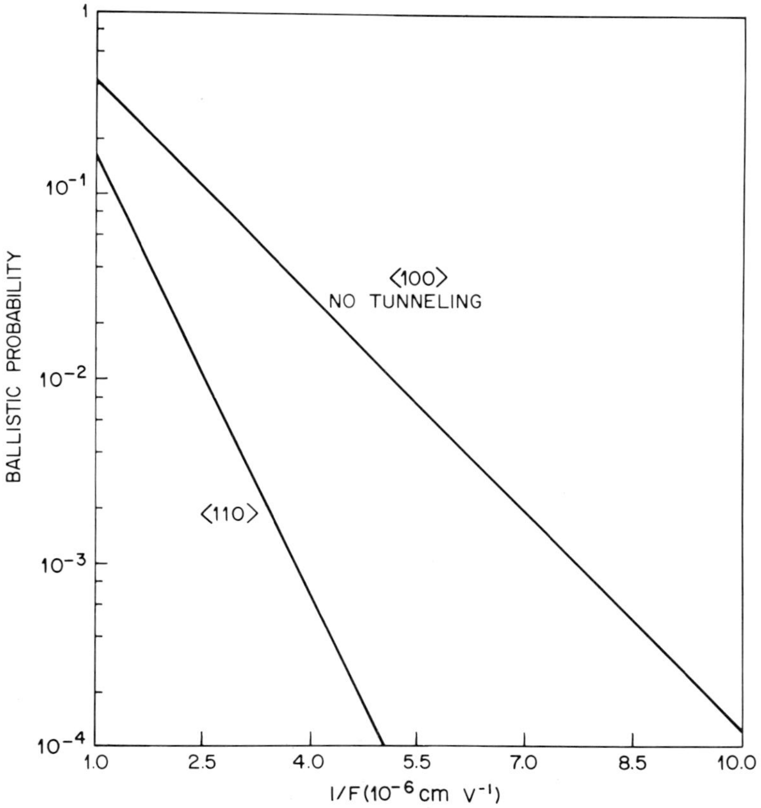

FIG. 22. Ballistic probability of reaching the wave vector corresponding to the ionization threshold versus reciprocal electric field for two different orientations of the electric field (⟨100⟩ and ⟨110⟩) in GaAs at 300 K; the initial electron energy is zero.

produces a marked orientation dependence of the ballistic ionization rate. Note that along the direction $\langle 111 \rangle$ in GaAs, there can be no impact ionization.

In Fig. 23 we have included the effect of tunneling into the upper conduction band for two different values of ΔE. We see that for $\Delta E = 0.2$ eV at an electric field $E = 4 \times 10^5$ V cm^{-1}, $P(E_i) = 2 \times 10^{-2}$. For the $\langle 110 \rangle$ direction, one finds instead $P(E_i) = 10^{-2}$. This shows that ballistic electrons starting from zero energy play a minor role in the ionization process at these values of the electric field in GaAs. At lower fields, such that the average energy gained between collisions $eF\lambda$ is smaller than the average energy lost per collision $\langle E_p \rangle$, $P(E_i)$ is even smaller, but only ballistic carriers will

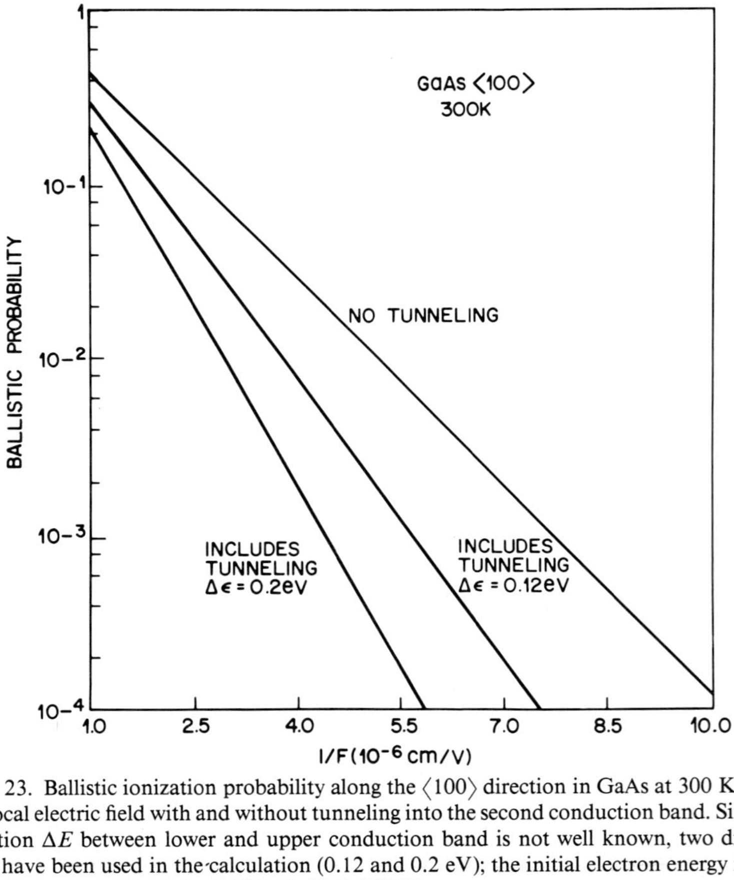

FIG. 23. Ballistic ionization probability along the $\langle 100 \rangle$ direction in GaAs at 300 K versus reciprocal electric field with and without tunneling into the second conduction band. Since the separation ΔE between lower and upper conduction band is not well known, two different values have been used in the calculation (0.12 and 0.2 eV); the initial electron energy is zero. [From Capasso and Bachelet (1980). © 1980 IEEE.]

contribute to the ionization rate. This should lead to a strong orientation dependence of α.

Note that if one used the correct band structure of Chelikowsky and Cohen also for the satellite valleys, the scattering rate would go up significantly since the density of states is higher than in parabolic bands. In fact, Chang *et al.* (1983) have found scattering rates substantially in excess of 10^{14} sec^{-1}, even using the correct nonperturbative calculation, which includes quantum many-body effects. Using these higher scattering rates would certainly decrease $P(E_i)$ considerably below 10^{-2}. Let us recall, however, that the values of the coupling constants D to the satellite valleys are not exactly known. Since $\tau^{-1} \propto D^2$, a value of D smaller than that used by Capasso *et al.* (1970a) and Chang *et al.* (1983) ($D = 10^9$ eV cm^{-1}) would result in a longer scattering time. This would enhance $P(E_i)$. From this it should be clear that an accurate knowledge of the coupling constants is needed for a calculation of the ionization rates.

Calculations of the electron ionization rate have made extensive use of the Monte Carlo (MC) method. The MC simulation technique has been extensively used in the investigation of steady-state and transient high-field transport properties in semiconductor devices. The principle of the method is to simulate the motion of electrons in momentum space according to the semiclassical particle description of the Boltzmann equation. This consists of a drift in the electric field followed by a scattering event. The drift phonon-scattering rates and final state are controlled by sequences of random numbers. The MC method is especially useful in studying transport properties when band-structure parameters and values of the phonon coupling constants are not well known and must be varied to achieve a best fit with experimental data. It has been applied to the ionization problem in GaAs by Shichijo and Hess (1981). It is important to analyze the approximations used by Shichijo and Hess. These are

(1) The pseudopotential band structure of Cohen and Bergstresser was used, which is qualitatively similar to that of Chelikowsky and Cohen.

(2) An isotropic threshold energy of 2.0 eV for electron-initiated impact ionization was used, and it was assumed that electrons could not reach this threshold along the $\langle 111 \rangle$ direction.

(3) An isotropic (only energy dependent) phonon scattering rate was taken, and it was assumed that electrons in the satellite valleys have the same scattering rate as in the central valley. This somewhat overestimates the scattering rate when electrons are in the satellite valleys.

(4) Interband tunneling and scattering were neglected, so that the second conduction band was not included.

(5) The Keldysh formula [Eq. (26)] was used for the ionization cross section with a large $P(> 50)$ as a parameter.

(6) It is tacitly assumed that electrons reach a spatial steady state and that the avalanche region is very long.

The most important result of Shichijo and Hess's work is the clarification of the mechanism by which electrons gain ionization energy and of Baraff's statement that the notion of ballistic electrons by Shockley and diffusing electrons by Wolff are complementary in determining the impact-ionization rate.

Other interesting results are the calculation of electric field dependence of the mean free path for phonon scattering and of the average energy. Figure 24 shows the average electron energy as function of the electric field, calculated from the MC simulation. Note that at low fields ($\leq 5 \times 10^3$ V cm^{-1}), the average energy (or the electron temperature) increases rapidly with the field because of carrier heating in the central valley, as described in the discussion of polar optical scattering in this section. Above 5×10^3 V cm^{-1}, the average energy increases with field at a lower rate because of intervalley scattering and then tends to saturate once the intervalley transfer is completed. The steep increase above 10^5 V cm^{-1} is not well understood and may be due to the band structure.

Figure 25 plots the phonon scattering mean free path versus electric field. In the electric field range where impact ionization occurs, the mean free path for phonon scattering ranges from 50 to 35 Å, in agreement with both Baraff's theory and the MC simulation using simple nonparabolic bands (Hauser, 1978). The reason for agreement with theories that do not include realistic band structures is that the average mean free path is mainly deter-

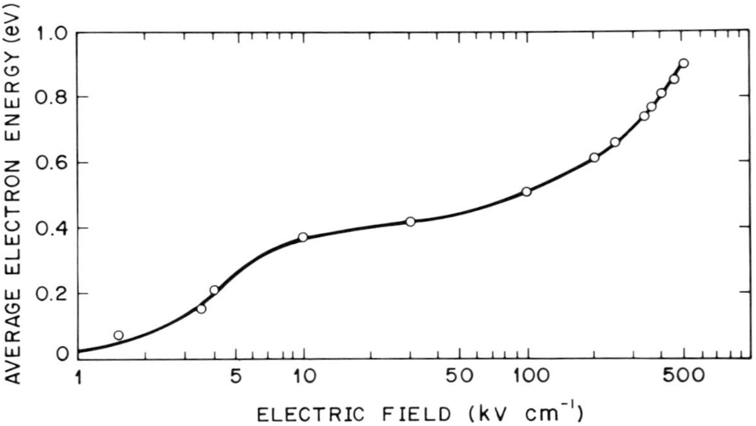

FIG. 24. Average electron energy in GaAs at 300 K as a function of electric field, calculated by the Monte Carlo simulation. [From Shichijo and Hess (1981).]

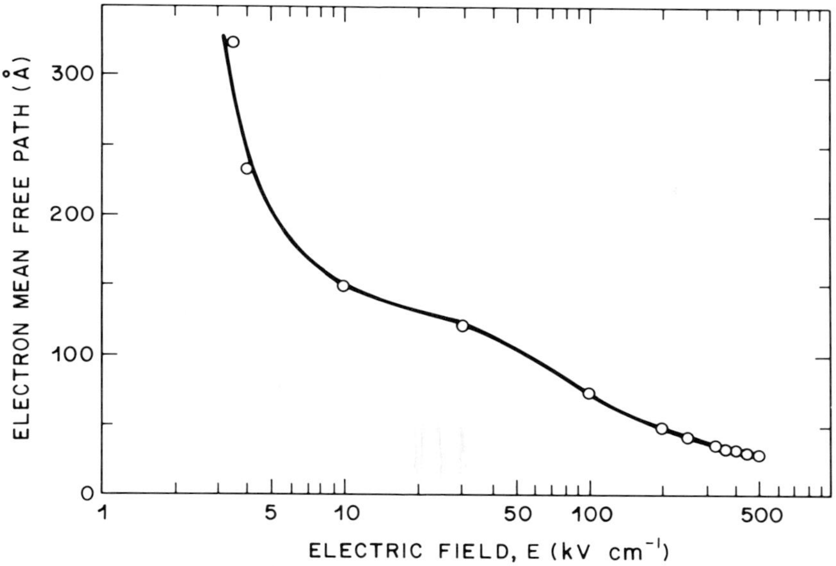

FIG. 25. Electron mean free path in GaAs at 300 K as a function of electric field, calculated by the Monte Carlo simulation (steady state). [From Shichijo and Hess (1981).]

mined by the average electron energy, which it still small enough (≤ 0.8 eV) for band-structure corrections to be important. Note also that λ is weakly field dependent for electric fields $> 10^5$ V cm^{-1}.

To understand how electrons gain the high energies needed for ionization, let us consider Fig. 26, which shows the variation of the electron energy after each scattering event for an electric field of 5×10 V cm^{-1}. The large reduction in the electron energy after some of the scattering events is due to phonon scattering against the direction of the electric field. The electron stays around the average energy most of the time, but occasionally escapes phonon collisions and moves up to high energies (the spikes in Fig. 26). When an electron reaches 2.0 eV, it ionizes. As Shichijo and Hess (1981) point out, "we can think of these electrons as the lucky electrons in Shockley's theory, and of those electrons around the average energy as the diffusing (in energy) electrons in Wolff's theory." This interpretation clarifies the physical basis of Baraff's distribution function, which represents a mixture of a spike in the direction of the field and a nearly spherically symmetric part. Thus, these "lucky" electrons can be pictured as "shooting out" from the nearly isotropic portion of the distribution, with an initial energy comparable to the average energy and reaching threshold ballistically or after a few collisions.

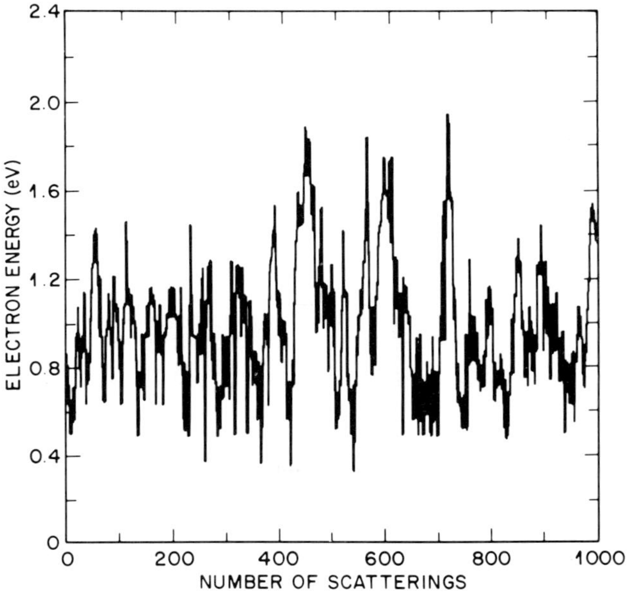

FIG. 26. Variation of electron energy after each scattering event for an electric field of 500 kV cm^{-1}, along the $\langle 100 \rangle$ direction, obtained via a Monte Carlo simulation. [From Shichijo and Hess (1981).]

Shichijo and Hess (1981) in their theory also calculated the electron-ionization rate in GaAs versus crystal orientation and concluded that within statistical uncertainty ($\pm 20\%$), there is no orientation dependence of impact ionization for electric fields $> 3.3 \times 10^5$ V cm^{-1}. They also find that the role of ballistic electrons is negligibly small.

The noninclusion of the anisotropy of the ionization energy should not be responsible for the lack of orientation dependence of α found in this calculation. This is due to the fact that at such high electric fields and at steady state, electrons are scattered all over the Brillouin zone before reaching threshold, so that any "memory" of the orientation of the electric field or the effect of the anisotropy of the conduction band is lost.

Capasso *et al.* (1981a, 1982a) criticized the use of the semiclassical Boltzmann picture inherent in the Monte Carlo simulation of Shichijo and Hess, in conjunction with the very high scattering rates at high energies. In fact, in the Boltzmann equation (for which the Monte Carlo method may be used as a solution technique), electrons are treated as semiclassical particles with well-defined energy and momentum, and phonon scattering is treated as a perturbation causing transitions between sharp momentum states (Barker,

1980). When the collision broadening cannot be neglected, as in the case of the scattering rates used by Shichijo and Hess (Fig. 15), the semiclassical Monte Carlo–Boltzmann model of electron transport should be used with caution.

Chang *et al.* (1983), in response to this criticism, described a quantum Monte Carlo approch to impact ionization in GaAs that goes beyond the semiclassical Boltzmann approach and includes nonperturbative quantum corrections to the phonon scattering rate. Although the quantitative results are different, the basic physical picture is unchanged. These quantum corrections were discussed in detail in Part II. The main effect is a reduction of the phonon scattering rate by $\simeq 20\%$ near the peak of the density of states. This enhances the electron-impact-ionization rate significatly. However, no orientation dependence of α was found even when a scattering rate of $\simeq 10^{14}$ $\sec^{-1}$ was used; in this case, in fact, α is very weakly dependent on the electric field, in disagreement with experiments. The authors concluded that the experimental anisotropy of α (Pearsall *et al.*, 1978) could be due to transient effects, as discussed by Capasso *et al.* (1982a) and by Hess *et al.* (1982).

The anisotropy of α $[\alpha(111) < \alpha(100) \simeq \alpha(110)]$ was found in abrupt $p-n$ junctions with relatively high doping [especially in the $\langle 111 \rangle$ orientation ($n \geq 8 \times 10^{16}$ cm^{3})] (Pearsall *et al.*, 1977b). In these $p-n$ junctions, the electric field decreases linearly with distance from the maximum field F_m. Thus, the ionization rate, which depends exponentially on F_m, is significant only in a small region near the peak field. The effective avalanche-region width is approximately 1000–2000 Å. Chynoweth and McKay (1957) were the first to point out that such junctions with narrow avalanche regions are the best candidates for observing the crystal-orientation dependence of impact ionization. This is because carriers that have suffered many collisions with phonons are unlikely to gain the ionization threshold and therefore exit the avalanche region before having ionized. This effect enhances the relative contribution of electrons that have suffered relatively few collisions to the total ionization rate. In more general terms, in short avalanche regions, carriers do not reach a spatial steady state, and the ionization coefficient should be considered a function of both field and position. The reduction of the randomizing effect of phonon collisions in narrow avalanche regions may thus reveal any crystal orientation or band-structure effect on α (Capasso *et al.*, 1982a)

On the other hand, it is extremely important to note that in any steady-state theory of impact ionization, including the Monte Carlo simulation of Shichijo and Hess (1981), even carriers that undergo many phonon collisions can still create electron–hole pairs, since a very large avalanche region is tacitly assumed. A simple calculation will clarify this point. For a field $F = 5 \times 10^5$ V cm^{-1}, Shichijo and Hess (1981) calculate an electron ionization rate of $\simeq 5 \times 10^4$ cm^{-1}. Using the definition of α as the reciprocal of the average distance between ionizations,

$$\frac{1}{\alpha} = \frac{eF}{E_{\rm i} + \langle n \rangle E_{\rm p}} \tag{129}$$

and substituting in Eq. (128) the values of F and α just given and taking $E_{\rm i} = 2.0$ eV, one finds that an electron is scattered on the average by $\langle n \rangle \simeq 400$ phonons before it ionizes. This is in agreement with Fig. 26. At this high electric field, the mean free path for phonon scattering is $\lambda \simeq 35$ Å (Fig. 25). Thus, a carrier has to travel several thousands of angstroms before it can ionize. If we now consider an abrupt $p^+ - n$ junction in GaAs with $n \simeq 5 \times 10^{16}$ cm^{-3} biased at a maximum field $\simeq 5 \times 10^5$ V cm^{-1}, the avalanche-region width is 2000 Å. Thus, electrons that have collided with many phonons do not contribute to the ionization, whereas they would in a diode with a uniform electric field $F = F_{\rm m}$ and long depletion width. This example shows that caution should be used in fitting experimental data taken in $p - n$ junctions, with steady-state models of α and β.

Brennan and Hess (1984a) have carried out a more refined Monte Carlo calculation of the electron ionization rate in GaAs extending over a wider range of electric fields and including all the quantum effects previously discussed. The results are shown in Fig. 27 for the $\langle 100 \rangle$ and $\langle 111 \rangle$ orientation, along with the data of Bulman *et al.* (1983) to be discussed in Section 12.

Note that at electric fields lower than 2.5×10^5 V cm^{-1}, an orientation dependence develops, and in the $\langle 111 \rangle$ direction α is about 40% smaller than in the $\langle 100 \rangle$ orientation. This anisotropy is connected with the fact that at such relatively low fields, the electron distribution is centered closer to $\mathbf{k} = 0$ than at high fields. Since the distribution is cooler, electrons reach the ionization energy only after gaining a lot of energy from the field. For this to occur, these electrons must not be scattered much from the direction of the field. In other words, the electrons are quasi-ballistic. Due to the anisotropy of the band structure, these electrons will gain energy at different rates along different crystal directions and cannot ionize along the $\langle 111 \rangle$ orientation. This will give rise to an orientation dependence of α and to a lower value of α when the field is along the $\langle 111 \rangle$ direction. In this latter calculation, it was again assumed that the ionization energy is isotropic and that it cannot be reached in the $\langle 111 \rangle$ direction.

In summary, there exists a range of electric fields in which the anisotropy of the conduction band produces an orientation dependence of α (Pearsall *et al.*, 1978); in this regime quasi-ballistic electrons are responsible for the anisotropy of α, similar to the suggestion of Capasso *et al.* (1979a) and Capasso and Bachelet (1980) but at significantly lower values of the field.

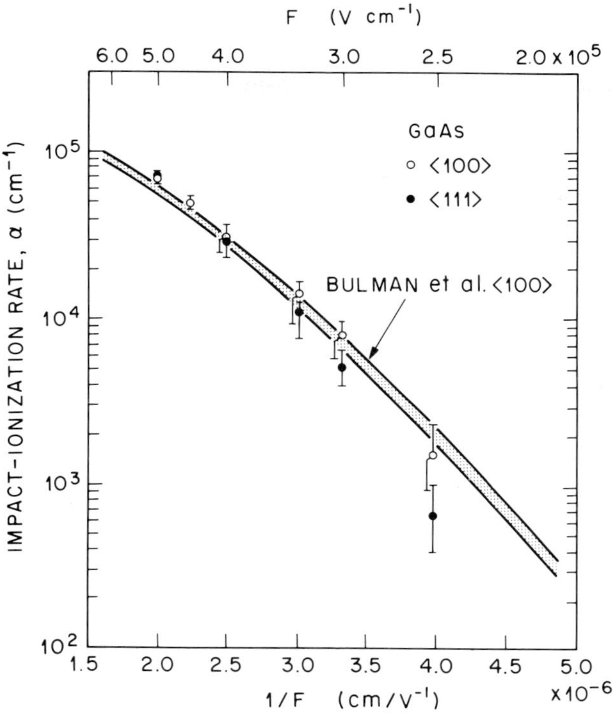

FIG. 27. Monte Carlo calculation of electron impact ionization rate in GaAs at 300 K for two crystal orientations (○, $\langle 100 \rangle$; ●, $\langle 111 \rangle$) versus reciprocal electric field. The error bars are based on convergence-error estimates from the calculation. The area between the solid curves represents experimental data. [From Brennan and Hess (1984a).]

The experimental anisotropy of α is observed at significantly higher electric fields than predicted by the Monte Carlo simulation. As discussed previously, any orientation dependence of α related to the anisotropy of the band structure will be enhanced by transient effects, and these may be present in the narrow p–n junctions in which that anisotropy has been reported (Pearsall *et al.*, 1978).

Brennan and Hess (1983a) have also reported Monte Carlo (MC) calculations of α in InP in the $\langle 100 \rangle$ and $\langle 111 \rangle$ orientations. These are shown in Fig. 28 together with the experimental data of Cook *et al.* (1982). Note that the electron-impact-ionization rate is much lower in InP than in GaAs, as can be seen from comparison with Fig. 27. In this calculation the same deformation potentials of GaAs were used, which is appropriate given the similarity of the two band structures. However, the density of states in InP at high energies is greater than in GaAs, giving rise to a higher phonon scattering rate and hence

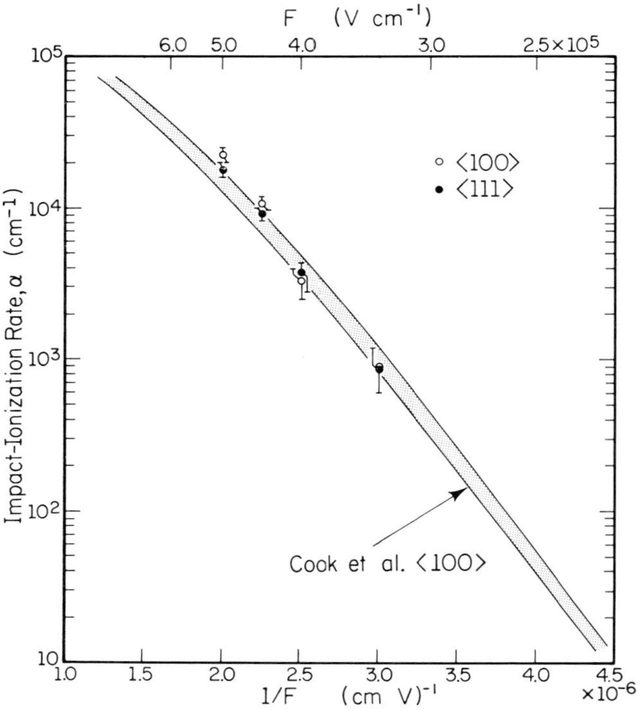

FIG. 28. Monte Carlo calculation of electron-impact ionization rate in InP at 300 K for two crystal orientations (○, ⟨100⟩; ●, ⟨111⟩) versus reciprocal electric field: error bars are based on convergence error estimates from the calculation. The area between the solid curves represents experimental data. [From Brennan and Hess (1984a).]

a lower ionization rate. Note that no orientation dependence of α is found, also at low electric fields. This fact has been experimentally confirmed by Armiento and Groves (1983).

The hole ionization rate β for InP and GaAs was also calculated by Brennan and Hess (1984b). It was found that $\beta(\mathrm{InP}) < \beta(\mathrm{GaAs})$, in agreement with the experimental data (see Figs. 42 and 48), and attributed to the higher total phonon scattering rate in InP, due to density of states considerations.

Brennan and Hess (1984a) also reported for the first time the MC simulation of α in InAs.

Tang and Hess (1983a) also performed a MC simulation of α in silicon. Their findings indicate that the second conduction band plays an important role, along with collision broadening and the "softness" of the ionization cross section above threshold, discussed in Section 2.

A study of hot-electron emission from Si into SiO_2 by Monte Carlo simulation has also been reported (Tang and Hess, 1983b). Impact ionization and collision broadening were included. Collision broadening in this process leads to an effective barrier lowering at the Si–SiO_2 interface. This is a consequence of the uncertainty principle. In fact, due to the high scattering rate the probability of finding electrons at extremely high energies (> 3 eV) is greater than if collision broadening is neglected.

Dmitriev *et al.* (1983) have presented an analytical theory of impact ionization for GaAs and other III–V materials. They conclude that along the $\langle 100 \rangle$ and $\langle 110 \rangle$ directions in GaAs, ballistic ionization is possible also at very high fields, whereas in the $\langle 111 \rangle$ direction a diffusion mechanism is predominant. They were able to produce a relatively good fit of Pearsall's data (1978) in the $\langle 100 \rangle$ and $\langle 111 \rangle$ orientations with $\exp(-b/F)$ and $\exp(-c/F^2)$ dependences, respectively. Their work is based on a theoretical scheme by Gribnikov (1978) who stated that ballistic ionization is also possible if $eF\lambda > \langle E_p \rangle$ and if the following conditions are satisfied. Carriers are heated by the field in a low-mass valley where the ionization threshold is located, and the dominant scattering mechanisms are intervalley transfer to a high-mass valley where carriers cool. Backscattering from this valley is much less probable than the reverse process. Thus, only carriers that have escaped intervalley collisions in the central valley can reach the ionization threshold.

One cannot apply this model to GaAs. At high energies, because of central valley nonparabolicities, the reverse-scattering process from the satellite valleys becomes much stronger and comparable in strength to Γ–L, Γ–X scattering.

Hess *et al.* (1983) have criticized Dmitriev *et al.* and showed that their expression for the ionization rate leads to a violation of the energy-conservation limitation at high fields. They also point out that their approach is not valid for nonparabolic bands.

To conclude this section, more experimental and theoretical work are needed to study the orientation dependence of impact ionization, especially in the transient regime. Observation of similar effects in p–n junctions of Ge, InGaAs, and AlGaSb (as discussed next in Part III) and in GaAs IMPATT diodes (Berenz *et al.*, 1979) have been reported. The anisotropy of the ionization threshold in GaAs and other III–V materials suggests that at low electric fields, the ionization coefficients may be anisotropic. This is confirmed by Monte Carlo calculations in the case of GaAs.

At very high electric fields ($> 2.5 \times 10^5$ V cm^{-1}) in GaAs, on the other hand, based on theoretical findings in steady-state conditions, α is isotropic, and ballistic electrons are not an important source of ionization.

III. Avalanche Multiplication and Measurement of Ionization Rates

10. Avalanche Rate Equations

The generally accepted theoretical model of impact ionization consists of two parts. The first, treated in Part II, relates the ionization rates to the distribution function via important physical quantities, such as ionization energies and phonon-scattering rates. The second relates the charge or current multiplication to ionization rates and is dealt with in this part. The measurement techniques, which are based on this analysis, are also briefly reviewed. Experimental ionization rate data for various materials are also presented, along with a discussion of important band-structure phenomena.

Figure 29 shows the typical geometry used in an avalanche-multiplication experiment. The spatial variation of the electric field is arbitrary, but the direction is assumed to be as shown so that the electrons drift in the positive direction with velocity v_n and the holes in the negative direction with velocity v_p. The current densities for electrons J_n and holes J_p are related to the carrier densities n and p by the usual relationships $J_n = -env_n$; $J_p = epv_p$, where e is the magnitude of the electronic charge.

The total current density $J_n + J_p$ is in the direction of the field, but in the multiplication process $|J_n|$ (and n) increases with increasing x, whereas $|J_p|$ (and p) increases with decreasing x; $J_n(W)$ and $J_p(0)$ are the injected electron and hole current densities, respectively; and W is the width of the field region and is generally a function of the reverse-bias voltage applied to the semiconductor. The differential equations describing avalanche multiplication in

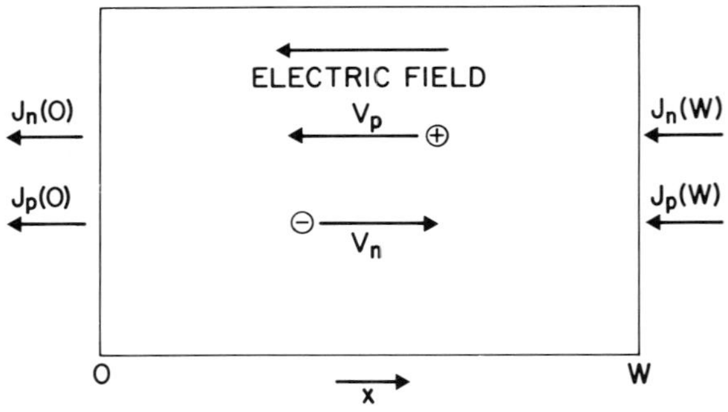

FIG. 29. Schematic of junction quantities and boundary conditions used in the avalanche rate equations: $0 < x < W$ is the field region; $v_{n,p}$ the electron and hole drift velocities; and $J_{n,p}$ the electron and hole current densities. Note that V_n is a positive quantity, whereas V_p is negative.

terms of ionization rates are given for the time-independent case by

$$dJ_n/dx = \alpha'(x)J_n(x) + \beta'(x)J_p(x), \tag{130}$$

$$-dJ_p/dx = \alpha'(x)J_n(x) + \beta'(x)J_p(x). \tag{131}$$

The reason for denoting the ionization rates by α' and β' rather than by α and β shall soon become apparent. In the past, several authors have used equations formally identical to Eqs. (129) and (130) but with carrier densities $n(x)$ and $p(x)$ in place of the current densities. Although in many practical situations this approach turns out to be equivalent, it is in general incorrect, since it assumes equal velocities for electron and holes (Beni and Capasso, 1979). In addition, in most avalanche multiplication experiments, it is the current rather than the charge that is measured. Equations (130) and (131) are very general and apply to situations of nonuniform fields and to cases where the carriers have not reached a spatial steady state; i.e., when the ionization coefficient depends *explicitly* on the position (Thornber, 1981).

We shall now consider the case of nonuniform electric fields as typically encountered in practical devices. If the electric field does not vary strongly over the ionization distance $E_{\rm i}/eF_{\rm m}$, where $F_{\rm m}$ is the maximum field, we may assume that α depends explicitly only on the electric field and not on the position; i.e.,

$$\alpha' = \alpha'[F(x)]. \tag{132}$$

In other words, if the spatial variation of F is sufficiently gradual, we can still assume a spatial steady state. Most ionization rate data are derived from avalanche-gain measurements via Eqs. (130) and (131), using the previous assumption. It is now worth asking whether the macroscopic ionization rates α' and β', defined through the current-density rate equations are the same as the microscopic ionization rates α and β calculated from the distribution function as the reciprocal of the average distance between ionizing collisions. Until recently, α' and β' have been acritically equated to α and β. Beni and Capasso (1979) have demonstrated that in general the two definitions are not equivalent, so that care must be used in fitting theoretically calculated ionization rates to experimental ones.

Microscopically, after obtaining the electron distribution through the Boltzmann equation, one calculates as follows: (1) the probability per unit time of creating a pair, $\tau_n^{-1}(\tau_p^{-1})$, and (2) the drift velocity $|v_n|(|v_p|)$. Then the ionization rates are identified with the inverse mean free paths between ionizing collisions and are calculated as

$$\alpha_{\rm cal} = (|v_n|\tau_n)^{-1}, \tag{133}$$

$$\beta_{\rm cal} = (|v_p|\tau_p)^{-1}. \tag{134}$$

Let us then consider a reference frame moving at the electron drift velocity v_n and a corresponding one moving at v_p with respect to the laboratory frame (the lattice). In the laboratory frame, because of stationarity the carrier densities n and p at point x are constant in time, whereas in the moving frames they are time-dependent quantities. Hence in the two moving frames, the rate equations for electrons and holes read

$$\left.\frac{dn}{dt}\right|_{v_n} = \frac{n}{\tau_n} + \frac{p}{\tau_p'}, \tag{135}$$

$$\left.\frac{dp}{dt}\right|_{v_p} = \frac{n}{\tau_n'} + \frac{p}{\tau_p}, \tag{136}$$

where $|_{v_{n(p)}}$ means "in the moving frame of reference of the electron (hole)"; τ_n^{-1} and τ_p^{-1} are the impact-ionization probabilities per unit time for electrons and holes; τ_n' and τ_p' are, in general, different from τ_n and τ_p, as is made clear later.

The relationships between the time derivatives of n and p in the moving frames and the time derivatives $dn/dt|_0$ and $dp/dt|_0$ in the laboratory frames are

$$\left.\frac{dn}{dt}\right|_{v_n} = \left.\frac{dn}{dt}\right|_0 + v_n \frac{dn}{dx}, \tag{137}$$

$$\left.\frac{dp}{dt}\right|_{v_p} = \left.\frac{dp}{dt}\right|_0 + v_p \frac{dp}{dx}. \tag{138}$$

From Eqs. (135)–(138), using the continuity equation

$$J_n + J_p = env_n + epv_p = \text{const}, \tag{139}$$

(which applies to the laboratory frame) and recalling that in steady state $dn/dt|_0 = dp/dt|_0 = 0$, one obtains

$$\frac{1}{\tau_p'} = \frac{1}{\tau_p} + \frac{dv_p}{dx}, \tag{140}$$

$$\frac{1}{\tau_n'} = \frac{1}{\tau_n} - \frac{dv_n}{dx}, \tag{141}$$

and finally,

$$\frac{dn}{dx} = \frac{n}{v_n\tau_n} + p\left(\frac{1}{v_n\tau_p} - \frac{1}{v_n}\frac{dv_p}{dx}\right), \tag{142}$$

$$\frac{dp}{dx} = \frac{p}{v_p\tau_p} + n\left(\frac{1}{v_p\tau_n} - \frac{1}{v_p}\frac{dv_n}{dx}\right). \tag{143}$$

Using the definition of α and β [Eqs. (133) and (134)], Eqs. (140) and (141) read

$$-\frac{dn}{dx} = \alpha n + \left(-\beta\frac{v_p}{v_n} - \frac{1}{v_n}\frac{dv_p}{dx}\right)p, \tag{144}$$

$$-\frac{dp}{dx} = \beta p + \left(-\alpha\frac{v_n}{v_p} + \frac{1}{v_p}\frac{dv_n}{dx}\right)n. \tag{145}$$

Substituting in Eqs. (130) and (131) the definitions of the current densities J_n and J_p and comparing with Eqs. (144) and (145) gives

$$\alpha' = \alpha - \frac{1}{v_n}\frac{dv_n}{dx}, \tag{146}$$

$$\beta' = \beta + \frac{1}{v_p}\frac{dv_p}{dx}. \tag{147}$$

These equations relate the measured quantities α' and β' to the microscopically calculated quantities, i.e., α and β.

We now discuss the physical significance of the second terms on the right-hand sides of Eqs. (146) and (147). These equations may be rewritten as

$$\alpha - \alpha' = \frac{1}{v_n}\frac{dv_n}{dF}\frac{dF}{dx}, \tag{148}$$

$$\beta - \beta' = -\frac{1}{v_p}\frac{dv_p}{dF}\frac{dF}{dx}. \tag{149}$$

If we consider a $p-i-n$ diode, which is constant-field structure, then $dF/dx = 0$, $\alpha = \alpha'$, and $\beta = \beta'$. Most measurements of ionization rates, however, have been performed in $p-n$ junctions, where $dF/dx \neq 0$. As we shall see later, in a $p-n$ junction, one measures the hole- and electron-initiated multiplications and derives α' and β' at the maximum fields F_m. In Si and GaAs, for impact ionization to occur, F_m must be greater than $\approx 10^5$ V cm^{-1}. Thus, in principle one must know the field dependence of the drift velocities for $F > 10^5$ V cm^{-1} in order to calculate the right-hand sides of Eqs. (148) and (149) and hence to evaluate how much the measured α' and β' differ from the microscopic ionization rates α and β.

Let us now discuss the magnitude of this correction. Consider the case of silicon, where the ionization rates and the drift velocities have been extensively studied. It is well established that in Si the drift velocities for electrons and holes reach a saturation value of $\approx 10^7$ cm sec^{-1} for electric fields between 10^4 and 10^5 V cm^{-1}. Unfortunately, for higher fields, where substantial avalanche multiplication occurs, there has been much less investigation of the behavior of the drift velocities.

However, existing data, although very limited, indicate an increase of the drift velocities above the saturated value for $F > 10^5$ V cm^{-1}. Duh and Moll (1967) measured the space-charge conductance of avalanching p^+-n-n^+ junctions in Si to obtain the electron drift velocity at electric fields $2 \times 10^5 \leq F \leq 4 \times 10^5$ V cm^{-1}. Their data suggest a smooth increase of the drift velocity with field, but the paucity of the experimental data led the authors to assign to the electron velocity a value of $(1.05 \pm 10\%) \times 10^7$ cm sec^{-1}.

Decker and Dunn (1975) have determined the field and the temperature dependence of the hole- and electron-drift velocities (and the carrier ionization rates) from microwave admittance and breakdown voltage data in Si avalanche diodes. An increase of the drift velocities above the saturated values was observed. At 300 K, the measured electron and hole drift velocities increase about 16% over the field range $5 \times 10^4 \leq F \leq 4 \times 10^5$ V cm^{-1}. Chen (1972) calculated the field dependence of the electron drift velocity in avalanching Si by solving the Boltzmann transport equation for the electron distribution function in the maximum anisotropy truncation (MAT) approximation of Baraff (1964). The calculated drift velocity increases from 1×10^7 to 2×10^7 cm sec^{-1} over the field range $1 \times 10^5 \leq F \leq 5 \times 10^5$ V cm^{-1}. Note that the magnitude of the increase depends on the choice of the fundamental microscopic parameters governing impact ionization, such as the ionization energy and the mean free path for optical phonon scattering. Although agreement with the data of Duh and Moll is not satisfactory, these parameters were not optimized for best fit because of the scarcity of the experimental data.

Roy and Ghosh (1975), using a simple energy-balance equation and assuming a drifted Maxwellian distribution for the electrons, predicted a smooth increase of the drift velocity with field above the scattering-limited value for $F > 10^5$ V cm^{-1}, indicating agreement with the data of Duh and Moll. Chwang *et al.* (1979) have calculated the drift velocity in Si in the avalanche regime using a rigorous Markov chain approach. They find that the average transport velocity increases from 10^7 to 1.5×10^7 cm sec^{-1} as the electric field is varied from 2.5×10^5 to 6.5×10^5 V cm^{-1}. The agreement with the data of Duh and Moll is excellent.

Thus, theoretical calculations and experiments in Si both suggest that the drift velocities increase with field for $F > 10^5$ V cm^{-1}. Let us now estimate for Si the relative corrections on α and β; i.e.,

$$\Delta_n = (\alpha' - \alpha)/\alpha', \qquad \Delta_p = (\beta' - \beta)/\beta'. \tag{150}$$

From Decker and Dunn (1975), we obtain

$$dv_n/dF \approx 6 \quad \text{cm}^2\,\text{V}^{-1}\,\text{sec}^{-1}, \qquad dv_p/dF \approx 4.2 \quad \text{cm}^2\,\text{V}^{-1}\,\text{sec}^{-1}, \tag{151}$$

in the range $2 \times 10^5 < F_m < 4 \times 10^5$ V cm^{-1} and $v_n \approx 1.15 \times 10^7$ cm sec^{-1}, $v_p \approx 8.2 \times 10^6$ cm sec^{-1}. Besides $dv_{n,p}/dF$, the relative correction $\Delta_{n,p}$ depends on the gradient of the electric field, which in turn is a function of the doping level and the type of structure used (abrupt, punchthrough, etc.). In a one-sided abrupt $p-n$ junction, the gradient of the electric field is given by

$$|dF/dx| = eN/\varepsilon_s, \tag{152}$$

where ε_s is the dielectric constant and N the doping of the lightly doped side of the junction.

Many groups have measured the impact-ionization rates in Si by using different diode structures and mathematical techniques to analyze the multiplication data. The experimental requirements that have to be met in such measurements will be carefully analyzed in Section 11. For silicon the data of Woods *et al.* are not affected by most of the experimental limitations present in previous work. Woods *et al.* (1973) used a Schottky-barrier one-sided $p-n$ junction doped with $N \approx 1.5 \times 10^{16}$ cm^{-3}, corresponding to $dF/dx \approx 2.3 \times 10^9$ V cm^{-2}. Their results for α' and β', in the field range $2 \times 10^5 < F_m < 4 \times 10^5$ V cm^{-1}, can be expressed in the form

$$\alpha' = 9.2 \times 10^5 \exp(-1.45 \times 10^6/F_m) \quad (\text{cm}^{-1}), \tag{153}$$

$$\beta' = 2.4 \times 10^5 \exp(-1.64 \times 10^6/F_m) \quad (\text{cm}^{-1}), \tag{154}$$

where F_m is in volts per centimeter.

From these numbers we can obtain Δ_n and Δ_p, which at low fields ($F \simeq 2.5 \times 10^5$ V cm^{-1}) turn out to be of the same order of the ionization rates themselves. Drift velocity corrections can therefore be significant.

Brennan and Hess (1984a) have calculated via the Monte Carlo method using the pseudopotential band structure, the drift velocity of electrons in InAs at 300 K over a wide range of electric fields including the avalanche regime. The results of this calculation are shown in Fig. 30. The velocity increases with electric field also after the onset of impact ionization ($\approx 10^4$ V cm^{-1}) and reaches very high values ($\simeq 8 \times 10^7$ cm sec^{-1}). This continuous increase can be interpreted as follows. As the field increases, electrons basically cannot reach an energy higher than the ionization threshold so that distribution is progressively compressed in energy and momentum in the low-mass high-velocity central valley. Note that no intervalley transfer occurs in this material because the energy separation from the satellite valleys is $> E_i$. It is clear that as a result of the velocity curve of Fig. 30, we expect that $\alpha \neq \alpha'$ and $\beta \neq \beta'$ in InAs.

In conclusion, the possible field dependence of the drift velocity in the avalanche regime may introduce a nonnegligible correction. This also illus-

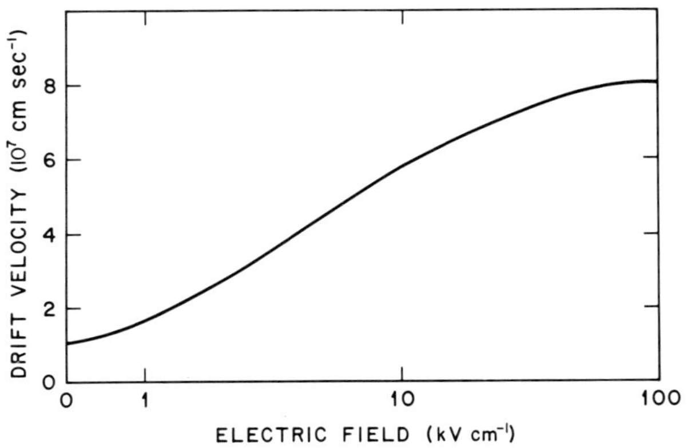

FIG. 30. Steady-state drift velocity in InAs as a function of electric field when impact ionization is present. The field is applied along the ⟨100⟩ direction at 300 K. [From Brennan and Hess (1984a).]

trates very clearly that in general $\alpha' \neq \alpha$, $\beta' \neq \beta$, so that it is necessary to know the field dependence of the drift velocities in order to fit the current microscopic models to the measured ionization rates α' and β'. This analysis also illustrates the convenience of using a $p-i-n$ diode (constant electric field) for a direct determination of the microscopic ionization rates. This also presents other important experimental advantages that are discussed shortly.

In response to some of the difficulties mentioned earlier, Thornber (1981) has introduced a much more general microscopic definition of the ionization coefficient in terms of the number of electron–hole pairs created per unit time per unit volume normalized to incident carrier density. This definition is also valid for situations of nonconstant field or when the ionization problem is nonlocal (i.e., when the ionization rate depends explicitly on position). It reduces to the usual definition of the reciprocal of average distance betwen ionizations when the electric field is constant and a spatial steady state has been achieved. Starting from his new microscopic definition of α and β, Thornber (1981) showed that the avalanche rate equations [Eqs. (130) and (131)] can be directly derived from the Boltzmann transport equation, so that $\alpha \equiv \alpha'$, $\beta \equiv \beta'$. In summary,

1. For uniform electric fields and spatial steady-state situations (e.g., p^+-i-n^+ diodes with long i regions), the microscopic ionization rates α and β can be correctly defined as the reciprocal of average distance between ionizing collisions. The ionization rates α' and β' appearing in the rate equations for the current densities are then identical to α and β (Beni and Capasso, 1979).

2. For nonuniform electric fields and nonlocal problems, one must change the microscopic definition of α and β to ensure that $\alpha = \alpha'$, $\beta = \beta'$. This new definition coincides with the old one for uniform fields and spatial steady-state situations (Thornber, 1981).

3. Theories of impact ionization that calculate α and β as the reciprocal of average distances between ionizing collisions in general cannot be used to fit ionization rates measured in devices with either nonuniform fields or very thin avalanche regions, since in these cases $\alpha' \neq \alpha$, $\beta' \neq \beta$. If, however, the variation of the electric field with distance is sufficiently gradual so that the carrier distribution can be assumed to be locally in equilibrium with the electric field, the previous impact-ionization models can still be used to analyze experimental data, provided that the field dependence of the drift velocity in the avalanche regime is known and accounted for (Beni and Capasso, 1979).

In the rest of this section, we shall denote the ionization rates in the rate equations as α and β, which is standard notation. The reader should be aware that these quantities may be different from the microscopic rates, as previously discussed.

The avalanche rate equations can be relatively simply solved for the currents or avalanche gains if it is assumed that α and β are either independent of position or depend on the space coordinate x only through the electric field $F(x)$. The general solutions of Eqs. (130) and (131) have been discussed by Stillman and Wolfe (1977) in a previous volume of this treatise, and the interested readers are referred to it. We shall concentrate here on two special cases of great interest for the experimental determination of α and β; i.e., pure electron and pure hole injection. These two cases can be understood by referring to Fig. 29 and arise respectively when there is no carrier generation other than impact ionization in the space-charge region and either $J_n(0) \neq 0$ and $J_p(W) = 0$ or $J_p(W) \neq 0$ and $J_n(0) = 0$.

Let us consider first the pure electron-initiation case. Electrons injected with a current density $J_n(0)$ undergo impact ionization so that J_n increases with distance. Under dc conditions, the total current is constant and given by

$$J = J_n(x) + J_p(x) = J_n(W) = \text{const}, \tag{155}$$

so that the hole current density increases from right to left. The amount of current depends on the values of α and β which are controlled by the electric field. The electron-initiated avalanche gain M_e is defined as the ratio of the current flowing through the device in the presence of avalanche gain to the current in the absence of gain under identical conditions of pure electron injection. Using Eq. (155), one has

$$M_n = J_n(W)/J_n(0). \tag{156}$$

Similarly, the hole-initiated avalanche gain is given by

$$M_p = J_p(0)/J_p(W). \tag{157}$$

The assumption of no generation of carriers in the space-charge region other than by ionization may seem unrealistic in view of the fact that in any device there is always thermal generation of carriers in the depletion layer, which gives rise to dark current. Since, however, avalanche gain is usually measured by using phase-sensitive detection, one only determines the multiplication of carriers injected externally (e.g., by photoexcitation).

Solution of the avalanche rate equations for the previous two cases yields

$$M_n = \left\{1 - \int_0^W \alpha \exp\left[-\int_0^x (\alpha - \beta)\, dx'\right] dx\right\}^{-1}, \tag{158}$$

$$M_p = \left\{1 - \int_0^W \beta \exp\left[\int_x^W (\alpha - \beta)\, dx'\right] dx\right\}^{-1}. \tag{159}$$

The avalanche breakdown voltage is defined as the voltage at which the multiplication goes to infinity and can be found by equating to zero the denominator of Eqs. (158) and (159). Although the breakdown voltage is independent of the injection conditions, the multiplication depends on where the primary carriers are injected. The reader is referred to the review article by Stillman and Wolfe (1977) for a detailed discussion of this point.

In the case of constant or linear field profiles, one can easily express, using Eqs. (158) and (159), α and β as a function of M_n and M_p. Thus, the determination of α and β reduces to the problem of measuring M_n and M_p.

11. Measurement Methods

To determine accurately M_n and M_p, several experimental conditions must be met (Stillman and Wolfe, 1977):

1. Pure electron and pure hole injection must be obtained in the same junction rather than in complementary structures. This is best achieved by using photoexcitation.
2. The photocurrent without avalanche gain must be measured precisely.
3. The electric field variation in the junction should be sufficiently gradual so that α and β can be assumed to be only functions of electric field.
4. The electric field profile must be accurately known.
5. The avalanche gain must be uniform across the active area of the device.

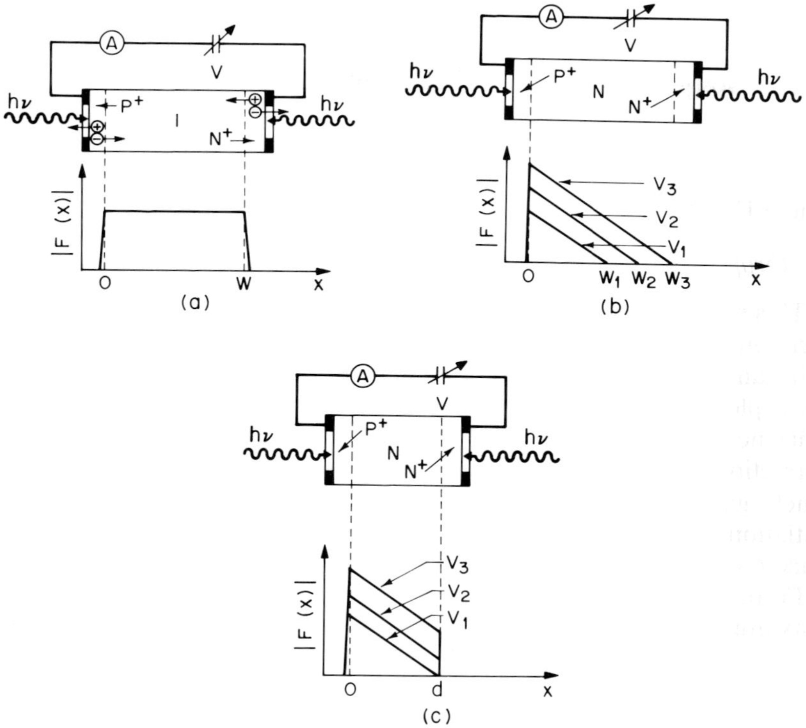

FIG. 31. Different diode configurations used in ionization rate measurements: (a) $p-i-n$ diode; (b) abrupt $p-n$ junction; (c) punchthrough diode.

Figure 31 shows a schematic of the device structures and field profiles that allow the most accurate determination of α and β.

a. p–i–n Diode

This is the best structure for the measurement of α and β since it satisfies all the requirements just listed. Pure electron and hole injection can be obtained by using light that is completely absorbed in the p^+ and n^+ regions, respectively (Fig. 31a). The collection efficiency is practically independent of bias voltage if either the p^+ and n^+ regions are very highly doped or the minority-carrier diffusion lengths are sufficiently long. In this case, the effect of the small widening of the depletion layer in the n^+ and p^+ regions is negligible, and the injected photocurrent can be accurately determined.

Since the electric field is constant in this structure, one easily obtains the following formulas related α and β to M_n and M_p (Stillman and Wolfe, 1977):

$$\alpha(F) = \frac{1}{W}\left[\frac{M_n(V) - 1}{M_n(V) - M_p(V)}\right] \ln\left[\frac{M_n(V)}{M_p(V)}\right], \tag{160}$$

$$\beta(F) = \frac{1}{W}\left[\frac{M_p(V) - 1}{M_p(V) - M_n(V)}\right] \ln\left[\frac{M_p(V)}{M_n(V)}\right], \tag{161}$$

where $V = FW$ is the bias voltage.

b. Abrupt p–n Junction

This structure (Fig. 31b) is more easily fabricated than the previous one (Section 11a). However, the hole photocurrent in the absence of avalanche gain cannot be determined in a straightforward way due to the widening of the depletion layer width with increasing voltage. This varies the collection efficiency of holes in the case of pure hole injection. Thus, appropriate corrections must be made to determine accurately the hole-initiated avalanche gain (Woods *et al.*, 1973). Another difficulty is the numerical differentiation of the multiplication data, which can introduce an additional uncertainty in the determination of α and β.

From the multiplication data one can determine the ionization rates at the maximum field F_m using the formulas

$$\alpha(F_m) - \beta(F_m) = F_m\left[\frac{1}{M_n(V)}\frac{dM_n(V)}{dV} - \frac{1}{M_p(V)}\frac{dM_p(V)}{dV}\right], \tag{162}$$

$$\beta(F_m) = F_m \frac{1}{M_n(V)M_p(V)}\frac{dM_p(V)}{dV}. \tag{163}$$

The maximum field F_m is related to the applied voltage by the equation

$$F_m = (2eN/\varepsilon)^{1/2}(V + V_{bi})^{1/2}, \tag{164}$$

where N is the net doping density in the n layer and V_{bi} the built-in voltage. Equations (162) and (163) were first derived by Dai and Chang (1971) and later discussed by Woods *et al.* (1973).

If the field gradient is too strong, α and β become explicit functions of the position. A simple though somewhat crude way to account for these nonlocal effects is to introduce a "dead-space" correction, as discussed in detail by Woods *et al.* (1973).

For the differentiation of the data, the following procedure can be used. The multiplication data can be fitted with the well-known empirical equation (Miller, 1955)

$$M = [1 - (V/V_B)^n]^{-1}. \tag{165}$$

The two adjustable parameters n and V_B are different for the case of hole and electron injection; V_B will usually be very close to the breakdown voltage of the diode. Equation (165) can then be differentiated and the resulting derivatives substituted in Eqs. (162) and (163).

c. Punchthrough Diode

In this structure (Fig. 31c), the drawback of the varying collection efficiency of holes can be eliminated, since the depletion layer punches through to the n^+ layer at low voltages so that the multiplication before punchthrough is negligible. The equations for α and β are (Stillman and Wolfe, 1977) as follows:

$$\alpha(F_m) = \frac{2d}{W_0^2}\frac{1}{M_n}\frac{dM_n}{dV} + \alpha(F(d))M_p - \beta(F_m)(M_n - 1), \tag{166}$$

$$\beta(F_m) = \frac{2d}{W_0^2}\frac{1}{M_n M_p}\frac{dM_p}{dV} + \frac{M_p}{M_n}\left\{\alpha(F(d)) - [\alpha(F(d)) - \beta(F(d))]\frac{1}{M_p}\right\}, \tag{167}$$

$$F(d) = F_m - F_W, \tag{168}$$

$$W_0 = (2\varepsilon/eN)^{1/2}, \tag{169}$$

$$F_W = 2d/W_0^2, \tag{170}$$

$$F_m = V/d + F_W/2, \tag{171}$$

where F_W is the maximum electric field at the punchthrough voltage and $F(d)$ the electric field at the edge of the N^+ region for a given applied voltage V. By making measurements of the multiplications before and after punchthrough, we can determine $\alpha(F(d))$ and $\beta(F(d))$ if they are significant.

Figure 32 shows two practical device structures that allow very pure injection of electrons and holes. These have become the standard device configurations for this type of measurements (Pearsall *et al.*, 1975; Bulman *et al.*, 1982). In the first (Fig. 32a), note that a hole is etched in the substrate to within a few tens of micrometers from the junction in order to obtain pure electron injection using strongly absorbed light. However, in materials with a short minority-carrier diffusion length, many of the electrons optically excited in the region of the substrate right above the etched hole recombine before being collected by the junction. The near-band-gap radiation generated by this recombination can be reabsorbed by the Franz–Keldysh effect in the space-charge region. If the width of the avalanche high-field region is appreciable, this reabsorption produces a considerable contamination of the electron-injected current. This effect was first observed by Bulman *et al.* (1982). These authors therefore introduced an important modification (Fig.

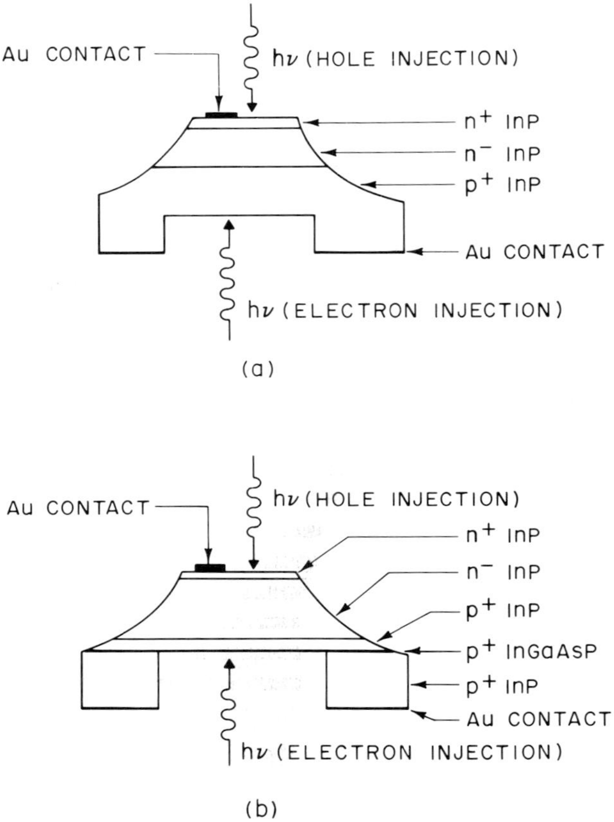

FIG. 32. Methods to achieve pure electron and pure hole injection in photomultiplication measurements: (a) standard method, and (b) improved method, to eliminate possible contamination from electroabsorption of recombination radiation from carriers diffusing toward the junction in the substrate.

32b) of the basic structure (Fig. 32a) to eliminate this source of contamination. A thin ($<1\ \mu$m) buffer layer with composition lattice matched to the substrate is inserted between the *n* layer and the substrate. The hole is etched with a selective etch that stops at the buffer layer; this layer completely absorbs the incident light, yet the thickness is substantially smaller than the minority-carrier diffusion length so that the amount of recombination is negligible.

To measure the ionization rates in an InP p^+-n diode, Bulman *et al.* (1982) used an InGaAsP buffer layer ($E_g = 0.90$ eV) lattice matched to InP ($E_g = 1.3$ eV) (Fig. 32b).

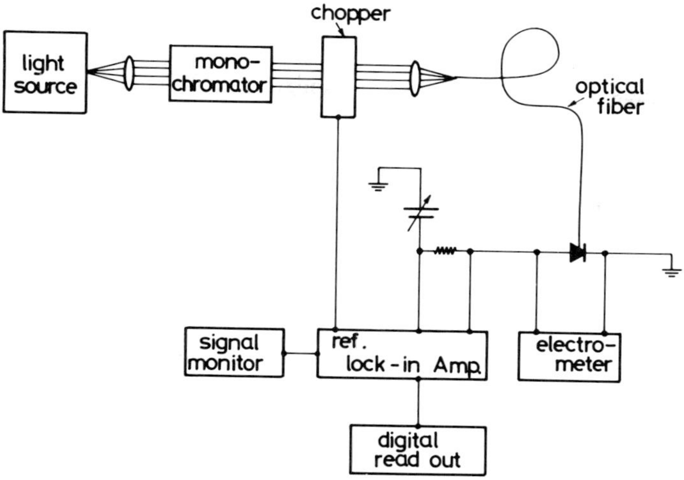

FIG. 33. Typical experimental setup used in photomultiplication measurements. [From Pearsall *et al.* (1978).]

Figure 33 shows a typical experimental setup for the measurement of the ionization rates (Pearsall *et al.*, 1978). The photocurrent is measured via phase-sensitive detection. Note also that the light was coupled to the diode via an optical fiber. This minimized any contamination of the electron photocurrent from stray light incident on the edge of the mesa.

12. EXPERIMENTAL IONIZATION RATES: Si, Ge, AND III–V SEMICONDUCTORS

The increasing and widespread research on III–V alloys suitable for long-wavelength optoelectronic devices has stimulated investigations of the ionization coefficients of these alloys and of their binary constituents. Section 12 focuses mainly on measurements of these materials; however, the experimental data for Si and Ge and other compounds of interest are also briefly reviewed.

Traditionally, measurements of impact-ionization rates have been affected by many experimental uncertainties. Substantial discrepancies among the results of various authors occur in the published literature. Only very recently have more accurate methods of analysis and measurement become available. For this reason, the following criteria have been used in reporting the experimental data. Only selected data, which, in the author's judgment, either represent an accurate measurement of α and β using one of the methods illustrated in Fig. 32 (when available) or are interesting from the point of view of physics are presented.

a. Ionization Rates in Si

There is an extensive literature on ionization coefficients in Si. It is now established beyond doubt that the electron ionization coefficient is significantly greater than the hole ionization coefficient for electric fields $<3 \times 10^5$ V cm^{-1}. As a consequence, Si APDs with very low excess avalanche noise are presently commercially available. These devices have effective ionization rates ratios between 20 and 100. The ionization rates that to date give best agreement with excess avalanche noise measurements and in general with the performance of Si APDs are those of Lee *et al.* (1964). These data, shown in Fig. 34, were obtained by illuminating a p^+-n junction with short- and long-wavelength radiation from the p^+ side to obtain pure electron and hole injection, respectively. Excellent agreement with Baraff's theory was found.

Measurements with one of the improved techniques illustrated in Fig. 32 are not yet available. The reader is referred to the review article of Stillman

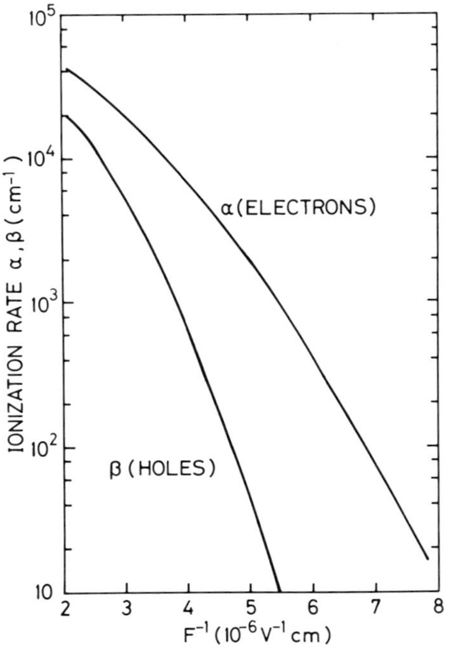

FIG. 34. Ionization rates in Si at 300 K versus reciprocal electric field; curves are obtained from the experimental data of Lee *et al.* (1964).

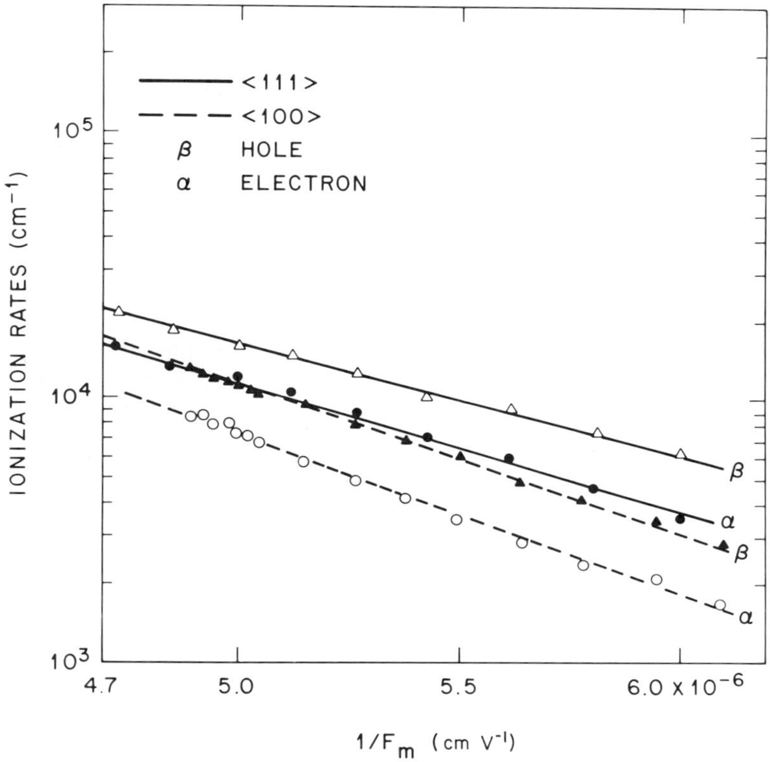

FIG. 35. Ionization rates versus reciprocal electric field for (- - -) ⟨100⟩ and (—) ⟨111⟩ Ge at 300 K: β, holes; α, electrons; lines are least square fits to the data. [From Mikawa *et al.* (1980).]

and Wolfe (1977) for a thorough analysis of Si ionization rate data from other authors.

b. Ionization Rates in Ge

Measurements of α and β using pure electron and hole injection in the same diode by the method of Fig. 32a have been reported by Mikawa *et al.* (1980). Their structure consists of a guard-ring planar-diffused $n^+ - p$ one-sided abrupt $p-n$ junction fabricated on both the ⟨100⟩ and ⟨111⟩ crystal orientations. The experimental arrangement is virtually identical to that of Fig. 32a, except that the diode is planar. Their results for the two orientations are shown in Fig. 35 for the electric field range $1.67 \times 10^5 \leq F \leq 2.13 \times 10^5$ V cm^{-1}. It is seen that $\beta > \alpha$ for both orientations; however, the ionization-rate ratio β/α of ⟨100⟩ Ge is found to be greater than that of ⟨111⟩

Ge. This finding is supported also by measurements of the excess noise factor for these two orientations (Kaneda *et al.*, 1979). A least squares fit gives

$$\alpha(F) = 2.72 \times 10^6 \exp(-1.1 \times 10^6/F), \tag{172}$$

$$\beta(F) = 1.72 \times 10^6 \exp(-9.37 \times 10^6/F), \tag{173}$$

for $\langle 111 \rangle$, and

$$\alpha(F) = 8.04 \times 10^6 \exp(-1.4 \times 10^6/F), \tag{174}$$

$$\beta(F) = 6.39 \times 10^6 \exp(-1.27 \times 10^6/F), \tag{175}$$

for $\langle 100 \rangle$ in the previous electric field range.

The magnitude of the β/α ratio is found to be comparable to that measured by Miller (1955) and by Dai and Chang (1971). Note that the latter authors used a $p-n-p$ transistor structure to achieve pure electron and hole injection and also reported measurement of α and β over the temperature range from 200 to 300 K. The orientation dependence of the ionization rates ratio obviously requires further investigation, since the experimental values of β/α in the two orientations differ by a modest amount.

c. *Ionization Rates in* GaAs

Initial measurements of α and β in this material assumed that these quantities were equal. Stillman *et al.* (1974) were the first ones not to use this assumption and employed a Schottky-barrier diode to attempt to obtain pure electron and hole injection, similar to work of Woods *et al.* (1973) in Si. They found that $\beta > \alpha$. However, both the magnitude of the ratio and the field dependence appear to be in error, as discussed in detail in a subsequent review by Stillman (1977). This is due to the many experimental uncertainties, especially the possible sources of contamination of the injected photocurrents. In addition the fact that the β/α ratio increases with increasing electric field appears to be unphysical.

Pearsall *et al.* (1977b, 1978) measured α and β for the symmetry directions $\langle 100 \rangle$, $\langle 110 \rangle$, and $\langle 111 \rangle$ using the method of Fig. 32a. Their measurements not only indicated that $\beta > \alpha$ but also showed a marked orientation dependence of the electron ionization rates. These data are shown in Figs. 36–38. It is seen that at high electric fields, β/α tends to unity along all orientations. The hole ionization rate shows negligible anisotropy, which was interpreted as a manifestation of the isotropic nature of the valence band and of the fact that holes, to impact ionize, have to scatter first to the spin–orbit split-off band. This tends to randomize any "memory" of the orientation.

In contrast, the orientation dependence of the electron ionization is clearly manifested by the rapid decrease of α at low fields along the $\langle 111 \rangle$ direction, compared to the two other orientations. Pearsall *et al.* (1977b, 1978) attrib-

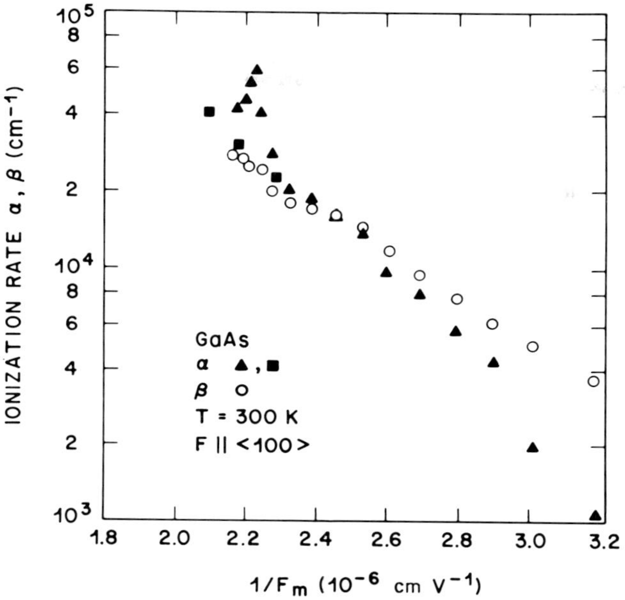

FIG. 36. Ionization rates versus reciprocal electric field for ⟨100⟩ GaAs at 300 K. [From Pearsall *et al.* (1978).]

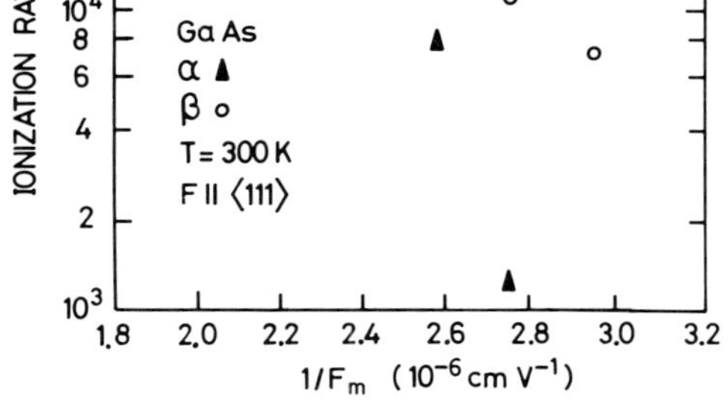

FIG. 37. Ionization rates versus reciprocal electric field for ⟨111⟩ GaAs at 300 K. [From Pearsall *et al.* (1978).]

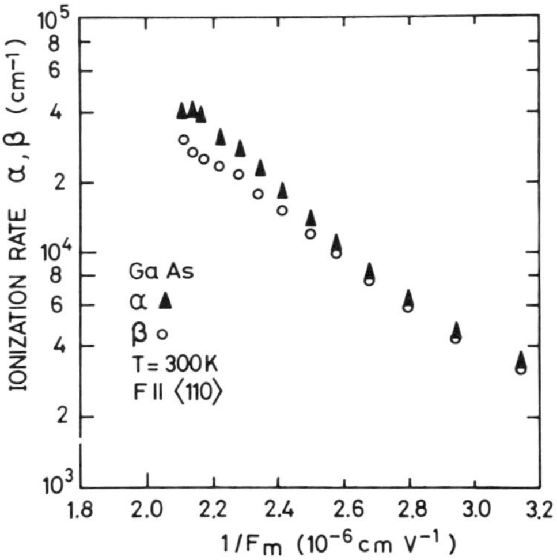

FIG. 38. Ionization rates versus reciprocal electric field for ⟨110⟩ GaAs at 300 K. [From Pearsall *et al* (1978).]

uted this effect to the fact that electrons cannot impact ionize while moving along the ⟨111⟩ orientation, since there is no electron-initiated threshold state along this direction. Thus, electrons can only ionize by first scattering to another orientation. In the ⟨100⟩ and ⟨111⟩ orientations instead, electron ballistic ionization is always possible, which increases the probability of ionizing along these directions with respect to the ⟨111⟩ orientation (Capasso *et al.,* 1979a).

This interpretation has been challenged by Shichijo and Hess (1981) via a Monte Carlo simulation, leading to lively controversy, which has already been discussed in Section 9. More experimental investigations of α, β, especially in the ⟨111⟩ orientation, are clearly needed.

The temperature dependence of α, β up to $T = 250°\text{C}$ was investigated by Capasso *et al.* (1977, 1979b) for the three principal symmetry orientations. Particularly interesting are the results for the ⟨100⟩ direction. Figure 39 shows α and β versus lattice temperature at an electric field of 4.3×10^5 V cm^{-1}. It is seen that both rates decrease with increasing temperature, a manifestation of increased phonon scattering. However, an additional effect is present in the behavior of α with temperature at $T = 200°\text{C}$. The same effect was observed over the complete field range from 4 to 5×10^5 V cm^{-1}. These experimental results were interpreted in terms of the temperature dependence of the threshold ionization energy, which is calculated from the band structure using the method of Anderson and Crowell (1972), taking

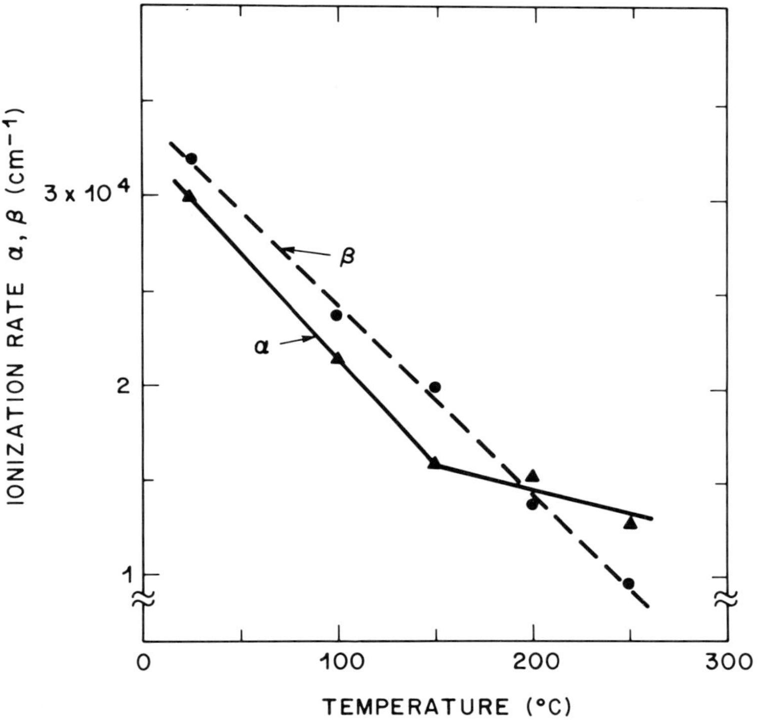

FIG. 39. Impact ionization rates versus temperature at $F_m = 4.3 \times 10^5$ V cm^{-1} in ⟨100⟩ GaAs. [From Capasso *et al.* (1977).]

into account the band-gap variation with temperature. The result of the calculation for the electron threshold energy in the ⟨100⟩ direction as a function of temperature is shown in Fig. 40. The abrupt jump in this energy near 240°C occurs because the threshold abruptly transfers from the upper to the lower conduction band. Thus, in this model, at temperatures above 240°C, electrons in the lowest conduction band can cause impact ionization, whereas at lower temperatures, the electrons must transfer into the upper conduction band to initiate multiplication. The measured change in α/β near 200°C is attributed to this abrupt change in electron threshold energy for ionization and is in good agreement with the calculation. The temperature dependence of α/β for the three main symmetry orientations is illustrated in Fig. 41.

Ito *et al.* (1978a), using $p^+ - n$ crater mesa junctions also observed $\beta > \alpha$ in ⟨100⟩ GaAs but found a stronger electric field dependence. Law and Lee (1978) performed measurements on Schottky-barrier diodes and found

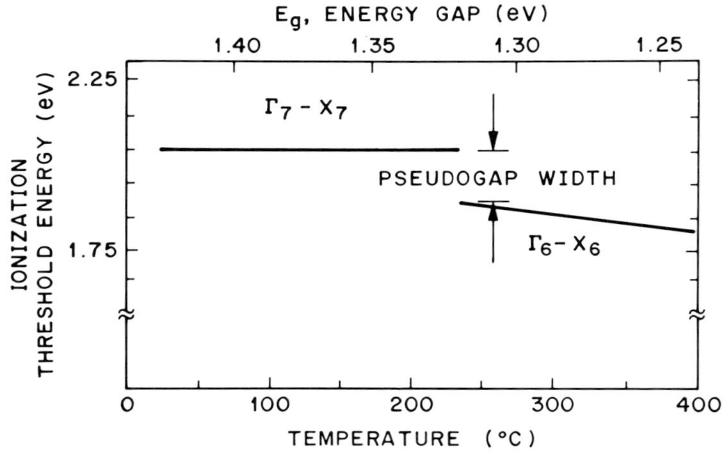

FIG. 40. Electron ionization threshold energy measured with respect to the bottom of the conduction band versus temperature calculated from the band structure of ⟨100⟩ GaAs. [From Capasso *et al.* (1977).]

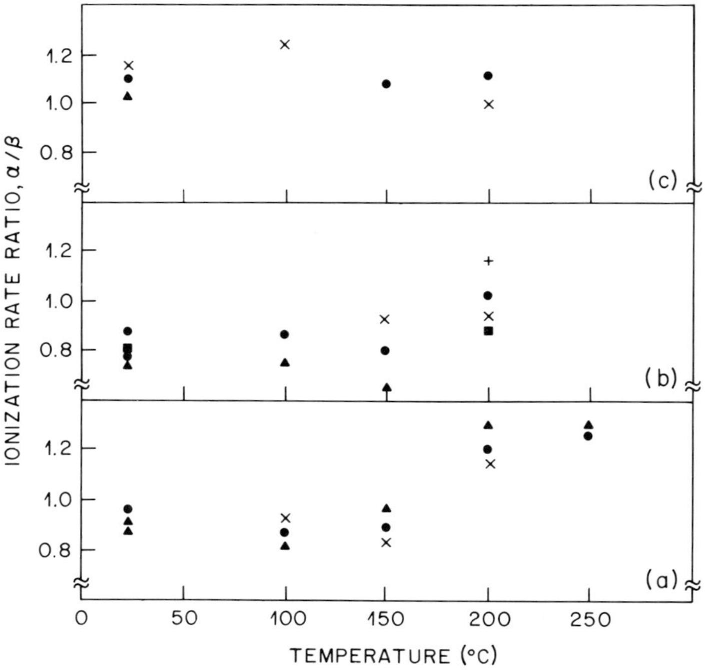

FIG. 41. Ionization rate ratio for GaAs versus temperature for $F = 4.3 \times 10^5$ V cm^{-1} and three orientations of the electric field: (a) ⟨100⟩, (b) ⟨111⟩, and (c) ⟨110⟩. [From Capasso *et al.* (1979a).]

$\beta > \alpha$ in heavily doped diodes but $\alpha > \beta$ in lightly doped ones. All these results were based on photocurrent multiplication data obtained under separate electron and hole injection. Ando and Kanbe (1981) performed photocurrent multiplication and noise measurements for electron injection in p^+-n diodes and found $\alpha/\beta \approx 2$.

These discrepancies prompted Bulman *et al.* (1983) to undertake a systematic and extremely detailed study of the ionization rates using the structure of Fig. 32b in a large number of wafers and diodes. For the etch-stop layer, an n^+-$Al_{0.35}Ga_{0.55}As$ layer was used. These measurements covered a very large range of electric fields and gave excellent reproducibility from wafer to wafer and from device to device. A total of 44 $p-n$ junction were measured. The results shown in Fig. 42 indicate that at 300 K in ⟨100⟩ GaAs, $\alpha > \beta$ in contrast with previous measurements and that the α/β ratio decreases from 2.5 to 1.3 as the electric field is increased from 2.2 to

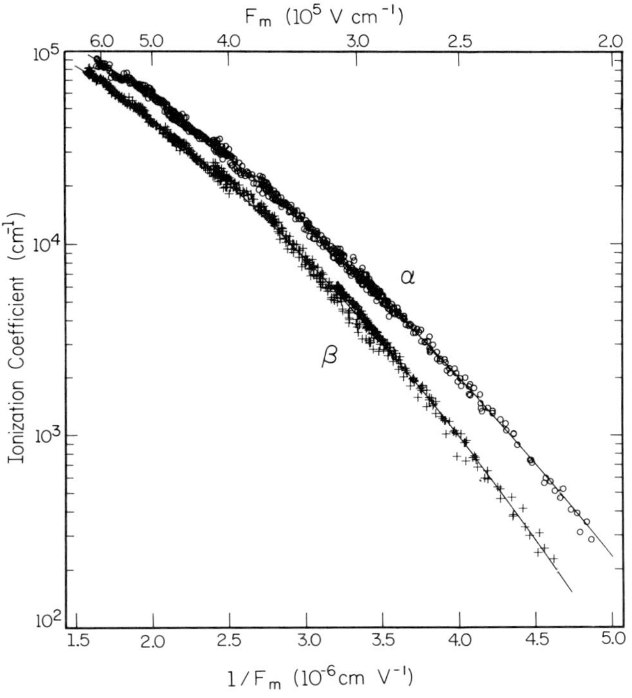

FIG. 42. Ionization rate versus reciprocal electric field in ⟨100⟩ GaAs at 300 K. [From Bulman *et al.* (1983). © 1983 IEEE.]

6.25×10^5 V cm^{-1}. The solid curves for α and β in Fig. 40 are given by

$$\alpha(F) = 2.994 \times 10^5 \exp[-(6.848 \times 10^5/F)^{1.6}], \tag{176}$$

$$\beta(F) = 2.215 \times 10^5 \exp[-(1.570 \times 10^5/F)^{1.5}]. \tag{177}$$

It is important to note that in this work dead-space corrections for electrons were carefully determined using a method suggested by Okuto and Crowell (1975). Since this correction on samples with heavily doped p regions results from the rapid field variation at the electron injection point, three wafers with less heavily doped p regions were grown to eliminate the necessity for this correction. In the diodes with the less heavily doped p regions, depletion into the p contacts cannot be ignored, so that the ionization rates cannot be derived from Eqs. (158) and (159) valid for one-sided p^+-n junctions. A numerical procedure due to Grant (1973), which allows for this effect, was used to extract α, β from the multiplication data. Excellent agreement between the ionization coefficients determined by the two techniques was found. The measured α is in good agreement with the calculations of Brennan and Hess (1983), discussed in Part II (Fig. 27).

To verify the measured values of α and β, avalanche noise measurements were made. These gave an excess noise factor for electrons F_n characterized by an effective ionization rate ratio $k_{\text{eff}} = (\beta/\alpha)_{\text{eff}}$ of 0.6, in good agreement with the data of Ando and Kanbe (1981). The effective ionization rate ratio k_{eff} is defined as (McIntyre, 1972)

$$k_{\text{eff}} = \frac{k_2 - k_1^2}{1 - k_2}, \tag{178}$$

where

$$k_1 = \frac{\int_0^W \beta M(x)\, dx}{\int_0^W \alpha M(x)\, dx}, \tag{179}$$

$$k_2 = \frac{\int_0^W \beta M^2(x)\, dx}{\int_0^W \alpha M^2(x)\, dx}, \tag{180}$$

where $M(x)$ is the position-dependent multiplication of an electron–hole pair injected at x. Note that k_{eff} is very weakly dependent on the electric field, unlike $k = \beta(x)/\alpha(x)$.

d. Ionization Rates in InAs

Reliable measurements of the ionization rates have only recently become available (Mikhailova *et al.*, 1976). These authors find that at 77 K, $\beta \approx 10\alpha$. The large enhancement of the hole ionization rate was explained by these authors in terms of the so called resonant enhancement of impact ionization.

At 77 K in InAs, the spin–orbit splitting becomes equal to the band gap ($E_g = 0.43$ eV). This reduces the hole-initiated ionization energy, which becomes equal to the band gap, as discussed previously in Section 1 on ionization energies [Eq. (19)]. The experimental data were well fitted with Baraff's theory. This gave $E_{ie} = 0.43$ eV for the electron ionization energy and $E_{ih} = 0.43$ eV for the hole threshold energy. The latter value coincides with the spin–orbit splitting at 77 K. The values of the mean free paths for phonon scattering derived from the fit were, respectively, $\lambda_e = 170$ Å for electrons and $\lambda_h = 400$ Å for holes. Note that the use of Baraff's theory for holes is legitimate because of the isotropy of phonon scattering for holes. However, it is not justified for electrons because electron–phonon scattering in InAs is predominantly small angle, as discussed in detail in Section 3 on theory.

e. Ionization Rates in $In_xGa_{1-x}As$ *and* $In_{0.52}Al_{0.48}As$

Pearsall *et al.* (1975) measured α and β in $\langle 111 \rangle$ $In_{0.14}Ga_{0.86}As$ using the configuration of Fig. 32a in a p^+-n junction. A buffer region was used that consisted of several layers of intermediate composition to relieve stress due to mismatch between the $In_{0.14}Ga_{0.86}As$ layers and the GaAs substrate. The results are shown in Fig. 43; it is seen that $\beta > \alpha$, with $\beta/\alpha = 2.5$ at $F = 3 \times 10^5$ V cm^{-1}. Least squares fits of the data in the electric field range $2.6 \times 10^5 - 3.5 \times 10^5$ V cm^{-1} gave

$$\alpha(F) = 1.0 \times 10^9 \exp(-3.6 \times 10^6/F), \tag{181}$$

$$\beta(F) = 1.3 \times 10^8 \exp(-2.7 \times 10^6/F). \tag{182}$$

Using the same experimental configuration, Pearsall (1979) also measured α and β for the alloy $In_{0.53}Ga_{0.47}As$ grown lattice matched to InP in the $\langle 100 \rangle$ orientation, which is very important for long-wavelength optoelectronic applications. He finds that $\alpha/\beta \simeq 2$, as illustrated in Fig. 44. Measurement of the ionization rates ratio with the electric field oriented along the $\langle 111 \rangle$ direction showed also that $\alpha > \beta$ but with $\alpha/\beta \simeq 5$ (Pearsall and Papuchon, 1978).

Capasso *et al.* (1984a) have measured the ionization rates for electrons and holes in $In_{0.52}Al_{0.48}As$, lattice-matched to InP, using a p^+-i-n^+ diode structure. $In_{0.53}Ga_{0.47}As$ was used as a buffer layer between the InP substrate and the n^+ $In_{0.52}Al_{0.48}As$ region to absorb 1.06-μm radiation for pure hole injection. For pure electron injection $\lambda = 6.328$ Å radiation absorbed in the p^+ $In_{0.52}Al_{0.48}As$ layer was used. The data are shown in Fig. 45 for seven different diodes. It is very clear that α is greater than β and that $\alpha/\beta \simeq 2.5-3$ in the electric field range $3.3 \times 10^5 - 4.3 \times 10^5$ V cm^{-1}. The data from the different diodes were averaged at each electric field and least squares fitted to

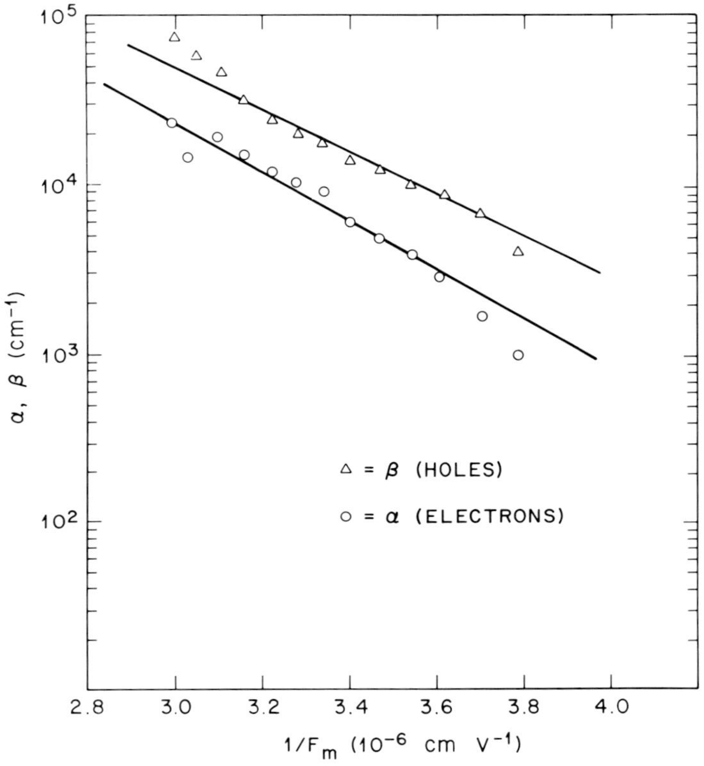

FIG. 43. Ionization rate versus reciprocal electric field in ⟨100⟩ $In_{0.14}Ga_{0.86}As$ at 300 K. [From Pearsall *et al.* (1975).]

give the solid lines in Fig. 45. These are represented by

$$\alpha(F) = 7.36 \times 10^4 \exp(-3.16 \times 10^{11}/F^2), \tag{183}$$

$$\beta(F) = 1.57 \times 10^4 \exp(-2.39 \times 10^{11}/F^2). \tag{184}$$

A long wavelength APD utilizing $In_{0.52}Al_{0.48}As$ as the avalanche region should therefore be designed for electron injection. Light would be absorbed in an adjacent $In_{0.58}Ga_{0.47}As$ layer.

f. Ionization Rates in $GaAs_{1-x}Sb_x$

Pearsall *et al.* (1976) investigated the α/β ratio of this alloy in the compositional range from $x = 0.05$ to $x = 0.12$ at 300 K using p^+-n junctions in the configuration of Fig. 32a. The results for $x = 0.12$ are reported in Fig. 46. It is seen that $\alpha/\beta > 1$. For the range of electric fields $2 \times 10^5 < F < 3 \times 10^5$ V

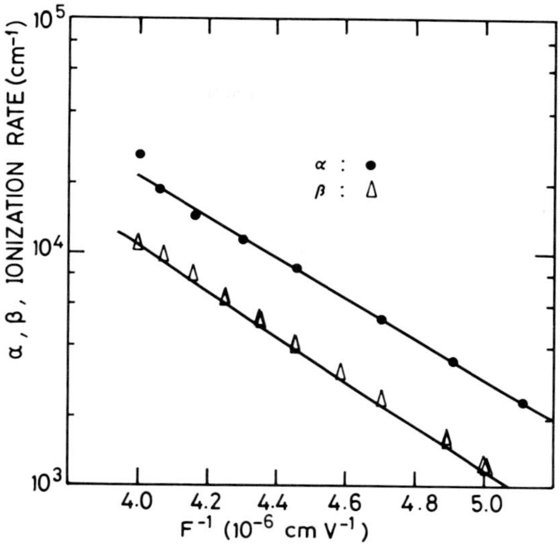

FIG. 44. Ionization rates versus reciprocal electric field in $In_{0.47}Ga_{0.53}As$ at 300 K. [From Pearsall (1980).]

cm^{-1}, the ionization rates were fitted with the equations

$$\alpha(F) = 1.5 \times 10^5 \exp(-6.4 \times 10^5/F), \tag{185}$$

$$\beta(F) = 1.1 \times 10^5 \exp(-7.2 \times 10^5/F). \tag{186}$$

It was also found the α/β varies with composition, as illustrated in Fig. 47.

g. Ionization Rates in InP

Several reports on the determination of α and β in $\langle 100 \rangle$ InP have been published. These experiments use either one-sided abrupt $p-n$ junctions or Schottky diodes to achieve pure hole and electron injection. All these measurements indicate that $\beta > \alpha$. Cook *et al.* (1982), using the improved geometry previously discussed (Fig. 32b), measured α and β in p^+-n and n^+-p junctions of $\langle 100 \rangle$ InP over a very wide range of fields, as shown in Fig. 48. It is seen that the β/α ratio decreases from 4.0 to 1.3 as the electric field is increased from 2.4×10^5 to 7.7×10^5 V cm^{-1}. Note that α and β vary exponentially with $1/F$ at low fields and with $1/F^2$ at high fields consistent with Shockley's and Wolff's laws and high-field limits, respectively. Excess noise measurements are also in good agreement with these data (Stillman *et al.*, 1982). These results are in fair agreement with Kao and Crowell (1980) for low fields and with Umebu *et al.* (1980) for high fields but are significantly higher than those reported by Armiento *et al.* (1979). The data of Fig. 48 are probably the most reliable data for this material to date. Table V gives

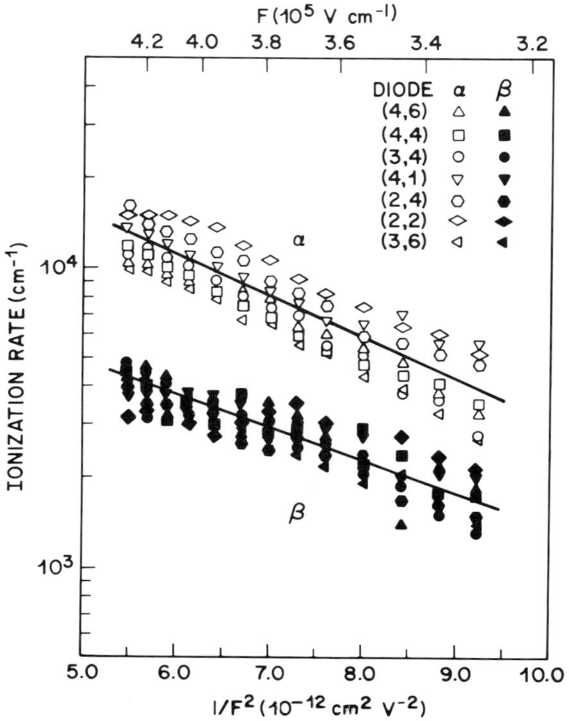

FIG. 45. Ionization rates versus reciprocal electric field squared in $In_{0.52}Al_{0.48}As$ at 300 K. [From Capasso *et al.* (1984a).]

the best fit curves to these data. The experimental values of α are in excellent agreement with the theory of Brennan and Hess (1984), as discussed in Part II (Fig. 28).

Armiento and Groves (1983) have studied experimentally impact ionization in $\langle 100 \rangle$, $\langle 110 \rangle$, and $\langle 111 \rangle$ InP. Their measurements indicate that no significant orientation dependence of the impact-ionization coefficients exists in InP, consistent with theoretical calculations of Brennan and Hess (1984a,b). Tabatabaie *et al.* (1983) reached independently the same conclusion by measuring α/β in $\langle 100 \rangle$- and $\langle 111 \rangle$-oriented InP diodes.

Takanashi and Horikoshi (1981) have also studied the temperature dependence of the effective ionization rates ratio k_{eff} in $\langle 100 \rangle$ InP by measuring the avalanche excess noise factor. They find that k_{eff} decreased with decreasing temperature from 2.3 at room temperature to slightly less than 1 at $-190°C$.

InP is routinely used as the avalanche layer in long-wavelength avalanche

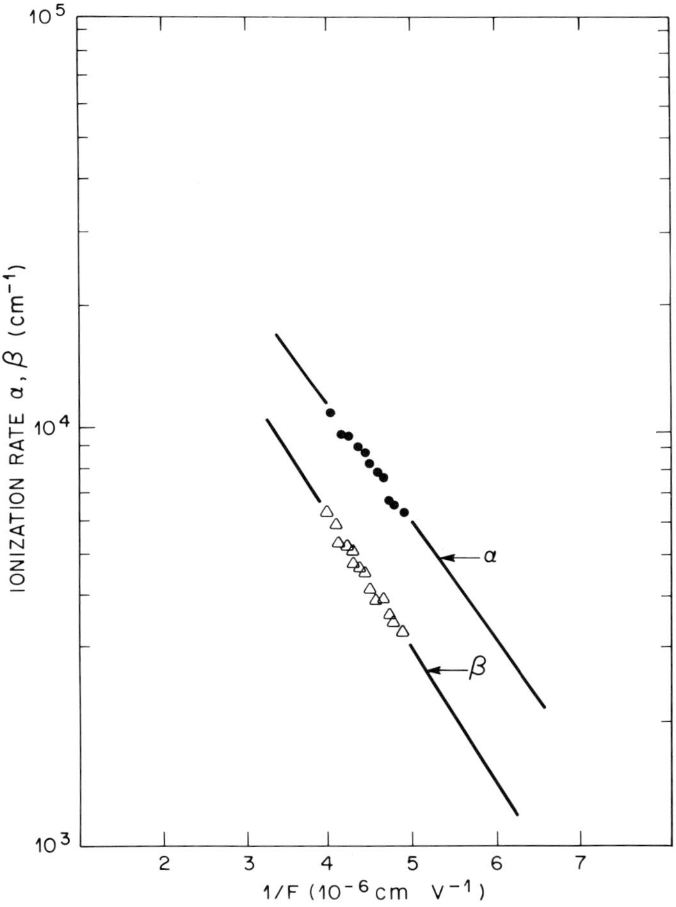

FIG. 46. Ionization rates versus reciprocal electric field in $GaAs_{0.89}Sb_{0.12}$ at 300 K. [From Pearsall *et al.* (1976).]

photodiodes with separate multiplication and absorption layers (Campbell *et al.*, 1983.)

h. Ionization Rates in InGaAsP

This alloy is of considerable interest for long-wavelength detectors and can be grown lattice matched to InP over a large compositional range spanning a band-gap range from 1.3 eV (InP) to 0.73 eV ($In_{0.53}Ga_{0.47}As$).

Takanashi and Horikoshi (1979) have measured the ionization rates in $In_{0.89}Ga_{0.11}As_{0.74}P_{0.26}$ ($E_g = 1.13$ eV) using the method illustrated in Fig.

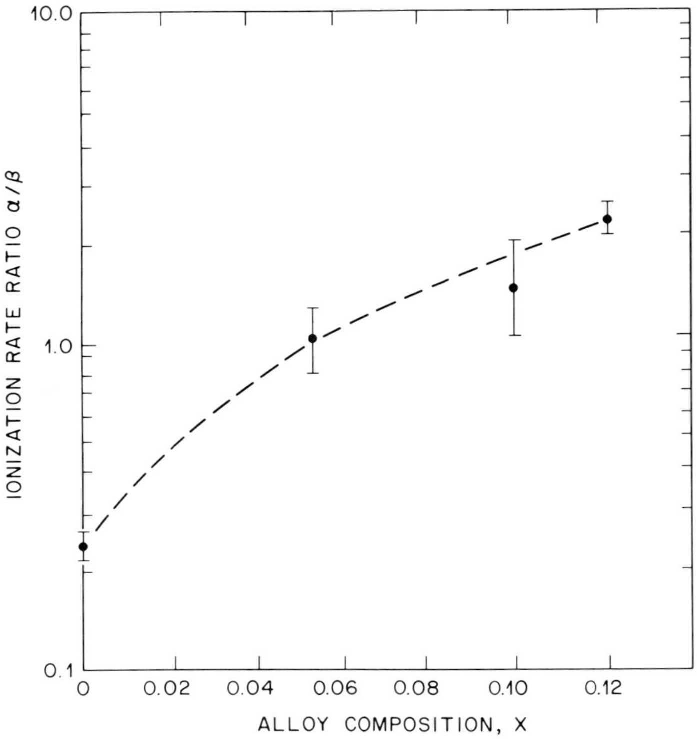

FIG. 47. Ionization rate ratio versus alloy composition in $GaAs_{1-x}Sb_x$ at 300 K in an electric field $\simeq 2.5 \times 10^5$ V cm^{-1} oriented along the $\langle 100 \rangle$ orientation. [From Pearsall *et al.* (1976).]

32a. Their results are shown in Fig. 49. The α/β ratio was found to be $\simeq 3.5$ in the electric field range from 3.85×10^5 to 2.95×10^5 V cm^{-1}.

The ionization rates can be expressed by the following formulas:

$$\alpha(F) = 2.46 \times 10^8 \exp(-3.2 \times 10^6/F), \tag{187}$$

$$\beta(F) = 2.15 \times 10^7 \exp(-3.07 \times 10^6/F). \tag{188}$$

Ito *et al.* (1978b) estimated the α/β ratio in $In_{0.73}Ga_{0.27}As_{0.43}$ ($E_g = 1.0$ eV) grown on $\langle 111 \rangle$ InP by measuring the wavelength dependence of the multiplication. They found $\alpha/\beta \simeq 3-4$.

Ionization rates data have also been reported for $\langle 100 \rangle$ $Ga_{0.33}In_{0.67}As_{0.70}P_{0.30}$ ($E_g = 0.92$ eV) and $\langle 100 \rangle$ $Ga_{0.18}In_{0.82}As_{0.39}P_{0.61}$ ($E_g = 1.11$ eV) by Osaka *et al.* (1984a,b). Abrupt mesa $p-n$ junctions and the injection method of Fig. 32a were used. For back illumination, a window was opened in the metallization, but the hole in the InP was not etched. Thus

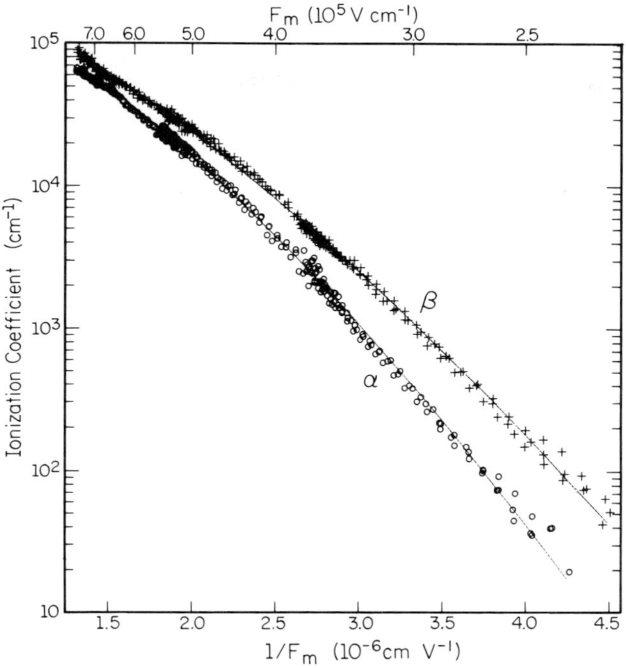

FIG. 48. Ionization rates versus reciprocal electric field in ⟨100⟩ InP at 300 K. [From Cook *et al.* (1982).]

there exists the possibility of electron contaminated hole injection due to electroabsorption of near band-gap recombination radiation generated by holes diffusing in the InP substrate towards the $p-n$ junction, as discussed in Section 11. The α and β data are presented in Figs. 50 and 51 for the larger and smaller band-gap respectively. In $Ga_{0.18}In_{0.82}As_{0.39}P_{0.61}$ ($E_g = 1.11$ eV)

TABLE V

BEST-FIT CURVES TO THE IONIZATION-COEFFICIENT DATA FOR ⟨100⟩ InP[a]

Doping level (cm^{-3})	Field range 10^5 ($V\ cm^{-1}$)	α (cm^{-1})	β (cm^{-1})
1.2×10^{15}	2.4–3.8	$1.12 \times 10^7 \exp(-3.11 \times 10^6/F)$	$4.79 \times 10^6 \exp(-2.55 \times 10^6/F)$
3.0×10^{16}	3.6–5.6	$2.93 \times 10^6 \exp(-2.64 \times 10^6/F)$	$1.62 \times 10^6 \exp(-2.11 \times 10^6/F)$
1.2×10^{17}	5.3–7.7	$2.32 \times 10^5 \exp(-7.16 \times 10^{11}/F^2)$	$2.48 \times 10^5 \exp(-6.23 \times 10^{11}/F^2)$

[a] After Cook *et al.* (1982).

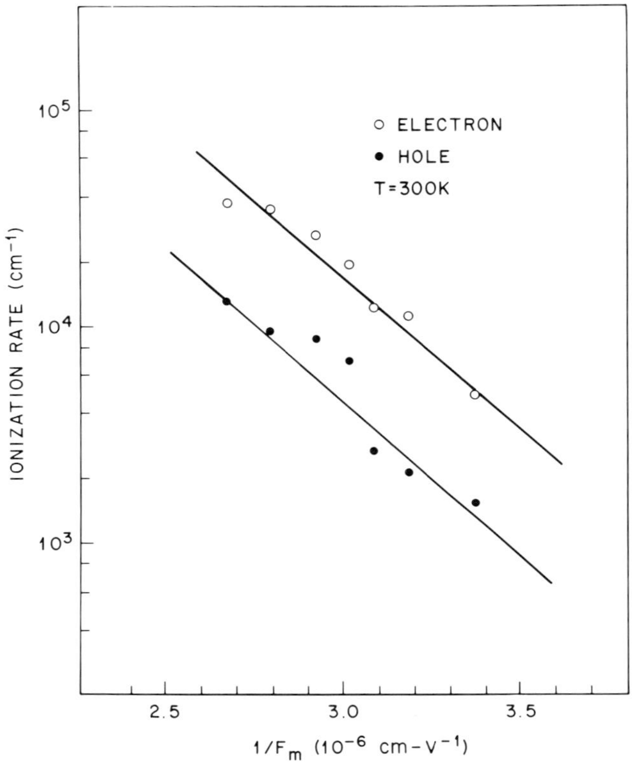

FIG. 49. Ionizaton rates versus reciprocal electric field in ⟨100⟩ $In_{0.89}Ga_{0.11}As_{0.74}P_{0.26}$ at 300 K: (○) electron, (●) hole. [From Takanashi and Horikoshi (1979).]

$\alpha/\beta = 1.1$ in the electric field range $4 \times 10^5 - 5 \times 10^5$ V cm^{-1}, whereas in $Ga_{0.33}In_{0.67}As_{0.70}P_{0.30}$ ($E_g = 0.92$ eV) $\alpha/\beta = 1.5$ in the field range $3.3 \times 10^5 - 4.3 \times 10^5$ V cm^{-1}. Note that the α/β value measured by Takanashi and Horikoshi (1979) for $E_g \approx 1.13$ eV (Fig. 49) is significantly higher than the value given by Osaka *et al.* (1984a) (Fig. 50) at a similar composition. A plot of the measured α/β ratio versus band gap for GaInAsP (including the InP and $Ga_{0.47}In_{0.53}As$ lattice-matched end points) shows that α/β decreases monotonically with increasing band gap except for the value measured by Takanashi and Horikoshi (1979). Osaka *et al.* (1984b) point out that there is no reason why α/β should have a maximum at some band gap of GaInAsP, since the band structure and scattering rates by optical phonons vary gradually with composition. They conclude therefore that their results are more reasonable than those by Takanashi and Horikoshi (1979).

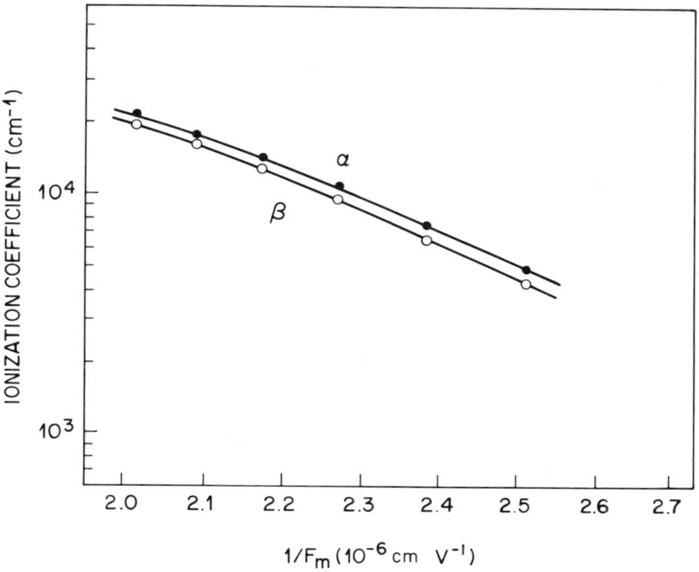

FIG. 50. Ionization rates versus reciprocal electric field in ⟨100⟩ $Ga_{0.18}In_{0.82}As_{0.39}P_{0.61}$, $E_g = 1.1$ eV, at 300 K. [From Osaka *et al.* (1984b).]

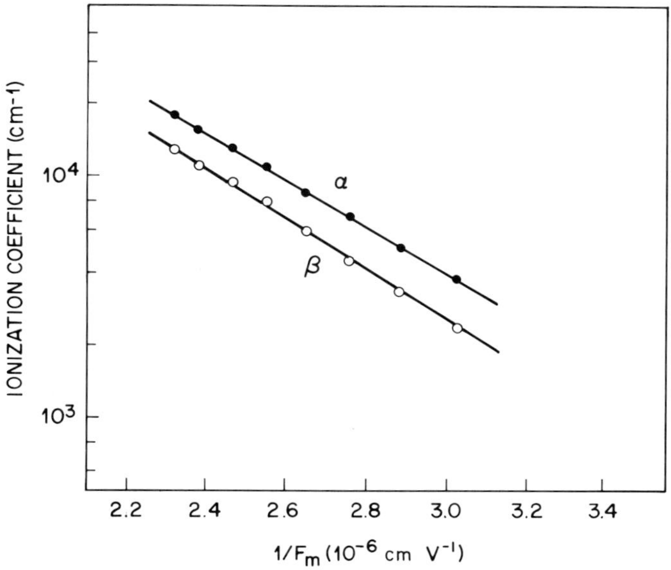

FIG. 51. Ionization rates versus reciprocal electric field in ⟨100⟩ $Ga_{0.33}In_{0.67}As_{0.70}P_{0.30}$, $E_g = 0.95$ eV, at 300 K. [From Osaka *et al.* (1984a).]

Takanashi and Horikoshi (1981) have determined the temperature dependence of $k_{eff} \equiv (\beta/\alpha)_{eff}$ in $In_{0.72}Ga_{0.78}As_{0.62}P_{0.38}$ ($E_g = 0.95$ eV) via excess noise measurements. They were not able to use temperatures higher than $-70°C$ because the tunneling dark currents produced extra noise, which made very difficult the measurements of the shot-noise power. A strong dependence on temperature was found, with k_{eff} varying from $\simeq 0.2$ to $\simeq 0.06$ in the range from -70 to $-190°C$.

i. Ionization Rates in GaP

The ionization coefficients for electrons and holes in GaP have been measured by Logan and Chynoweth (1962) and by Logan and White (1965). They found that $\alpha = \beta$ and that the rates can be fitted with Baraff's theory.

j. Ionization Rates in $GaSb/Al_xGa_{1-x}Sb$

Hildebrand *et al.* (1980, 1981) have carried out a systematic study of α and β in GaSb and in the $Al_xGa_{1-x}Sb$ alloy for various values of x. Their work provides conclusive evidence of the resonant enhancement of impact ionization. The ionization coefficients were measured in p^+-i-n^+ diodes. As discussed previously, this is the ideal geometry for the determination of α and β. To achieve pure injection of electrons and holes, these authors used an electron beam focused on the p^+ and n^+ layers of the structure.

Figure 52 shows the ionization threshold energies for electrons and holes

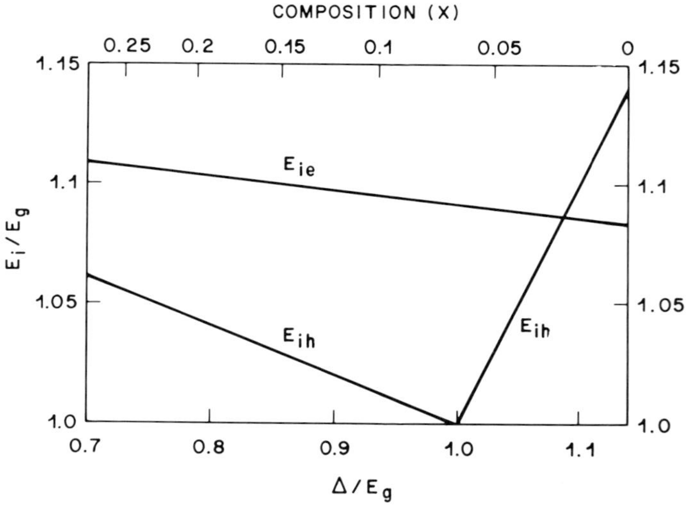

FIG. 52. Hole threshold energy E_{ih} and electron threshold energy E_{ie} for $Ga_{1-x}Al_xSb$ plotted versus the spin–orbit splitting-to-band-gap ratio Δ/E_g. [From Hildebrand *et al.* (1981). © 1981 IEEE.]

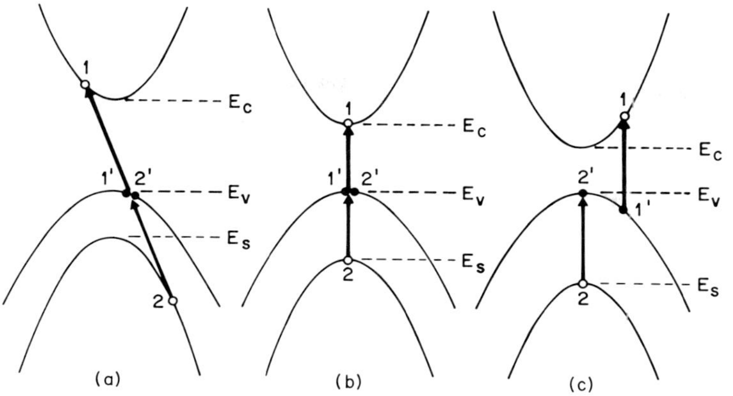

FIG. 53. The process of impact ionization initiated by holes from the split-off band for the three cases (a) $E_g > \Delta$, (b) Δ, $E_g = \Delta$, and (c) $E_g < \Delta$ in $Al_xGa_{1-x}Sb$. The initiating hole is labeled with a "2." The curvature of the bands is not in scale with the real band structure of this alloy. [From Hildebrand *et al.* (1981). © 1981 IEEE.]

normalized to the band gap of this alloy system versus composition x and versus the spin–orbit-splitting–band-gap ratio and calculated according to Eqs. (15) and (19). Note that as x approaches 0.065, Δ/E_g tends to unity, and the hole ionization energy goes through a pronounced minimum. Thus, an enhancement of β/α is expected at or near this critical composition. The process of impact ionization initiated by hot holes in the spin–orbit split-off light hole band is shown schematically in Fig. 53 for three cases with $E_g > \Delta$, $E_g = \Delta$, and $E_g < \Delta$. For $E_g > \Delta$, conservation of energy and momentum requires a threshold energy for holes given by Eq. (19). As Δ/E_g approaches unity, the threshold energy decreases to E_g. For $\Delta > E_g$, the threshold energy is always equal to Δ since no momentum transfer is necessary.

The hole ionization rate is probably enhanced most by the fact that a very small momentum is transferred in the impact-ionization collision ($\Delta k = 0$ for $\Delta = E_g$), as can be seen from Fig. 53b. The impact-ionization collision is Coulombic in nature. The square of the matrix element of the Coulomb interaction is inversely proportional to $(\Delta k)^4$, where Δk is the wave vector transferred in the interaction (for simplicity, we have neglected screening). Thus, for $\Delta k \simeq 0$, the impact-ionization probability above threshold will be strongly enhanced.

The experimental β/α is plotted in Fig. 54 versus Δ/E_g for two values of the electric field. As Δ/E_g approaches unity β becomes much larger than α. For $\Delta/E_g = 1.02$, $\beta/\alpha \gtrsim 20$ at $F = 3.3 \times 10^4$ V cm^{-1}.

The electric field dependence of α and β in GaSb and $Al_xGa_{1-x}Sb$ is illustrated in Fig. 55. The previous enhancement of β/α leads to a very low

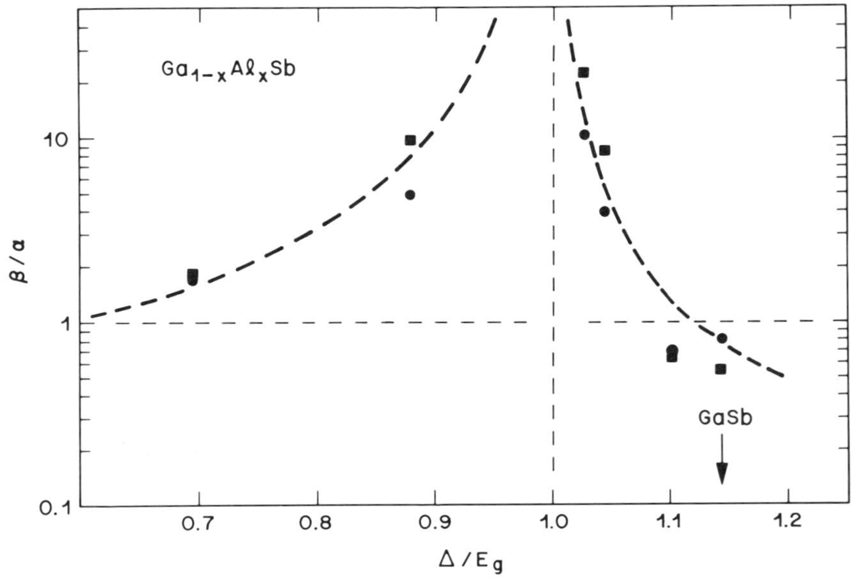

FIG. 54. The ionization rate ratio β/α plotted versus Δ/E_g: (●) $F = 4 \times 10^4$ V cm^{-1}; (■) $F = 3.3 \times 10^4$ V cm^{-1}. For $\Delta/E_g = 1.1$, β/α is independent of the field F. [From Hildebrand *et al.* (1981). © 1981 IEEE.]

excess noise factor for long-wavelength 1.55-μm avalanche photodiodes having a $Ga_{0.935}Al_{0.065}Sb$ avalanche region.

Zhingarev *et al.* (1980) independently confirmed the spin–orbit split-off band resonance effect in AlGaSb alloys. They also found the same effect in GaSb at 77 K, since at this temperature $E_g = \Delta = 0.8$ eV.

In a more recent study, Zhingarev *et al.* (1981) have systematically investigated the ionization rates in this alloy system at 77 K in the $\langle 100 \rangle$ and $\langle 111 \rangle$ orientation. The most interesting finding is that the electron ionization rate in the $\langle 111 \rangle$ orientation is smaller than in the $\langle 100 \rangle$ direction over the compositional range from $x = 0$ to $x = 0.15$. The hole ionization rate undergoes a strong resonant enhancement for $x = 0$ (Figs. 56 and 57), but as x increases, the resonance condition is no longer satisfied (unlike at 300 K), and β decreases very rapidly. Figures 56 and 57 show the electric field dependence of α and β for the $\langle 111 \rangle$ and the $\langle 100 \rangle$ directions, respectively. The ratio of ionization coefficients is higher in the $\langle 111 \rangle$ direction with a maximum value of $\beta/\alpha = 115$. The fact that $\alpha(\langle 111 \rangle) < \alpha(\langle 100 \rangle)$ has been interpreted along lines similar to those used by Pearsall *et al.* (1978) in the case of GaAs, such as the lack of an electron ionization threshold in the $\langle 111 \rangle$ direction.

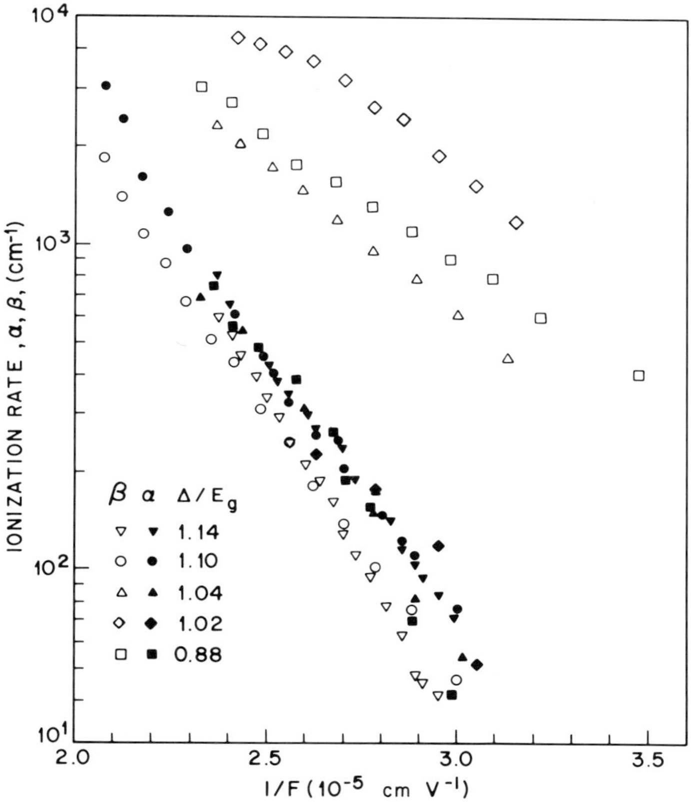

FIG. 55. Experimental results for the ionization coefficients α and β with various values of Δ/E_g for $Ga_{1-x}Al_xSb$. The open symbols correspond to β; closed symbols correspond to α. The measurements were performed with the following compositions x and temperatures: □, ■, $x = 0.052/80$ K, $\Delta/E_g = 0.88$; ○, ●, $x = 0.052/300$ K, $\Delta/E_g = 1.02$; △, ▲, $x = 0.052/320$, K, $\Delta/E_g = 1.04$; $x = 0.015/300$ K ($\Delta/E_g = 1.1$); $x = 0/300$ K ($\Delta/E_g = 1.14$). [From Hildebrand *et al.* (1981). © 1981 IEEE.]

Law *et al.* (1978) have measured α and β in $Al_{0.28}Ga_{0.72}Sb$. They find $\beta/\alpha = 3$ independent of the electric field; Fig. 58 shows their data. The solid lines are least squares fits, given by

$$\alpha(F) = 1.04 \times 10^5 \exp[-(4.1 \times 10^5/F^2)], \tag{189}$$

$$\beta(F) = 1.91 \times 10^5 \exp[-(4.12 \times 10^5/F^2)]. \tag{190}$$

Brennan and Hess (1985a) have performed a Monte Carlo simulation of β in $Al_xGa_{1-x}Sb$ (for compositions such that $\Delta = E_g$ and $\Delta < E_g$) and in GaSb at 300 K. They find that in order to fit the hole ionization data at resonance

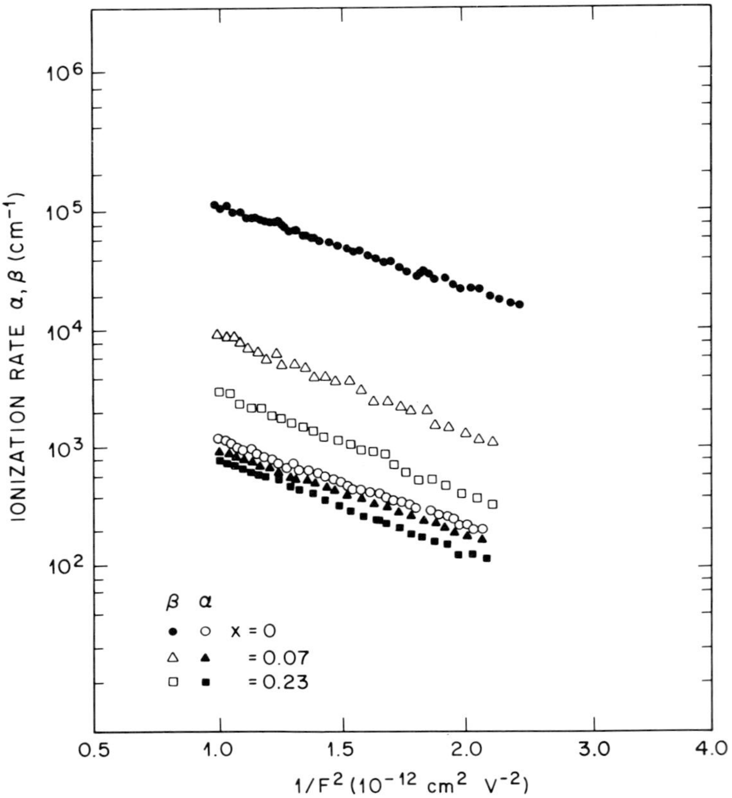

FIG. 56. Electric field dependence of the ionization rates in ⟨111⟩ $Al_xGa_{1-x}Sb$ at 77 K. [From Zhingarev *et al.* (1981).]

($\Delta = E_g$) of Hildebrand *et al.* (1981) (Fig. 55), a very high value of the P coefficient ($\simeq 10^5$) in Keldysh's formula (Eq. 26) for the impact ionization cross section must be assumed at the bottom of the split-off band, over a narrow energy range of the order of the collision broadening of the state. Above this energy range, a much smaller value of P must be used ($\simeq 0.2$). Physically this model suggests a resonance in the ionization cross section, at $k \simeq 0$. This is a consequence of the negligible momentum transfer by a hole ionizing at the bottom of the split-off band and is responsible for most of the observed resonant enhancement of β.

Brennan *et al.* (1985a) also found that the minimum in the hole ionization threshold when $\Delta = E_g$, accounts only for a small part of the resonant enhancement of β. Another interesting finding is that the hole ionization

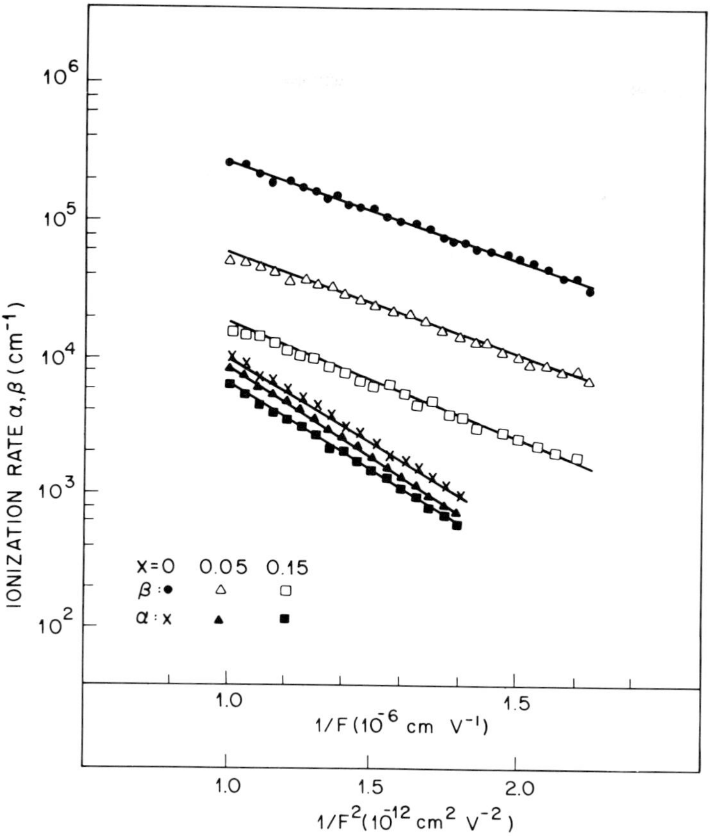

FIG. 57. Electric field dependence of the ionization rates in ⟨100⟩ $Al_xGa_{1-x}Sb$ at 77 K: the upper horizontal scale is for holes; the lower for electrons. [From Zhingarev *et al.* (1981).]

threshold energy must be very high (1.4 eV) in both the light and heavy hole band.

Physically this is related to the fact that most of the holes reside in these two bands, because their density of states is much higher than in the split-off band. Yet, for a resonance to occur, most of the holes must impact ionize in the split-off band. As pointed out by Brennan *et al.* (1985a), this can only be accomplished if enough holes are scattered into the split-off band from which they can impact ionize before they ionize from either the heavy or light hole bands. For that, it is necessary for the holes to drift to an energy suffciently high that the density of states within the split-off band is appreciable before the holes can be scattered into it. Therefore in order for the

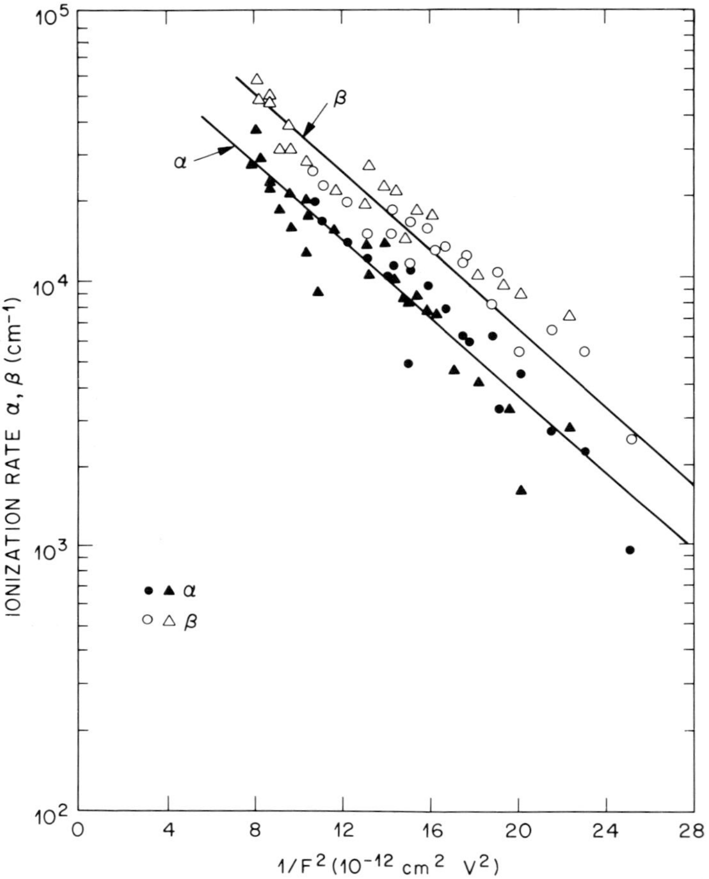

FIG. 58. Electric field dependence of the ionization rates in ⟨100⟩ $Al_{0.28}Ga_{0.72}Sb$ at 300 K. [From Law *et al.* (1978).]

resonance to be possible, the impact ionization threshold energy must be high within the heavy and light hole bands ($\simeq 1.4$ eV).

k. Ionization Rates in InSb–$In_xGa_{1-x}Sb$

Baertsch (1967) has studied avalanche multiplication in InSb, and from the variation of multiplication with wavelength he concluded that $\alpha \gg \beta$ at 77 K. Neglecting hole-initiated impact ionization, he found that α is nearly independent of electric field for $F \geq 5 \times 10^3$ V cm^{-1} and $\simeq 2 \times 10^3$ cm^{-1}.

More recently Gavrusko *et al.* (1978) have reported multiplication studies in InSb $p-n$ junctions. Their conclusions are generally in agreement with those of Baertsch; namely, that $\alpha \gg \beta$ and that there is a range of electric fields for which α is constant. They find that $\alpha \simeq 2 \times 10^3\ \mathrm{cm}^{-1}$ for $F \leq 10^4\ \mathrm{V\ cm}^{-1}$. However, for higher fields α was found to increase exponentially. From the onset of multiplication an ionization threshold energy of 0.24–0.25 eV was found, in agreement with the theoretical value obtained by Van Welzenis (1981).

Zhingarev *et al.* (1980) have investigated avalanche multiplication in $In_xGa_{1-x}Sb$ alloys with $0 \leq x \leq 0.10$ at 77 K. They find that electron-initiated multiplication is negligible compared to hole-initiated multiplication, implying that $\beta \gg \alpha$. At these compositions the spin–orbit splitting is $> E_g$, so that the observed high β/α ratio is due to the resonant mechanism previously discussed.

l. Ionization Rates in $Hg_xCd_{1-x}Te$

Avalanche gain in photodiodes made of this alloy has been observed by Vèrié *et al.* (1982). Measurements indicate that for $x = 0.73$, $\beta/\alpha \simeq 10$. At this composition, the spin–orbit splitting is very nearly equal to the band gap, and a large enhancement of hole-impact ionization is expected. Thus, this alloy system also is a promising candidate for low-noise long-wavelength 1.3-μm APDs.

IV. Avalanche Photodiodes with Enhanced Ionization Rate Ratios and Solid-State Photomultipliers

13. Avalanche Excess Noise

Knowledge of the ionization rate ratios and of their electric field dependence is essential for the design of a low-noise APD (McIntyre, 1966, 1972). For a detailed description of this work, the reader is referred to the review by Stillman and Wolfe (1977); we shall restrict ourselves here to a brief summary.

The noise of an APD per unit bandwidth can be described by the formula (McIntyre, 1966)

$$\langle i^2 \rangle = 2eI_{\mathrm{ph}} \langle M \rangle^2 F, \tag{191}$$

where e is the electron charge, I_{ph} the unmultiplied photocurrent (signal), $\langle M \rangle$ the average avalanche gain, and F the excess noise factor. For simplicity, we have neglected the dark current of the device.

If the avalanche process were deterministic, in other words, if every injected photocarrier would undergo the same gain M, the factor F would be unity, and the resulting noise would only be the multiplied input shot noise due to the random arrival of signal photons. The avalanche process is instead intrinsically statistical in nature so that individual carriers in general have different avalanche gains characterized by a distribution with an average $\langle M \rangle$. This causes additional noise so that we can write (McIntyre, 1966)

$$\langle i^2 \rangle = 2eI_{\text{ph}}\langle M \rangle^2 + 2eI_{\text{ph}}\sigma^2, \tag{192}$$

where σ^2 is the variance of the distribution of gains

$$\sigma^2 = \langle M^2 \rangle - \langle M \rangle^2. \tag{193}$$

Rearranging Eq. (192) gives

$$\langle i^2 \rangle = 2eI_{\text{ph}}\langle M \rangle^2[1 + \sigma^2/\langle M \rangle^2], \tag{194}$$

where the quantity in square brackets is the excess noise factor

$$F = \langle M^2 \rangle/\langle M \rangle^2 = 1 + \sigma^2/\langle M \rangle^2. \tag{195}$$

Photocurrent multiplication experiments always measure $\langle M \rangle$, which is now simply denoted as M. It has been shown that F is strongly dependent on the ratio of ionization coefficients for electrons and holes α/β (McIntyre 1966, 1972). McIntyre's results can be qualitatively summarized as follows. Low avalanche noise is obtained when the ratio of ionization coefficients β/α is either very large or very small and when the multiplication process is initiated by the carrier with the highest ionization coefficient. This can be intuitively understood as follows. If $\alpha = \beta$, for every electron–hole pair created, let us say, by an electron, there will be another pair generated by the newly created hole traveling in the opposite direction. This strong feedback effect "amplifies" noise fluctuations. The opposite case is the one of no feedback, when neither electrons nor holes impact-ionize. The latter is the situation of minimum excess noise, whereas the $\alpha = \beta$ case is that of maximum excess noise. It should be clear, then, that for low-noise APDs, a large difference between α and β is required.

The F factor is generally dependent on the avalanche gain, on where the current is injected or generated, and on the ionization rate ratios. For the case of a constant (i.e., not dependent on the electric field) ratio of ionization coefficients $k = \beta/\alpha$ and electron-initiated multiplication, the excess noise factor is given by (McIntyre, 1966)

$$F_n = M_n\{[1 - (1 - k)((M_n - 1)/M_n)^2]\}, \tag{196}$$

whereas for hole-initiated gain, the excess noise factor can be expressed by (McIntyre, 1966)

$$F_p = M_p\{1 - [1 - (1/k)][(M_p - 1)/M_p]^2\}, \tag{197}$$

where M_n and M_p are the electron and hole multiplication factors, respectively. The excess noise factors calculated from these expressions are plotted in Fig. 59 for various values of $k = \beta/\alpha$. Since Eqs. (196) and (197) are

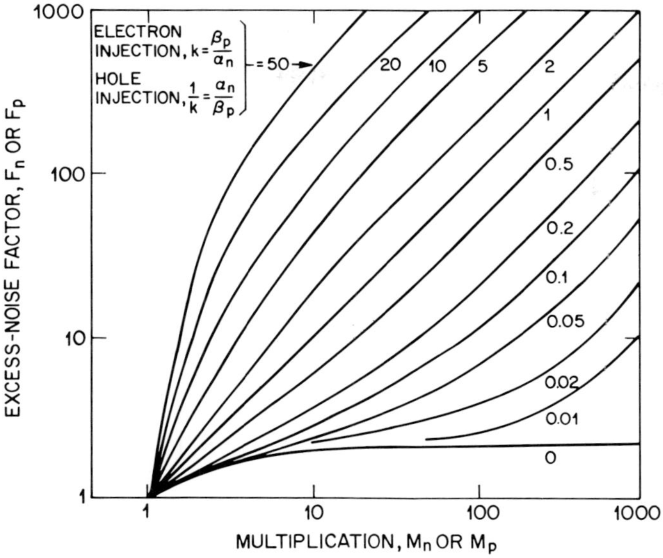

FIG. 59. Excess-noise factors for various values of multiplicaton and ratios of electron- and hole-ionization coefficients.

symmetric in n, p, k, and $1/k$, only one set of curves is required. It can be seen that for low F, the electron and hole-ionization coefficients must be very different.

Note that for the case $\alpha = \beta, F \simeq M$ so that the noise power increases as the cube of the gain. In the other limiting case in which either $\beta = 0$ or $\alpha = 0$, it is seen that for $M \gtrsim 10, F = 2$. Thus, at high gain the excess noise can never be made arbitrarily small, even in the ideal case where only one type of carrier ionizes. This means that a conventional APD is intrinsically more noisy than a photomultiplier tube where the avalanche process is instead nearly noise free (i.e., $F \simeq 1$). This was thought to be a fundamental limitation of avalanche multiplication in semiconductors; however, this is not the case. A novel APD structure (staircase APD) has been disclosed (Capasso *et al.*, 1983). This device represents the solid-state analog of a photomultiplier and has virtually noise-free gain ($F \simeq 1$); it is discussed in detail in Section 17.

A material with a large ratio of ionization coefficients is also desirable to maximize the gain–bandwidth product and to achieve APDs with uniform high gain (Stillman and Wolfe, 1977). As discussed before in Part III most materials sensitive in the low-loss low-dispersion window of optical fibers ($1.3 \leq \lambda \leq 1.55\ \mu m$) have comparable ionization coefficients for electrons and holes. It is therefore of considerable practical interest to explore the possibility of artificially increasing or decreasing β/α by using new device structures. Part IV shall focus on these novel avalanche detector structures with enhanced α/β ratio. Several principles have either been used or proposed for enhancing the ionization rate ratio α/β:

1. elimination of the "positive feedback" of holes by trapping them in a potential well formed between two heterojunctions (Gordon *et al.*, 1979);
2. difference between the ionization energies and the quasi-electric fields for electrons and holes in graded-gap materials (graded-gap APD) (Capasso *et al.*, 1982b);
3. large asymmetry between conduction- and valence-band discontinuities at many III–V heterojunctions [structures using this principle include the superlattice or multiquantum-well APD (Chin *et al.*, 1980; Capasso *et al.*, 1981b) and the staircase APD (Capasso *et al.*, 1982); in the multi-quantum-well APD, the α/β ratio can be further enhanced by introducing a period doping profile (Blauvelt *et al.*, 1982)];
4. resonant enhancement of impact ionization in superlattices, induced by the zone folding effect (Mon and Hess, 1982);
5. spatial separation of electrons and holes in materials of differing band gaps (Capasso, 1982a,b; Tanoue and Sakaki, 1982).

14. Multiple *p–n*-Junction Heterostructure Avalanche Detector

This structure was proposed by Gordon *et al.* (1979). Figure 60 is a schematic diagram of such a device with two stages of avalanche multiplication. The device consists of two $p-n$ junctions, reverse biased and separated by the heterojunction B–C, with $E_g^B > E_g^C$. The junctions A–B and C–D can be either homojunctions or heterojunctions. Figure 60 illustrates, for example, a case with two $p-n$ homojunctions and one heterojunction, where the energy step occurs entirely in the valence band, creating the desired hole trap.

A photon incident through the window layer generates an electron–hole pair in region A. After diffusing through this layer, the electron falls through the first accelerating voltage in region B, producing new pairs, which have themselves the possibility of further production of pairs. Region B should be lightly n doped, and thin enough so that under bias it is entirely depleted of carriers. We define M_1 as the resultant multiplication: M_1 electrons enter region C, and $M_1 - 1$ holes plus the original photoproduced hole pass out of the device to the left. The electrons diffuse through region C whereupon they avalanche anew while falling into region D. In this second avalanche, each of the M_1 electrons produces M_2 electrons in region D as well as $M_2 - 1$ holes trapped in region C. The result is a totally integrated current pulse of $M_1 M_2$ electrons leaving the device through the contract at the right. The trapped holes are removed from region C to the external circuit. What has been avoided in this structure is the reentry of the $M_1(M_2 - 1)$ holes into the first

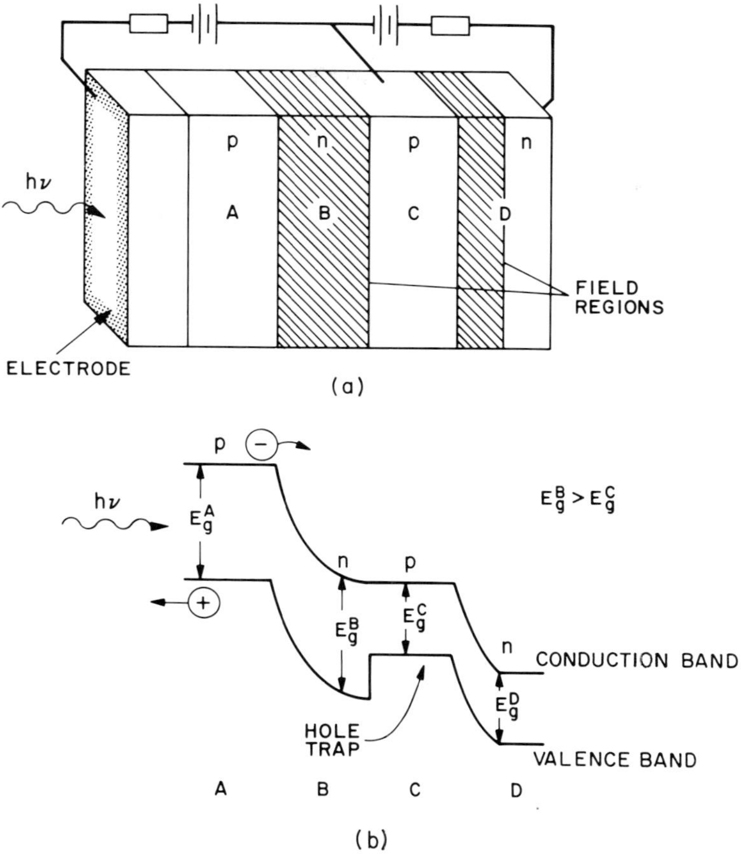

FIG. 60. (a) Schematic of the $p-n-p-n$ avalanche-detector structure; (b) electronic band structure for the device with appropriate bias applied; region C provides the desired hole trap. [From Gordon *et al.* (1979). © 1979 IEEE.]

avalanche region. This drastically reduces hole feedback and therefore excess noise.

In a one-stage device with average multiplication M, for electron injection, the variance of the gain previously defined by Eq. (191) is obtained from Eqs. (193) and (194):

$$\sigma^2 = M[(M-1) + (\beta/\alpha)(M-1)^2]. \tag{198}$$

Now, consider the two-stage device of Fig. 60, consisting of stages 1 and 2 separated by a junction that passes electrons without loss, but blocks holes. The multiplication and variances for the two stages are M_1, σ_1^2 and M_2, σ_2^2, whereas the overall values are M, σ^2. Evidently, $M = M_1 M_2$, and for this device it can be shown that

$$\sigma^2 = M_1 \sigma_2^2 + M_2^2 \sigma_1^2. \tag{199}$$

If we assume that β/α is the same for the two stages and use Eq. (199) to evaluate σ_1^2 and σ_2^2, we find the result for the variance of the two-stage device:

$$\sigma^2 = M\{(M-1) + (\beta/\alpha)[(M_2-1)^2 + M_2(M_1-1)^2]\}. \tag{200}$$

Comparing Eq. (200) with Eq. (198), we see that the term proportional to β/α has been reduced. For example, if we arbitrarily take $M_1 = M_2 = \sqrt{M}$, then this term becomes $(\beta/\alpha)(M-1) \times (\sqrt{M}-1)$. It can be further minimized for fixed M by taking $2M_2 = M_1^2 + 1$. The generalization of Eq. (200) to n stages for the case where the ratio β/α may be different for each stage, gives

$$\sigma^2 = M\{(M-1) + [(\beta/\alpha)_n(M_n-1)^2 + M_n(\beta/\alpha)_{n-1}(M_{n-1}-1)^2 + \cdots + M_n M_{n-1} \cdots M_2(\beta/\alpha)_1(M_1-1)^2]\}. \tag{201}$$

The reduction in excess noise obtainable with such a multistage APD is shown graphically in Fig. 61. The excess noise efficiency γ is related to the standard excess noise factor F by

$$F = 1 + \gamma[(M-1)/M]. \tag{202}$$

The excess noise efficiency γ has been plotted as a function of the total multiplication factor M for several cases including two- and three-stage devices. The horizontal line in the lower part of Fig. 61 corresponds to the ideal case of $\beta/\alpha = 0$, which gives the lowest excess noise obtainable in an electron-multiplying APD. The worst case, a single-stage device with $\beta/\alpha = 1$, is also shown in Fig. 61. One can observe that a two- or three-stage device can reduce the excess noise in this worst case ($\beta/\alpha = 1$), by more than an order of magnitude, to a value several times greater than that for the ideal case.

Unfortunately, there is one consideration that may make this structure impractical. To ensure high speed, the trapped carrier somehow has to be drained out of the potential wells. For this to occur, additional electrodes are needed. Another potential speed problem is due to the fact that electron motion in the A and C regions is by diffusion rather than by drift.

15. Graded-Gap Avalanche Detectors

This device was disclosed in a patent (Capasso, 1983c) and was first experimentally demonstrated by Capasso *et al.* (1982b). Moderate enhancements of the measured α/β ratio with respect to bulk GaAs and $Al_xGa_{1-x}As$,

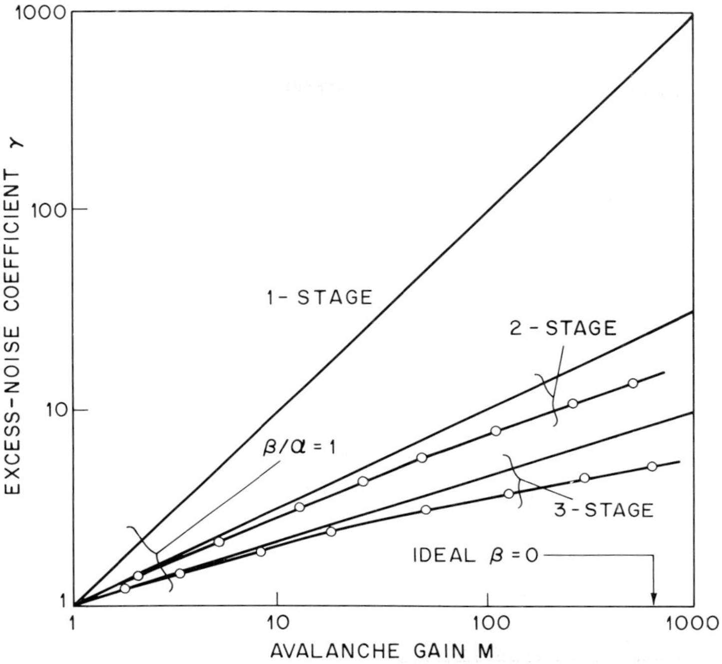

FIG. 61. Excess-noise coefficient $\gamma = \sigma^2/M(M-1)$ versus overall avalanche gain M. Curves are shown for one-, two-, and three-stage devices under the condition $\beta/\alpha = 1$. For the multistage devices, the solid curves are for equal gain per stage, and the dashed curves are for devices with gain ratios optimized as described in the text. The lower horizontal line $\gamma = 1$ represents the ideal case $\beta = 0$. The coefficient γ is related to the standard excess-noise factor F by $F = 1 + \gamma[(M-1)/M]$. [From Gordon *et al.* (1979).]

have been reported by these authors in a $p^+\text{-}i\text{-}n^+$ diode with a graded-gap $Al_xGa_{1-x}As$ region.

Kroemer (1957) first considered the problem of transport in a graded-gap semiconductor. As a result of compositional grading, electrons and holes experience "quasi-electric" fields of different intensities. In addition, the forces resulting from these fields push electrons and holes in the same direction (Fig. 62a).

Let us assume that the graded material (assumed to be $\leq 1\ \mu m$ wide and intrinsic) is sandwiched between a p^+ and an n^+ region to form a $p\text{-}i\text{-}n$ diode. If reverse bias is then applied to the device, electrons will experience a higher total electric force than will holes. This is shown in Fig. 62b, where $eF_{\mathrm{g,e(h)}}$, eF_{b} respectively represent the intensities of the electrical forces resulting from the band-gap grading and from the applied voltage. Above a certain bias voltage, impact ionization occurs. Since the electron experiences

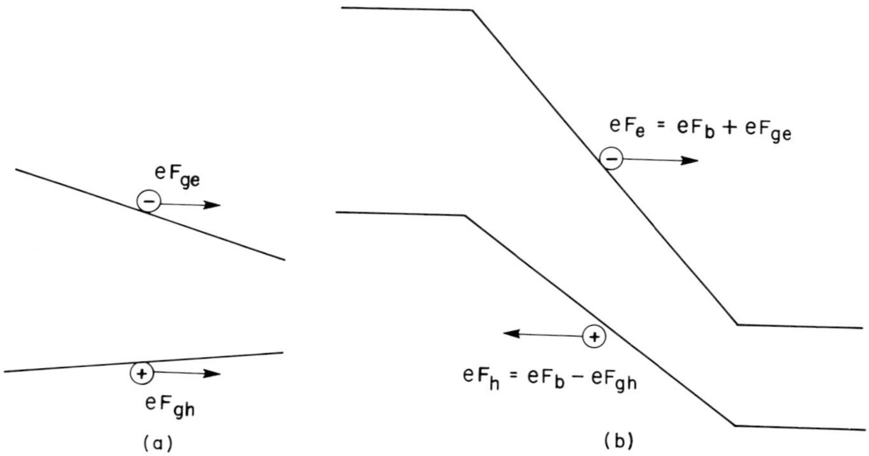

FIG. 62. (a) Effect of quasi-electric field in a graded-gap material; (b) combined effect of real and quasi-electric fields. [From Capasso *et al.* (1982b).]

a total effective electric field higher than that acting on the holes, the net effect is an increase in the α/β ratio. The difference between the quasi-electric fields acting on electrons and holes can amount to as much as 10% of the applied bias field. The exponential dependence of α and β on the electric field can thus produce increase of α/β.

The major effect of band-gap grading is, however, on the ionization threshold energies for electrons and holes. This is illustrated in Fig. 63. Let us assume that the avalanche is initiated by an electron–hole pair (1–1′) excited, for example, by a photon inside the graded region. Electron 1 is accelerated by the electric field toward regions of lower band gap and generates an electron–hole pair (2–2′) by impact ionization after an average distance $1/\alpha$. Hole 1′, on the other hand, drifts in the opposite direction toward a higher band-gap region, thus requiring less energy (with respect to electron 1) to create an electron–hole pair (3–3′). Thus, the effective ionization threshold energy for electrons is smaller than that for holes. This argument can be repeated for every electron–hole pair subsequently generated by impact ionization. Since the ionization rates increase exponentially with decreasing ionization energy, the ionization rates ratio is expected to be greatly enhanced. Obviously, to achieve minimum excess noise the avalanche should always be initiated only by the carrier with the highest ionization coefficient.

Finally, we would like to discuss another important property of graded-gap diodes. It can be shown that a graded-band-gap diode has a "softer" breakdown than a nongraded diode, meaning a more gradual slope of the IV

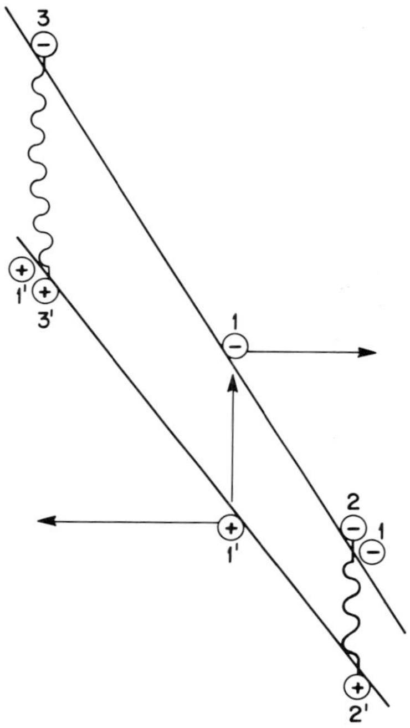

FIG. 63. Schematic of impact ionization in a graded-gap semiconductor. The initial electron–hole pair is denoted by 1 – 1′. Note that the electron sees a lower ionization energy than the hole, since it is moving toward a lower-gap region. [From Capasso *et al.* (1982b).]

characteristic very close to the breakdown voltage (Arutyunyan and Petrosyan, 1980; Capasso *et al.,* 1982b). The physical reason for this effect is that avalanche multiplication starts first in the low-gap region and then gradually proceeds toward higher gap regions as the voltage is increased. Thus, the graded-gap avalanche diode is expected to have an improved gain stability (with respect to fluctuations of the applied voltage) compared to a homogeneous diode.

For a quantitative estimate of the enhancement of α/β, a Shockley model shall be used. Although this approach is not a complete discription of impact ionization, nevertheless, it has the advantage of providing a simple heuristic description of the essential physical aspects. Consider a graded intrinsic region of width W in a spatially uniform applied electric field of intensity F. The field is directed toward the higher-gap region. We assume an i region linearly graded from $Al_{0.45}Ga_{0.55}As$ ($E_g = 2$ eV) to GaAs ($E_g = 1.42$ eV). It is well known that in both GaAs and $Al_xGa_{1-x}As$ ($x \leq 0.45$), $\alpha \lesssim \beta$; thus, the effect of grading is expected to be an increase of α/β above unity. The

Shockley model assumes that the carriers that impact-ionize are those which escape phonon collisions and reach the ionization energy. The ionization rates are then given by [Eq. (47)].

$$\alpha = D_e^{-1} \exp[-D_e/\lambda_e]; \qquad \beta = D_h^{-1} \exp[-D_h/\lambda_h], \tag{203}$$

where λ_e, λ_h are the mean free paths for phonon scattering and D_e, D_h are the distances required to ballistically gain the ionization threshold energy $E_{i,e}$ and $E_{i,h}$. It is easy to extend Eq. (203) to the case of a graded-band-gap material. In this case, D_e, D_h satisfy the equations

$$\left(eF + \frac{dE_c}{dz}\right) D_e = E_{i,e}(z) - \frac{dE_{i,e}}{dz} d_e, \tag{204}$$

$$\left(eF - \frac{dE_v}{dz}\right) D_h = E_{i,h}(z) + \frac{dE_{i,h}}{dz} D_h, \tag{205}$$

where dE_c/dz, dE_v/dz are the quasi-electric force intensities expressed as gradients of the band-edge energies. The right-hand sides of Eqs. (204) and (205) express the physical fact that the electron initially at z has to gain an ionization energy smaller than $E_{i,e}(z)$ since it is moving toward lower-gap regions; the opposite occurs for holes.

We now recall that in an $Al_xGa_{1-x}As/Al_yGa_{1-y}As$ heterojunction $\Delta E_c = 0.60\ \Delta E_g$ and $\Delta E_v = 0.40\ \Delta E_g$, implying that $dE_c = 0.60\ dE_g$, $dE_v = 0.40\ dE_g$. Next we assume that $E_{i,e} = E_{i,h} = \frac{3}{2}E_g(z)$ and $\lambda_e = \lambda_h$ throughout the graded region. These are reasonable approximations also in view of the fact that experimentally $\alpha \simeq \beta$ in $GaAs/Al_xGa_{1-x}As$ ($x \leq 0.45$). Equations (204) and (205) and can then be solved with respect to D_e, D_h, which substituted in Eq. (203) give

$$\frac{\alpha(z)}{\beta(z)} = \frac{eF + 2.10 dE_g/dz}{eF - 1.40 dE_g/dz} \frac{\exp[-1.5E_g(z)/(eF + 2.10 dE_g/dz)\lambda]}{\exp[-1.5E_g(z)/(eF - 1.40 dE_g/dz)\lambda]}. \tag{206}$$

Note that for $W < 1\ \mu m$, the terms in dE_g/dz cannot be neglected with respect to F, so that α/β is enhanced. Instead, for $W \gg 1\ \mu m$, $\alpha/\beta \rightarrow 1$, and α, β depend only on the local band gap. In Fig. 64, $\alpha(z)/\beta(z)$ has been plotted versus position inside the device using Eq. (204) for three different applied electric fields. We have assumed that $W = 0.4\ \mu m$ and $\lambda = 50$ Å. For $F = 3 \times 10^5\ V\ cm^{-1}$ the α/β ratio is extremely large; but it can be shown that the gain is small. However, at higher fields ($\geq 4 \times 10^5\ V\ cm^{-1}$) at which the gain is sizeable, it is seen that the ratio α/β is still favorable in terms of device noise.

The experimental device consists of an n^+-p junction grown by molecular beam epitaxy (MBE) on a high-quality Si-doped GaAs (100) n^+ substrate. A 1.5-μm-thick Sn-doped n^+ buffer layer was then grown, followed by the

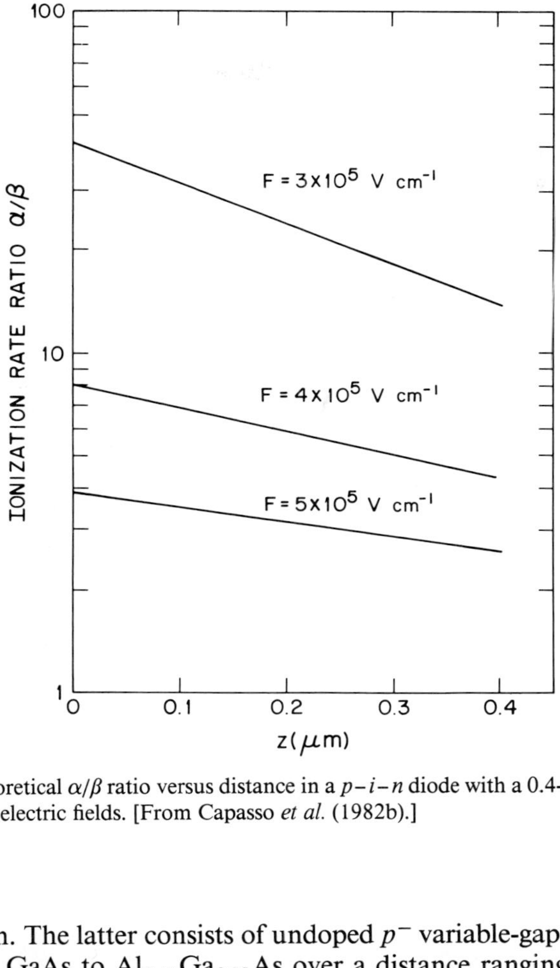

FIG. 64. Theoretical α/β ratio versus distance in a $p-i-n$ diode with a 0.4-μm graded-gap i layer at various electric fields. [From Capasso *et al.* (1982b).]

graded region. The latter consists of undoped p^- variable-gap $Al_xGa_{1-x}As$ graded from GaAs to $Al_{0.45}Ga_{0.55}As$ over a distance ranging from 0.3 to 0.7 μm. Subsequently, a Be-doped ($\simeq 5 \times 10^{17}$ cm^{-3}) p^+-$Al_{0.45}Ga_{0.55}As$ layer $\simeq 1.0-1.5$ μm thick is grown, followed by a 1000-Å p^+-GaAs contact layer.

To determine the ionization rate ratio it is necessary to measure separately the electron- and hole-initiated avalanche gain M_n and M_p, as discussed in Part III. The technique of Fig. 32a was used, and 4500 and 6238 Å lights were respectively used to inject electrons and holes. Base-line corrections for the electron-initiated gain (due to depletion-layer widening effects in the layer with higher background doping) were accurately determined using Wood's *et al.* (1973) method. No correction is needed for injected holes.

Two types of structures were tested. In the first the undoped graded region was 0.7 μm wide. The background doping was graded from 5×10^{16} to 8×10^{16} cm^{-3}, and the breakdown voltage was $\simeq 20$ V. The second structure had a 0.4-μm-thick graded region with background doping estimated at 10^{15} cm^{-3} and a breakdown voltage of 32 V. The diode dark current was $\simeq 1$ nA at $0.9 V_B$. Figure 65 shows M_n and M_p measured versus reverse voltage in the diodes having 0.4- and 0.7-μm-wide graded regions, respectively. We note a significant difference in the electron and hole multiplications for the strongly graded diodes. This is evidence of a large effective ionization rate ratio. The turn-on voltage of the electron-initiated multiplication corresponds to a field of $\simeq 2 \times 10^5$ V cm^{-1}. For the diodes with the wider graded region, it is seen that M_n and M_p are comparable (Fig. 65b), indicating that the electron and hole ionization rates are comparable. This is not only due to the smaller band-gap gradient but also to the fact that due to the high doping, the avalanche region extends only to a small fraction of the graded region.

Because of the variable gap and the intrinsic nonlocal nature of the ionization rates in our structure, the data of Fig. 65 cannot be used to determine the individual rates α, β as is commonly done in nongraded structures. However, from the data we can extract an effective α/β, which is the important quantity in noise considerations. Starting from the avalanche rate equations, with arbitrary ionization rates $\alpha(z)$, $\beta(z)$, one can derive the equation (McIn-

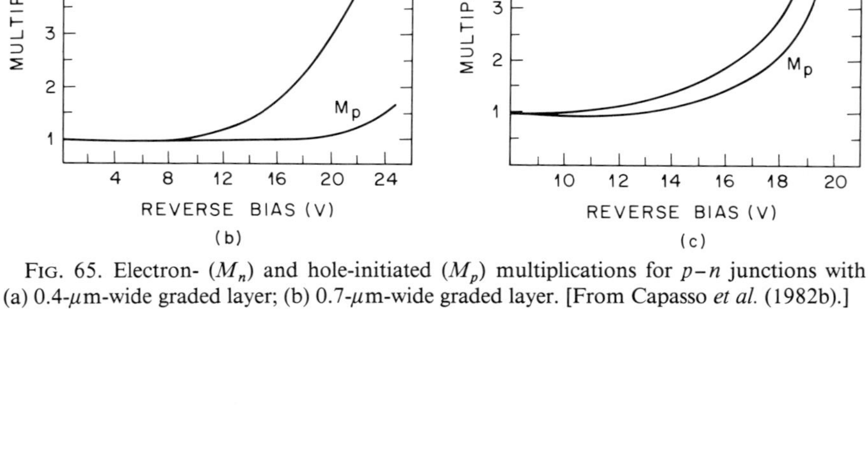

FIG. 65. Electron- (M_n) and hole-initiated (M_p) multiplications for p–n junctions with (a) 0.4-μm-wide graded layer; (b) 0.7-μm-wide graded layer. [From Capasso *et al.* (1982b).]

tyre, 1972) for the effective k_1 ionization rate ratio previously defined [Eq. (179)]:

$$\frac{M_p - 1}{M_n - 1} = \frac{\int_0^W \beta M(z)\, dz}{\int_0^W \alpha M(z)\, dz} \equiv k_1, \tag{207}$$

where $M(z)$ is the multiplication that an electron–hole pair generated at z will undergo. The weighted ionization rate ratio k_1 is directly related to the effective β/α ratio k_{eff}, which determines the noise. Note also that $k_1 \cong k_{\text{eff}}$ (McIntyre, 1972). From the data of Fig. 65a we find an ionization rate ratio $k_1 \simeq 7.5$ at a gain of 5 and even higher values (10) at lower gains. On the other hand, for the other structure with a smaller band-gap gradient, we find (Fig. 65b) at a gain of 5 an ionization rate ratio $k_1 \simeq 1.7$.

There is an obvious trade-off between gain and ionization rate ratio. The α/β ratio can be made very large (>10) but at the expense of gain because of the reduced length of the device. We can also obtain $\beta/\alpha \gg 1$ instead of $\alpha/\beta \gg 1$ by interchanging the p^+ and n^+ regions in Fig. 62b. Extremely high ($\simeq 1000$) spatially uniform gains were measured in our structures by biasing the devices very close to breakdown.

16. Superlattice Avalanche Detectors

a. *The Multi-Quantum-Well Avalanche Photodiode*

Capasso *et al.* (1981b, 1982c) first demonstrated experimentally that in an AlGaAs/GaAs multi-quantum-well structure, the effective impact-ionization rates for electrons and holes are very different ($\alpha/\beta \simeq 8$), although they are comparable in the basic bulk materials. This effect is attributed to the difference between the conduction- and valence-band-edge discontinuities at the $Al_{0.45}Ga_{0.55}As$/GaAs interface, a feature common to several lattice-matched heterojunctions used in long-wavelength (1.3-μm) detectors ($Al_{0.48}In_{0.52}As/Ga_{0.47}In_{0.53}As$; AlSb/GaSb, HgCdTe/CdTe). This makes possible the development of low-noise APDs in these materials. The enhancement of α/β in such structures was first predicted by Chin *et al.* (1980).

To understand the superlattice APD, consider (Fig. 66) the energy-band diagram of the structure implemented by Capasso *et al.* (1981b, 1982c). Because of the very low-doped material, the field is constant across the 2.5-μm-long depletion layer. This consists of alternating GaAs (450 Å) and $Al_{0.45}Ga_{0.55}As$ (550 Å) layers, for a total of 25 periods. To illustrate, assume a field $F = 2.7 \times 10^5$ V cm^{-1}; at this value a sizable multiplication was observed. For $F > 10^5$ V cm^{-1}, electrons gain between collisions an energy greater than the average energy lost per phonon-scattering event $\langle E_p \rangle$ ($\simeq 21$ meV). Thus, carriers are strongly heated by the field and can gain the ionization energy.

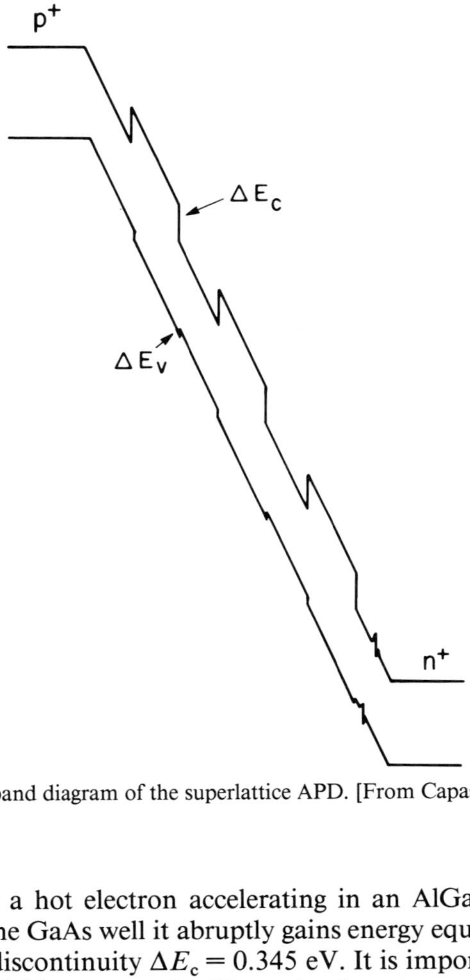

FIG. 66. Energy-band diagram of the superlattice APD. [From Capasso *et al.* (1982c).]

Consider now a hot electron accelerating in an AlGaAs barrier layer. When it enters the GaAs well it abruptly gains energy equal to the conduction-band-edge discontinuity $\Delta E_c = 0.345$ eV. It is important to stress the ballistic nature of this energy gain. Given the abruptness of the heterointerface, it occurs over a distance much smaller than the phonon scattering mean free path so that the average distance required to reach the ionization threshold is greatly reduced (α enhanced). Since $\Delta E_c > \Delta E_v \simeq 0.23$ eV, electrons enter the GaAs well with higher kinetic energy than do holes and are therefore much more likely to produce electron–hole pairs in the GaAs (impact-ionization by band-edge discontinuities). Therefore a steplike band structure, as shown in Fig. 66, where the discontinuity in the conduction band is greater than that in the valence band, will enhance the ionization rate ratio α/β.

Another way to express this concept is to say that when it enters the well, the electron "sees" an ionization energy ($E_i \simeq 1.65$ eV) reduced by ΔE_c with respect to the threshold energy in bulk GaAs ($E_i = 2.0$ eV). Since the impact-ionization rate α increases exponentially with decreasing E_i, a large increase in the effective α with respect to bulk GaAs is expected. When the electron enters the next AlGaAs barrier region, the threshold energy in this material is increased by ΔE_c, thus decreasing α in the AlGaAs layer. However, since $\alpha_{GaAs} \gg \alpha_{AlGaAs}$, the exponential dependence on the threshold energy ensures that the average α

$$\bar{\alpha} = \alpha_{GaAs} L_{GaAs} + \alpha_{AlGaAs} L_{AlGaAs} / L_{GaAs} + L_{AlGaAs}, \tag{208}$$

is largely increased (L denotes layer thickness).

Electrons that have impact ionized in the GaAs easily get out of the well; the voltage drop across the well is > 1 V. In addition, at fields $\geq 10^5$ V cm^{-1} in GaAs, the average electron energy is ≥ 0.6 eV (Shichijo and Hess, 1981; Fig. 24) so that electron trapping effects in the wells are negligible.

The above values of ΔE_v and ΔE_c for the $Al_{0.45}Ga_{0.55}As$/GaAs heterojunction are based on several recent measurements of band offsets, which indicate that for AlGaAs/GaAs heterointerfaces ΔE_c is approximately equal to 60% of the band-gap difference (Miller *et al.,* 1984; Wang *et al.,* 1984; Arnold *et al.,* 1984).

It is important to note that in the superlattice APD, a given electron will not ionize at every step, and the process is statistical in nature. Once the electron has gained the energy ΔE_c, it may not yet have the energy required to ionize. When it enters the next barrier layer it loses this energy ΔE_c, but on the average it has a kinetic energy larger than in the previous barrier layer, since $eF\lambda > \langle E_p \rangle$. Thus, on average, electrons go through several quantum wells without ionizing until they acquire an energy within ΔE_c from the ionization threshold. At this point they will ionize in the next well. This greatly reduces the average distance $1/\alpha$ required to create an electron–hole pair. For the superlattice APD reported by Capasso *et al.* (1981b, 1982c), $1/\alpha \simeq 1\ \mu$m at $F = 2.7 \times 10^5$ V cm^{-1}, corresponding to roughly 10 periods of the superlattice.

Note, however, that since multiplication is an intrinsically statistical process, some electrons will ionize after only a few wells, whereas others will create an electron–hole pair after a distance $> 1/\alpha$. A few "lucky" or ballistic electrons can gain sufficient energy ($\simeq 1.6$ eV) in only one barrier layer without colliding with phonons and impact-ionize in the next well. These considerations illustrate the need of maintaining an electric field as constant as possible over the entire length of the superlattice and thus of very low background doping (Chin *et al.* 1980).

The superlattice-detector structure in which the enhancement of α/β was

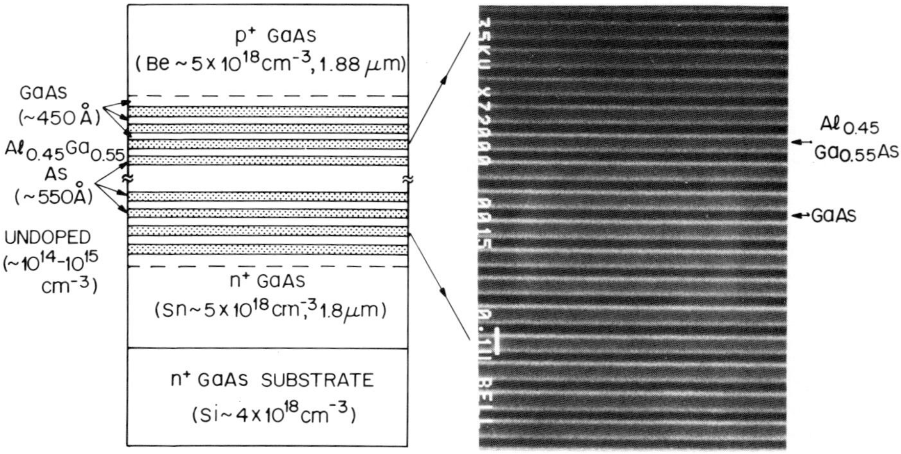

FIG. 67. Schematic diagram showing the layer structure and doping concentrations of the superlattice APD grown by MBE. The scanning electron microscope photograph shows the stained cross section of the superlattice. The bright images of the interfaces result from the chemical staining. [From Capasso *et al.* (1982c).]

observed is shown in Fig. 67 and was grown by MBE. A buffer layer of n^+–GaAs (1.8 μm) doped with Sn to $\sim 5 \times 10^{18}$ cm^{-3} was first grown on a high-quality Si-doped ($\sim 4 \times 10^{18}$ cm^{-3}) (100) GaAs substrate. The Sn-doping beam was then shuttering off $\sim$450 Å prior to the growth of the first $Al_{0.45}Ga_{0.55}As$ barrier to avoid any doping of the initial layers due to Sn segregation. The undoped 50-layer GaAs ($\simeq$450 Å)/$Al_{0.45}Ga_{0.55}As$ ($\simeq$550 Å) superlattice structure was grown at a substrate temperature of $\simeq$695 °C. The Al/Ga flux ratio was optimized by shuttling the Al shutter. The net acceptor concentration of this multilayer structure is in the 10^{14}–10^{15}-cm^{-3} range. At the end of the last $Al_{0.45}Ga_{0.55}As$ layer, an undoped GaAs layer of $\sim$600 Å was grown prior to the growth of the p^+–GaAs (Be $\simeq$ 5×10^{18} cm^{-3}, 1.88 μm) contact layer.

The quality of the $Al_{0.45}Ga_{0.55}As$ material ($E_g \simeq 2.0$ eV) and the heterointerfaces were independently checked by growing under similar conditions GaAs/$Al_{0.5}Ga_{0.5}As$ single and double quantum wells with 40–100-Å GaAs wells and 15–40-Å $Al_{0.5}Ga_{0.5}As$ barriers. This was done using the same source charges and without breaking the UHV of the growth chamber. Subsequent low-temperature photoluminescence (PL) and excitation spectroscopy measurements on these quantum-well wafers show that the heterointerfaces are smooth to within one atomic layer and that there are *no alloy clusters* in the $Al_{0.5}Ga_{0.5}As$ down to a scale of 15 Å (Miller and Tsang, 1981).

The low-temperature PL and excitation spectra also indicate very low impurity levels in the undoped GaAs and $Al_xGa_{1-x}As$. Such high-quality material and heterointerfaces are essential for superlattice detectors because of the large number (50) of heterointerfaces and the high electric field ($>2 \times 10^5$ V cm^{-1}).

The samples were processed into 3×10^{-4}-cm^2 mesas using photolithographic techniques and a $H_2O:H_2O_2:H_2SO_4$ (20, 16, 2 ml) etching solution at $\simeq 10°C$. The etched depth was 6–7 μm. Alloyed Au–Zn and Au–Sn were used for the p and n contacts, respectively.

For the experiment the samples with the lowest doping density ($\sim 2 \times 10^{14}$ cm^{-3}) were used. Capacitance–voltage profiles of these diodes showed that the entire superlattice region is depleted at nearly zero bias, ensuring that the field profile is constant at voltages above the avalanche turn-on voltage ($\simeq 50$ V). This considerably simplifies the determination of α and β and eliminates base-line corrections due to depletion-layer-widening effects. The breakdown voltages of our diodes are in the 75–80-V range, and the dark current is $\simeq 1$ nA at $0.9 V_B$.

To determine α and β, the electron- and hole-initiated avalanche gains M_n and M_p must be measured separately. This was done using the standard configuration of Fig. 32a. Pure electron injection was achieved by coupling highly absorbed 6328-Å laser light via a microscope into the p^+ side. The 5-μm light spot was scanned over the sensitive area to verify the absence of microplasmas, edge breakdown, or field-enhancement response at the edge. The gain uniformity was $\Delta M/M \simeq 1\%$ at $M = 15$.

Both electron- and hole-injected photocurrents were constant over a substantial voltage range before the onset of avalanche, giving a horizontal gain baseline. This ensures a precise determination of M_n and M_p. The large measured difference between M_n and M_p (Fig. 68) is direct evidence that the effective α is much greater than β.

In general, α and β can be derived from M_n and M_p after solving the avalanche rate equations Eqs. (130) and (131). This solution was previously discussed and is represented by Eqs. (158) and (159).

In our constant field case, we assume that $\alpha(x)$ and $\beta(x)$ are periodic with the superlattice period, taking the values α_1, β_1 in the GaAs layers of thickness L_1 and the values α_2, β_2 in the AlGaAs layers of thickness L_2. Equation (158) then reduces to

$$1 - M_e^{-1} = \{[\alpha_1/(\alpha_1 - \beta_1)][1 - \exp(\beta_1 - \alpha_1)L_1] + [\alpha_2/(\alpha_2 - \beta_2)][\exp(\beta_1 - \alpha_1)L_1][1 - \exp(\beta_2 - \alpha_2)L_2]\} \times \frac{1 - \exp(\bar{\beta} - \bar{\alpha})W}{1 - \exp(\bar{\beta} - \bar{\alpha})L}, \qquad L = L_1 + L_2, \tag{209}$$

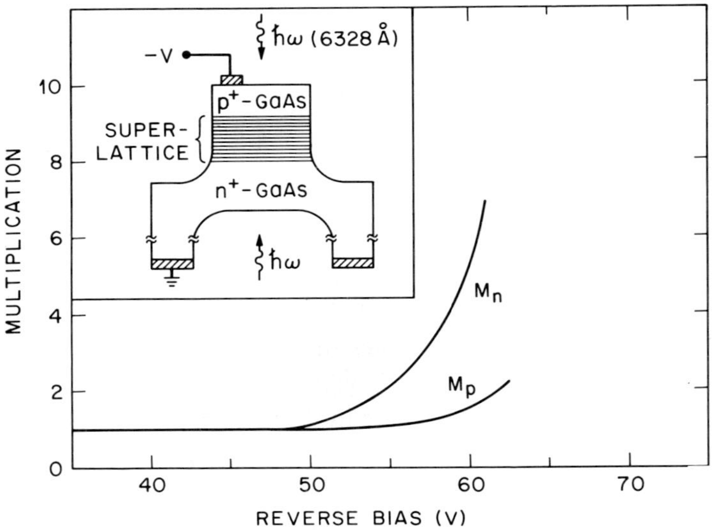

FIG. 68. Electron- (M_n) and hole-(M_p) multiplications initiated versus voltage; inset shows schematic diagram of the experimental configuration. [From Capasso *et al.* (1982c).]

where $\bar{\alpha}$ and $\bar{\beta}$ are the average ionization rates

$$\bar{\alpha} = \frac{\alpha_1 L_1 + \alpha_2 L_2}{L_1 + L_2}, \qquad \beta = \frac{\beta_1 L_1 + \beta_2 L_2}{L_1 + L_2}. \tag{210}$$

Equation (159) reduces to an equation identical to Eq. (209) with exchanged α and β and exchanged indices 1, 2.

We now recall that the superlattice region consists of many (50) thin (500 Å) layers and that the devices are operated at moderate gains (≤ 15). Under these conditions we have $1/\alpha_1 \gg L_1$, $1/\alpha_2 \gg L_2$, $1/\beta_1 \gg L_1$, $1/\beta_2 \gg L_2$; and by making use of a Taylor expansion to first order and keeping first-order terms overall, Eq. (209) and the corresponding one for M_p then reduce to

$$\bar{\alpha}(F) = \frac{1}{W}\frac{M_n - 1}{M_n - M_p} \ln \frac{M_n}{M_p}, \tag{211}$$

$$\bar{\beta}(F) = \frac{1}{W}\frac{M_p - 1}{M_p - M_n} \ln \frac{M_p}{M_n}, \tag{212}$$

which are the well-known equations of a homogeneous $p-i-n$ diode with ionization rates $\bar{\alpha}$, $\bar{\beta}$ [Eqs. (160) and (161)]. These can be used to derive $\bar{\alpha}$ and $\bar{\beta}$ from the experimental data $M_n(V)$ and $M_p(V)$.

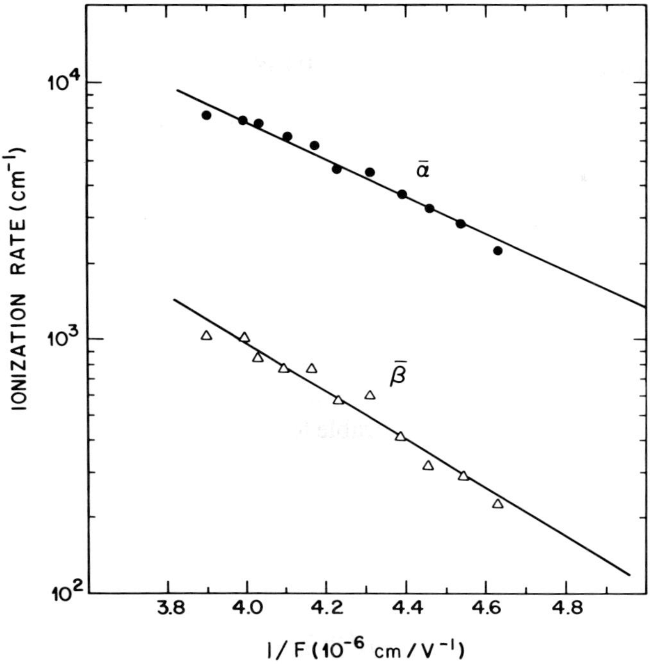

FIG. 69. Measured effective ionization rates for electrons and holes in the superlattice APD. The solid lines represent least square fits to the data. [From Capasso *et al.* (1982c).]

The effective ionization rates are plotted in Fig. 69 versus reciprocal electric field $1/F$ in the range $2.1-2.7 \times 10^5$ V cm^{-1}. The straight lines are the least square fits given by

$$\bar{\alpha} = 5.7 \times 10^6 \exp(-1.7 \times 10^6/F), \qquad (213)$$

$$\bar{\beta} = 5.5 \times 10^6 \exp(-2.17 \times 10^6/F). \qquad (214)$$

For example, at $F = 2.5 \times 10^5$ V cm^{-1}, $\alpha/\beta = 8$ and $M_n = 10$. This gives, according to McIntyre's theory, an excess-noise factor $F = 2.9$ for electron-initiated multiplication.

Again, the large $\bar{\alpha}/\bar{\beta}$ ratio is mainly the result of impact ionization in the GaAs layers. This is because the band gap of GaAs is substantially smaller than that of $Al_{0.45}Ga_{0.55}As$, so that α_1 and β_1 give the largest contribution to the averages $\bar{\alpha}$ and $\bar{\beta}$. Measurements of α and β in low-doped bulk GaAs give in the same field range $\alpha/\beta \simeq 2$ (Bulman *et al.*, 1983). Note that the hole data in Fig. 67 are close to the values of β reported by Bulman *et al.* (1983) for bulk GaAs. Thus, the measured high α/β ratio is largely due to enhanced electron impact ionization.

Brennan *et al.* (1985b) have reported on a Monte Carlo calculation of α and β in a superlattice of AlGaAs/GaAs identical to the one used in the superlattice APD of Capasso *et al.* (1982). The material parameters, and the band structure were identical to those previously used (Brennan and Hess, 1984) in the Monte Carlo simulation of α and β in GaAs to fit the data of Bulman *et al.* (1983). They find an excellent fit to the electron ionization rate data of Capasso *et al.* if ΔE_c is taken to be 75% of the band-gap difference. The calculated β instead is comparable to the bulk value, which is agreement with experimental data. It is important to note that the observed enhancement of α cannot be explained with the old "Dingle rule" $\Delta E_c = 0.85\ \Delta E_g$, now experimentally demonstrated to be wrong. Using the latter band lineup, the Monte Carlo simulation gives a much larger enhancement than the experimental one. On the other hand, if one uses the recent experimental value $\Delta E_c = 0.6\ \Delta E_g$, one gets $\alpha/\beta \simeq 5$.

b. Superlattice APD with Graded-Gap Sections

The multi-quantum-well APD configuration of Fig. 66 is only appropriate for moderate values of ΔE_c. In general, it is desirable for maximum enhancement of electron ionization to maximize ΔE_c. This may lead in some heterojunctions to electron-trapping effects in the wells. In these situations it is desirable to compositionally grade a region of the material after the well. The corresponding energy-band diagram is shown in Fig. 70 (Capasso *et al.*, 1983).

This modified superlattice APD structure is useful for lattice-matched heterojunctions, such as $Al_{0.48}In_{0.52}As/In_{0.53}Ga_{0.47}As$ ($\Delta E_c = 0.5$ eV, $\Delta E_v = 0.2$ eV), (People *et al.*, 1983) and $AlAs_{0.03}Sb_{0.93}/GaSb$ ($\Delta E_c = 0.72$ eV, $\Delta E_v = 0.14$ eV), useful for long-wavelength detectors (1.3 – 1.6 μm). In these cases, the graded section would be of AlInGaAs and AlGaAsSb lattice matched respectively to InP and GaSb substrates.

There is another important consideration in designing a long-wavelength superlattice APD with materials, such as AlInGaAs/InGaAs, with low effective masses. Band-to-band tunneling in the wells may cause high leakage currents before the onset of avalanche. It is thus necessary to prevent tunneling by reducing the thickness of the wells to the point where the valence and conduction bands no longer overlap. For this to occur, the potential drop across a well of thickness L_w should not exceed the band gap; i.e. $eFL_w \leq E_g$. For example, for a superlattice APD such as that of Fig. 70 with $Al_{0.48}In_{0.52}As$ barrier layers and $In_{0.53}Ga_{0.47}As$ wells (the graded regions would consist of lattice-matched AlInGaAs) with a constant electric field of 2×10^5 V cm^{-1}, the well thickness L_w should be smaller than 365 Å.

It is interesting to note that for this particular heterojunction, $\Delta E_c \simeq 0.5$ eV, whereas the electron ionization energy in $In_{0.53}Ga_{0.48}As$ is

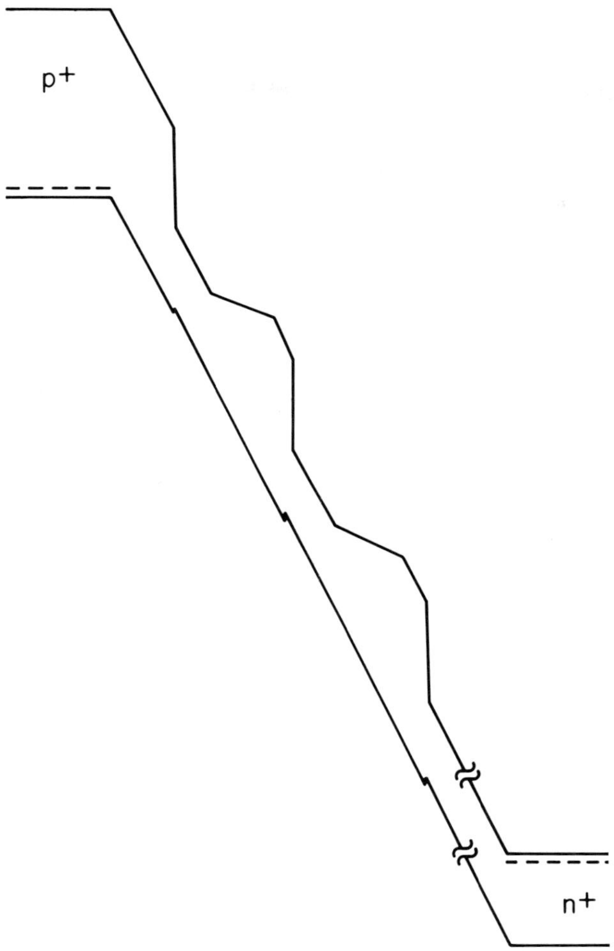

FIG. 70. Energy-band diagram of superlattice APD with graded sections to avoid electron trapping in the wells. [From Capasso *et al.* (1983). © 1983 IEEE.]

estimated to be $\simeq 1.0$ eV. Thus, if electrons gain on the average an energy ≥ 0.5 eV in the $Al_{0.48}In_{0.52}As$ barrier layer, a large fraction of them are likely to ionize at nearly every well. This can easily be achieved with bias electric fields of $\simeq 2 \times 10^5$ V cm^{-1}. For this reason, this particular superlattice detector should yield large α/β enhancements, usable in the 1.3–1.6-μm region. It shall later become clear to the reader that this structure in fact represents the conceptual link between the superlattice multi-quantum-well APD and the staircase solid-state photomultiplier.

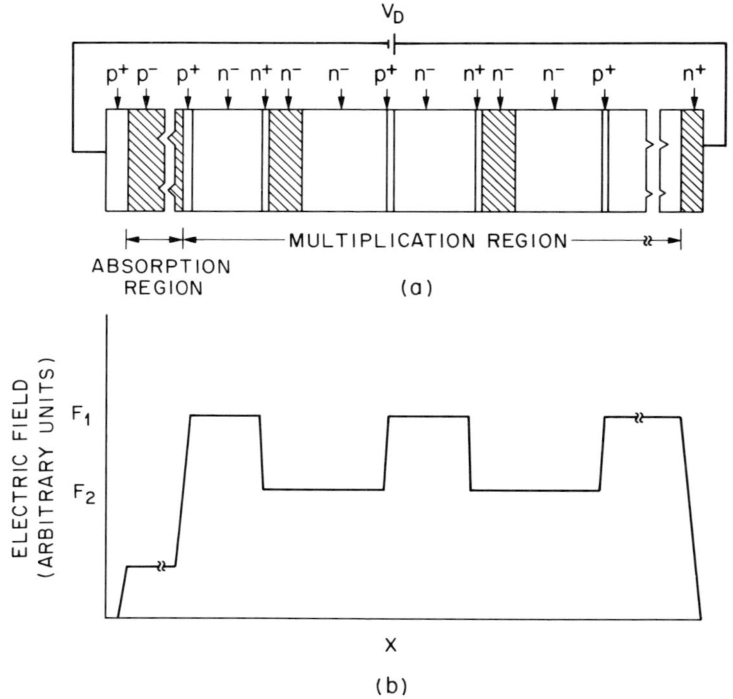

FIG. 71. (a) Schematic diagram of superlattice APD structure with periodic doping profile: material A (shaded region); material B (white region). (b) Electric field in each layer of multilayer structure. [From Blauvelt *et al.* (1982).]

c. *Superlattice APD with Periodic Doping Profile*

Blauvelt *et al.* (1982) proposed a modification of the superlattice detector that should yield much higher α/β ratios. Their structure is schematically shown in Fig. 71a. The shaded areas are the regions where the ionization process occurs, whereas the white areas indicate layers with significantly higher band gaps. The ionization occurs in the former regions. Typically, the structure is completely depleted, even at low voltage. By inserting a periodic doping profile as in Fig. 71a the electric field will also vary periodically within the depletion region as shown in Fig. 71b. The rapid variations from high to low electric field are achieved by means of the thin highly doped p^+ and n^+ layers, since the gradient of the electric field in the depleted layer is proportional to the doping.

Note that electrons are injected in the low-gap layers (shaded areas) from a high-electric-field region, whereas holes enter the same layers from the right coming from a low-electric-field region. Thus, the fraction of the electrons that are injected with energies above the ionization threshold can be significantly larger than the fraction of holes that are injected with energies above the ionization threshold. The two materials can be, for example, GaAs ($E_g = 1.43$ eV) and $Al_{0.45}Ga_{0.55}As$ ($E_g = 2.0$ eV). Typical values of the thicknesses of the GaAs, high-field AlGaAs, and low-field AlGaAs layers are 400, 700, and 900 Å, respectively. With these values, Blauvelt *et al.* (1982) estimated the expected α/β ratio using a simple Shockley model. The results are shown in Fig. 72. The ionization rate is plotted for various values of the differences between the electric fields in the high- and low-field layers. The curve for $\Delta F = 0$ corresponds to a detector similar to that reported by Capasso *et al.* (1981b). Note that for $\Delta F \neq 0$, significantly greater enhancements of α/β can be obtained.

Finally, several practical considerations are in order. To minimize secondary ionization by holes, the low-field layers have to be sufficiently thick that holes can lose (by phonon collision) the kinetic energy gained in the preceding high-field layer. In addition, the difference between the electric fields in the high- and low-field regions should be as large as is practical. A 50-Å-thick n^+ layer with a doping of 2×10^{18} cm^{-3} will result in a change ΔF in the electric field of approximately 1.6×10^5 V cm^{-1}. It is also desirable to have the total number of donors in a unit cell nearly equal to the total number of

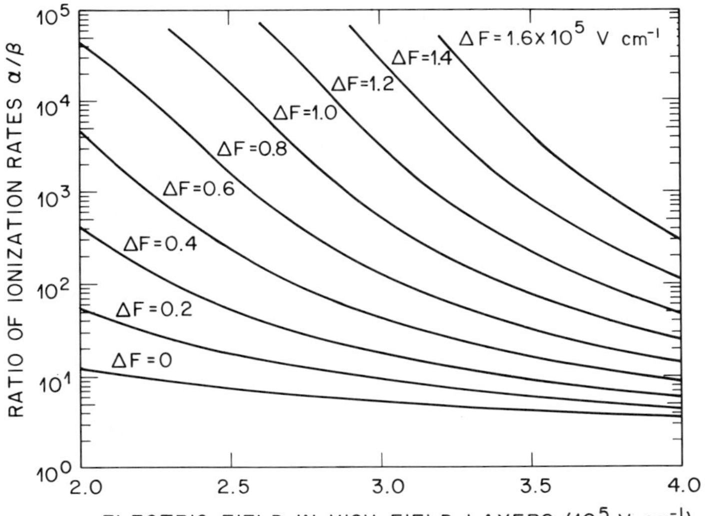

FIG. 72. Ionization rate ratio for various values of the electric fields in high- and low-field layers. [From Blauvelt *et al.* (1982).]

acceptors, so that the electric-field pattern repeats itself in each unit cell. Finally, it may be desirable to grade the interfaces between the GaAs and the n-$Al_{0.45}Ga_{0.55}As$ layers to avoid trapping.

It is important in the design of a superlattice detector, similar to any APD, to initiate the avalanche with the carrier having the highest ionization rate. To achieve that without sacrificing speed, the incident high should be absorbed in a lower electric field drift region adjacent to the superlattice avalanche region. A possible way to achieve this is illustrated in Fig. 71 and can be applied to the other detectors described in this section as well.

d. Enhancement of the α/β Ratio in III–V Alloys with Clustering

It is worth mentioning that the notion of impact ionization assisted by band-edge discontinuities, on which the superlattice detector is based, has been used to model impact ionization in III–V alloys proven to clustering, such as InGaAsP (Burroughs *et al.*, 1982). Clustering is a form of disorder in ternary and quaternary compounds associated with the irregular distribution of column III and/or column V atoms in their respective lattice sites.

Burroughs *et al.* (1982) start from a one-dimensional model of clustering in $Al_xGa_{1-x}As$. The effect of this type of disorder is to introduce random insertion of thin GaAs regions (10–50 Å) in the otherwise periodic alternation of Al and Ga in the column-III lattice site. In other words, it is assumed that in a ternary tetrahedral zinc blende crystal represented by $A_xB_{1-x}C$, such as $Al_xGa_{1-x}As$, the C atoms completely fill their face-centered-cubic (fcc) sublattice sites. This leaves the other fcc sublattice for the A and B atoms. It is assumed that these A and B atoms are randomly placed in their fcc sublattice in proportion to their respective compositions x and $1 - x$. Thus, the effect of clustering is to produce a random type of quantum-well superlattice. Fig. 73 shows a linear array of A, B, and C atoms showing a cluster of compound BC. The dotted lines define the cluster boundaries. Also shown are the conduction- and valence-band edges. Clearly, the asymmetry of the band-edge discontinuities in this random type of superlattice will tend to produce an enhancement of the α/β in alloys prone to clustering, analogous to that observed in multi-quantum-well detectors. Burroughs *et al.* (1982) have used a procedure similar to that of Chin *et al.* (1980) to model this enhancement based on a polynomial fit (Sutherland, 1980) to Baraff's theory. Burroughs *et al.* note that experimental data for InGaAsP, InGaAs, and GaAsSb, reviewed in Part III support their model.

e. Resonance Impact Ionization in Superlattices

If the barrier-layer thicknesses in a multi-quantum-well superlattice are made comparable or smaller than the tunneling length of the carriers, the quantum states of neighboring wells become coupled, leading to formation

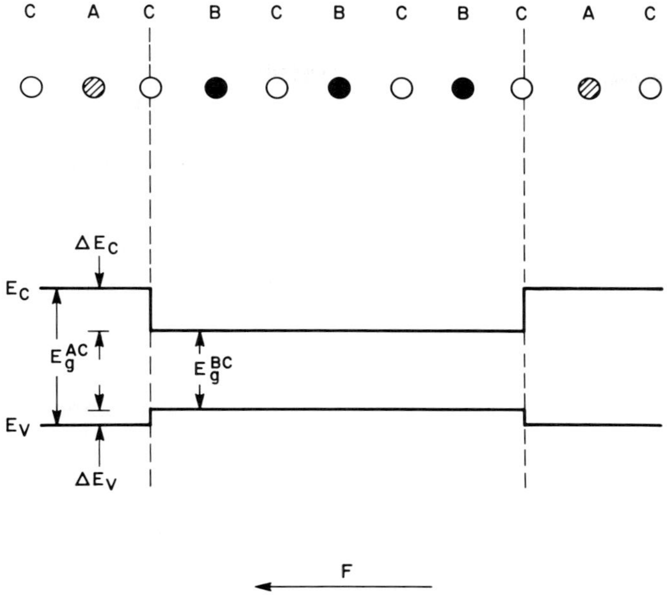

FIG. 73. Linear array of A, B, and C atoms showing a cluster of compound C. The dotted lines define the cluster boundaries. Also shown are the conduction- and valence-band edges. [From Burroughs *et al.* (1982).]

of minibands in the conduction and valence bands and in general to large variations of the band structure. Both the energy gap E_g and the spacing between minibands can be altered over a wide range by varying the layer thickness and the composition of the constituent alloys. In particular, one can introduce resonance states in the valence band at $k = 0$ (Mon and Hess, 1982). These authors have shown that, for an AlAs/GaAs superlattice with a 1:2-layer ratio, hole resonance states at $\simeq$2.03 eV from the bottom of the conduction band are introduced in the valence band at $k = 0$. The energy band gap in this case is nearly equal to this separation ($E_g = 1.93$ eV). Thus, one can enhance the β/α ratio by a resonance effect similar to that discussed in Part III in the case of near equality between the band gap and the spin–orbit splitting.

f. Enhancement of the Ionization Rate Ratio via Superlattice Minigaps

Another mechanism of enhancement of the ionization rate ratio in a superlattice has been proposed by Vodakov *et al.* (1981) and by Dmitriev *et al.* (1983). A superlattice splitting of either the conduction or valence band along a given crystal orientation may reduce the ionization probability if the electric field is oriented along this direction and the minigap widths are of the order of a few tenths of an electron volt. If the superlattice potential affects, for example, only the conduction band, then the electron ionization rate will be reduced, whereas β is not affected.

This is apparently the situation of the semiconductor 6H Silicon Carbide, where a natural superlattice structure exists along the c axis (Dubrovski, 1972). Ionization-rate measurements in 6H SiC ($E_g = 2.9$ eV) p–n junctions grown on this orientation indicate that $\beta/\alpha = 10^2$ (Vodakov *et al.*, 1981; Dmitriev *et al.*, 1983). These researchers attribute this large ratio of ionization coefficients to the low tunneling probability of electrons between the first two minibands. This clearly reduces the probability of reaching the ionization energy and α. Holes, on the other hand, are not affected by this mechanism, because the superlattice structure in this compound modifies only the conduction band.

17. The Staircase Solid-State Photomultiplier

This device is a multistage graded-gap structure. It is based on the same principle of the multi-quantum-well APD (impact ionization assisted by band discontinuities); however, here the electron ionization energy is entirely provided by conduction-band steps (Capasso *et al.*, 1983; Williams *et al.*, 1982; Capasso and Tsang, 1982; Capasso, *et al.*, 1982d, 1984b). Figure 74a shows the energy-band diagram of the graded-gap multilayer material (assumed intrinsic) at zero applied field. Each stage is linearly graded in composition from a low (E_{g1}) to a high (E_{g2}) band gap, with an abrupt step back to low-band-gap material. The conduction-band discontinuity shown accounts for most of the band-gap difference, as is typical of many III–V heterojunctions. The materials are chosen for a conduction-band discontinuity equal to or greater than the electron-ionization energy E_{ie} in the low-gap material immediately following the step.

The band structure of the complete staircase detector under bias is shown in Fig. 74b. Consider a photoelectron generated next to the p^+ contact. Under the combination of the bias field F and the grading field $\Delta E_c/l$, it drifts toward the first conduction-band step. The combined field $F - (\Delta E_c/l)$ is sufficiently small so that the electron does not impact-ionize before it reaches the step. After the step, since $\Delta E_c \simeq E_{ie}$, the electron impact-ionizes; this ballistic ionization process is repeated in each stage.

Ideally, the avalanche gain per stage is exactly two; each electron impact-ionizes once after each conduction-band step. In practice, the gain is $2 - \delta$, where δ is the fraction of electrons that do not impact-ionize. The total gain of the structure is then $\langle M \rangle = (2 - \delta)^N$, where N is the number of stages.

Any hole-initiated ionization is caused by the applied electric field F only;

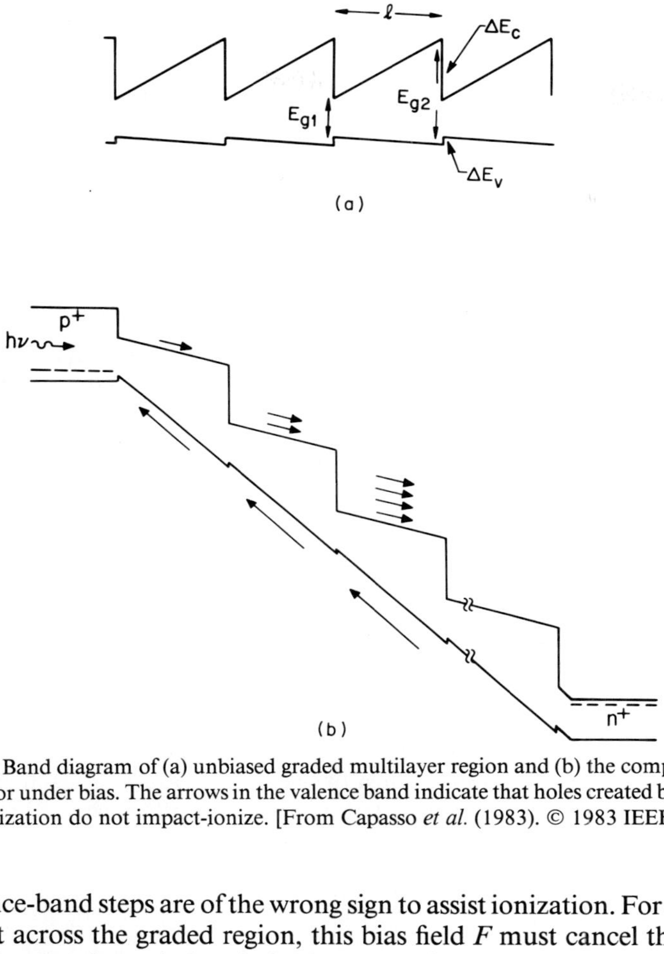

FIG. 74. Band diagram of (a) unbiased graded multilayer region and (b) the complete staircase detector under bias. The arrows in the valence band indicate that holes created by electron impact ionization do not impact-ionize. [From Capasso *et al.* (1983). © 1983 IEEE.]

the valence-band steps are of the wrong sign to assist ionization. For electron transport across the graded region, this bias field F must cancel the $\Delta E_c/l$ conduction-band quasi-electric field and provide a small extra component to assure drift rather than diffusion transport. The device is then designed so that the hole ionization rate at F ($\gtrsim \Delta E_c/l$) is negligible. The arrows in the valence band of Fig. 74b simply indicate that holes do not impact-ionize; obviously, they will always undergo multiplication due to electron-initiated ionization. Because only electrons cause ionization, this device mimics a photomultiplier; the conduction-band steps correspond to the dynodes.

a. Multiplication Noise of the Staircase APD

We shall now present a simple derivation of the multiplication noise for the staircase APD (Capasso, 1983c; Capasso *et al.*, 1983). The noise on the output signal per unit bandwidth, neglecting dark current, is

$$\langle i^2 \rangle = 2eI_0 \langle M \rangle^2 F(N, \delta), \tag{215}$$

where I_0 is the primary photocurrent. The excess-noise factor $F(N, \delta)$ is calculable by noting that at each stage, a fraction of electrons (characterized by a random variable with average δ) do not ionize. This causes fluctuations of the average gain. The noise contribution of the nth stage $(1 \leq n \leq N)$ is multiplied by the rest of the stages by the factor $(2 - \delta)^{2(N-n)}$. We write the mean-square noise current generated by the nth stage as

$$\langle i^2 \rangle_n = 2eI_{n-1}(\langle m^2 \rangle - \langle m \rangle^2) = (\delta - \delta^2)(2eI_{n-1}), \tag{216}$$

where $I_{n-1} = I_0(2 - \delta)^{n-1}$ is the average current into the nth stage and m the gain of the nth stage. The total noise per unit bandwidth on the output signal is

$$\begin{aligned} \langle i^2 \rangle &= 2eI_0(2 - \delta)^{2N} + \sum_{n=1}^{N} 2eI_0(2 - \delta)^{n-1}(\delta - \delta^2)(2 - \delta)^{2(N-n)} \\ &= 2eI_0(2 - \delta)^{2N}\left[1 + \sum_{n=1}^{N}\left(\frac{1}{(2 - \delta)^n}\frac{\delta(1 - \delta)}{2 - \delta}\right)\right]. \end{aligned} \tag{217}$$

Therefore the excess-noise factor is

$$F(N, \delta) = 1 + \{\delta[1 - (2 - \delta)^{-N}]/(2 - \delta)\}. \tag{218}$$

It is useful to plot $F(N, \delta)$ as a function of the multiplication per stage $1 + \gamma$ (where $\gamma = 1 - \delta$ is the ionization yield per stage) for varying number of stages (Fig. 75). Note that if most of the carriers ionize at each step (i.e., $1 + \gamma \simeq 2$), the excess-noise factor is nearly unity and independent of the number of stages, implying that the multiplication process is virtually noise free even at high gain (large N), similar to a photomultiplier. This cannot be achieved in a conventional APD at high gain even if one of the ionization coefficients is zero. In fact, the McIntyre theory predicts that for $\alpha/\beta = 0$ or $\beta/\alpha = 0$, $F = 2$ at large gains (Fig. 56). Note that in Eq. (198), $F(N, \delta) < 2$ for any δ and N. In the opposite limit where no carrier ionizes $(\gamma = 0)$, one obviously has also $F = 1$, thus recovering the well-known shot-noise expression.

The intrinsically lower noise of this structure is due to the fact that the avalanche process is much less random than in a conventional APD. The multiplication occurs only at well-defined positions in space (at the conduction-band steps), and if the device is properly designed, most carriers ionize at each step; thus, the statistical variations of the gain are very small with a resulting near-unity excess-noise factor.

In a conventional APD, carriers can instead ionize essentially anywhere

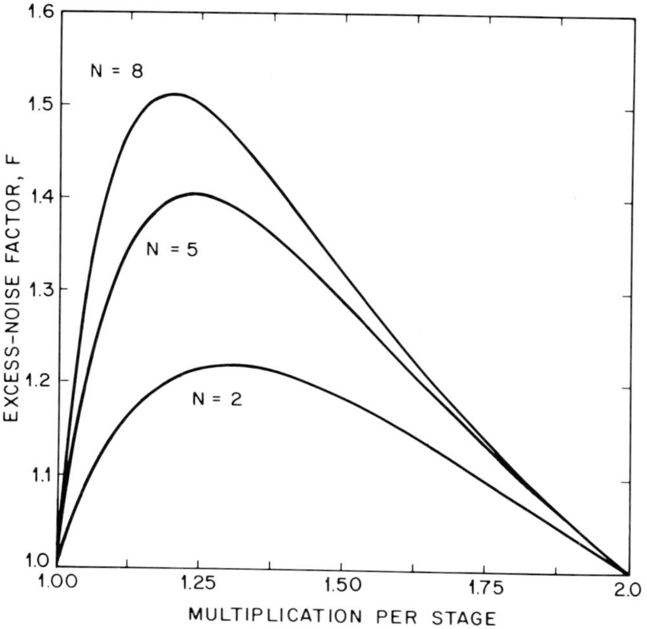

FIG. 75. Excess-noise factor of the staircase APD as a function of the gain per stage for varying number of stages. Note that if most electrons ionize at each step, the multiplication process is nearly noise free even at high gain. [From Capasso *et al.* (1983). © 1983 IEEE.]

within the avalanche region so that the avalanche process is more random. This causes much larger gain fluctuations and, consequently, more excess noise. Thus, the noise of the staircase APD cannot be described by the McIntyre theory.

Note that also in a photomultiplier, the multiplication process is nearly noise free with $F \simeq 1$. However, there is an important difference. In the staircase APD, electrons at the conduction-band step can either create one electron–hole pair or none, whereas in the phototube, electrons can create one, two, and, in general, many new electrons. Thus, for the same total gain, the variance of the multiplication, i.e., the noise fluctuations, will be greater in the phototube, giving an F factor slightly larger than in a staircase APD.

It is interesting to compare the excess noise of the staircase APD with that of a photomultiplier tube. This is given by (Anderson and McMurtry, 1966)

$$F = (G^{N+1} - 1)/[G^N(G - 1)], \tag{219}$$

where G is the multiplication per stage. This expression is plotted versus G with N as a parameter in Fig. 76. Note that, for example, for a typical

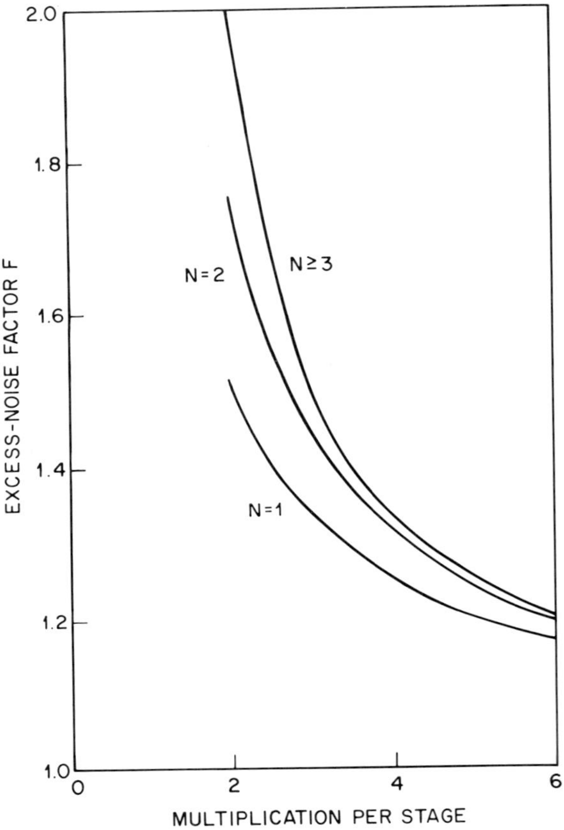

FIG. 76. Excess-noise factor F of a photomultiplier tube for varying number of diodes. Each dynode has been assumed to have the same gain. [From Capasso *et al.* (1983). © 1983 IEEE.]

multiplication per stage of 6, the excess-noise factor is $\simeq 1.2$ at high gain (large N). This formula is only valid for $G \geq 2$; note, in fact, that in a practical phototube, electrons always arrive at the dynodes with an energy sufficient to create at least one electron.

To achieve the small δ required for low noise and high gain, the transition from high- to low-gap material should be over a distance smaller than the mean free path for phonon scattering (50–100 Å). This is achievable using molecular beam epitaxy (MBE).

For long-wavelength photodetectors, two material systems are presently of interest: AlGaAsSb/GaSb on GaSb substrate and $Hg_xCd_{1-x}Te$ lattice matched to CdTe or InSb substrates. For the AlGaAsSb system, the minimum electron ionization energy is 0.80 eV (GaSb); the maximum heterointerface band-gap difference ($AlAs_{0.03}Sb_{0.97}$ to GaSb) is 0.85 eV, of which 85% appears in the conduction band (Naganuma *et al.,* 1981), giving a conduction-band step of 0.72 eV. Thus, $\Delta E_c \approx E_i$; the 0.08 eV deficit is easily furnished by the electron drift field $F - \Delta E_c/l$.

Consider a five-stage AlGaAsSb alloy detector ($\langle M \rangle \approx 32$), having each layer 3000 Å thick. If the bias field is 3×10^4 V cm^{-1}, the average hole ionization rate is negligible, including even the resonant ionization at the $Al_{0.065}Ga_{0.935}As_{0.005}Sb_{0.995}$ composition. The effective electric field is reduced by the grading field to 10^4 V cm^{-1}; the electron ionization rate in the graded section is negligible. However, the transport is at the saturation velocity, and the average electron energy is a few tenths of an electron volt (~0.2 eV). Therefore, the majority of the electrons ionize at the conduction-band steps, despite the 0.08-eV deficit.

Note that for Al concentrations greater than 20%, in the AlGaAsSb alloy, the band gap becomes indirect. Thus, electrons reside in the satellite valleys of GaSb immediately after injection over the conduction-band step, their energy being roughly equal to ΔE_c plus the average kinetic energy before the step. Since the ionization threshold state in GaSb lies in the central valley, electrons must scatter into this valley to impact-ionize. This intervalley phonon process has a high scattering rate $1/\tau \geq 10^{14}$ sec^{-1} because of the high electron energy. Thus, if the average kinetic energy before the step is greater than the optical phonon energy, a single intervalley scattering event will make ionization possible.

The operating voltage of the previous AlGaAsSb device is ≃5 V. The physical reason for this low operating voltage is that the ionization energy is delivered abruptly to the electrons by the conduction-band steps (ballistic ionization) rather than gradually via the applied field as in a conventional APD. Thus, the competing energy losses by phonon emission are much smaller, and most of the applied voltage is used to create electron–hole pairs.

The most interesting material for the implementation of the staircase APD is probably $Hg_xCd_{1-x}Te$. This alloy is lattice matched to CdTe and InSb substrates for $0 \leq x \leq 1$. Its band gap varies with x from 0 eV (HgTe) to 1.6 eV (CdTe) and stays direct over all the compositional range. In addition, there are good theoretical reasons to believe that in heterojunctions made of this alloy, $\Delta E_v = 0$ (Shulman and McGill, 1979). Thus, if CdTe/$Hg_xCd_{1-x}Te$ heterojunctions were to be used as "dynodes" in the staircase APD, all the band-gap difference would be used to impact-ionize. Furthermore, if the band gap of the material after the conduction-band step is chosen sufficiently small (0.1–0.3 eV), more than one electron–hole pair per dynode will be created.

Although the MBE growth of this alloy is still in its infancy, HgTe/CdTe

MBE superlattices of good quality were grown by Faurie *et al.* (1982). Magneto-optical data from HgTe/CdTe heterojunctions led to the conclusion that $\Delta E_v \simeq 40$ meV (Guldner *et al.*, 1983).

It is important to note that the staircase-APD energy-band diagram (Fig. 74) can be obtained from that of a conventional superlattice APD (Fig. 66) by starting to grade the heterointerface at the exit of the well (Fig. 70) and finally by extending the grading up to the preceding interface.

Thus, with compositional grading, the energy-band diagram and the associated transport properties can be tuned and varied continuously. This procedure, named by Capasso (1983a) "band-gap engineering," can be extended to other electronic devices as well (Capasso 1984).

b. Ionization Yield per Dynode

From the previous considerations, it is clear that in order to achieve nearly noise-free multiplication and high gain, the ionization yield γ must be close to 1. In this section, we present the theory of the ionization yield and discuss how a high γ can be achieved (Capasso *et al.*, 1983). For simplicity, we shall assume ΔE_c equal to the electron ionization energy immediately after the step. It will be shown that to ensure a high yield, electrons have to be "heated" by the field in the graded region before the step to an average energy much higher than the optical phonon energy.

We assume that hot electrons approaching the conduction-band step can be described by a Maxwell–Boltzmann distribution $f(E)$ with an electron temperature T_e larger than the lattice temperature; i.e.,

$$f(E) = A \exp(-E/kT_e), \tag{220}$$

where A is a normalization factor such that $\int_0^\infty f(E)\, dE = 1$. Although this assumption represents a somewhat crude description of carrier dynamics, it is a useful tool for a heuristic investigation of the problem. When electrons go over the conduction-band step, they can either impact-ionize or emit a phonon. We neglect for simplicity absorption processes. Phonon absorption will slightly increase the ionization yield.

Following Baraff (1962), we phenomenologically describe ionizing collisions and phonon collisions above the ionization threshold with energy-independent cross sections σ_i and σ_{ph}. In many practical cases, σ_i and σ_{ph} are comparable. The ionization yield depends on T_e, σ_i, and σ_{ph}.

Consider now a hot electron approaching the conduction-band step with an energy E measured with respect to the conduction-band edge before the step. When it reaches the step, it can either ionize directly or it can emit one phonon and ionize, or two consecutive phonons and ionize, and so on, depending on the relative magnitude of E and the optical phonon energy E_p. For example, for an electron with incoming energy $0 < E < E_p$, the ionization probability is

$$\frac{\sigma_i}{\sigma_T}\int_0^{E_p} A\exp(-E/kT_e)\,dE = \frac{\sigma_i}{\sigma_T}[1-\exp(-E_p/kT_e)], \tag{221}$$

where $\sigma_T = \sigma_i + \sigma_{ph}$. An electron with incoming energy $E_p < E < 2E_p$ instead can either ionize directly or after having emitted a phonon. The total probability for these processes is clearly

$$\frac{\sigma_i}{\sigma_T}\int_{E_p}^{2E_p} A\exp\left(-\frac{E}{kT}\right)dE + \left(1-\frac{\sigma_i}{\sigma_T}\right)\frac{\sigma_i}{\sigma_T}\int_{E_p}^{2E_p} A\exp\left(-\frac{E}{kT_e}\right)dE$$
$$= \left[\frac{\sigma_i}{\sigma_T} + \frac{\sigma_i}{\sigma_T}\left(1-\frac{\sigma_i}{\sigma_T}\right)\right]\left[\exp\left(-\frac{E_p}{kT_e}\right) - \exp\left(-\frac{2E_p}{kT_e}\right)\right]. \tag{222}$$

In general, for an electron with incoming energy $(m-1)E_p < E < mE_p$, the ionization probability is

$$\sum_{n=0}^{m-1}\frac{\sigma_i}{\sigma_T}\left(1-\frac{\sigma_i}{\sigma_T}\right)^n \int_{(m-1)E_p}^{mE_p} A\exp\left(-\frac{E}{kT_e}\right)$$
$$= [1-(1-\sigma_i/\sigma_T)^m][\exp(-(m-1)E_p/kT_e) - \exp(-mE_p/kT_e)]. \tag{223}$$

The ionization yield γ is obtained by summing Eq. (223) over all possible incoming electron energies; i.e.,

$$\gamma = \sum_{m=1}^{\infty}\left[1-\left(1-\frac{\sigma_i}{\sigma_T}\right)^m\right]\times\left[\exp\left(-(m-1)\frac{E_p}{kT_e}\right) - \exp\left(-\frac{mE_p}{kT_e}\right)\right]. \tag{224}$$

This gives

$$\gamma = 1 + [1-\exp(E_p/kT_e)]\{[1-(1-\sigma_i/\sigma_T)\exp(-E_p/kT_e)]^{-1} - 1\}. \tag{225}$$

It is useful to consider the two limiting cases where $E_p/kT_e \gg 1$ or $kT_e/E_p \gg 1$. In the first case,

$$\gamma \simeq \sigma_i/(\sigma_i + \sigma_{ph}), \tag{226}$$

which is an obvious and expected result, since electrons injected with such small energy in excess of the ionization energy can only ionize directly or emit a phonon and not ionize at all.

The other case is more interesting. In this case, γ reduces to

$$\gamma \simeq \frac{\sigma_i}{\sigma_i + (E_p/kT_e)\sigma_{ph}}. \tag{227}$$

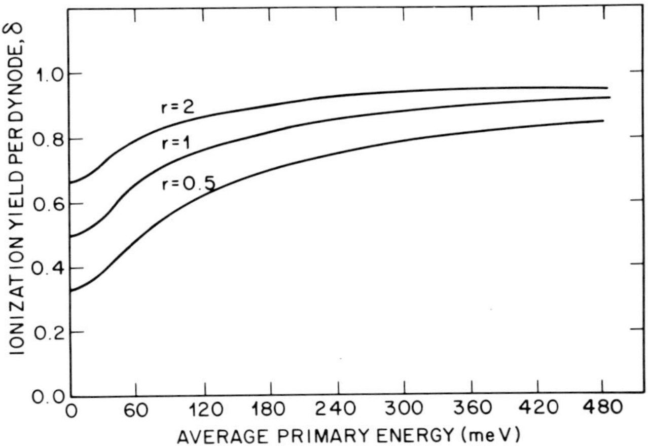

FIG. 77. Ionization yield per dynode as a function of average primary electron kinetic energy before the step for various values of the ratio of the phonon scattering cross sections to the impact-ionization cross section $r = \sigma; 1\sigma_{ph}. E_p = 30$ meV. [From Capasso *et al.* (1983). © 1983 IEEE.]

The physical interpretation of this formula is simple. When electrons are injected with high energy ($\approx kT_e$), the only way they cannot ionize is by emitting many consecutive phonons ($\approx kT_e/E_p$). Thus, the phonon emission cross section that competes with ionization is effectively reduced by the factor kT_e/E_p. Note that since σ_i and σ_{ph} are comparable, the probability of emitting consecutively many phonons is very small so that $\gamma \simeq 1$.

Thus, in order to achieve a high ionization yield per dynode, electrons approaching the step should have an energy well in excess of the phonon energy (typicaly, a few tenths of an eV). In Fig. 77 we have plotted Eq. (225) as a function of the average carrier energy $\frac{3}{2}kT_e$ before the step for different values of σ_i/σ_{ph}. A phonon energy of 30 meV has been assumed, which is typical of several III–V semiconductors. Note that the average energy should exceed 0.3 eV to achieve a high yield. This energy can be obtained with effective fields for electrons in the graded region $\simeq 10^4$ V cm^{-1}, using AlGaAsSb alloys.

Finally, a few points should be mentioned here for the sake of completeness. We have so far tacitly assumed that electrons impact-ionize as soon as they reach the conduction-band step. However, due to the finite ionization mean free path, λ_i electrons will impact-ionize after a distance of order of λ_i. Typically, $\lambda_i = 50–100$ Å in most III–V semiconductors. Thus, it is advisable that ungraded layers having a thickness corresponding to a few λ_i be inserted between the steps and the graded regions. This will maximize the ionization probability (Fig. 78).

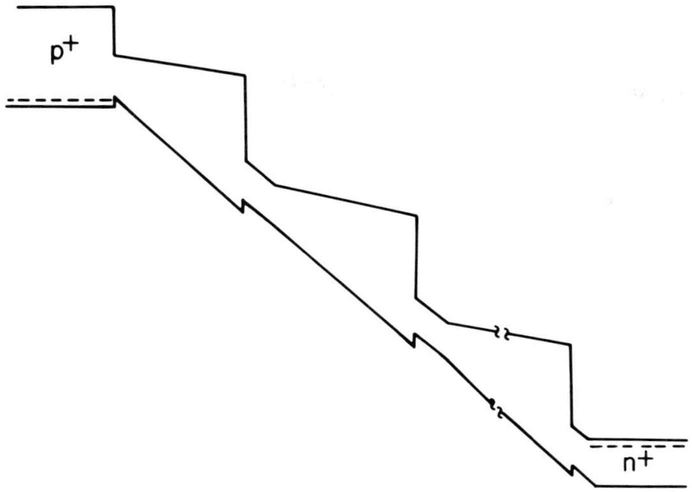

FIG. 78. Modified energy-band diagram of the staircase APD. The finite ionization mean free path above the ionization threshold is taken into account by introducing ungraded regions (≈ 100 Å) after the steps. [From Capasso *et al.* (1983). © 1983 IEEE.]

Previous discussions have assumed that $\Delta E_c = E_i$. In practical situations, the exact equality is seldom achieved, and the small energy deficit can be compensated by the effective field in the graded region. This is, for example, the case of the AlGaAsSb/GaSb system previously discussed, where $E_i - \Delta E_c = 0.08$ eV. For larger energy deficits, the field must be further increased. One way to do that is also to add an ungraded region before the step so that by applying the same voltage, one obtains a higher field acting on electrons before the step and therefore a higher electron energy.

Another important point worth discussing, is the valence-band discontinuity ΔE_v. Holes created by electron impact ionization after the conduction-band step must surmount this barrier to avoid trapping effects, which may affect the device speed. This conditions is

$$eF\lambda_i + E_{f,h} = \Delta E_v, \tag{228}$$

where $E_{f,h}$ is the hole energy immediately after its creation by electron-initiated impact ionization, which depends on the energy at which the incoming electron has ionized. For a staircase detector with stages graded from GaAb to $AlAs_{0.08}Sb_{0.92}$ followed by a thin GaSb ionization region, $\Delta E_v \simeq 0.14$ eV, and $E_{f,h}$ is typically in the range 0.02–0.1 eV, depending on whether the incoming electron had an energy very close to the ionization threshold or a few tenths of an electron volt higher. Assuming $E_{f,h} = 0.05$ eV and $\lambda =$

100 Å, then $F \simeq 10^5$ V cm^{-1}. At this field, hole-initiated ionization starts to become a problem in the low-gap regions of the detector.

One way to eliminate altogether, or at least to minimize, trapping effects without producing hole-initiated impact ionization is to introduce dopants on both sides of the conduction-band step very near the interface to locally enhance the electric field over a distance smaller than the minimum hole-ionization length. For example, a sheet of ionized acceptors on the wide-gap side of the step and a corresponding sheet of ionized donors on the opposite low-gap side will create an additional local electric field capable of accelerating holes over the valence-band step and further enhancing electron-initiated ionization (Capasso *et al.*, 1983). These sheets of charges can relatively easily be introduced by MBE as in the case of "planar-doped barriers." Assume that the two sheets are at the same distance $d/2$ from the step and that they have the same surface-charge density σ. The additional field created by the "plane capacitor" is $e\sigma/\varepsilon_s$ and should be such that

$$\frac{e\sigma}{\varepsilon_s}\frac{d}{2} + eF\frac{d}{2} \geq \Delta E_v, \tag{229}$$

where F is the "background" electric field due to the external bias, to promote holes over the barrier of height ΔE_v. Since holes are created by impact ionization at a distance λ_i from the step, $d/2$ should be $\simeq \lambda_i$. The minimum charge density required is then [neglecting $F\,(\simeq 10^4$ V cm$^{-1})$ with respect to $e\sigma/\varepsilon_s$]

$$\sigma = (2\varepsilon_s\,\Delta E_v)/e\lambda_i. \tag{230}$$

Choosing $\Delta E_v = 0.14$ V, $\lambda_i = 100$ Å, and $\varepsilon_s \simeq 15\varepsilon_0$ for the AlGaAsSb case previously discussed gives $\sigma \simeq 2.7 \times 10^{12}$ cm^{-2}, which can easily be achieved by MBE.

Note that in the limit of a few atomic layers separation between the two charge sheets one can say that the band-edge discontinuities have been modified. This is because of the dipolar nature of the electrostatic potential associated with the doping interface dipole (DID). The DID concept may have important implications for many heterojunction devices, since their performance is usually a sensitive function of the band discontinuities. Capasso *et al.* (1985) have succeeded for the first time in artificially tuning the conduction and valence band barrier heights at an abrupt intrinsic semiconductor–semiconductor heterojunction via a DID.

A near one order of magnitude enhancement in the photocollection efficiency of an abrupt AlGaAs/GaAs heterojunction has been observed as a result of the conduction-band barrier lowering induced by the DID.

c. Derivation of the Ionization Yield from Avalanche Gain Measurements

One common situation encountered in impact-ionization experiments is the derivation of the ionization coefficients α and β from the electron- and hole-initiated multiplications M_n and M_p. It was shown in Part II that α and β can be obtained from a knowledge of M_n and M_p versus voltage. For simple field profiles ($p-i-n$ diode, abrupt onesided $p-n$ junction), simple analytical formulas relating α and β to M_n and M_p have been derived.

It is the object of this section to show that a similar situation exists in the case of the staircase APD. Namely, that from the knowledge of M_n and M_p, one can obtain via simple formulas the ionization yield $1-\delta$ and the average hole-ionization coefficients in the graded region $L^{-1}\int_0^L \beta\, dz \simeq \bar{\beta}$, where L is the length of each stage. We have assumed some residual hole-initiated ionization, which may occur under the conditions discussed in Section 17b.

Let us consider first the case of a one-stage device. Consider a single electron injected into the device and approaching the step; we want to calculate the total number of electrons after the step at the device output. This number is the electron-initiated gain M_n, which is given by

$$M_n = (2-\delta) + (2-\delta)(1-\delta)\left[\exp\left(\int_0^L \beta\, dz\right) - 1\right] + (2-\delta)(1-\delta)^2\left[\exp\left(\int_0^L \beta\, dz\right) - 1\right]^2 + \cdots . \qquad (231)$$

This simple infinite geometrical series is obtained by a very simple physical argument. The injected electron creates $1-\delta$ electrons at the step, giving a total of $2-\delta$ electrons after the step (first term). The $(1-\delta)$ holes created by electron impact ionization travel in the opposite direction and ionize generating $(1-\delta)[\exp(\int_0^L \beta\, dz) - 1]$ electrons, which are "fed back" into the input and impact-ionize to give additional $(2-\delta)[\exp(\int_0^L \beta\, dz) - 1]$ electrons at the output (second term). This procedure can be repeated for each additional, higher order "feedback loop." Adding the infinite number of terms of the series gives

$$M_n = \frac{2-\delta}{1-(1-\delta)[\exp(\int_0^L \beta\, dz) - 1]} . \qquad (232)$$

The breakdown condition,

$$(1-\delta)\left[\exp\left(\int_0^L \beta\, dz\right) - 1\right] = 1, \qquad (233)$$

is simply understood by noting that the left-hand side of Eq. (233) represents the number of electrons created in a single loop by the feedback effect of holes per injected electron. A similar procedure leads to the hole-initiated avalanche gain

$$M_p = \frac{\exp(\bar{\beta}L)}{1-(1-\delta)[\exp(\bar{\beta}L)-1]}. \tag{234}$$

From Eqs. (232) and (234), it is straightforward to derive δ and $\bar{\beta}$ in terms of M_n and M_p; i.e.,

$$\delta = 1 - \frac{M_n - 1}{M_p}, \tag{235}$$

$$\bar{\beta} = \frac{1}{L}\ln\left[\left(1 + \frac{M_n - 1}{M_p}\right)\left(\frac{M_n}{M_p}\right)\right]. \tag{236}$$

Thus, from a simple measurement of the electron- and hole-initiated photocurrent multiplication in a single-stage staircase APD, one can directly obtain the important physical quantities δ and $\bar{\beta}$. It is easy to generalize the previous procedure to an N-stage device to give

$$\delta = 1 - \frac{(M_n - 1)[(M_n/M_p)^{1/N} - 1]}{M_n - M_p}, \tag{237}$$

$$\bar{\beta} = \frac{1}{L}\left[\ln(2-\delta)\left(\frac{M_p}{M_n}\right)^{1/N}\right]. \tag{238}$$

Note that in this derivation we have assumed that holes ionize in the graded regions, whereas electrons impact-ionize only at the steps. In other words, the field F is sufficiently high to cause hole ionization but is such that the effective field $F - \Delta E_c/l$ does not cause electron ionization in the graded regions. This is a reasonable assumption in view of the exponential dependence of α and β on the electric field.

In conclusion, in this section we have derived useful formulas for experimentalists that relate measured quantities, such as M_n and M_p, to basic quantities such as δ and $\bar{\beta}$.

Obviously, the excess-noise factor is increased above the values given by Eq. (218) when hole-initiated ionization is present. The formulas are much more complicated and cumbersome and are not given here. However, it can be shown that if the residual hole-initiated ionization can be reduced, a significant improvement in noise performance results in most cases with respect to conventional APDs.

18. Channeling Avalanche Detector

This new structure was first proposed by Capasso (1982a,b). After publication of this work, a similar concept was independently put forward by Tanoue and Sakaki (1982).

This device uses a novel interdigitated $p-n$-junction structure, where electrons and holes are spatially separated by a transverse electric field in layers of different band gap. A parallel electric field subsequently channels the carriers along the layers where they undergo ionization. Since electrons and holes impact ionize in layers of different band gap, the ionization rate ratio can be made extremely large by properly choosing the band-gap difference while maintaining a very high gain.

a. The Interdigitated p–n-Junction Structure and its Electrical Characteristics

Figure 79 shows a schematic diagram of the structure. This consists of several abrupt $p-n$ heterojunctions with alternated p and n layers of band-gap E_{g1} and E_{g2} ($E_{g1} > E_{g2}$), respectively. Note, however, that for most applications other than APDs, the band gaps can be equal. The layers are lattice matched to a semi-insulating substrate. The p^+ and n^+ regions, which extend perpendicular to the layers, can be obtained by ion implantation or by etching and epitaxial regrowth techniques. The voltage source supplying

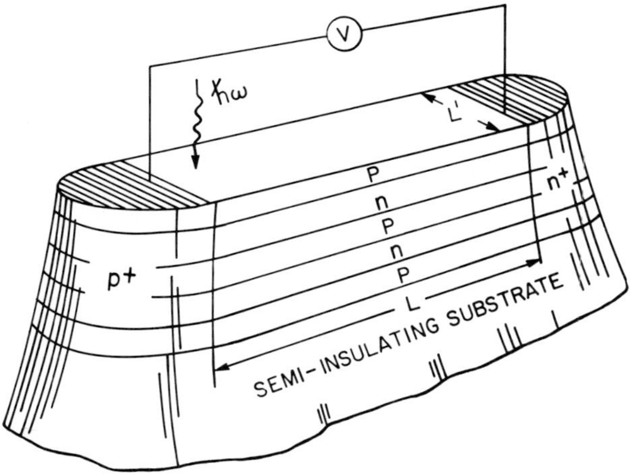

FIG. 79. Schematic diagram of the channeling APD (not to scale). The p layers have a wider band gap then the n layers. See text for layer thicknesses and doping levels. [From Capasso (1982b). © 1982 IEEE.]

the reverse bias is connected between the p^+ and n^+ regions. We assume, to illustrate the principle of this structure, equal doping levels ($n = p = n$) for the n and p layers. The three center layers have thickness d, whereas the topmost and bottommost p layers have thicknesses $d/2$; the layer length is L ($\gg d$) and the sensitive area width L'.

For zero bias, the p and n layers are in general only partially depleted on both sides of the heterojunction interfaces, as shown in Fig. 80a, which represents a cross section of the structure. The shaded areas denote the space-charge regions. The undepleted portions of the p and n layers (white areas) are at the same potential of the p^+ and n^+ end regions, respectively, so that the structure appears as a single interdigitated $p-n$ junction. Because of this geometry, when a reverse bias is applied between the p^+ and n^+ regions, this potential difference will appear across every $p-n$ heterojunction, thus increasing the space-charge width on both sides of the heterojunction interface (Fig. 80b). The bias is then further increased until all the p and n layers are completely depleted at a voltage $V = V_{\text{pth}}$ (where V_{pth} is the punchthrough voltage) (Fig. 80c). At this point, analogously to a p^+-i-n^+ diode, any further increase in the reverse bias will only add a constant electric field F_{p} parallel to the length L of the layers. This field is then increased to values such that avalanche multiplication takes place.

The capacitance of this novel structure has an interesting dependence on the applied voltage. For voltages $< V_{\text{pth}}$, the capacitance is essentially that of the four $p-n$ junction capacitors in parallel (since $L \gg d$). Thus,

$$C' = 4\varepsilon_{\text{s}}(LL'/W), \tag{239}$$

where W is the depletion-layer width of each $p-n$ junction (the dielectric constants of the n and p layers of different band gap have been assumed equal). As the reverse voltage is increased and approaches V_{pth}, the capacitance decreases toward the value

$$C_{\text{pth}} = 4\varepsilon_{\text{s}}(LL'/d). \tag{240}$$

At the punchthrough voltage V_{pth}, the capacitance drops abruptly from this relatively large value to

$$C_{V>V_{\text{ph}}} = 4\varepsilon_{\text{s}}L'd/L, \tag{241}$$

since the layers have been completely depleted, and the residual capacitance is that of the "p^+-i-n^+ diode" formed by the p^+ and n^+ regions. The capacitance is thus reduced by a factor $(L/d)^2$, which is $\gtrsim 100$ for typical dimensions. For $L \simeq L'$ and $d \simeq 1\ \mu\text{m}$, $C_{V>V_{\text{ph}}} \leq 10^{-3}$ pF. Note that in the general case of different acceptor and donor concentrations N_{D} and N_{A}, the

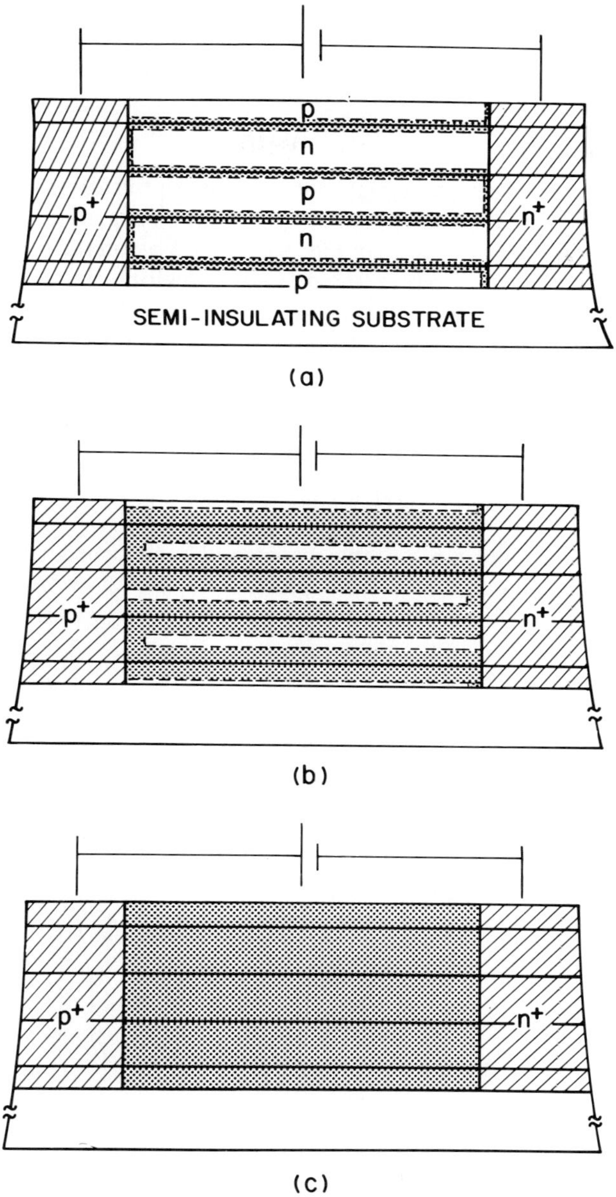

FIG. 80. Cross section of the channeling diode at various reverse-bias voltages. (a) Zero bias. (b) $0 < V < V_{pth}$. The shaded portions of the layers represent the depleted volume, which increases with increasing bias. (c) $V = V_{pth}$. At this voltage the layers are completely depleted. Any further increase in bias adds a constant field F_p parallel to the device length. Note that to ensure a parallel electric field constant along most of the device length L, this dimension is made much greater than the layer thicknesses. For simplicity of illustration, equal doping levels ($p = n$) have been assumed. [From Capasso (1982b). © 1982 IEEE.]

thicknesses of the center p layer and of the n layers (d_n and d_p) should be in the ratio

$$d_n/d_p = N_A/N_D, \tag{242}$$

and the top- and bottommost p layer should have a thickness $d_p/2$ to ensure complete depletion of all the p and n layers. The structure may have less or more than the number of layers in Fig. 80, depending on the type of application. For a three-layer structure, the p^+ and n^+ regions may be obtained via ion implantation of relatively light ions, such as Be, Mg, and Si. For devices with more layers, schemes using etching and epitaxial regrowth techniques are more practical.

b. Photodetector Operation

To understand the photodetector, it is important to consider the three-dimensional picture of the APD band diagram. This is illustrated in Fig. 81 for voltages $> V_{pth}$.

Suppose that radiation of suitable wavelength is absorbed in the lower-gap layer, thus creating electron–hole pairs. The two p–n heterojunctions formed at the interfaces between the relatively narrow band gap and the

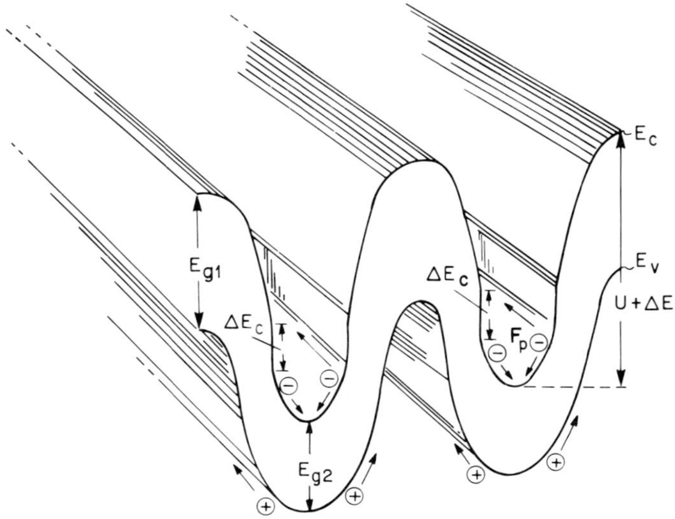

FIG. 81. Band diagram of the channeling APD under operating conditions; F_p is the parallel electric field causing carriers to impact ionize; U is the sum of the punchthrough voltage and the built-in potential. The valence-band discontinuity has been assumed negligible with respect to ΔE_c. Also shown is the electron–hole separation mechanism. [From Capasso (1982b). © 1982 IEEE.]

surrounding higher-band-gap layers serve to confine electrons to the narrow-band-gap layers while sweeping holes out into the contiguous wider-band-gap p layers, where they are confined by the potential. The parallel electric field F_p causes electrons confined to the narrow-band-gap layers to impact-ionize. Holes generated in this way are swept out in the surrounding higher-gap layers before undergoing ionizing collisions in the narrower-gap layers, since the layer thickness is made much smaller than the hole-ionization distance $1/\beta$. In conclusion, electrons and holes impact-ionize in spatially separated regions of different band gap. The holes in the wider-gap layers impact ionize at a much smaller rate than electrons in the relatively low-gap material, due to the exponential dependence of α, β on the band gap, so that α/β can be made extremely large.

In order for the device to operate in the described mode, several conditions should be met. First the potential well confining the electrons in the narrow-gap layer should be equal to or greater than the electron-ionization energy, so that electrons do not escape the potential wells before impact ionizing. A similar condition is required for holes in the wider-gap layers. Assuming a net donor concentration N_D in the n layers of thickness d_n and a net acceptor concentration N_A in the p layers of thickness d_p, these conditions read

$$eV_n + \Delta E_c \geq E_{ie,2}, \qquad eV_p - \Delta E_v \geq E_{ih,1}, \tag{243}$$

where ΔE_c and ΔE_v are the conduction- and valence-band discontinuities of the heterojunction; $E_{ie,2}$ and $E_{ih,1}$ are, respectively, the electron- and hole-ionization energies in the low- and high-gap layers; V_n and V_p are the portions of the voltage that at punchthrough are supported in the n and p layers and are given by

$$V_n = \tfrac{1}{4} F_{tm} d_n = [V_{pth} + V_{bi}][N_A/(N_A + N_D)], \tag{244}$$

$$V_p = \tfrac{1}{4} F_{tm} d_p = [V_{pth} + V_{bi}][N_D/(N_A + N_D)], \tag{245}$$

where V_{pth} is the punch-through bias voltage required to completely deplete the n and p layers, V_{bi} is the built-in potential, and F_{tm} is the maximum field perpendicular to the plane of the layers at punchthrough and is given by

$$F_{tm} = eN_D d_n / 2\varepsilon_s. \tag{246}$$

Note that we have assumed for simplicity equal dielectric constants for the two materials. Using Eqs. (244) and (245), the electron- and hole-confinement conditions, Eqs. (243), may be expressed as

$$d_n \geq 4(E_{ie,2} - \Delta E_c)/eF_{tm}, \tag{247}$$

$$d_p \geq 4(E_{ih,1} + \Delta E_v)/eF_{tm}, \tag{248}$$

Recalling the charge neutrality condition [Eq. (242)], Eq. (247) reads

$$d_n \geq \frac{4(E_{\text{ih},1} + \Delta E_{\text{v}})N_{\text{A}}}{eF_{\text{tm}}N_{\text{D}}}. \tag{249}$$

It is easily seen that for

$$N_{\text{D}}/N_{\text{A}} \geq (E_{\text{ih},1} + \Delta E_{\text{v}})/(E_{\text{ie},2} - \Delta E_{\text{c}}), \tag{250}$$

Eq. (247) rather than Eq. (249) determines the confinement of both electrons and holes (note that $E_{\text{ie},2} - \Delta E_{\text{c}} < E_{\text{ih},1} + \Delta E_{\text{v}}$). Equation (250) is important, as will be seen later for the optimum design of the structure (highest α/β ratio).

In addition, we also require that the transverse field F_{tm} at punchthrough be smaller than the avalanche or tunneling threshold field F_{th}, so that no avalanche multiplication or tunneling perpendicular to the layers occurs prior to punchthrough. This condition reads

$$F_{\text{tm}} < F_{\text{th}}. \tag{251}$$

Finally, it is also necessary that holes created by incident photons or by electron impact ionization be swept out of the lower gap layers before undergoing impact ionization. This condition may be expressed as

$$d_n/2 \ll 1/\beta_{2\text{m}}, \tag{252}$$

where $\beta_{2\text{m}}$ is the maximum value of the hole-ionization rate in the low-gap material. Note that the hole-ionization coefficient β_2 depends on the geometrical sum of the parallel and transverse components of the field $(F_{\text{p}}^2 + F_{\text{tm}}^2)^{1/2}$; F_{p} is independent of position, whereas F_{t} increases linearly with distance from the center of the channel and attains the maximum value F_{tm} at the interface between the low- and high-gap material. Thus, β_2 increases exponentially with distance from the center of the channel, reaching the maximum value $\beta_{2\text{m}}$ corresponding to the field $(F_{\text{p}}^2 + F_{\text{tm}}^2)^{1/2}$ at the heterojunction interface. For practical purposes, we shall require

$$d_n/2 \leq 1/10\beta_{2\text{m}}. \tag{253}$$

Equations (247), (249), (251), and (253) determine the operating-point diagram of the structure, as is illustrated by several examples in Subsection 18a. For example, for a specified donor-to-acceptor concentration ratio $N_{\text{D}}/N_{\text{A}}$, it is easily shown that the layer-thickness range, for which carriers are confined and no multiplication occurs prior to punchthrough, is given by

$$\left[\frac{8\varepsilon_s(E_{ih,1}+\Delta E_v)N_A}{e^2N_D^2}\right]^{1/2} < d_n < \frac{2\varepsilon_s F_{th}}{eN_D}, \tag{254}$$

for

$$N_D < \frac{\varepsilon_s F_{th}^2 N_D}{2(E_{ih,1}+\Delta E_v)N_A}, \tag{255}$$

or

$$\left[\frac{8\varepsilon_s(E_{ie,2}-\Delta E_c)}{e^2N_D}\right]^{1/2} < d_n < \frac{2\varepsilon_s F_{th}}{eN_D}, \tag{256}$$

for

$$N_D < \frac{\varepsilon_s F_{th}^2}{2(E_{ie,2}-\Delta E_c)}, \tag{257}$$

depending on whether N_D/N_A is smaller or greater than $(E_{ih,1}+\Delta E_v)/(E_{ie,2}-\Delta E_c)$. Note that the corresponding range for d_p is obtained by multiplying the right-hand and left-hand sides of Eqs. (254) and (256) by N_D/N_A. Once d_n and N_D are chosen, the transverse electric field F_{tm} is also determined, and one has to verify that the electron–hole spatial separation condition Eq. (252) is satisfied.

The parallel field F_p is constant along the length of the layers (except for a negligible distance, comparable to the layer thickness, at both ends of the layers), so that

$$F_p = (V - V_{pth})/L. \tag{258}$$

Let us now discuss the α/β ratio of this structure. In the limit of negligible hole ionization in the low-band-gap layers, the ionization rate ratio is given by

$$k = \alpha_2(F_p)/\beta_1(F_p), \tag{259}$$

where α_2 and β_1 are the electron and hole ionization rates, respectively, in the low- and high-gap material. Note that the k ratio is evaluated at the parallel field F_p; the transverse field F_t is expected to affect k only marginally, since carriers are confined near the center of the layers where $F_t \ll F_p$. Since Eq. (259) is a good approximation when the majority of holes are swept out of the low-gap channels before undergoing ionization, the difference between the maximum hole ionization distance $1/\beta_{2m}$ and the low-gap layer thicknesses should be made as large as possible.

From the previous considerations, it is clear that the minimum layer thickness compatible with carrier confinement is obtained, for a specified transverse electric field F_{tm}, by choosing $N_{\text{D}}/N_{\text{A}} \geq (E_{\text{ih},1} + \Delta E_{\text{v}})/(E_{\text{ie},2} - \Delta E_{\text{c}})$ and is given by [Eq. (247)]

$$d_{n,\min} = 4(E_{\text{ie},2} - \Delta E_{\text{c}})/eF_{\text{tm}}. \tag{260}$$

The optimum transverse electric field F_{tm} is found by maximizing the ratio $1/\beta_{2\text{m}} d_{n,\min}$. The optimum donor density is then obtained from Eq. (246). The function $\beta_{2\text{m}}$, which depends on $(F_{\text{p}}^2 + F_{\text{tm}}^2)^{1/2}$, can be derived by fitting the experimental hole-ionization-rate data. In practice, the device can be easily optimized using the graphical method discussed in Subsection 18c.

Note that if the conduction-band discontinuity ΔE_{c} is comparable to the electron ionization energy, the minimum layer thickness can be made very small (a few thousand angstroms). The physical reason for this fact is that the large conduction-band discontinuity provides a significant portion of the confining potential $eV_n + \Delta E_{\text{c}}$, so that the voltage drop across the layer and therefore the layer thickness can be maintained small. In several heterojunctions of interest for 1.3 – 1.6-μm photodiodes, the ratio of the hole ionization distance in the low-gap material to the low-gap layer thickness can thus be made very large ($\geq$20), so that Eq. (259) is a good approximation to the effective ionization rate ratio.

To achieve minimum noise, the avalanche should be initiated by electrons. This is done by shining light near the p^+ region in Fig. 79. This ensures that holes created by the incident light exit from the device before ionizing. In the limit of negligible hole impact ionization in the low-gap material, the electron-initiated gain is given by

$$M_n = \frac{[1 - (\beta_1/\alpha_2)] \exp\{\alpha_2 L[1 - (\beta_1/\alpha_2)]\}}{1 - (\beta_1/\alpha_2) \exp\{\alpha_2 L[1 - (\beta_1/\alpha_2)]\}}, \tag{261}$$

which reduces, in the limit of $\alpha_2/\beta_1 \gg 1$, to $M_n \simeq \exp(\alpha_2 L)$. Thus, in this limit, the channeling APD closely mimics a channeltron photomultiplier.

Note that space-charge effects due to electron – hole separation are negligible for the range of photocurrents typical of APDs.

c. Operating-Point Diagram: Application to the Design of AlGaAs/GaAs, InP/InGaAs, *and* AlGaAsSb/GaSb *Channeling APDs*

In this section, we shall discuss three examples of the channeling-APD structure using AlGaAs/GaAs, InP/InGaAs, and AlGaAsSb/GaSb heterojunctions. Note that GaAs, InGaAs, and GaSb have all comparable α and β.

In the first example, let us assume the lower gap n layers to be GaAs ($E_{\text{g2}} = 1.42$ eV) and the p layers $Al_{0.45}Ga_{0.55}As$ ($E_{\text{g1}} = 2.0$ eV). For the con-

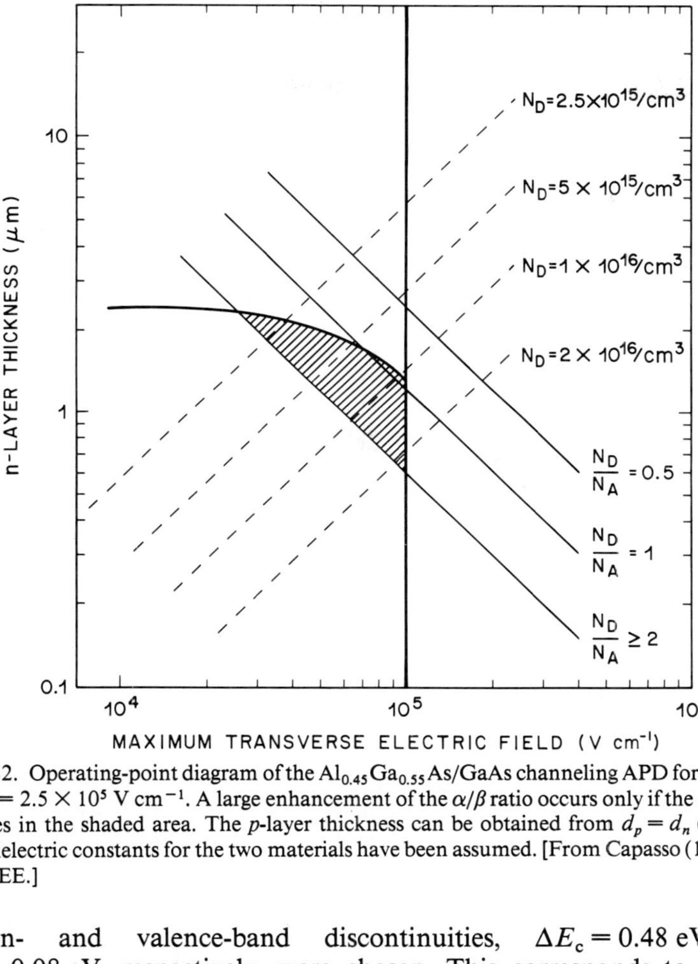

FIG. 82. Operating-point diagram of the $Al_{0.45}Ga_{0.55}As$/GaAs channeling APD for a parallel field $F_p = 2.5 \times 10^5$ V cm^{-1}. A large enhancement of the α/β ratio occurs only if the operating point lies in the shaded area. The p-layer thickness can be obtained from $d_p = d_n\ (N_D/N_A)$. Equal dielectric constants for the two materials have been assumed. [From Capasso (1982b). © 1982 IEEE.]

duction- and valence-band discontinuities, $\Delta E_c = 0.48$ eV and $\Delta E_v = 0.08$ eV, respectively, were chosen. This corresponds to the old "Dingle rule" ($\Delta E_c = 0.85\ \Delta E_g$) thought to be valid for this heterojunction. The new correct alignment is $\Delta E_c = 0.6\ \Delta E_g$. This, however, does not change significantly the operating point diagram of the channeling APD. The hole ionization energy in $Al_{0.45}Ga_{0.55}As$ is $\simeq \frac{3}{2} E_g = 3.0$ eV, whereas the electron ionization energy in GaAs is $E_{ie,2} = 2$ eV; the permittivity is $\varepsilon_s = 13\varepsilon_0$, and we set $F_{th} \simeq 10^5$ V cm^{-1}, since below this field impact ionization in GaAs is negligible. By setting the n-layer thickness equal to the right-hand side of Eq. (249), we obtain a family of solid straight lines (Fig. 82), the so-called confinement lines. For a given N_D/N_A, the operating point should be above the corresponding line to achieve confinement of both electrons and holes in the low- and high-gap layers, respectively. Note that for $N_D/N_A > 2$, the confinement line is independent of this ratio and is given by the equal sign in Eq. (247).

The vertical solid line is the so-called transverse avalanche multiplication line and is obtained by equating F_{tm} to F_{th}. In order to avoid avalanche multiplication perpendicular to the layers prior to punchthrough, the operating point should be on the left side of this vertical line.

The curved line is the electron–hole spatial separation line and is obtained by equating $d_n/2$ to the right-hand side of Eq. (253). The operating point should stay below this line to minimize residual hole ionization in the low-gap material. The curvature represents the effect of the transverse electric field. In calculating β_{2m}, the following expression for the hole ionization rate

$$\beta = 5.5 \times 10^6 \exp(-2.2 \times 10^6/F), \tag{262}$$

was used, which fits hole ionization data in GaAs (Ando and Kanbe, 1981). The total field $F = (F_{tm}^2 + F_p^2)^{1/2}$ was obtained by assuming $F_p = 2.5 \times 10^5$ V cm^{-1} and varying F_{tm}.

In conclusion, the device operates in the described mode when the operating point lies in the shaded area in Fig. 82. The dashed lines, described by Eq. (246), can be used [along with Eqs. (247), (249), and (250)] to determine for a given N_D and N_D/N_A the appropriate layer thicknesses and transverse electric field ranges.

As an example, let us assume $N_D/N_A = 2$ and $N_D = 1 \times 10^{16}$ cm^{-3}. The operating point should lie on the $N_D = 1 \times 16$ cm^{-3} dashed line between the electron–hole spatial separation curve and the $N_D/N_A = 2$ confinement line. Thus, the n-layer thickness should be between $\simeq 0.95$ and $\simeq 1.4$ μm, whereas the p-layer thickness $[d_p = d_n(N_D/N_A)]$ varies from $\simeq 1.9$ to $\simeq 2.8$ μm. The corresponding transverse electric field is between 6×10^4 and 9.5×10^4 V cm^{-1} and the reciprocal of the maximum hole ionization rate in the low-gap material $1/\beta_{2m}$ between 9 and 7 μm, thus satisfying the electron–hole spatial separation condition. The confining potential V_n ranges from 1.54 to 3.4 V for electrons and from 3.14 to 6.8 V for holes, whereas the total voltage across the layers at punchthrough $V_{pth} + V_{bi}$ varies between 4.7 and 10.2 V.

Note that for $N_D/N_A \geq 2$ and for $N_D > 2.4 \times 10^{16}$ cm^{-3}, the device cannot operate as a channeling APD, since avalanche multiplication perpendicular to the layers occurs at a voltage lower than that required for carrier confinement. On the other hand, for $N_D/N_A \geq 2$ and for $N_D \leq 1.5 \times 10^{15}$ cm^{-3}, the required layer thickness becomes sufficiently large that the number of holes not swept out of the lower-gap layer before ionization can no longer be neglected, with consequent signal-to-noise degradation. The optimum operating point is found by choosing $N_D/N_A \geq 2$ and by maximizing the vertical separation between the carrier-confinement line $N_D/N_A \geq 2$ and the $d_n = 1/5\beta_{2m}$ electron–hole spatial separation curve. This is done by choosing an n-layer thickness of $\simeq 0.6-0.7$ μm and a transverse electric field in the $9 \times 10^4 - 10^5$ V-cm^{-1} range corresponding to $n_D \simeq 2 \times 10^{16}$ cm^{-3}. The hole-ionization-distance–layer-thickness ratio is then $\simeq 11$. This choice of the operating point minimizes residual hole ionization in the n channels, so that $\kappa = \alpha(\text{GaAs})/\beta(\text{Al}_{0.45}\text{Ga}_{0.55}\text{As})$ is a reasonable approximation for the ionization rate ratio. In GaAs at 2.5×10^5 V cm^{-1}, $\alpha = 1.66 \times 10^3$ cm^{-1}. There are no data for β in $Al_{0.45}Ga_{0.55}As$/GaAs, but an estimate can be obtained by scaling the constant A in $\beta = a \exp(-A/F)$, which fits the hole data in GaAs (Ando and Kanbe, 1981) by the ratio of the ionization energies in the two materials. This gives at $F = 2.5 \times 10^5$ V cm^{-1}, $\beta = 4.7$ cm^{-1}. Thus, $\alpha/\beta \simeq 350$. With such high k, the gain M is $\approx \exp(\alpha L)$, since the hole feedback is greatly reduced. At $F = 2.5 \times 10^5$ V cm^{-1} and $L = 25$ μm, $M = 63$.

From the previous considerations, it is seen that for a practical device that detects radiation with $\lambda \geq 8300$ Å, a detector with two GaAs layers ($n \simeq 2 \times 10^{16}$ cm^{-3}) $\simeq 0.6$ μm thick, one $Al_{0.45}Ga_{0.55}As$/GaAs center layer ($p = 1 \times 10^{16}$ cm^{-3}) $\simeq 1.2$ μm thick, and two cladding $Al_{0.45}Ga_{0.55}As$/GaAs layers ($p \simeq 1 \times 10^{16}$ cm^{-3}) 0.6 μm thick is appropriate. The semi-insulating substrate is Cr-doped GaAs.

Let us now consider the case where the wider-gap layer is InP and the lower-gap layer consists of the lattice-matched composition $Ga_{0.47}In_{0.53}As$, which can be grown lattice matched to an Fe-doped InP semi-insulating substrate. This heterojunction has been widely used for APDs with separated multiplication and absorption regions (see, e.g., Capasso *et al.*, 1984c), since it is a very attractive material combination for 1.3–1.6-μm fiber communication systems. However, k in these structures is at best 0.3, which is far from satisfactory. By using the channeling APD concept, the k ratio can be made very high in this particular heterostructure also.

For the InP/$In_{0.53}Ga_{0.47}As$ heterojunction (Forrest and Kim, 1981), $\Delta E_c = 0.24$ eV, $\Delta E_v = 0.38$ eV; the hole ionization energy in InP is $E_{ih,1} = 1.67$ eV, whereas the electron ionization energy in $In_{0.53}Ga_{0.47}As$ is $E_{ie,2} = 0.85$ eV (Pearsall, 1979, 1980). Tunneling in the InGaAs layers is negligible at maximum fields $\leq 10^5$ V cm^{-1}, so that we set F_{th} equal to this value. The relatively large valence-band discontinuity, which could produce storage of holes being swept out of the InGaAs, can be conveniently eliminated by intentionally grading the interface, as demonstrated by Capasso *et al.* (1984c), using a graded-gap InP/GaInAs pseudoquaternary superlattice. Note that because of the lower band gap, this long-wavelength-channeling APD requires considerably less voltage than the AlGaAs/GaAs APD.

Assuming a parallel field $F_p = 1.9 \times 10^5$ cm^{-1}, this corresponds to a

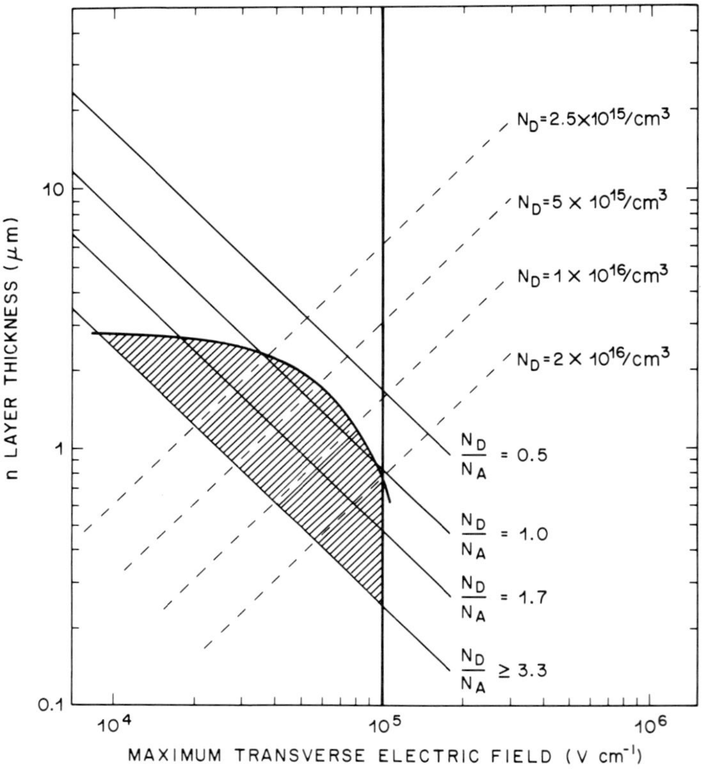

FIG. 83. Operating-point diagram of the $InP/In_{0.53}Ga_{0.47}As$ channeling APD for a parallel field $F_p = 1.9 \times 10^5$ V cm^{-1}. [From Capasso (1982b). © 1982 IEEE.]

voltage in the 360–900-V range for $20 \leq L \leq 50$ μm. The operating-point diagram of this detector can easily be obtained, as in the case of the AlGaAs/GaAs detector, and is shown in Fig. 83 for a field $F_p = 1.9 \times 10^5$ V cm^{-1}. Note that the area of the shaded region is substantially larger than that of the AlGaAs/GaAs diagram. This is due to the fact that the $(E_{ih} + \Delta E_v)/(E_{ie} - \Delta E_c)$ ratio is more than twice as large. This allows one to make the hole-ionization-distance–n-layer-thickness ratio higher than in the $Al_{0.45}Ga_{0.55}As/GaAs$ detector. The optimum operating point is found by choosing $N_D/N_A \geq 3.3$ and setting $d_n \simeq 0.45$ μm and $F_{tm} \simeq 5.7 \times 10^4$ V cm^{-1}, which corresponds to a net donor density in the n layers of $\simeq 2 \times 10^{16}$ cm^{-3}. The corresponding electron-confining potential is $V_n = 0.7$ V. Assuming $N_D/N_A \simeq 3.6$, the p-layer thickness is then $d_p \simeq 1.5$ μm, and the hole-confining potential $V_p \simeq 2.30$ V. The $1/\beta_{2m} d_n$ ratio is $\simeq 20$. The hole ionization rate data for $In_{0.53}Ga_{0.47}As$ are obtained from Pearsall (1979). The ionization rate ratio can be approximated by $k = \alpha(\mathrm{InGaAs})/\beta(\mathrm{InP})$. At $F_p \simeq 1.9 \times 10^5$ V cm^{-1}, we estimate $\beta(\mathrm{InP}) = 7.1$ cm^{-1} and $\alpha(\mathrm{InGaAs}) \simeq 2 \times 10^3$ cm^{-1}, so that $k = 280$. The electron-initiated gain is $\simeq 150$ for $L = 25\ \mu$m.

Note that this structure could be grown lattice matched to a semi-insulating InP iron-doped substrate using either MBE or vapor-phase epitaxial growth. A device optimized as previously described would consist of two *n*-InGaAs layers to increase photon absorption at the longer wavelengths, a center *p* channel, and two external *p* layers of the same doping level but half the thickness of the center channel. Note that instead of $InP/In_{0.53}Ga_{0.47}As$ heterojunctions, one could use $Al_{0.48}In_{0.52}As/In_{0.53}Ga_{0.47}As$ heterojunctions, which can be grown by MBE lattice matched to InP. Measurements indicate that a large fraction of the band-gap difference ($\simeq 70\%$) should be concentrated in the conduction band (People *et al.*, 1983). This fact, along with the higher band gap of $Al_{0.48}In_{0.52}As$ ($\simeq 1.50$ eV), with respect to InP ($\simeq 1.29$ eV), indicates that an $Al_{0.48}In_{0.52}As/In_{0.53}Ga_{0.47}As$ APD will be superior in terms of α/β ratio to an $InP/In_{0.53}Ga_{0.47}As$ channeling APD. The α/β ratio would be expected to be virtually infinite.

Finally, we shall consider an $Al_xGa_{1-x}As_ySb_{1-y}GaSb$ channeling APD. The $Al_xGa_{1-x}As_ySb_{1-y}$ alloy can be grown lattice matched to GaSb over a wide range of compositions along the $y = 0.08x$ lattice-matched line. The corresponding band gap also spans a large range from 0.7 (GaSb) to 1.58 eV ($AlAs_{0.08}Sb_{0.92}$). In addition, measurements suggest that in the AlSb/GaSb heterojunction, 85% of the band-gap difference is concentrated in the conduction band (Naganuma *et al.*, 1981). These features immediately indicate that a channeling APD consisting of GaSb *n* channels and $AlAs_{0.08}Sb_{0.92}$ *p* channels will have an extremely high α/β (virtually infinite). This is because at the fields typical of avalanche multiplication in GaSb ($\approx 10^4$ V cm^{-1}), no impact ionization occurs in the wide-gap $AlAs_{0.08}Sb_{0.92}$. The threshold field for ionization in this material is $\gtrsim 10^5$ V cm^{-1}. In addition, as a result of the large conduction-band discontinuity comparable to the electron ionization energy in GaSb, a very small additional potential ($E_{ie} - \Delta E_c$) is needed to completely confine the electrons in the GaSb. Since this potential is proportional to $N_D d_n^2$, the *n* layers can be made very thin ($d_n \simeq 1500$ Å) by a suitable choice of N_D and N_D/N_A. This ensures that residual hole ionization in the GaSb is completely negligible. (The hole ionization distance in this material exceeds several micrometers at typical operating voltages.) Finally, the fact that avalanche multiplication in GaSb occurs at substantially lower fields than in $In_{0.47}Ga_{0.53}As$ means that the operating voltage of this device will be typically $\simeq 150-200$ V. Figure 84 represents the operating-point diagram for the previous detector. Note that the shaded area defining the

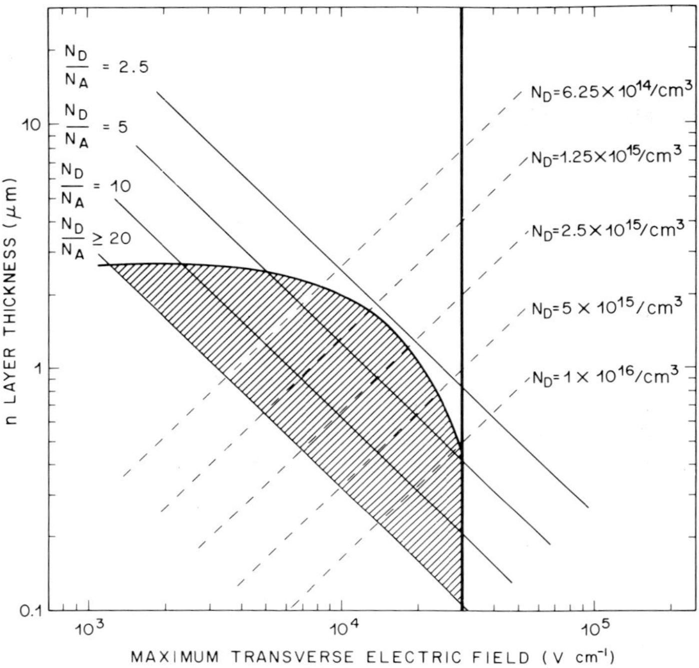

FIG. 84. Operating-point diagram of the $AlAs_{0.08}Sb_{0.02}/GaSb$ channeling APD for a parallel field $F_p = 5.35 \times 10^4$ V cm^{-1}. [From Capasso (1982b). ©1982 IEEE.]

layer thickness and transverse electric field ranges of operation is much larger than the previously considered cases. This is due to the large band-gap difference between the two materials and to the large conduction-band discontinuity ($\simeq E_{ie,2}$). For this particular detector, we have modified the hole-confinement condition by requiring

$$d_p > 4(E_{g,1} + \Delta E_v)/eF_{tm}, \tag{263}$$

instead of Eq. (246), where $E_{g,1}$ is the band gap of $AlAs_{0.08}Sb_{0.92}$. The physical reason is that at the operating fields of this detector ($\simeq 5 \times 10^4$ V cm^{-1}), holes acquire in the wide-gap $AlAs_{0.08}Sb_{0.92}$ an average energy of a few tenths of a volt. Thus, the probability that a hole gains energies in excess of the band gap ($\simeq 1.58$ eV) is negligibly small. Consequently, hole transfer back into the GaSb layers can be prevented by a confining potential smaller than the hole ionization energy in $AlAs_{0.08}Sb_{0.92}$ ($\simeq 3.2$ eV) (of the order of the band gap). We have taken $F_{th} = 3 \times 10^4$ V cm^{-1} for the avalanche threshold field in GaSb (Hildebrand *et al.*, 1981).

The optimum operating point can again be chosen graphically by maximizing the vertical separation between the lowest carrier-confinement line ($N_D/N_A \geq 20$) and the electron–hole spatial separation line $1/5\beta_{2m}$, calculated from measurements of β in GaSb (Hildebrand *et al.*, 1981). We find that an optimized device should have $N_D/N_A \geq 20$, an n-layer thickness $d_n \simeq 2000$ Å, a transverse electric field $F_{tm} = 1.3 \times 10^4$ V cm^{-1}, and an n-doping level $N_d \simeq 1 \times 10^{16}$ cm^{-3}. Setting $N_D/N_A = 20$, then $d_p = 4\ \mu$m and $N_A = 5 \times 10^{14}$ cm^{-3}. The ratio of the hole ionization distance $1/\beta_{2m}$ to the n-layer thickness d_n is then $\simeq 40$. For a parallel field, $F_p = 4.35 \times 10^4$ V cm^{-1}, $\alpha = 10^3$ cm^{-1} in GaSb, which provides a gain of $\simeq 150$ for a 50-μm-long device. This corresponds to a voltage of $\simeq 200$ V. Note that at this field, β in $AlAs_{0.08}Sb_{0.92}$ is completely negligible because of the wide gap; thus, the k ratio is infinite, and the device mimics a channeltron photomultiplier.

If the high dark current characteristic of GaSb can be reduced, the AlGaAsSb/GaSb alloy system then appears to be one of the most promising materials for the implementation of the channeling APD. For the semi-insulating substrate, one could use AlSb (Shaw and McKell, 1963), followed by matching layers of intermediate compositions. A structure optimized according to the previous discussion will consist of a plurality (3–5) of thin n-GaSb layers to increase the quantum efficiency, alternated with $AlAs_{0.08}Sb_{0.92}$ p layers.

The main advantage of the channeling APD with respect to conventional APDs is that a very high ionization rate ratio (> 100) and thus a very small effective noise factor can be maintained even at high gains. In conventional APDs, this is not possible because at very high fields approaching breakdown, the difference between the ionization rates is reduced, a feature typical of all semiconductors. The reason that the channeling APD does not have this limitation is that electrons and holes ionize in different materials.

Finally, we shall consider the speed of this device. The capacitance will not affect the response time because it is completely negligible, as discussed previously. For avalanche gains smaller than the α/β ratio, the bandwidth is not affected by the multiplication process and is only limited by the sum of the electron and hole transit times, which is a well known result (Emmons, 1967). Since the channeling-APD α/β is very high (100 to infinity), the bandwidth will then not be limited by avalanche multiplication up to very high gains. The total transit time is on the order of 500 psec–1 nsec for carrier velocities in the $5 \times 10^6 - 10^7$-cm sec^{-1} range and layer lengths between 25 and 50 μm. Thus, the device can be operated at signal bit rates as high as 500 Mbit/sec.

Note that throughout this chapter, we have assumed charge neutrality $N_D d_n = N_A d_p$. This control of the doping and of the layer thickness can be achieved by MBE.

d. Device Fabrication

The fabrication and the $C-V$ measurements of the channeling diode have been reported by Capasso *et al.* (1982e,f). For the experiments, three-layer $p-n-p$ structures were grown by liquid-phase epitaxy (LPE) and MBE (Fig. 85a). The GaAs n layer has thickness and carrier concentration in the 1.0–1.5-μm and $5 \times 10^{15}-10^{16}$-cm^{-3} ranges, respectively. The two $Al_{0.45}Ga_{0.55}As$ p layers have the same doping concentration approximately equal to that of the n layer and half the n-layer thickness. The transverse n^+ and p^+ regions are defined by etching and LPE regrowth of $Al_{0.2}Ga_{0.8}As$.

Fifty-μm stripes with 500-μm center to center were defined and mesa etched down to the substrate (Fig. 85b). The sample was then chemically cleaned, anodized to grow a native oxide and cleaned to remove the native oxide, and inserted in the growth furnace. Prior to the regrowth, approximately 0.2 μm of the exposed surface was removed by a meltback step. This cleanup of the etched surfaces reduces nonradiative recombination at the mesa-regrown layer interface. The n^+-$Al_{0.45}Ga_{0.5}As$ was then regrown, as shown in Fig. 85c. The surface was then masked with 100-μm-wide stripes, 500-μm center-to-center spacing in Shipley 1350J photoresist. The mask was positioned to cover about 40 μm of the mesa top and 60 μm of the n^+ regrowth. This ensures that on etching away the exposed crystal, one side of the mesa will contain the originally grown layers, as shown in Fig. 85d.

The wafer was then cleaned as after the first regrowth, and in a procedure identical to the first regrowth, an $Al_{0.2}Ga_{0.8}As$ layer doped with Ge to a level $\geq 10^{18}$ cm^{-3} was grown, as shown in Fig. 85e. Note that since the first regrown n^+ layer was $Al_{0.2}Ga_{0.8}As$, no special masking was again needed to prevent regrowth on the top of the mesas. The remaining steps included the formation of ohmic contacts and the electrical isolation of the devices (Fig. 85f). For the isolation, 5-μm-wide slots were etched between and orthogonal to the mesas. The sensitive area of the detector between the n^+ and p^+ regions was typically in the $5 \times 10^{-5}-10^{-4}$-cm^2 range. Metal contact was provided by evaporation of 200 Å of Ti followed by 2000 Å of Au. To ensure stable contacts, the wafer was heated for a few seconds in a H_2 flow at 500°C.

e. Capacitance Measurements

The capacitance of the diodes was measured at 1 MHz as a function of reverse bias for various incident light intensities (Fig. 86). A 2-mW He–Ne laser attenuated with neutral density filters was used as the light source. The top $Al_{0.45}Ga_{0.55}As$ is transparent to the $\lambda = 6328$-Å radiation. Let us first discuss the dark $C-V$ curve (absence of light). The curve has three distinct regions. First is a decrease of the capacitance with voltage, characteristic of a reverse-biased diode, followed by a one-order-of-magnitude drop in capacitance over a small voltage range, and a final region of nearly constant ultra-

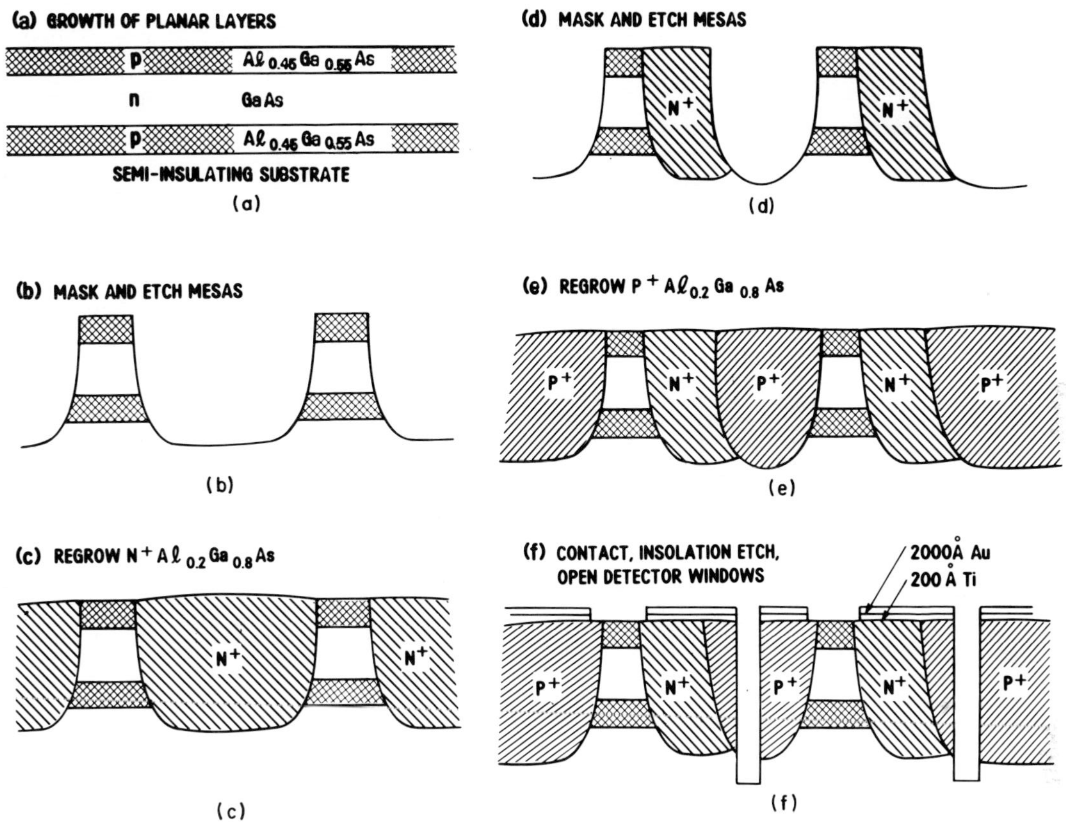

FIG. 85. Illustration of the fabrication procedure (not to scale: the distance between the p^+ and n^+ regions is much larger than the layer thicknesses): (a) growth of planar layers; (b) mask and etch mesas; (c) regrow $N^+ Al_{0.2}Ga_{0.8}As$; (d) mask and etch mesas; (e) regrow $P^+ Al_{0.2}Ga_{0.8}As$; (f) contact, insolation etch, open detector windows. [From Capasso *et al.* (1982f).]

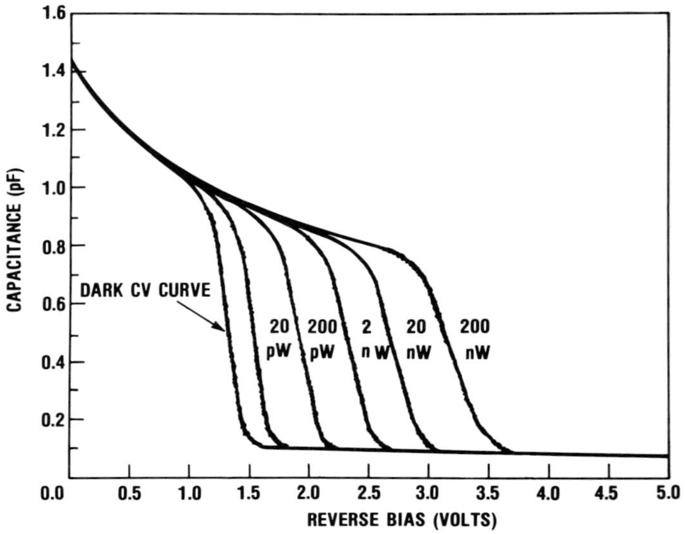

FIG. 86. $C-V$ characteristics of the channeling photodiode for optical power levels in the range from 20 pW to 200 nW. The scales are 0.1 mA div^{-1} and 10 V div^{-1}. [From Capasso *et al.* (1982f).]

small ($\leq$0.1 pF) capacitance. The overall behavior agrees well with the previous predictions. It is very important to note that for a large change in capacitance, complete depletion of all the layers is not required. Assume, for example, that for the device in Fig. 79, the acceptor concentration is greater than the donor concentration; in this situation, only the n layers will be completely depleted. The residual capacitance is that formed by the two undepleted sections of the p layers with the n^+ region. This capacitance is still much smaller than that before punchthrough.

Doping measurements were made on the top AlGaAs p layer using a Hg probe; the net acceptor concentration is $N_A \simeq 5 \times 10^{15}$ cm^{-3}. The decrease in capacitance before punchthrough gives a linear $1/C^2$-versus-voltage plot, implying abruptness of the junctions. By assuming equal acceptor concentrations for the top and bottom p layers, from the slope of the $1/C^2$-versus-V line, we find $(N_D N_A)/(N_D + N_A) \simeq 2.5 \times 10^{15}$ cm^{-3}. The measured value of N_A then implies that $N_D \simeq N_A \simeq 5 \times 10^{15}$ cm^{-3}. The layer thicknesses were $d_n = 1.3$ and $d_p \simeq 0.7$ μm for the n and p layers, respectively. The measured punchthrough voltage is consistent with the previous dimensions and doping levels.

The other $C-V$ curves in Fig. 86 were obtained by varying the incident laser power over four orders of magnitude from 20 pW to 200 nW. Note the increase of the punchthrough voltage with increasing power and the large variations in capacitance (0.6–1.0 pF) (with respect to the "dark capacitance") produced by the low optical power levels used. It is clear that this device can be used as an ultrahigh-sensitivity photocapacitive detector. The essential features of this novel photocapacitive phenomenon can be easily interpreted with the aid of Fig. 80.

Let us assume that the device is biased at or slightly above the punchthrough voltage, so that the layers are completely depleted. When light is shined on the device, the photogenerated electrons and holes are spatially separated and collected in the depleted n and p layers, respectively, thereby partially neutralizing the ionized donor and acceptor space charge. The net effect is that the width of the depletion layer is reduced; this produces a large increase in capacitance. An additional voltage is thus required to completely deplete the layers. This explains the shift of the punchthrough voltage with increasing optical power. The ultrahigh sensitivity of the structure is due to two factors. First, even a small ($\leq 10\%$) reduction of the depletion-layer width produced by a very low incident power is sufficient to cause a large change in capacitance. Second, the spatial separation of optically generated electrons and holes greatly increases their recombination lifetime so that substantial quasi-stable excess densities of electrons and holes are present in the layers to compensate the ionized space charge. The increase in lifetime due to the spatial separation of electrons and holes was first discussed by Dohler (1972) in the context of $n-i-p-i$ superlattices.

It is important to stress that this detector can also be used as an ultralow capacitance, low-punchthrough-voltage $p-i-n$ photodiode, provided it is biased at voltages such that the capacitance is not affected by the incident light. For example, assume that for the particular application of interest (e.g., fiber optics receivers), the optical power does not exceed 200 nW. From Fig. 84 it is clear that for $p-i-n$ photodetector operation, the reverse bias should be greater than 4 V. Note the very small capacitance (≤ 0.1 pF) and the low operating voltage of this structure. These characteristics are extremely important for very low noise operation and for power-supply dimensions and reliability considerations in $p-i-n$ field-effect transistor (FET) receivers. The insert shows a typical $I-V$ characteristic of the channeling photodiode. We expect to be able to reduce further the dark current by improving the fabrication technology. The external quantum efficiency at $\lambda = 6328$ Å was $\sim 60\%$.

A final comment on the speed of the device is in order. Used as a photocapacitive detector, the device is expected to be slow, due to the long recombination time of electrons and holes. Operated instead as a $p-i-n$ photodetector, the speed is limited by the transit time along the layers. If light is absorbed in the depletion region near the p^+ contact, only electrons will contribute to the transit time. For a separation between the p^+ and n^+ regions of $\simeq 50\ \mu m$ and an applied voltage $V = 11.5$ V, the parallel field is 2×10^3 V cm^{-1}. [Note that a portion of the voltage (1.5 V) is used to deplete the layers and thus does not contribute to the parallel field.] At this field, the electron drift velocity in GaAs is close to the peak value of 2×10^7 cm sec^{-1}. This corresponds to a transit time of 250 psec.

It is worth stressing the unique features and important differences of this structure with respect to conventional $p-i-n$ diodes. The novel interdigitated $p-n$-junction scheme allows one to achieve an ultrasmall capacitance, largely independent of the detector area between the p^+ and n^+ regions and of the doping of the layers. Thus, the sensitive area can be maintained reasonably large and the doping moderately high. Note that conventional $p-i-n$ diodes require very low doping levels. The other very interesting aspect of this device is that the interdigitated scheme allows depletion of a large thickness of material while keeping the punchthrough voltage small.

Avalanche multiplication was not observed in the structure of Fig. 85 because of the relatively high dark current at voltages of $\gtrsim 30$ V. This is a technological problem presumably related to the quality of the interface between the etched and regrown regions.

f. Spatial Separation by Band-Edge Discontinuities

This structure, proposed by Tanoue and Sakaki (1982), uses the band-edge discontinuities rather than the built-in potentials to separate spatially electrons and holes in layers of different band gap. For this to occur, however, the electron affinities must satisfy the condition

$$\chi_A > \chi_B; \qquad E_{gA} + \chi_A > E_{gB} + \chi_B, \tag{264}$$

illustrated in Fig. 87. The condition given by Eq. (264) can be met by the lattice-matched combination of $GaAs_{1-y}Sb_y/In_{1-x}Ga_xAs$. By adding Al to each of the two constituent materials, one can control the band gap and create a variety of InGaAlAs/AlGaAsSb combinations without disturbing the lattice-matched condition. As specific examples, consider two types of quaternary–ternary alloy combinations, consisting of $In_{1-y}(Ga_{1-x}Al_x)_yAs$/GaAsSb and InGaAs/$Ga_{1-z}Al_zAs_{1-w}Sb_x$, which are lattice matched to InP.

Tanoue and Sakaki calculated the α/β ratio for these alloy systems using a simple Shockley model (1982). This gives, assuming equal pre-exponential factors for electrons and holes and the 3/2-band-gap rule for the ionization energies,

$$\alpha/\beta = \exp[-1.5(E_{gA} - E_{gB}) - \Delta E_v]/eF\lambda]. \tag{265}$$

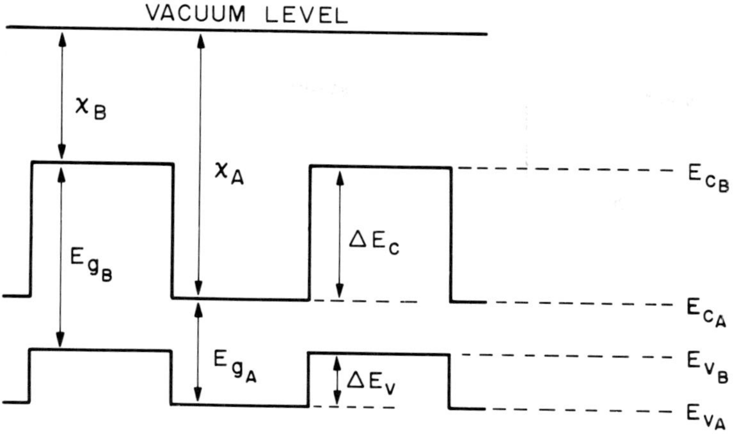

FIG. 87. Alternative scheme to spatially separate electrons from holes via band discontinuities in a multilayer avalanche photodiode.

The ΔE_v term takes into account the fact that when holes enter the wide-gap layer from material A, they are shifted upward in energy by an amount ΔE_v. Figure 88 shows the band-edge discontinuity situations of three extreme cases [(a) $x = 1$, $y = 0.48$; (b) $x = 0$, $y = 0.47$; and (c) $z = 1$, $w = 0.56$] and the results of the calculations for an electric field $F = 1 \times 10^5$ V cm^{-1}. Also shown in Fig. 88 is the calculation of α/β as a function of Al content x or z on the basis of Eq. (265). From Figs. 88d,e one can readily see that the α/β ratio can be controlled by the present scheme over an extremely wide range (10^2-10^3).

The dashed lines in Figs. 88d,e represent the three regions where some of the basic conditions are no longer satisfied; in the first region where $x > 0.9$, $\chi_A > \chi_B$ is no longer met, as shown in Fig. 88a. In the second and third regions, where $0 < z \lesssim 0.3$ and $0 < x \simeq 0.3$, the condition $1.5E_{gA} - 1.5E_{gB} + \Delta E_v \gg eF\lambda_0$ is no longer satisfied. Hence, the predominance of impact ionization caused by electrons (or holes) over that by holes (or electrons) is no longer guaranteed. In the latter two cases, where E_{gA} and E_{gB} are close to each other, electrons and holes are likely to play similarly important roles, and the α/β ratio becomes close to unity, as shown by the dotted lines in Figs. 88d,e.

19. Pseudoquaternary Semiconductors: A New Graded-Gap Superlattice High-Speed Avalanche Photodiode

In this last section we show how a superlattice can be used also to enhance the device speed. Capasso *et al.* (1984b) have demonstrated a new superlat-

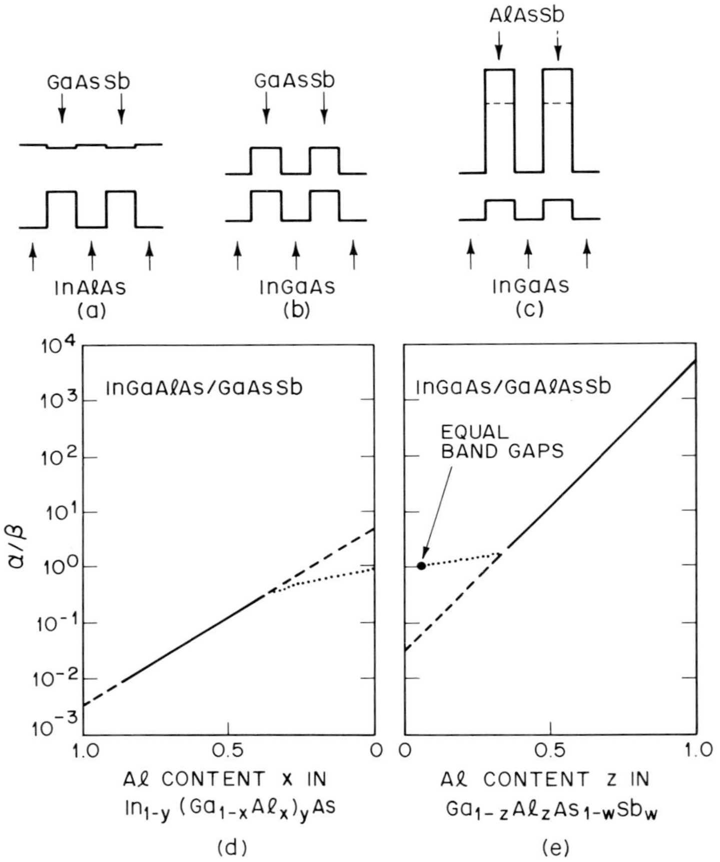

FIG. 88. Energy-band diagrams for the combinations of (a) GaAsSb/InAlAs, (b) GaAsSb/InGaAs, and (c) AlAsSb/InGaAs lattice matched to InP substrates; (d) and (e) show the ionization rate ratio α/β calculated for various ternary–quaternary combinations lattice matched to InP substrates. [From Tanoue and Sakaki (1982).]

tice (pseudoquaternary GaInAsP) capable of conveniently replacing conventional GaInAsP semiconductors in a variety of device applications. It is important to recall that these quaternary materials play a key role in optoelectronic devices for the 1.3–1.6 μm low-loss, low dispersion window of silica fibers. Here we shall discuss the application of pseudoquaternary semiconductors to long-wavelength avalanche photodiodes with separated absorption and multiplication regions (SAM APDs). SAM APDs are extensively discussed in the chapter by Pearsall and Pollack in this volume.

The concept of a pseudoquaternary GaInAsP semiconductor is easily explained. Consider a multilayer structure of alternated $Ga_{0.47}In_{0.53}As$ and

InP. If the layer thicknesses are sufficiently thin (typically a few tens of angstroms) one is in the superlattice regime. One of the consequences is that this novel material has now its own band gap, intermediate between that of $Ga_{0.47}In_{0.53}As$ and InP. In the limit of layer thicknesses of the order of a few monolayers the energy band gap can be approximated by the expression

$$\overline{E}_g = \frac{E_g(Ga_{0.47}In_{0.53}As)L(Ga_{0.47}In_{0.53}As) + E_g(InP)L(InP)}{L(Ga_{0.47}In_{0.53}As) + L(InP)}, \quad (266)$$

where the L's are the layer thicknesses.

These superlattices can be regarded as novel pseudoquaternary GaInAsP semiconductors. In fact, similarly to $Ga_{1-x}In_xAs_{1-y}P_y$ alloys, they are grown lattice matched to InP and their band gap can be varied between that of InP and that of $Ga_{0.47}In_{0.53}As$. The latter is done by adjusting the ratio of the $Ga_{0.47}In_{0.53}As$ and InP layer thicknesses. Pseudoquaternary GaInAsP is particularly suited to replace variable gap $Ga_{1-x}In_xAs_{1-y}P_y$. Such alloys are very difficult to grow since the mole fraction x (or y) must be continuously varied while maintaining lattice matching to InP.

Figure 89(a) shows a schematic of the energy-band diagram of undoped (nominally intrinsic) graded-gap pseudoquaternary GaInAsP. The structure consists of alternated ultrathin layers of InP and $Ga_{0.47}In_{0.53}As$ and was grown by a new vapor-phase epitaxial growth technique (Levitation Epitaxy) (Cox, 1984). Other techniques such as molecular-beam epitaxy or metal-loorganic chemical vapor deposition may also be suitable to grow such superlattices. From Fig. 89(a) it is clear that the duty factor of the InP and $Ga_{0.47}In_{0.53}As$ layer is gradually varied, while the period of the superlattice is kept constant. As a result the average composition and band gap [dashed lines in Fig. 89(a)] of the material are also spatially graded between the two extreme points (InP and $Ga_{0.47}In_{0.53}As$). In our structure both ten and twenty periods (1 period = 60 Å) were used. The InP layer thickness was linearly decreased with distance from ≃ 50 Å to ≃ 5 Å while correspondingly increasing the $Ga_{0.47}In_{0.53}As$ thickness to keep the superlattice period constant (= 60 Å).

The graded gap superlattice was incorporated in a long-wavelength InP/$Ga_{0.47}In_{0.53}As$ avalanche photodiode as shown in Fig. 89(b). This device is basically a photodetector with separate absorption ($Ga_{0.47}In_{0.53}As$) and multiplication (InP) layers and a high–low electric field profile (HI-LO SAM APD). This profile [(Fig. 89(c)] is achieved by a thin doping spike in the ultralow-doped InP layer and considerably improves the device performance compared to conventional SAM APDs. The $Ga_{0.47}In_{0.53}As$ absorption layer is undoped ($n \approx 1 \times 10^{15}/cm^3$) and 2.5 μm thick. The n^+ doping spike thickness and carrier concentration were varied in the 500–200 Å and 5×10^{17} cm^{-3} ranges respectively (depending on the wafer), while maintaining the same carrier sheet density (≃ 2.5×10^{12} cm^{-2}). The n^+ spike was

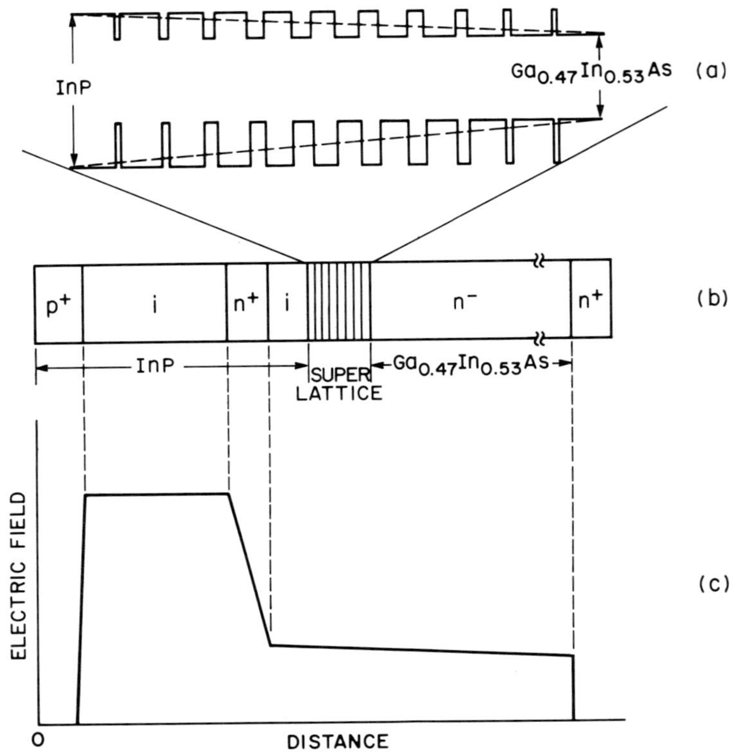

FIG. 89. (a) Energy band gap of the graded gap pseudo-quaternary GaInAsP. The thicknesses of the InP and of the $Ga_{0.47}In_{0.53}As$ are gradually varied between 5 and 55 Å while keeping constant the period of the superlattice (=60 Å). The dashed lines represent the average band gap. (b) Schematics of the HI-LO heterojunction avalanche photodiode incorporating the superlattice. (c) Electric field profile.

separated from the superlattice by an undoped 700–1000-Å-thick InP spacer layer. The p^+ region was defined by Zn diffusion in the 3-μm-thick low-carrier-density ($n^- \approx 10^{14}$ cm^{-3}) InP layer. The junction depth was varied from 0.8 to 2.5 μm. Similar devices, but without the superlattice region, were also grown.

Previous pulse response studies of conventional SAM APDs with abrupt $InP/Ga_{0.47}In_{0.53}As$ heterojunctions found a long (> 10 nsec) tail in the fall time of the detector due to the pile-up of holes at the heterointerface (Forrest *et al.*, 1982). This is caused by the large valence-band discontinuity ($\simeq$0.45 eV). It has been shown that this problem can be eliminated by inserting between the InP and $Ga_{0.47}In_{0.53}As$ region a $Ga_{1-x}In_xAs_{1-y}P_y$ layer of intermediate band gap (Campbell *et al.*, 1983). This quaternary layer is replaced, in our structure, by the $InP/Ga_{0.47}In_{0.53}As$ variable gap superlattice. This not only offers the advantage of avoiding the growth of the critical, independently lattice-matched, GaInAsP quarternary layer, but also may lead to an optimum "smoothing out" of the valence-band barrier for reproducible high-speed operation. This feature is essential for HI-LO SAM APDs since the heterointerface electric field is lower than in conventional SAM devices.

For the pulse response measurement we used a 1.5-μm GaInAsP driven by

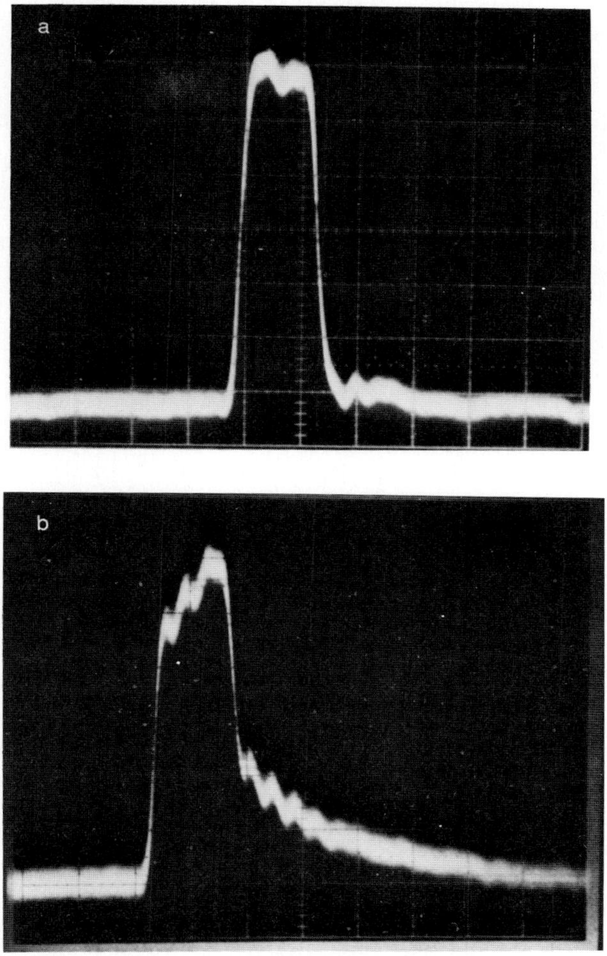

FIG. 90. Pulse response of HI-LO SAM APD at unity gain (a) with graded-gap superlattice and (b) without to $a \simeq 2$ nsec, $\lambda = 1.55$ μm laser pulse. The bias voltage is -65.5 V for both devices. The horizontal scale is 2 nsec/div.

a pulse pattern generator. Figure 90 shows the response to a 2-nsec laser pulse of a HI-LO SAM APD with (a) and without (b) a 1300-Å-thick superlattice. Both devices had similar doping profiles and breakdown voltages ($\simeq 80$ V) and were biased at -65.5 V. At this voltage the ternary layer was completely depleted in both devices and the measured external quantum efficiency was about 70%. The results of Fig. 90 were reproduced in many devices on several wafers. The long tail in Fig. 90(b) is due to the pile-up effect of holes associated with the abruptness of the heterointerface.

In the devices with the graded-gap superlattice [Fig. 90(a)] there is no long tail. In this case the height of the barrier seen by the holes is no more the valence-band discontinuity ΔE_V but

$$\Delta E = \Delta E_V - e\varepsilon_i L, \tag{267}$$

where ε_i is value of the electric field at the InP/superlattice interface and L the thickness of the pseudoquaternary layer.

The devices are biased at voltage such that $\varepsilon_i > \Delta E_V / eL$ so that $\Delta E = 0$ and no trapping occurs.

In the devices with no superlattice instead $\Delta E = \Delta E_V$ for every ε_i so that relatively long tails in the pulse response are observed at all voltages.

Acknowledgments

The author is very grateful to Prof. Karl Hess of the University of Illinois for many stimulating conversations. These have been of conseiderable help in preparing the theoretical section of the chapter. Useful discussion with K. K. Thornber, T. P. Pearsall, R. E. Nahory, M. A. Pollack, G. F. Williams, W. T. Tsang, G. B. Bachelet, G. A. Baraff, J. Chelikowsky, G. E. Stillman, B. Ridley and H. Sakaki are also gratefully acknowledged.

References

Anderson, C. L., and Crowell, C. R. (1972). *Phys. Rev. B* **5,** 2267.
Anderson, L. K., and McMurty, B. J. (1966). *Proc. IEEE.* **54,** 1335.
Ando, H., and Kanbe, H. (1981). *Solid-State Electron.* **24,** 629.
Antončik, H. (1967). *Czech. J. Phys. B* **17,** 735.
Armiento, C. A., Groves, S. H., and Hurwitz, C. E. (1979). *Appl. Phys. Lett.* **35,** 333.
Armiento, C. A., and Groves, S. H. (1983). *Appl. Phys. Lett.* **43,** 198.
Arnold, D. A., Ketterson, T., Henderson, T., Klem, J., and Morkoc, H. (1984). *Appl. Phys. Lett.* **45,** 1237.
Arora, V. K. (1983). *J. Appl. Phys.* **54,** 824.
Arutyunyan, V. M., and Petrosyan, S. G. (1980). *Sov. Phys. Semicond. (Engl. Transl.)* **14,** 1188.
Aspnes, D. E. (1976). *Phys. Rev. B* **14,** 5331.
Baertsch, R. D. (1967). *J. Appl. Phys.* **38,** 4267.
Balliger, R. A., Major, K. G., and Mallinson, J. R. (1973). *J. Phys. C.* **6,** 2573.
Baraff, G. A. (1962). *Phys. Rev.* **128,** 2507.
Baraff, G. A. (1964). *Phys. Rev.* **133,** A26.
Berenz, J. J., Kinoshita, J., Hierl, R. L., and Lee, C. A. (1979). *Electron. Lett.* **15,** 150.

Barker, J. R. (1973). *J. Phys. C* **6,** 2663.
Barker, J. R. (1980). *In* "Physics of Nonlinear Transport in Semiconductors" (D. K. Ferry, J. R. Barker, and C. Jacoboni, eds.), Chap. 5, p. 424. Plenum, New York.
Beni, G., and Capasso, F. (1979). *Phys. Rev. B* **19,** 2197.
Blauvelt, H., Margalit, S., and Yariv, A. (1982). *Electron Lett* **18,** 375.
Brennan, K., and Hess, K. (1984a). *Solid-State Electron.* **27,** 347.
Brennan, K., and Hess, K. (1984b). *Phys. Rev. B* **29,** 5581.
Brennan, K., Hess, K., and Chang, W. C. (1985a). *J. Appl. Phys.* **57,** 1971.
Brennan, K., Wang, T., and Hess, K. (1985b). *IEEE Electron Device Lett.* **EDL-G,** 199.
Bulman, G. E., Cook, L. W., and Stillman, G. E. (1982). *Appl. Phys. Lett.* **40,** 589.
Bulman, G. E., Robbins, V. M., Brennan, K. F., Hess, K., and Stillman, G. E. (1983). *IEEE Electron Dev. Lett.* **EDL-4,** 181.
Burroughs, M. S., Hess, K., Holonyak, N., Jr., Laidig, W. D., and Stillman, G. E. (1982). *Solid-State Electron.* **25,** 161.
Campbell, J. C., Dentai, A. G., Holden, W. S., and Kasper, B. L. (1983). *Electron. Lett.* **9,** 818.
Camphausen, D. L., and Hearn, C. J. (1972). *Phys. Status Solidi B* **50,** K139.
Capasso, F. (1982a). *Electron. Lett.* **18,** 12.
Capasso, F. (1982b). *IEEE Trans. Electron Devices* **ED-29,** 1388.
Capasso, F. (1983a). *J. Vac. Sci. Technol.* **B1,** 457.
Capasso, F. (1983b). *Surf. Sci.* **132,** 527.
Capasso, F. (1983c). U.S. Patent 4383269.
Capasso, F. (1983d). *IEEE Trans. Nucl. Sci.* **NS-30,** 424.
Capasso, F. (1984). *Surf. Sci.* **142,** 513.
Capasso, F., and Bachelet, G. B. (1980). *Tech. Dig.—Int. Electron Devices Meet., 1980, Washington, D.C.,* p. 633.
Capasso, F., and Tsang, W. T. (1982). *Tech. Dig.—Int. Electron Devices Meet., 1982, San Francisco,* 334.
Capasso, F., Nahory, R. E., Pollack, M. A., and Pearsall, T. P. (1977). *Phys. Rev. Lett.* **39,** 723.
Capasso, F., Nahory, R. E., and Pollack, M. A. (1979a). *Solid-State Electron.* **22,** 977.
Capasso, F., Nahory, R. E., and Pollack, M. A. (1979b). *Electron. Lett.* **15,** 117.
Capasso, F., Pearsall, T. P., and Thornber, K. K. (1981a). *IEEE Electron Device Lett.* **EDL-2,** 1295.
Capasso, F., Tsang, W. T., Hutchinson, A. L., and Williams, G. F. (1981b). *Tech. Dig.—Int. Electron Devices Meet., 1981, Washington, D.C.,* p. 284.
Capasso, F., Pearsall, T. P., Thornber, K. K., Nahory, R. E., Pollack, M. A., Bachelet, G. B., and Chelikowski, J. R. (1982a). *J. Appl. Phys.* **53,** 3324.
Capasso, F., Tsang, W. T., Hutchinson, A. L., and Foy, P. W. (1982b). *Conf. Ser.—Inst. Phys.* **63,** 473.
Capasso, F., Tsang, W. T., Hutchinson, A. L., and Williams, G. F. (1982c). *Appl. Phys. Lett.* **40,** 38.
Capasso, F., Tsang, W. T., and Williams, G. F. (1982d). *Proc. Soc. Photo-Opt. Instrum. Eng.* **340,** 50.
Capasso, F., Logan, R. A., and Tsang, W. T. (1982e). *Electron Lett.* **18,** 760.
Capasso, F., Logan, R. A., Tsang, W. T., and Hayes, J. R. (1982f). *Appl. Phys. Lett.* **41,** 944.
Capasso, F., Tsang, W. T., and Williams, G. F. (1983). *IEEE Trans. Electron Devices* **ED-30,** 38.
Capasso, F., Mohammed, K., Alavi, K., Cho, A. Y., and Foy, P. W. (1984a). *Appl. Phys. Lett.* **45,** 968.
Capasso, F., Tsang, W. T., and Williams, G. F. (1984b). U.S. Patent 44764777.
Capasso, F., Cox, H. M., Hutchinson, A. L., Olsson, N. A., and Hummel, S. G. (1984c). *Appl. Phys. Lett.* **45,** 1193.

Capasso, F., Cho, A. Y., Mohammed, K., and Foy, P. W. (1985a). *Appl. Phys. Lett.* **46,** 664.
Chang, Y. C., Ting, D. Z. Y., Tang, J. Y., and Hess, K. (1983). *Appl. Phys. Lett.* **42,** 76.
Chelikowsky, J. R., and Cohen, M. L. (1976). *Phys. Rev. B* **14,** 556.
Chen, P. C. Y. (1972). *IEEE Trans. Electron Devices* **19,** 285.
Chin, R., Holonyak, N., Jr., Stillman, G. E., Tang, J. T., and Hess, K. (1980). *Electron Lett.* **16,** 467.
Chynoweth, A. G. (1968). *In* "Semiconductors and Semimetals" (R. K. Willardson and A. C. Beer, eds.), Chap. 4, p. 263. Academic Press, New York.
Chynoweth, A. G., and McKay, K. G. (1957). *Phys. Rev.* **108,** 29.
Chuenkov, V. A. (1967). *Sov. Phys.—Solid State (Engl. Trans.)* **9,** 35.
Chwang, R., and Crowell, C. R. (1976). *Solid State Commun.* **20,** 169.
Chwang, R., Kao, C. W., and Crowell, C. R. (1979). *Solid-State Electron.* **22,** 599.
Cohen, M. L, and Bergstresser, T. K. (1966). *Phys. Rev.* **141,** 789.
Conwell, E. M. (1967). "High Field Transporting Semiconductors." Academic Press, New York.
Cook, L. W., Bulman, G. E., and Stillman, G. E. (1982). *Appl. Phys. Lett.* **40,** 589.
Cox, H. M. (1984). *J. Crystal Growth* **69,** 641.
Crowell, C. R. (1983). (Private communication.)
Crowell, C. R., and Sze, S. M. (1966). *Appl. Phys. Lett.* **9,** 242.
Dai, B. T., and Chang, C. Y. (1971). *J. Appl. Phys.* **42,** 5198.
Decker, D. R., and Dunn, C. N. (1975). *J. Electron. Mater.* **4,** 527.
Devreese, J. T., van Welzenis, R. G., and Errard, R. P. (1982). *Appl. Phys.* **A29,** 125.
Dmitriev, A. P., Mikhailova, M. P., and Yassievich, I. N. (1982). *Phys. Status Solidi B* **113,** 125.
Dmitriev, A. P., Konstantinov, A. O., Litvin, D. P., and Sankin, V. I. (1983). *Soviet Phys.—Semicond. (Engl. Transl.)* **17,** 685.
Dohler, G. (1972). *Phys. Status Solidi B* **52,** 533.
Dubrovskii, G. B. (1972). *Sov. Phys.—Solid State (Engl. Trans.)* **13,** 2107.
Duh, C. Y., and Moll, J. L. (1967). *IEEE Trans. Electron Devices* **14,** 46.
Dumke, W. P. (1968). *Phys. Rev.* **167,** 783.
Emmons, R. B. (1967). *J. Appl. Phys.* **38,** 3705.
Faurie, J. P., Million, A., and Piaguet, J. (1982). *Appl. Phys. Lett.* **41,** 713.
Fawcett, W., Boardman, A. D., and Swain, S. (1970). *J. Phys. Chem. Solids* **31,** 1963.
Ferry, D. K., Barker, J. R., and Jacoboni, C. (1980). "Physics of Nonlinear Transport in Semiconductors." Plenum, New York.
Forrest, S. R., and Kim, O. K. (1981). *J. Appl. Phys.* **52,** 5838.
Forrest, S. R., DiDomenico, M., Jr., Smith, R. G., and Stocker, H. J. (1980). *Appl. Phys. Lett.* **36,** 580.
Forrest, S. R., Kim, O. K., and Smith, R. G. (1982). *Appl. Phys. Lett.* **41,** 95.
Gavrusko, V. V., Kosogov, O. V., and Lebedeva, V. D. (1978). *Sov. Phys.—Semicond. (Engl. Transl.)* **12,** 1398.
Gordon, J. D., Nahory, R. E., Pollack, M. A., and Warlock, J. M. (1979). *Electron. Lett.* **15,** 518.
Grant, W. N. (1973). *Solid-State Electron.* **16,** 1189.
Gribnikov, Z. S. (1978). *Sov. Phys.—JETP (Engl. Trans.)* **47,** 1099.
Guldner, Y., Bastard, G., Vieren, J. P., Voss, M., Faurie, J. P., and Million, A. (1983). *Tech. Dig.—Int. Conf. Electron Properties Two-Dimensional Systems, 5th, Oxford, England, 1983,* p. 618.
Hauser, J. R. (1978). *Appl. Phys. Lett.* **33,** 351.
Hess, K. (1981). *IEEE Electron Dev. Lett.* **EDL-2,** 297.
Hess, K., Tang, J. Y., Brennan, K., Shichijo, H., and Stillman, G. E. (1982). *J. Appl. Phys.* **53,** 3324.

Hess, K., Brennan, K., and Stillman, G. E. (1983). Private communication.
Hildebrand, O., Kuebart, W., and Pilkuhn, M. H. (1980). *Appl. Phys. Lett.* **37,** 801.
Hildebrand, O., Kuebart, W., Benz, K. W., and Pilkuhn, M. H. (1981). *IEEE J. Quantum Electron.* **QE-17,** 284.
Ito, M., Kagawa, S., Kaneda, T., and Yamaoka, T. (1978a). *J. Appl. Phys.* **49,** 4607.
Ito, M., Kaneda, T., Nakajima, K., Toyama, Y., Yamaoka, T., and Kotani, T. (1978b). *Electron. Lett.* **14,** 418.
Kafka, H. J., and Hess, K. (1981). *IEEE Trans. Electron Devices* **ED-28,** 831.
Kane, E. O. (1967). *Phys. Rev.* **159,** 624.
Kaneda, T., Mikawa, T., and Toyama, Y. (1979). *Appl. Phys. Lett.* **34,** 692.
Kao, C. W., and Crowell, C. R. (1980). *Solid-State Electron.* **23,** 881.
Kash, K., and Wolff, P. A. (1983). *Appl. Phys. Lett.* **42,** 173.
Keldysh, L. V. (1958). *Sov. Phys. JETP (Engl. Transl.)* **34,** 655.
Keldysh, L. V. (1960). *Sov. Phys. JETP (Engl. Transl.)* **10,** 509.
Keldysh, L. V. (1965). *Sov. Phys. JETP (Engl. Transl.)* **21,** 1135.
Kroemer, H. (1957). *RCA Rev.* **18,** 332.
Kyuregyan, A. S. (1976). *Sov. Phys.—Semicond. (Engl. Transl.)* **10,** 410.
Law, H. D., and Lee, C. A. (1978). *Solid-State Electron.* **21,** 331.
Law, H. D., Nakano, K., Tomasetta, L. R., and Harris, J. S. (1978). *Appl. Phys. Lett.* **33,** 948.
Lee, C. A., Logan, R. A., Batdorf, R. L., Keimack, J. J., and Wiegmann, W. (1964). *Phys. Rev.* **134,** A761.
Littlejohn, M. A., Hauser, J. R., and Glisson, T. H. (1977). *Appl. Phys. Lett.* **48,** 4587.
Littlejohn, M. A., Glisson, T. H., and Hauser, J. R. (1982). *In* "GaInAsP Alloy Semiconductors" (T. P. Pearsall, ed.), p. 243. Wiley, New York.
Logan, R. A., and Chinoweth, A. G. (1962). *J. Appl. Phys.* **33,** 1649.
Logan, R. A., and White, H. G. (1965). *Appl. Phys. Lett.* **36,** 3945.
Logan, R. A., and Sze, S. M. (1966). *J. Phys. Soc. Jpn Suppl.* **21,** 434.
McIntyre, R. J. (1966). *IEEE Trans. Electron Devices* **ED-13,** 164.
McIntyre, R. J. (1972). *IEEE Trans. Electron Devices* **ED-19,** 703.
Meyer, N. I., and Jorgensen, M. H. (1965). *J. Phys. Chem. Solids* **26,** 823.
Mikawa, T., Kagawa, S., Kaneda, T., Toyama, Y., and Mikami, O. (1980). *Appl. Phys. Lett.* **37,** 387.
Mikhailova, M. P., Smirnova, N. N., and Slobodchikov, S. V. (1976). *Sov. Phys.—Semicond. (Engl. Transl.)* **10,** 509.
Miller, S. L. (1955). *Phys. Rev.* **99,** 1234.
Miller, R. C., and Tsang, W. T. (1981). *Appl. Phys. Lett.* **4,** 334.
Miller, R. C., Kleinmann, D. A., and Gossard, A. C. (1984). *Phys. Rev. B* **29,** 7085.
Mon, K. K., and Hess, K. (1982). *Solid-State Electron.* **25,** 491.
Mönch, W. (1969). *Phys. Status Solidi* **36,** 9.
Naganuma, M., Suzuki, Y., and Okamoto, H. (1981). *Conf. Ser.—Inst. Phys.* **63,** 125.
Okuto, Y., and Crowell, C. R. (1972). *Phys. Rev. B* **6,** 3076.
Okuto, Y. and Crowell, C. R. (1976). *Solid-State Electron.* **18,** 161.
Osaka, F., Mikawa, T., and Kaneda, T. (1984a). *Appl. Phys. Lett.* **45,** 654.
Osaka, F., Mikawa, T., and Kaneda, T. (1984b). *Appl. Phys. Lett.* **45,** 292.
Pearsall, T. P. (1979). *Appl. Phys. Lett.* **35,** 168.
Pearsall, T. P. (1980). *Appl. Phys. Lett.* **36,** 218.
Pearsall, T. P., and Papuchon, M. (1978). *Appl. Phys. Lett.* **33,** 640.
Pearsall, T. P., Nahory, R. E., and Pollack, M. A. (1975). *Appl. Phys. Lett.* **27,** 330.
Pearsall, T. P., Nahory, R. E., and Pollack, M. A. (1976). *Appl. Phys. Lett.* **28,** 403.
Pearsall, T. P., Nahory, R. E., and Chelikowski, J. R. (1977a). *Conf. Ser.—Inst. Phys.* **33b,** 331.

Pearsall, T. P., Nahory, R. E., and Chelikowski, J. R. (1977b). *Phys. Rev. Lett.* **39,** 295.
Pearsall, T. P., Capasso, F., Nahory, R. E., Pollack, M. A., and Chelikowski, J. R. (1978). *Solid-State Electron.* **21,** 297.
People, R., Wecht, K. W., Alavi, K., and Cho, A. Y. (1983). *Appl. Phys. Lett.* **43,** 118.
Reuter, H., and Hübner, K. (1971). *Phys. Rev. B* **4,** 2572.
Ridley, B. K. (1977). *J. Appl. Phys.* **48,** 754.
Ridley, B. K. (1983a). *J. Phys. C* **16,** 3373.
Ridley, B. K. (1983b). *J. Phys. C.* **16,** 4733.
Robbins, D. J. (1980a). *Phys. Status Solidi B* **97,** 9.
Robbins, D. J. (1980b). *Phys. Status Solidi B* **97,** 387.
Robbins, D. J. (1980c). *Phys. Status Solidi B* **98,** 11.
Roy, S. K., and Ghosh, R. (1975). *Solid-State Electron.* **18,** 1035.
Shaw, D., and McKell, A. D. (1983). *Br. J. Appl. Phys.* **14,** 295.
Shichijo, H., and Hess, K. (1981). *Phys. Rev. B.* **23,** 4197.
Shichijo, H., Hess, K., and Stillman, G. E. (1980). *Electron Lett.* **16,** 208.
Shichijo, H., Hess, K., and Stillman, G. E. (1981). *Appl. Phys. Lett.* **38,** 89.
Shockley, W. (1961). *Solid-State Electron.* **2,** 35.
Shulman, J. N., and McGill, T. C. (1979). *Appl. Phys. Lett.* **34,** 663.
Stillman, G. E. (1977). *Conf. Ser.—Inst. Phys.* **33a,** 185.
Stillman, G. E., and Wolfe, C. M. (1977). *In* "Semiconductors and Semimetals" (R. K. Willardson and A. C. Beer, eds.), Vol. 12, 291. Academic Press, New York.
Stillman, G. E., Wolfe, C. M., Foyt, A. G., and Lindley, W. T. (1974). *Appl. Phys. Lett.* **24,** 8.
Stillman, G. E. Cook, L. W., Bulman, G. E., Tabatabaie, N., Chin, R., and Dapkus, P. D. (1982). *IEEE Trans. Electron Devices* **ED-29,** 1355.
Sutherland, A. D. (1980). *IEEE Trans. Electron Devices* **ED-27,** 1299.
Sze, S. (1969). "Physics of Semiconductor Devices," Wiley (Interscience), New York.
Tabatabaie, N., Robbins, V. M., Brennan, K. F., Hess, H., and Stillman, G. E. (1983). *IEEE Trans. Electron Devices* **ED-30,** 1608.
Takanashi, Y., and Horikoshi, Y. (1979). *Jpn. J. Appl. Phys.* **18,** 2173.
Takanashi, Y., and Horikoshi, Y. (1981). *Jpn. J. Appl. Phys.* **20,** 1907.
Tang, J. Y., and Hess, K. (1983a). *J. Appl. Phys.* **54,** 5139.
Tang, J. Y., and Hess, K. (1983b). *J. Appl. Phys.* **54,** 5145.
Tanoue, T., and Sakaki, H. (1982). *Appl. Phys. Lett.* **41,** 67.
Thornber, K. K. (1978). *Solid-State Electron.* **21,** 259.
Thornber, K. K. (1981). *J. Appl. Phys.* **52,** 279.
Umebu, I., Choudhury, A. N. M. M., and Robson, P. N. (1980). *Appl. Phys. Lett.* **36,** 302.
Van Overstraeten, R., and DeMan, H. (1970). *Solid-State Electron.* **13,** 583.
Van Welzenis, R. G. (1981). *Appl. Phys.* **A26,** 157.
Van Welzenis, R. G., and de Zeew, W. C. (1983). *Appl. Phys.* **A30,** 151.
Vèrié, C., Raymond, F., Besson, F., and Nguyen, Duy (1982). *J. Cryst, Growth* **59,** 342.
Vodakov, A. Y., Konstantinov, A. O., Litvin, D. P., and Sankin, V. I. (1981). *Sov. Tech. Phys. Lett. (Engl. Transl.)* **7,** 301.
Wang, W. I., Mendez, E. E., and Stern, F. (1984). *Appl. Phys. Lett.* **45,** 639.
Williams, G. F., Capasso, F., and Tsang, W. T. (1982). *IEEE Electron Device Lett.* **EDL-3,** 71.
Woods, M. H., Johnson, W. C., and Lampert, M. A. (1973). *Solid-State Electron.* **16,** 381.
Wolff, P. A. (1954). *Phys. Rev.* **95,** 1415.
Zhingarev, M. Z., Korol'kov, V. I., Mikhailova, M. P., Sazonov, V. V., and Tretyakov, D. N. (1980). *Sov. Phys.—Semicond. (Engl. Transl.)* **14,** 801.
Zhingarev, M. Z., Korol'kov, V. I., Mikhailova, M. P., and Sazonov, V. V. (1981). *Sov. Tech. Phys. Lett. (Engl. Transl.)* **7,** 637.

CHAPTER 2

Compound Semiconductor Photodiodes

T. P. Pearsall

AT&T BELL LABORATORIES
MURRAY HILL, NEW JERSEY

M. A. Pollack

AT&T BELL LABORATORIES
CRAWFORD HILL LABORATORY
HOLMDEL, NEW JERSEY

† The nomenclature in current use for the alloy system GaInAsP varies from author to author. In this chapter, the element ordering is written to conform with the current rules established by the International Union of Pure and Applied Chemistry (IUPAC) and with the current practice of the Chemical Abstracts Service of the American Chemical Society. The element ordering for HgCdTe conforms instead to almost universal usage.

ISBN 0-12-752153-4

I. Introduction

1. Optical Communications: Considerations for Photodetector Technology

The advent of optical-fiber transmission in telecommunications means that vast increases in information capacity are now available to the end user of the communications network. By the 1990s we can expect to see applications covering the entire telecommunications spectrum from undersea long-distance cable to new high-bandwidth services, such as digital video and high-speed data transmission. *Bell Telephone Magazine* (1979) reported that the capacity to carry 10% of all telephone traffic in the United States by glass fibers would be in place at the end of 1982—almost entirely the result of the installation of optical fibers in high-traffic, intercity trunks (Pearsall, 1982). Each different application places special demands on detector technology and performance. For submarine cable, stringent requirements to assure device reliability are of primary concern. In wide-band, user-premises equipment, robust and inexpensive devices are needed. In intercity trunking, high speed and sensitivity are the primary objectives. It would be ideal to develop a single detector technology that could satisfy all of these needs. The current state of the art in compound-semiconductor photodiodes suggests that this goal may indeed be achieved.

If compromises have to be made, however, it is clear that photodetector technology will be driven by the requirements of the wide-band user-premises network, because it is here that the vast majority of components will be used. Whether separate technologies and manufacturing capabilities are developed to satisfy a small number of specialized applications, such as undersea cable where fewer than 10^5 detectors will be needed, is essentially an economic question that is too complex to be answered simply and definitively here. In addition to technical considerations alone, the time and cost of developing specialized photodetectors must be weighed against the marginal increase in revenues that would be generated by a system equipped with these devices—a figure that depends as much on the likely level of traffic as on the promise of competitive pressure. In this chapter we review the performance and technology-related considerations that would be used to make this decision.

The appropriate wavelength for optical fiber telecommunications is determined primarily by the absorption spectrum of the fiber. Wavelength dispersion of the fiber is an additional consideration in wide-band, long-distance transmission. Happily, thanks to the compensating effects of mode-versus-material dispersion (White and Nelson, 1979), it is possible to minimize dispersive effects over much of the wavelength range of interest for monomode silica fibers; i.e., from 1.3–1.6 μm. It is to be stressed that the

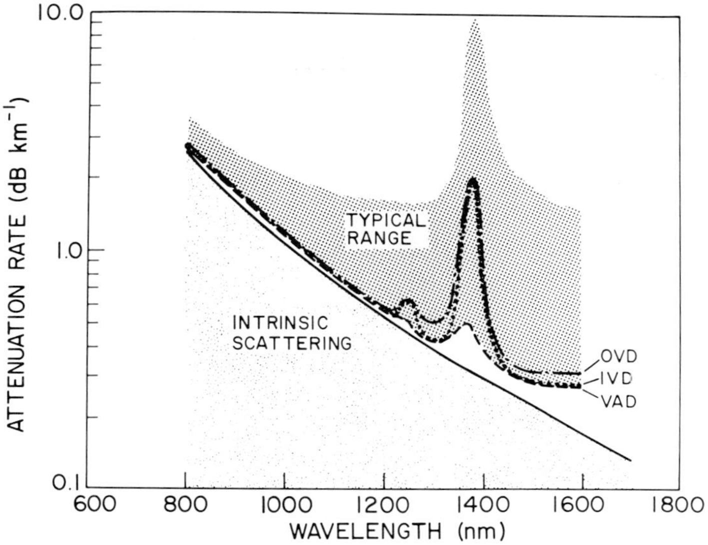

FIG. 1. Attenuation versus transmission wavelength for multimode optical fibers grown by various growth processes: vapor-phase axial deposition (VAD), inside vapor deposition (IVD), and outside vapor deposition (OVD). [From Keck (1982).]

situation for multimode fibers is different. Dispersion-free transmission is possible only near 1.32 μm. In Figs. 1 and 2, we show the attenuation spectrum for some state-of-the-art multimode and monomode optical fibers. It can be seen that the scattering loss is close to the fundamental Rayleigh limit (Ishida *et al.,* 1982; Keck, 1982). In the region of minimum attenuation near 1.55 μm, the fiber loss is less than 0.2 dB/km—one order of magnitude below that at 0.82 μm, the wavelength of GaAs lasers and light-emitting diodes (LEDs). This difference is crucial for practical optical fiber transmission, so much so that no more very long-distance optical-fiber links are now being planned using GaAs sources and Si detectors. The development of compound semiconductors whose optical properties can be manipulated to make both photodetectors and light sources for the 1.0–1.6-μm region of the spectrum is an essential element of this new direction.

2. Compound Semiconductors for Optical-Fiber Communications

With the exception of Ge, no naturally occurring semiconductors are suitable either as detectors or emitters in the 1.0–1.6-μm spectral region. Germanium is an indirect-gap semiconductor (with band-gap energy $E_g = 0.6$ eV) and is therefore unsuitable for emitter applications. The small

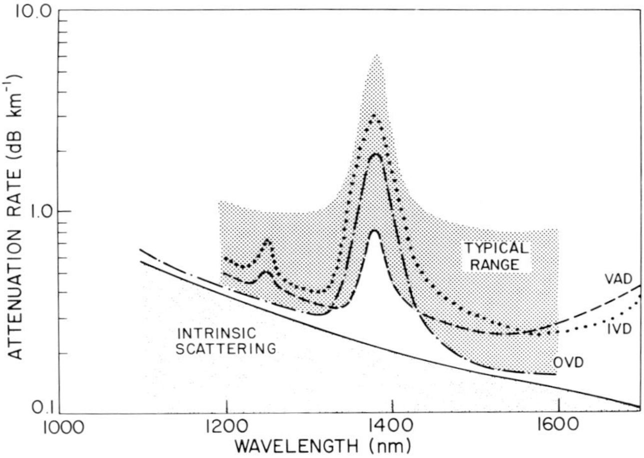

FIG. 2. Attenuation versus transmission wavelength for monomode optical fibers (compare Fig. 1). [From Keck (1982).]

value of its band-gap energy may lead to excessive noise in detector applications at room temperature and above. Promising alternatives to Ge are the new compound semiconductor alloys of known binary compounds, most of which are direct-gap materials. The resulting semiconductor alloys can be used in applications both as light emitters and as detectors. By adjusting the alloy composition, the band gap can be made to coincide with the wavelength of interest for optical-fiber transmission.

There are three principal alloy systems that are appropriate for this application: AlGaSb, GaInAsP, and HgCdTe. These alloys share the common property that each can be grown nearly lattice matched on a commercially available binary substrate. In Fig. 3 we show how the composition of these alloys can be varied to match the regions of optimum optical transmission by fibers at 1.3 and 1.6 μm. In the case of the two ternary alloys, there is a slight lattice mismatch with respect to the binary substrate. This mismatch can be eliminated by the addition of a small amount of a fourth component; e.g., in AlGaSb, arsenic may be added to keep the lattice parameter fixed. The third system is a true quaternary alloy that bridges InP ($E_g = 1.35$ eV) to $Ga_{0.47}In_{0.53}As$ ($E_g = 0.75$ eV) while simultaneously maintaining a fixed lattice parameter ($a_0 = 5.869$ Å). Of these three systems, only for GaInAsP has a substantial technology been developed that permits the fabrication of high-performance lasers, LEDs, and photodiodes.

Of the three principal alloy systems for lightwave applications that are shown in Fig. 3, HgCdTe has long been known as a material for photodetec-

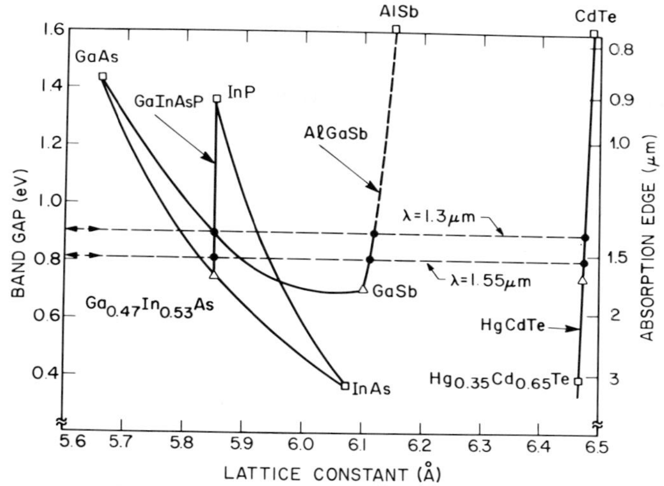

FIG. 3. Band gap versus lattice parameter for the principal compound semiconductors of interest for optical-fiber communication applications. AlGaSb and HgCdTe are ternary alloys whose lattice parameter changes very little with band gap. For the quaternary GaInAsP, the lattice parameter can in principle be maintained constant over the entire band-gap range between InP and $Ga_{0.47}In_{0.53}As$; (△) detector materials; (●) emitter materials.

tor applications in the 8–14-μm-wavelength range (e.g., Long and Schmit, 1970). A considerable industrial capability exists worldwide to fabricate detectors and detector arrays from this material. Although the present fabrication technology involves the processing of bulk-grown crystals, heteroepitaxial structures of HgCdTe grown on CdTe substrates are a subject of current materials research. Such an effort may lead to the development of double-heterostructure lasers and LEDs, as well as photodiodes (Pichard *et al.,* 1982) with band gaps appropriate for optical-fiber telecommunications.

Fast and responsive photodetectors at 1.3 and 1.55 μm have been demonstrated using $Al_xGa_{1-x}Sb$ grown epitaxially on GaSb substrates. The lattice mismatch present in this system is small, as shown in Fig. 3, and it may be eliminated totally by the addition of a small amount of arsenic. The $Al_xGa_{1-x}Sb$ system also has an electronic energy-level structure with favorable consequences for avalanche photodiodes. For the composition with $x \approx 0.065$, the band gap becomes equal in size to the energy splitting in the valence band. This peculiar feature, also present in InAs (Mikhailova *et al.,* 1976a) and HgCdTe, reduces the threshold energy for hole-initiated impact ionization to its minimum value (Pearsall *et al.,* 1977) and increases the probability of an impact-ionization collision to unity (Mikhailova *et al.,* 1976b; Hildebrand *et al.,* 1980). These features improve the signal-to-noise ratio of the avalanche gain process. As we shall show in Section 4, the sensitivity improvement of the receiver increases approximately as the $\frac{1}{3}$ power of the ratio of the individual ionization rates of holes and electrons. For the specific case of AlGaSb, this feature, if fully exploited, would lead to a doubling of the receiver sensitivity.

With the demonstration of the first long-wavelength photodiode made from the GaInAsP alloy system (Pearsall and Hopson, 1978), it was apparent that the performance was far superior to existing alternatives, such as Ge or PbS. The common technology for both optical sources and detectors leads to considerable cost reductions for any serious commercial exploitation of this material. In addition, the economic benefits of common technology impose no compromise on device performance. For these reasons the development of the GaInAsP alloys has been rapid, and this system appears to be the preferred choice at the present time for components for optical-fiber telecommunications.

3. Specifications for $p-i-n$ Photodiode Performance

The central issue for photodetectors in optical-fiber telecommunications is the detection of weak optical signals. In some applications, the length of an optical link will be determined by the minimum optical signal that can be detected at the transmitted bandwidth or bit rate. The minimum detectable signal level is determined entirely by the system noise. The photodiode operating with unity gain itself contributes shot noise both from the photocurrent and from the dark current. In addition, thermal noise from the preamplifier, referred to the photodiode, contributes a noise term that depends on the photodiode capacitance. The goal in photodiode design is to reduce these contributions to a level below that of the other noise contributions of the optical receiver. In the case of avalanche gain, the goal is to further reduce the dark-current contribution so that it is about one order of magnitude less than the primary photocurrent. Other desirable operating characteristics, such as high speed and near-100% quantum efficiency, must also be achieved.

In Fig. 4 we show the measured receiver sensitivity of state-of-the-art preamplifiers for optical communications as a function of the bit rate of a digitally modulated bit stream (Yamada *et al.*, 1978, 1982; Abbott and Muska, 1979; Smith *et al.*, 1979; Williams and Leblanc, 1983). The amplifiers are all of a high impedance front-end design, although the details of the equalization and feedback impedance schemes vary with the individual receiver design. At bit rates below 100 Mbit/sec, field-effect transistors (FETs) appear to be preferred for transistors in these preamplifiers, whereas above this bit rate, both Si bipolars and GaAs FETs are used. Despite the variety in receiver design, the performance figures describe a reasonably

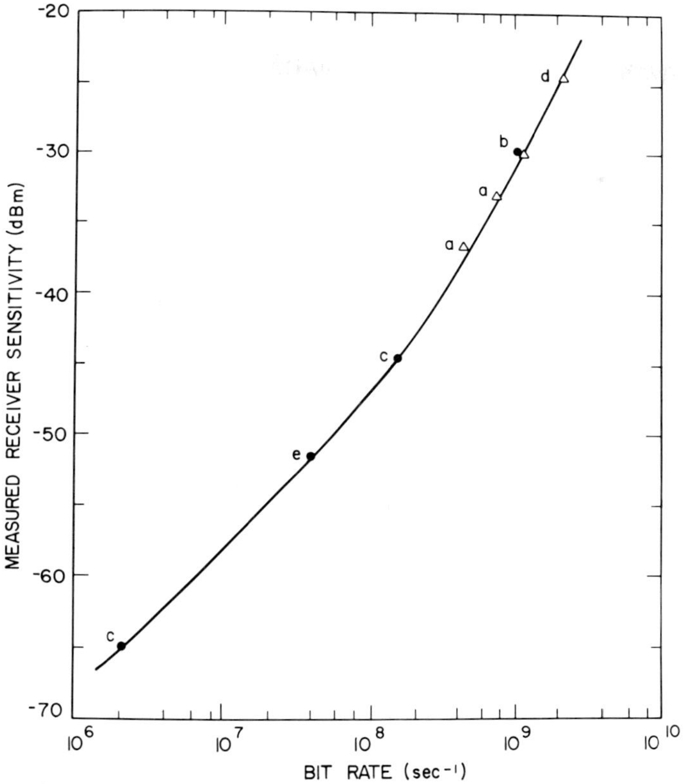

FIG. 4. Receiver sensitivity for state-of-the-art preamplifiers used in optical-fiber telecommunications. All receivers are used to detect a digital bit stream with a bit-error rate of 10^{-9}: (a) Yamada *et al.* (1978); (b) Abbott and Muska (1979); (c) Smith *et al.* (1979); (d) Yamada *et al.* (1982); (e) Williams and Leblanc (1983). ●, FET; △ Si bipolar.

clear picture of receiver performance: -70 dBm (decibels above 1 mW) at 1 Mbit/sec, -50 dBm at 100 Mbit/sec, and -30 dBm at 1 Gbit/sec.

In Fig. 5 we show the equivalent diode current (photocurrent I_p plus reverse dark current I_d) whose shot-noise power would equal the preamplifier noise power (Pearsall, 1981). The shot-noise power depends linearly on bandwidth B:

$$\langle i_n^2 \rangle = 2q(I_p + I_d)B, \tag{1}$$

where q is the electronic charge. The rapid increase in the noise-equivalent dark current above 100 Mbit/sec reflects the rapid rise in amplifier-generated noise above this bit rate.

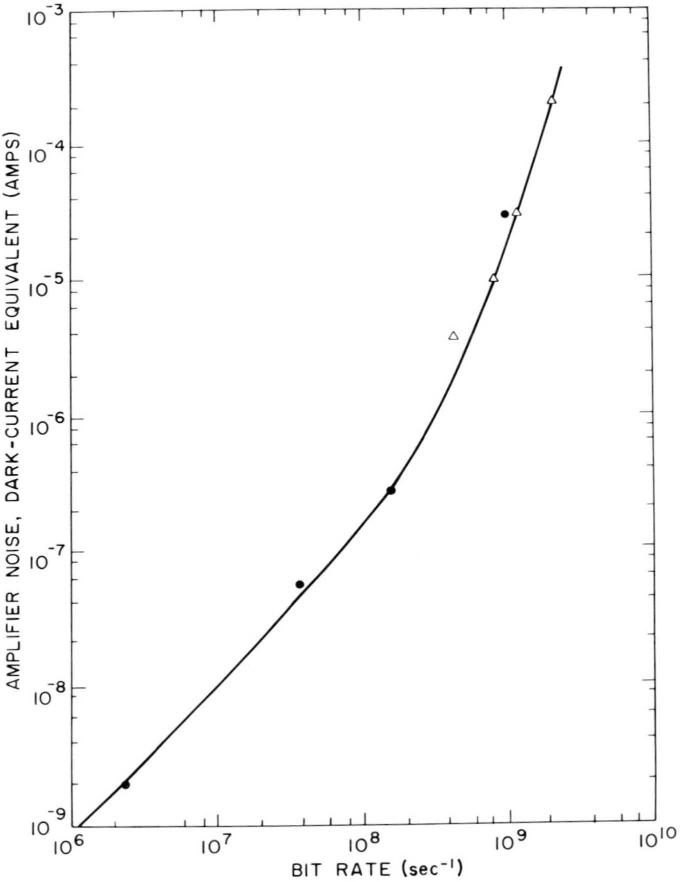

FIG. 5. Noise-equivalent photodiode dark current for the amplifiers shown in Fig. 4. This curve gives the upper limit on photodiode dark current as a function of bit rate when the amplifier noise dominates the system sensitivity. Because of the effect of temperature on the dark current, the photodiode dark current must be specified at the highest temperature of operation. ●, FET; △, Si bipolar.

The dark current is a strong function of temperature, being thermally activated in origin. In a GaInAs homojunction photodiode, for example, it increases by an order of magnitude for a temperature rise of 50°C (see Section 10b). This means that the photodiode specification on dark current must be met at the highest receiver-operating temperature. In land-based telephone networks, this temperature limit often reaches 85°C. Under these conditions, the room-temperature dark current must be much lower than that indicated in Fig. 5.

If avalanche gain is to be used, the photodiode-generated noise must be substantially less than that of the amplifier. The optimum avalanche gain is determined when the multiplied photocurrent and dark current reach the limits shown in Fig. 5. As a general rule for compound semiconductor avalanche photodiodes, the unmultiplied dark current should be at least a factor of 20 less than the upper limit shown in Fig. 5 at the reverse bias and maximum temperature of operation.

Receiver sensitivity also depends on the total capacitance of the photodiode–preamplifier system. The contribution of capacitance to receiver performance is discussed in detail in Chapter 4 by Forrest. Following this treatment, the noise of an FET preamplifier originating from the channel conductance and referred to the input is give by

$$\langle i_c^2 \rangle = (4kT\Gamma/g_m)[(2\pi C_T)^2 I_3 B^3], \tag{2}$$

$$= 1.9 \times 10^{-18} C_T^2 B^3. \tag{3}$$

In Eq. (2), kT is the thermal energy at temperature T, $\Gamma \sim 1$ for GaAs FETS, g_m is the channel transconductance of the FET, I_3 (=0.085) a Personick integral, B the bit rate, and C_T the total input capacitance of the amplifier and photodiode. The strong dependence of the noise power on the bit rate means that the capacitance of the photodiode will eventually become the dominant noise term at high bit rates.

In order to place this effect in perspective, we show in Fig. 6 the upper limit on photodiode capacitance, as a function of bit rate, which would be required to keep the capacitance-dependent noise term [Eq. (3)] below the noise of the amplifier. The amplifier noise power depends on the Johnson noise of the input resistor, the gate-leakage current, and the sum of the amplifier and lead capacitances. At bit rates greater than 200 Mbit/sec, the lowest achieved amplifier noise power is equivalent to that associated with a 2-pF capacitance. In order not to increase that noise power significantly, the photodiode capacitance would have to be about one order of magnitude less than this value, or about 0.2 pF under operating conditions. The reduction in sensitivity for a photodiode with 0.5-pF capacitance is about 1 dB. This specification represents not only a more reasonable goal from the standpoint of fabrication but also a measurable compromise in receiver sensitivity. In long-distance optical fiber links using state-of-the-art fibers, a 1-dB penalty in receiver performance costs 5 km in transmission distance. Such a penalty may have important economic consequences in certain applications, such as undersea cable where maximizing the distance betwen repeater stations is an important consideration.

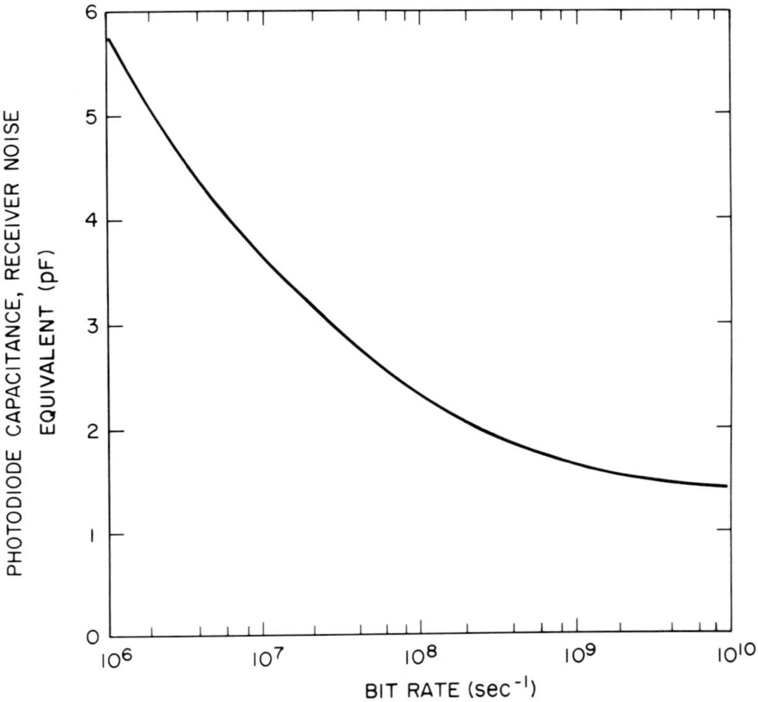

FIG. 6. Equivalent capacitance for amplifier-generated noise. A photodiode whose capacitance is equal to that shown on this curve would result in an increase of 3 dB in the receiver noise power; bit-error rate, 10^{-9}.

4. Specifications for Avalanche Photodiode Performance

The sensitivity of a receiver employing an avalanche photodiode (APD) may be expressed in terms of the signal-to-noise ratio (SNR) in the form (Pearsall, 1981)

$$\mathrm{SNR} = \frac{(P_0 R)^2 M_0^2}{N_S M_0^2 F(M_0) + N_A}, \tag{4}$$

for a digitally modulated signal in return-to-zero format with equal numbers of 1's and 0's. In Eq. (4), P_0 is the optical power, R the photodiode absolute responsivity ($I_p = RP_0$), N_S the square of the shot-noise current, N_A the square of the amplifier noise current, M_0 the average current multiplication from avalanche gain, and $F(M_0)$ the excess-noise factor. It can be seen immediately in Eq. (4) that avalanche gain can be used to increase the SNR if the photodiode-generated noise is less than the amplifer noise N_A.

Avalanche gain occurs when the electric field in the depletion region of the photodiode accelerates the electrons and holes to energies above the fundamental band-gap energy so that they can create additional secondary carriers in ionizing collisions with the lattice. The statistical nature of these collisions leads directly to the generation of excess noise. The excess-noise factor depends in a complex way on the multiplication M_0 and on the ratio of the ionization rates of the electrons and holes (Part II). However, for a number of practical avalanche photodiodes, it has been found experimentally that the excess-noise factor can be expressed approximately as

$$F(M_0) = kM_0, \tag{5}$$

where k is less than 1.

An expression for the optimum avalanche gain $\hat{M}$ can be found by differentiating Eq. (4) with respect to multiplication M_0 (Pearsall, 1981):

$$\hat{M} = (2N_A/kN_S)^{1/3}. \tag{6}$$

Thus, the optimum avalanche gain becomes larger if the amplifier noise is increased or if the dominant contribution to the shot noise is decreased. When an APD is operated at the optimum gain, the increase in the SNR is always much less than the optimum gain. This result is a direct consequence of the excess-noise factor. The sensitivity improvement or the "real gain" G is given approximately by

$$G \simeq \hat{M}/F(\hat{M}). \tag{7}$$

By using Eqs. (4)–(6), we can derive an expression for the minimum detectable optical power:

$$P_{\min} = R^{-1}[\tfrac{3}{2}\,\mathrm{SNR}(2N_A N_S^2 k^2)^{1/3}]^{1/2}. \tag{8}$$

Although Eq. (8) must be solved numerically because the optical power appears on both sides of the expression, several qualitative points concerning receiver sensitivity can be readily seen. The minimum detectable optical power, when avalanche gain is used, is relatively insensitive to the level of the amplifier noise; that is, if the amplifier noise is high, then the optimum gain will be high, but the minimum detectable signal will not change very much. The minimum detectable power is more dependent on the dark current. An order of magnitude increase in dark current will increase the minimum detectable power by a factor of 2. On the other hand, the factor k does not vary very much ($k \sim 0.4$ for InP-based compound semiconductor APDs to about 0.8 for Ge). Germanium APDs are generally recognized to represent the worst case. The smaller value of k for compound-semiconductor APDs reduces the minimum detectable power by 20%, which is of course useful. New kinds of APD structures and materials that would exploit features of the electronic energy-level structure of compound semiconductors have been proposed. One of these structures is an APD in which the avalanche region is deliberately grown as an alternating sequence of high- and low-band-gap semiconducting materials. As originally proposed (Chin, *et al.,* 1980) this structure supposedly leads to reduced excess noise because of a large asymmetry in the ionization rate ratio that is induced by the structure. Despite initial promising results (Capasso *et al.,* 1982) showing such a large asymmetry, subsequent careful measurements have shown no reduction in the actual excess noise factor or improvement in sensitivity of avalanche photodiodes made in this way (Susa and Okamoto, 1984).

There are a few cases of specific semiconductor materials whose electronic structure enhances the probability of hole-initiated ionization over that of electrons. Use of these materials, which are discussed in Part III, may reduce the value of k to less than 0.08. This would result in a factor of 2 improvement over currently achieved sensitivities using Ge. The improvement achieved by developing these devices is equivalent to a reduction in the receiver capacitance of 0.5 pF, from 0.7 to 0.2 pF, for example.

In Fig. 7 we show the optimum avalanche gain and increase in sensitivity as a function of bit rate, determined using Eqs. (4) and (8) and using the amplifier sensitivities shown in Fig. 4. It is a general property of avalanche gain, as illustrated in Fig. 7, that the actual improvement in receiver sensitivity is much less than the value of the gain itself. In this example, the excess-noise factor is obtained from Eq. (5) using $k = 0.4$, which is typical of the value measured in III–V APDs made from GaAs or InP, including the GaInAs/InP APD with separate absorption and multiplication regions (Section 10c), which is currently regarded as a promising candidate for optical fiber telecommunications. The sensitivity improvement for this photodiode is shown by the dashed line in Fig. 7 and is only about $\frac{1}{2}$ the optimum avalanche gain. In the case of Si APDs, where k is 0.02, a more advantageous value, the improvement in sensitivity would be about $\frac{2}{3}$ the optimum multiplication.

In Fig. 8 we use the results of this section to show the increased transmission distance that avalanche gain would permit as a function of bit rate for a 1-dB/km fiber. At 100 Mbit/sec, about 5 km of additional transmission is expected. Using a laser-transmitter/$p-i-n$-receiver combination, the maximum transmission distance at 100 Mbit/sec is abut 40 km for a 1-dB km^{-1} fiber; hence avalanche gain gives about a 12% increase in transmission

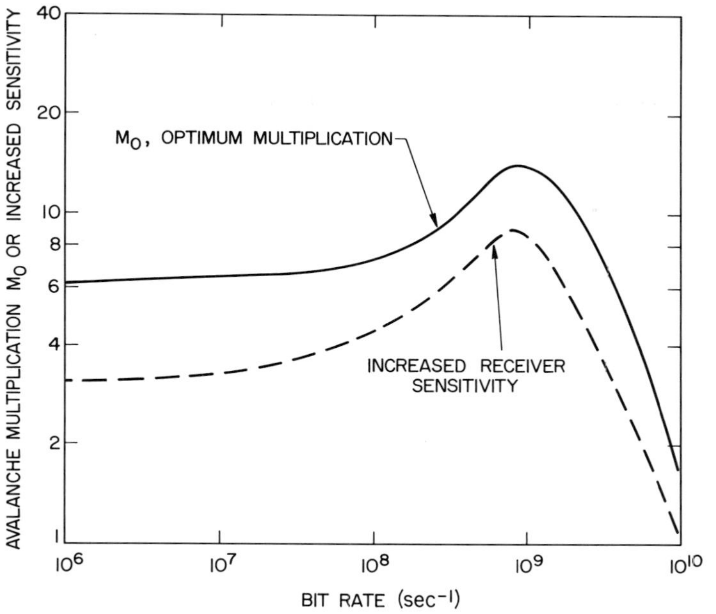

FIG. 7. Optimum avalanche multiplication and receiver sensitivity improvement for a GaInAsP/InP SAM–APD at $T = 23°C$ with $k = 0.4$ as a function of bit rate. Significant improvement in receiver performance is obtained above 1 Gbit/sec because of the increase in amplifier noise at higher frequencies. Above 2 Gbit/sec, the receiver sensitivity is limited by the gain–bandwidth product for SAM-APDs of ~ 8 GHz. Below 100 Mbit/sec, the improvement afforded by avalanche gain is small but useful.

distance. If a fiber cable is installed with an attenuation equal to that shown in Figs. 1 or 2, i.e., 0.2 dB km^{-1}, the maximum transmission distance for a laser/$p-i-n$ system should approach 200 km at 100 Mbit/sec. Since few applications require this length (more than 90% of the link requirements call for less than 15 km), it is questionable that the additional 24-km transmission made possible by avalanche gain could be used in any but a small number of specialized applications. In Fig. 8 it can be seen that at bit rates above 500 Mbit/sec, the optimum gain increases dramatically. In addition, the minimum detectable optical power decreases with gain. At 1 Gbit/sec, the transmission distance over a 1-dB/km fiber is about 8 km, which could be doubled to about 16 km using an avalanche detector. Avalanche photodetectors may therefore be a component of optical fiber systems operating at 1 Gbit/sec.

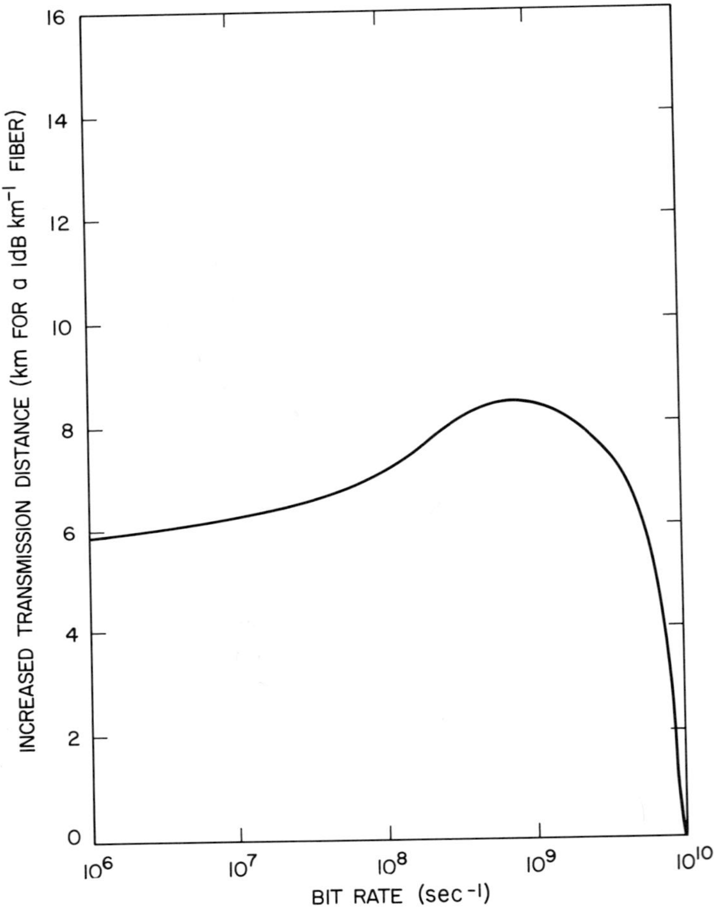

FIG. 8. Increased transmission distance in kilometers for GaInAs/InP SAM–APD at $T = 23°C$ as a function of bit rate using the optimum gain APD of Fig. 7. The distances shown are calculated for a 1-dB/km fiber. For a 0.5-dB/km fiber, multiply this distance by 2.

II. Compound Semiconductor Photodiode Principles

In Part I of this chapter we described some of the photodiode characteristics that are required for high receiver sensitivity. In Part II, we outline the physical principles that determine these characteristics. In Part III, the properties of particular photodiode structures are described.

A high-performance photodiode first must be efficient in converting optical signal power at the operating wavelength to electrical signal current. This requires the correct choice of semiconductor band-gap wavelength and proper device design. These matters are treated in Section 5. To meet the speed requirements of high-bit-rate systems, the photodiode must have an adequately wide bandwidth, i.e., a fast temporal response. This also requires attention to the photodiode configuration as well as the achievement of low device capacitance. These factors are discussed in Section 6. To contribute the smallest amount of noise to the receiver, a good photodiode must have low dark current, especially for low-bit-rate systems, and low capacitance, particularly for high-bit-rate systems. The factors affecting dark current are treated in Section 7. The concepts developed in Sections 5 – 7 apply not only to $p-i-n$ photodiodes, but to APDs as well. A high-performance APD must have in addition, appropriate gain and bandwidth, and must not contribute substantial excess noise. The principles determining these performance factors are outlined in Sections 8 and 9, respectively. Additional material on photodiode principles can be found in several general review articles (Stillman and Wolfe, 1977; Lee and Li, 1979a; Matsushima and Sakai, 1982; Stillman *et al.,* 1982; Lee, 1983).

5. Quantum Efficiency and Spectral Response

The basic $p-i-n$ photodiode, illustrated schematicaly in Fig. 9, consists of three semiconductor regions. The p and n regions are usually doped to high carrier concentrations. The i (intrinsic) region of width d is unintentionally doped in the typical compound semiconductor photodiode. It may have a small, residual n- or p-type background carrier concentration (assumed in what follows to be n type). Under the application of a reverse bias of magnitude V, a depletion region of width W and internal field $E(x)$ is formed in the i region. In normal operation, V is made sufficiently large to cause the depletion region to "punchthrough" to the n layer so that $W = d$.

Above-band-gap light shining on the photodiode results in the absorption of photons and in the creation of an electron – hole carrier pair for each absorbed photon. The carriers photogenerated in the depletion region are separated by the field and collected, generating a current in the external load. In general, the quantum efficiency η_Q is the ratio of carriers collected to photons incident on the photodiode. Absorption of all the photons within the depletion region and collection of all the photogenerated carriers yields a unity quantum efficiency.

In an actual photodiode, a fraction r of the incident power P_0 is reflected at the air – photodiode interface (Fig. 9). Also, the amount of optical power absorbed in the depletion region is a function of its width W and its wavelength-dependent absorption coefficient $\alpha_0(\lambda)$. Thus, ignoring any absorption in the p region,

$$\eta_Q = (1 - r)(1 - e^{-\alpha_0 W}). \tag{9}$$

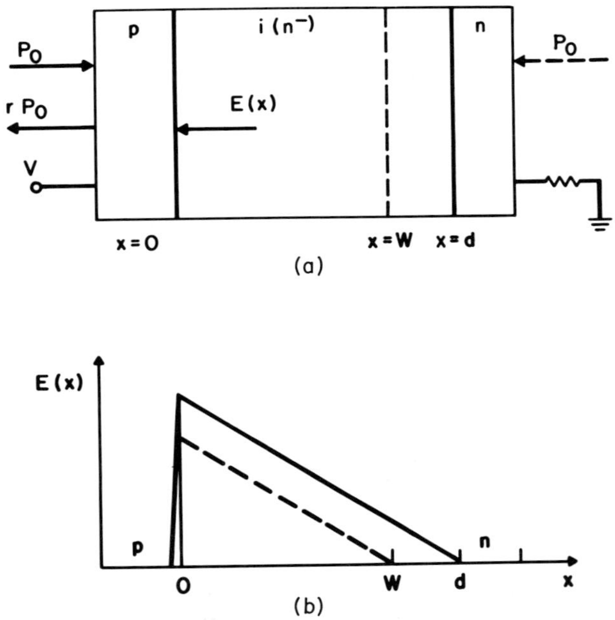

FIG. 9. Basic (a) $p-i-n$ photodiode structure and (b) internal electric field $E(x)$. The i layer, of width d, has an n^- residual carrier concentration. The depletion region extends from $x = 0$ to $x = W$ (dashed) at bias voltage V and to $x = 0$ at punchthrough. Incident power P_0 impinges on the photodiode from the left (or from the right, dashed).

In order to maximize η_Q, the depletion layer must be wide enough to ensure that $\alpha W \gg 1$. The direct-band-gap semiconductor alloys have absorption coefficients in excess of 10^4 cm^{-1} at energies above the band gap, so that W need only be a few micrometers wide to absorb nearly all the incident light.

In most of the materials suitable for photodiodes, the discontinuity in the refractive index going from air, $n_a = 1$, to the semiconductor, $n_s = 3.5$, leads to a Fresnel reflection coefficient $r = 0.3$ at normal incidence. This reflection loss, which would otherwise reduce the maximum value of η_Q to about 70%, can be eliminated almost completely by depositing an appropriate transparent antireflection film coating on the photodiode surface (Baumeister and Picus, 1969). Silicon nitride, with a refractive index in the range $n = 1.8$–2.0, can be used as the antireflection film by choosing $n_{ar} = (n_a n_s)^{1/2}$. The film thickness must correspond to a quarter wavelength of light in the dielectric.

Any absorption of light outside the depletion region has a negative effect on the device quantum efficiency, because carriers photogenerated there must diffuse to the depletion region in order to contribute to the photocurrent. For example, in the geometry of Fig. 9 with light incident from the left, some of the electrons generated by absorption in the *p* region will not reach the depletion region by diffusion and will recombine within the *p* region or at its surface. The details of the device geometry and the surface-recombination properties of the semiconductor material determine the resulting effect on the quantum efficiency. This may not be a very serious problem for light incident on a *p* layer because the diffusion of electrons can be rapid in *p*-type materials. The effects of the much slower diffusion of holes on quantum efficiency, as well as on speed, may be more severe in geometries where photons are absorbed in an undepleted *n* region, as might be the case for light incident from the right in Fig. 9.

Absorption outside the fully depleted *i* region can be eliminated by making either the *p* or *n* region transparent to the incident light. This can be accomplished with lattice-matched alloy semiconductors by designing the photodiode with a wide-band-gap "window" layer. This layer may serve either as the *p* or *n* region in Fig. 9. Carrier recombination at the heterojunction thus formed is negligible in comparison with recombination at a free surface.

To the photodiode user, quantum efficiency is not as important as responsivity, the ratio of photocurrent I_p to incident power P_0. The responsivity is wavelength dependent through the energy per incident photon and is related to the quantum efficiency by

$$R = I_P/P_0 = q\eta_Q/h\nu, \tag{10}$$

where h is Planck's constant and ν the photon frequency. For a wavelength λ given in micrometers, $h\nu = 1.24/\lambda$ eV, so that a convenient relationship for responsivity is

$$R = \eta_Q\lambda/1.24. \tag{11}$$

6. Temporal Response

The speed of response of a photodiode depends both on the circuit time constant and on the internal time duration from carrier generation to collection. The circuit or *RC* time constant is the product of the diode load, often 50 Ω for high-speed detection, and the photodiode capacitance. Achieving the smallest value of C requires minimizing the junction, packaging, and stray capacitances. (As noted in Section 4, low capacitance also results in reduced receiver noise contributions.) The junction capacitance of the diode of Fig. 9 is simply $C_j = \varepsilon A/W$, where ε is the dielectric permittivity (typically, $\varepsilon/\varepsilon_0 \sim 13$ for most compound semiconductors) and A the junction area. Minimizing C_j demands the smallest junction area consistent with the ac-

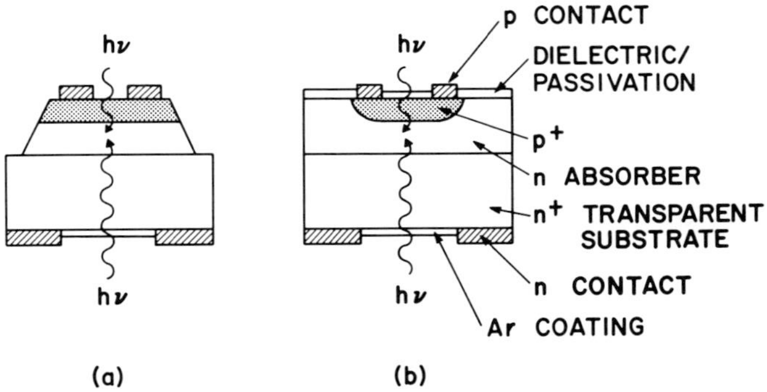

FIG. 10. Junction-area restriction in a $p-i-n$ photodiode: (a) mesa geometry; (b) planar geometry. Either geometry may be rear illuminated through a transparent substrate, or top illuminated through a thin or transparent p region.

ceptance of all of the incident light. The junction area may be restricted in several ways, two of the most common of which are shown schematically in Fig. 10. The mesa geometry of Fig. 10a is formed by etching away material to leave a mesa-shaped junction region. In the planar diode example of Fig. 10b, the junction region is defined by the implantation or diffusion of an acceptor species into n-type material. In either case, light may be admitted through a transparent substrate layer or through a thin or transparent top p region.

The junction capacitance is reduced by designing the largest practical depletion width. This requires a low i-layer background-carrier concentration, in order to be able to deplete it at a modest bias voltage. For an i layer with a net donor concentration N_D, the depletion-layer width for the resulting one-sided junction is given by

$$W = [2\varepsilon(V - V_{bi})/qN_D]^{1/2}, \qquad (12)$$

where V is the reverse bias voltage and V_{bi} the junction built-in voltage (Sze, 1981). For a photodiode with $N_D = 10^{15}$ cm^{-3}, the depletion width W is about 2 μm at a few volts bias. A junction diameter of 100 μm and a fully depleted i layer 2 μm wide results in a capacitance of about 1 pF and an RC time constant of 50 psec for a 50-Ω load.

The total carrier transit time is a combination of the time for carriers photogenerated in the depletion region to drift across it under the influence of the applied field E and the time for any carriers photogenerated outside the depletion region to diffuse to it. Scattering-limited drift in typical fields of 10^4 V cm^{-1} or more takes place at a velocity of about 10^7 cm sec^{-1} in materials commonly used in photodiodes. Thus, in the previous example with $W = 2$ μm, the transit time would be 20 psec.

Carriers generated outside the fully depleted *i* region diffuse to it with a diffusion time τ_d. For electrons diffusing through a *p* region, for example, τ_d is given by

$$\tau_d = L_n^2/2.4D_n, \tag{13}$$

where L_n is the electron diffusion length or the diffusion distance, whichever is smaller, and D_n the electron diffusion constant, which is proportional to electron mobility. Typically, $\tau_d = 100$ psec for $1-2\ \mu$m of diffusion distance. Thus, if a very fast photodiode with a bandwidth greater than a few gigahertz is required, it is necessary to minimize the thickness of any absorbing *p* region. Absorption directly in the depletion region, for example, by means of a wider band-gap window layer as described previously for improved quantum efficiency, can effectively eliminate the diffusion component of the carrier transit. In the case of designs where absorption may take place in *n* regions, the effects of the much slower hole diffusion can lead to much more severe device speed limitations. This would be the case if light were admitted to the diode of Fig. 9 from the right or if photons were absorbed in a part of the *i* region which was not fully depleted.

7. Dark Current

The noise power generated in a $p-i-n$ photodiode is well described by the shot-noise expression [Eq. (1)]. Minimization of noise requires that the photodiode have dark current I_d, consisting in general of contributions from bulk and surface effects, that is substantially less than the detected photocurrent. In imperfect diodes, I_d may be dominated by $p-n$-junction defects and surface leakage. In properly designed and processed $p-i-n$ photodiodes, the sum of the generation–recombination current I_{g-r} and the diffusion current I_{dif} represents the minimum dark current at low bias voltages.

The value of I_{g-r} is proportional to the volume of the depletion region and inversely proportional to the effective carrier lifetime; the highest quality defect-free material is required for the smallest values of I_{g-r}. The dependence of I_{g-r} on bias voltage V, effective carrier lifetime τ_e, and temperature T is given by (Grove, 1967) as

$$I_{g-r} = (qn_iAW/\tau_e)[1 - \exp(-qV/kT)], \tag{14}$$

where n_i is the intrinsic carrier concentration. At any reverse-bias V above a few tenths of a volt, the exponential term in Eq. (14) may be neglected. The temperature and band-gap dependence of I_{g-r} comes about through n_i, which is given by

$$n_i = n_i' \exp[-E_g(T)/2kT]. \tag{15}$$

Because the magnitude of I_{g-r} is proportional to the volume of the depletion region, it increases as the square root of the applied bias until the i region is fully depleted [Eq. (12)].

The diffusion current I_{dif} arises from thermally generated minority carriers diffusing into the depletion region from the surrounding undepleted p and n regions. In the geometry of Fig. 9, contributions to I_{dif} come from carriers in the p region as well as from carriers diffusing from the undepleted part of the i region if $W < d$ or the n region if $W = d$. In addition, in a planar photodiode of finite cross section (Fig. 10b), a diffusion current due to carriers generated around the edges of the depletion region is also present.

The contributions from the different regions surrounding the depleted volume depend specifically on the details of the photodiode geometry and may be calculated from (Sze, 1981)

$$I_{dif} = I'_{dif}[1 - \exp(-qV/kT)], \tag{16}$$

where, for the diffusion current generated in a p region,

$$I_{dif,n} = qn_i^2(D_n/\tau_n)^{1/2}(A_p/N_A), \tag{17}$$

and for the diffusion current generated in an n region,

$$I_{dif,p} = qn_i^2(D_p/\tau_p)^{1/2}(A_n/N_D). \tag{18}$$

In Eqs. (17) and (18), D_n and D_p are the minority-carrier diffusion constants and τ_n and τ_p the minority-carrier diffusion lifetimes in the p and n regions, respectively. In most of the III–V compound semiconductors, $D_n/D_p \approx$ 10–20, and $\tau_n \simeq \tau_p$. The areas of the depletion region bounding the p and n regions are A_p and A_n and the acceptor and donor concentrations of those regions are N_A and N_D. For a diode with very high p-layer carrier concentration, the dominant diffusion current will be from the n region or, in the planar case, from the undepleted side regions (Fig. 10b). Because the diffusion current contribution to I_d is proportional to n_i^2 and the generation–recombination contribution is proportional to n_i, I_{dif} becomes relatively more significant in materials with small band gaps and at high temperatures.

At sufficiently high fields, tunneling and avalanche processes control the dark current. As will be seen, the successful operation of compound semiconductor APDs requires that tunneling contributions to I_d be minimized. Tunneling can take place as a band-to-band process or via traps or defect levels in the forbidden energy gap. The expression for band-to-band tunneling in a uniformly doped depletion region may be written as follows (Kane, 1959; Moll, 1964; Forrest *et al.*, 1980a):

$$I_{tun} \approx \gamma_i A \exp[-\theta m_0^{1/2} E_g^{3/2}/q\hbar E_m], \tag{19}$$

where m_0 is the free-electron mass, E_m the maximum junction field, and $\hbar$ Planck's constant divided by 2π. The parameter $\theta = \alpha(m_e/m_0)^{1/2}$, where m_e is the electron effective mass and $\alpha \sim 1$, depends on the detailed shape of the tunneling barrier. The factor γ_i is different for tunneling via traps in the band gap and for band-to-band tunneling. In the latter case, $\gamma_i = (2m_e/E_g^{1/2})q^3 E_m V/4\pi^2\hbar^2$.

From the exponential term in Eq. (19), it can readily be seen that the tunneling current contribution to I_d will be more significant for semiconductor materials with narrow band gaps E_g. Because of the decrease of E_g with increasing temperature, tunneling currents increase with temperature. Tunneling is especially significant at the high fields found in APDs, rather than those typical of $p-i-n$ photodiodes, and in heavily doped materials, in which high fields are reached at low applied bias voltages.

8. Multiplication and Bandwidth for APDs

Electrons and holes moving in the depletion region of a reverse-biased $p-i-n$ photodiode gain kinetic energy from the electric field. If the electric field is made sufficiently high, these carriers will gain enough kinetic energy before being collected to create additional electron–hole pairs by the process of impact ionization. The energy that is lost upon impact (inelastic collision) with the lattice is sufficient to promote a valence electron to the conduction band, leaving behind an additional hole in the valence band. This process may continue, producing an avalanche effect accompanied by a multiplication of the original number of photogenerated carriers. Impact ionization is characterized by the rates α for electron- and β for hole-initiated ionization. These rates, in inverse centimeters, give the number of secondary-electron–hole pairs created by a single initiating carrier per centimeter of travel. The reciprocal of the ionization rate is the average distance a carrier will travel before an impact-ionization event. The rates α and β, generally unequal, are strong functions of the local electric field, and depend on the electronic band structure of the semiconductor.

The multiplication, bandwidth, and noise properties of an APD depend critically on the relative magnitudes of the rates α and β. Although the general case of an APD with arbitrary carrier-injection conditions and arbitrary ionization rates has been analyzed in detail (McIntyre, 1966, 1972), it is useful to consider two limiting cases as illustrations. Figure 11a shows the time sequence of the avalanche process for the first case, where $\beta = 0$, and Fig. 11b shows it for the second case, where $\beta = \alpha$. We assume that α and β are independent of position (uniform electric field) and that only electron injection takes place (from the left.) The same results would be obtained for $\alpha = 0$ in the first case if hole-initiated ionization were treated instead.

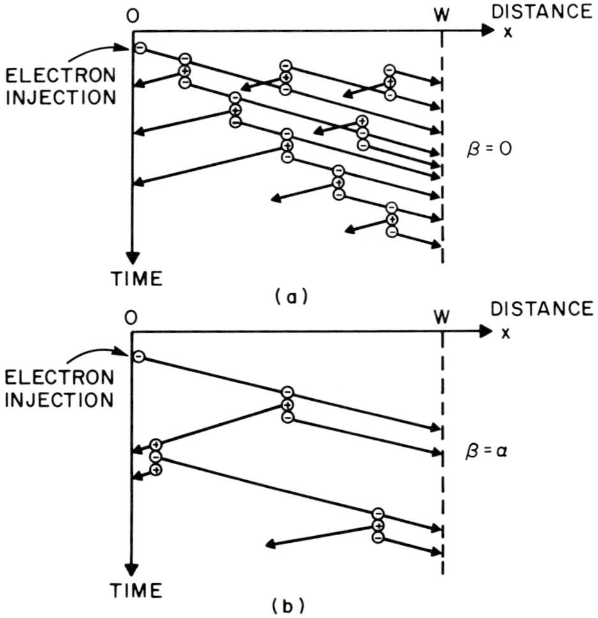

FIG. 11. Representation of the time evolution of avalanche multiplication in a uniform high-field depletion region of width W for pure electron injection at $x = 0$: (a) $\beta = 0$; (b) $\beta = \alpha$. [From Stillman and Wolfe (1977).]

Consider first the case of $\beta = 0$ (Fig. 11a). An injected electron is accelerated in the field until it creates an electron–hole pair by impact ionization. The original electron plus the new electron are then accelerated to repeat the process until such time as all the electrons have traversed the depletion region, i.e., one-electron transit time. Because $\beta = 0$, the holes created by the process travel to the left without further interaction. The maximum time that all carriers remain in the depletion region in response to a short pulse of photoinjected electrons is limited to the sum of the electron and hole transit times. The time response, and therefore the bandwidth, are independent of the multiplication, which in this case can be simply written as

$$M_n = \exp(\alpha W). \tag{20}$$

This expression shows that the current increases in a smooth, exponential manner with α. Because α is a function of reverse bias (through the applied electric field), a photodiode in which $\beta = 0$ (or $\alpha = 0$) will exhibit a gradual rather than a sharp breakdown, making the multiplication easy to control externally.

Next, consider the case of equal ionization rates, $\beta = \alpha$, shown in Fig. 11b. Here the initiating electron starts out as before and creates the first electron–hole pair. Now, however, the hole created upon impact ionization may initiate ionizing collisions itself during acceleration (to the left), resulting in a positive feedback process in which the multiplication is described by

$$M_n = M_p = (1 - \alpha W)^{-1}. \tag{21}$$

Breakdown under these conditions is sharp and occurs when $\alpha W = 1$ or when an injected carrier produces one electron–hole pair on the average while traversing the depletion region. Such a sharp breakdown is more difficult to control by adjusting the bias voltage than the smooth breakdown obtained when only one carrier type takes part in the multiplication process. In addition, the entire avalanche process now lasts much longer than the sum of the electron and hole transit times, which is the case for $\beta = 0$. In fact, the avalanche duration increases with multiplication, and the APD exhibits a limited gain–bandwidth product.

The frequency response of an APD with uniform electric field in the multiplication region has been calculated by Emmons (1967) as a function of the low-frequency multiplication M_0. It depends strongly on the ratio of ionization rates, $k = \beta/\alpha$ for electron injection or $k = \alpha/\beta$ for hole injection. The 3-dB bandwidth is shown in Fig. 12 normalized to $2\pi T_{av}$, where T_{av} is the average of the electron and hole transit times in the avalanche region. The curves are extended only to $M_0 = 200$, for useful gains beyond this value are uncommon in compound semiconductor APDs. Above the dashed $M_0 = 1/k$ curve, the bandwidth varies little with gain, with the case of ionization by a single carrier (e.g., $\beta = 0$) giving essentially constant bandwidth, as discussed before. Below the line, one approaches the case of equal ionization rates and a constant gain–bandwidth product. For almost all of the III–V compound-semiconductor APDs known to date, $k \sim 0.3$–0.5, and the constant gain–bandwidth product case applies, even for the smallest multiplications. In all cases, avalanche gain *reduces* the bandwidth compared to p–i–n diodes of the same material.

9. Excess Noise in APDs

The description of multiplication just given applies to ensemble-average quantities and ignores the statistical nature of the impact-ionization process. The statistical fluctuations in the multiplication process produce a noise current in excess of the p–i–n photodiode shot noise [Eq. (1)] multiplied by the square of the average multiplication M_0. The mean-squared noise current is now given by

$$\langle i_{\mathrm{n}}^2 \rangle = 2q(I_{\mathrm{p}} + I_{\mathrm{d}})\langle M^2 \rangle B, \tag{22}$$

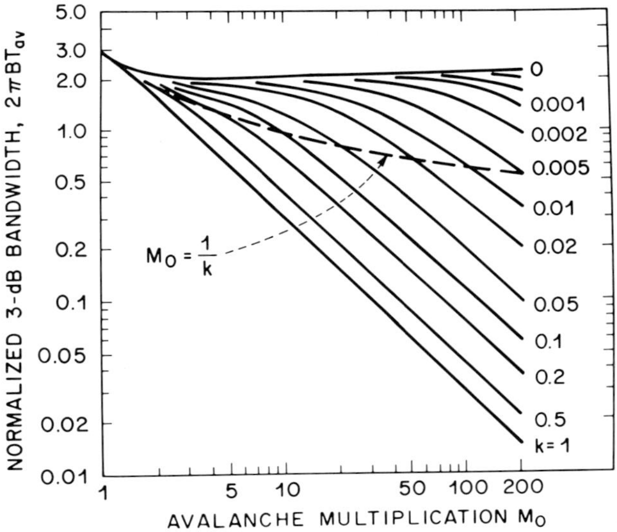

FIG. 12. Normalized 3-dB bandwidth as a function of average avalche multiplication M_0 for several values of the ionization rate ratio k. For pure electron injection, $k = \beta/\alpha$; for pure hole injection, $k = \alpha/\beta$; T_{av} is the average of the electron and hole transit times in the avalanche region. [From Emmons (1967).]

where $\langle M^2 \rangle$ is the mean square value of the multiplication. The excess-noise factor is defined as

$$F(M_0) = \langle M^2 \rangle / M_0^2, \tag{23}$$

and is a function of the average multiplication M_0 and the ionization rate ratio k. For the case of pure electron injection (Fig. 11), the general expression for $F(M_0)$ with $k = \beta/\alpha$ reduces to the approximate expression (McIntyre, 1966):

$$F(M_0) = kM_0 + [2 - (1/M_0)](1 - k). \tag{24}$$

For modest values of M_0 ($\gtrsim 10$) and common values of k ($\sim 0.3-0.5$), this reduces further to the relation given by Eq. (5). The same expression may be used to describe the case of pure hole injection by defining $k = \alpha/\beta$.

The two limiting cases of Fig. 11 may now be examined. In the first case, $\beta = 0$, so that $k = 0$ and $F(M_0) \to 2$ in the limit of large values of multiplication. For the case of $\beta = \alpha$, $k = 1$, and $F(M_0) = M_0$. That equal ionization rates for electrons and holes should result in a larger excess noise than the case of only one carrier taking part in the multiplication can be seen intui-

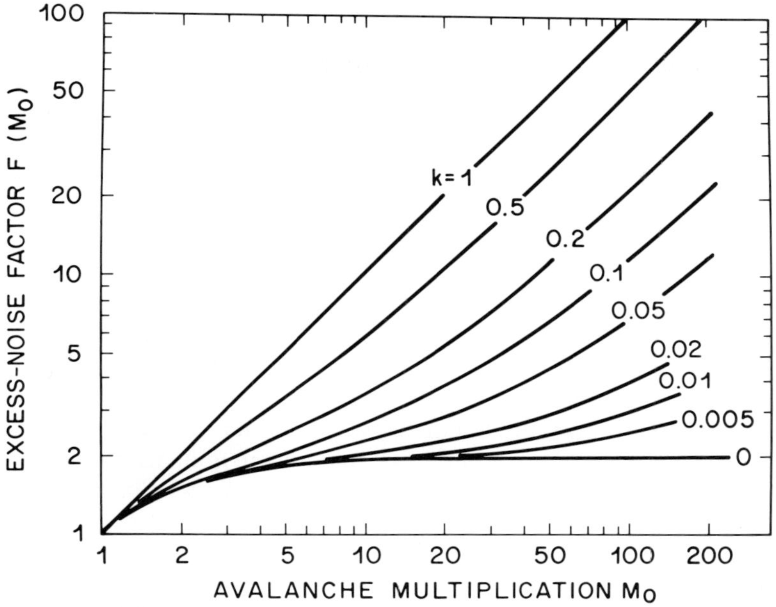

FIG. 13. Excess noise factor $F(M_0)$ as a function of average avalanche multiplication M_0 for several values of the ionization rate ratio k. For pure electron injection, $k = \beta/\alpha$; for pure hole injection, $k = \alpha/\beta$. [From Webb *et al.* (1974).]

tively from Fig. 11. The exponential dependence of multiplication in the case of $\beta = 0$ [Eq. (20)] requires a large value of αW to achieve large values of M_0. This implies a large total number of electrons present in the avalanche region. A small statistical variation in the ionization process will cause only a small relative variation in the total number of carriers or multiplication and a small excess noise. In the case of $\alpha = \beta$ [Eq. (21)], large gain is obtained for $\alpha W = 1$, and relatively few carriers take part in the avalanche process. The same small statistical variation in the ionization process in this case will result in large variations in the multiplication and in a large excess noise.

All intermediate cases of $\alpha \neq \beta$ have been treated as well (McIntyre, 1966), and the results for $F(M_0)$ are summarized in Fig. 13 for various values of $k = \beta/\alpha$ and pure electron injection. Figure 13 also describes the case of pure hole injection if we define $k = \alpha/\beta$. The excess-noise factor for mixed injection has also been treated in detail elsewhere (Webb *et al.*, 1974; Nishida, 1977). In all cases, the lowest excess-noise factor is obtained by injection of only the carrier with the larger ionization rate and for conditions yielding the greatest difference in α and β.

III. Compound Semiconductor Photodiode Properties

The use of compound semiconductor photodiodes for lightwave communications applications is based on new and rapidly evolving technology. The following sections therefore describe not only current $p-i-n$ and APD designs and characteristics, but also touch on some of the material and processing technologies that are likely to be part of future photodiode development. Central to the material technology is the ability to prepare high-quality defect-free epitaxial or bulk materials with low residual carrier concentrations. From such materials, photodiodes can be designed with low leakage current, low capacitance, high speed, and high quantum efficiency.

At present, compound-semiconductor photodiodes are most important in the $1-1.6$-μm spectral region. Germanium $p-i-n$ and APDs are also sensitive in this region and are treated extensively in Chapter 3 by Kaneda. Among the compound semiconductors from which potentially useful devices at these wavelengths can be fabricated are the quaternary mixed crystals GaInAsP and AlGaAsSb; the ternaries GaInAs, AlGaSb, GaAsSb, and HgCdTe; and the binary GaSb (see Part I). For wavelengths shorter than 1.0 μm, photodiodes of GaAs, AlGaAs, and InP are possible and may serve some special application. However, silicon $p-i-n$ and APDs offer superior performance at these wavelengths and are used extensively in short-wavelength lightwave communication systems. Photodiodes for optical communications at wavelengths beyond about 1.6 μm have received limited attention to date, but the achievement of lower-loss fibers at these wavelengths would surely change this situation.

10. GaInAs and GaInAsP Photodiodes

a. Material and Processing Technology

The quaternary alloy $Ga_xIn_{1-x}As_yP_{1-y}$ can be grown epitaxially and lattice matched on InP substrates with any room-temperature band-gap wavelength from 0.92 μm (InP) to 1.65 μm ($Ga_{0.47}In_{0.53}As$, hereafter abbreviated GaInAs) by proper adjustment of the composition and by keeping $y/x \approx 2.2$ (Fig. 3). The most extensive efforts on crystal growth in this system have been by liquid-phase epitaxy (LPE) and vapor-phase epitaxy (VPE), although considerable progress has also been made in organometallic chemical vapor deposition (OM-CVD) and molecular beam epitaxy (MBE). The material-growth technology of the GaInAsP system based on these four methods is treated elsewhere in these volumes; here only the subject of material purity, central to high-quality photodiodes, is considered.

Common to all methods of crystal growth is the difficulty of handling phosphorus in a completely reproducible way. In LPE, the distribution coefficient of phosphorus is very large, meaning that only minute quantities are needed to obtain sizable atomic percentages in the solid (Pollack *et al.*, 1978). Inaccuracies in preparing the growth solution, or small inhomogeneities in the solution during growth, may lead to sizable variations in solid composition. In vapor-phase growth, including MBE, phosphorus vapor tends to form many complexes, only some of which are active in crystal growth itself. There are problems with all methods of growth in producing phosphorus-bearing epitaxial films of uniform composition and of sufficient thickness ($t \geq 2\ \mu m$) for photodetector applications. For these reasons, GaInAs has emerged as a clear materials choice for photodiodes in the 1–1.6-μm region of the spectrum. Its absorption edge occurs at a wavelength sufficiently long ($\lambda = 1.65\ \mu m$) to produce high-quantum-efficiency photodiodes at both 1.3 and 1.55 μm, the two low-loss transmission windows. Experience so far has shown that it is less difficult to grow high-purity low-defect material from this ternary than from a quaternary whose composition is tuned to match a particular wavelength. This result may be directly related to the absence of phosphorus, whose high volatility and small atomic radius could lead to vacancy concentrations that would affect both the crystalline quality and the free-carrier concentration.

Very high-purity InP and GaInAs have been prepared by both LPE and VPE techniques. These materials are *n* type when grown without intentional dopants. Net residual carrier concentrations $N_D - N_A$ in the low 10^{14}-cm^{-3} range have been reported for GaInAs (Amano *et al.*, 1981; Cook *et al.*, 1982b). The room-temperature mobility of GaInAs with $N_D - N_A$ below 10^{15} cm^{-3} is typically 10,000 cm^2 V^{-1} sec^{-1} (Cook *et al.*, 1982b). The preparation of such high-purity low-carrier-concentration GaInAsP compositions has generally not received as much attention, although values of $N_D - N_A$ below 10^{15} cm^{-3} have been obtained (Kuphal and Pocker, 1981).

Growth of low-background carrier-concentration material by LPE requires the use of highly purified graphite boats and high-purity source materials for the growth solutions. The indium, GaAs, InAs, and InP starting materials must also be baked in a pure hydrogen ambient for tens of hours or more at temperatures in the neighborhood of the growth temperature, typically 600–700°C. Baking in this way reduces the residual carrier concentration of LPE GaInAs, for example, from the 10^{16}–10^{17}-cm^{-3} range to the 10^{15}-cm^{-3} range (Amano *et al.*, 1981). Such baking is thought to influence the concentration of Si impurities in the growth solution (Kuphal and Pocker, 1981). Schemes incorporating water vapor in the baking ambient to modify the Si equilibrium concentration have been claimed to reduce successfully residual carrier concentrations (Oliver and Eastman, 1980), although the subject remains controversial. The presence of sulfur as a residual donor in LPE-grown InP, GaAs, and GaInAs has also been confirmed (Stillman *et al.*, 1981, 1982). Both S and Si have fairly high distribution coefficients, and great care in their removal from the growth solution, boat, and ambient are required to achieve low residual carrier concentrations. For this reason, the details of the construction of the LPE apparatus may be critical, and this may explain the varying degrees of success different workers have achieved in obtaining the very purest materials.

Growth of the very highest purity VPE InP and GaInAs also requires careful attention to system cleanliness and use of very pure source gases. For GaInAs growth, high-purity In, Ga, H_2, and AsH_3 are used, together with an HCl carrier gas (Yamauchi *et al.,* 1982). The purest HCl currently is obtained by thermal decomposition of $AsCl_3$ at 850°C (Susa and Yamauchi, 1981a). Etching of the growth system with HCl prior to growth at temperatures slightly above the growth temperature is considered to be essential for obtaining InP and GaInAs epitaxial layers with the lowest residual carrier concentrations (Susa, 1982). As with LPE growth, the details of the VPE apparatus appear to be critically important in achieving these results. Carrier concentrations in the $10^{12}-10^{13}$-cm^{-3} range for GaInAs and in the $10^{13}-10^{14}$-cm^{-3} range for InP have been obtained (Susa, 1982), although residual concentrations in the 10^{15}-cm^{-3} range are more typical (Yamauchi *et al.,* 1982).

An important advantage of the entire GaInAsP material system is the ease with which it can be doped either *n* or *p* type in any of the epitaxial growth methods to carrier concentrations required in photodiode structures. For *n*-type materials, Sn is the most common donor, although S, Se, and Te have been used with success. For *p*-type materials, Zn is the most commonly used acceptor dopant in LPE and VPE, with special applications using Be, Cd, Mg, or Mn. For MBE growth and for ion implantation, Be is the most useful acceptor dopant.

b. p–i–n Photodiode Performance

Excellent performance has been achieved with *p–i–n* photodiodes incorporating GaInAsP alloys for detection in the 1–1.6-μm-wavelength band. Use of particular quaternary alloy compositions for absorption permits the photodiode spectral response to be tailored to the source wavelength, whereas use of the lattice-matched GaInAs composition permits coverage of the entire band. An advantage of tailoring the composition to the largest usable band-gap energy E_g is that the dark current may be reduced, as can be seen from the dependence of the dark current on band gap given in Section 7.

Low dark-current low-capacitance GaInAsP *p–i–n* photodiodes have been reported by a number of workers (e.g., Burrus *et al.,* 1979). The variation of quantum efficiency τ_Q with wavelength for photodiodes prepared from five different GaInAsP-alloy compositions, including GaInAs, is shown in Fig. 14 (Washington *et al.,* 1978). These photodiodes are hetero-

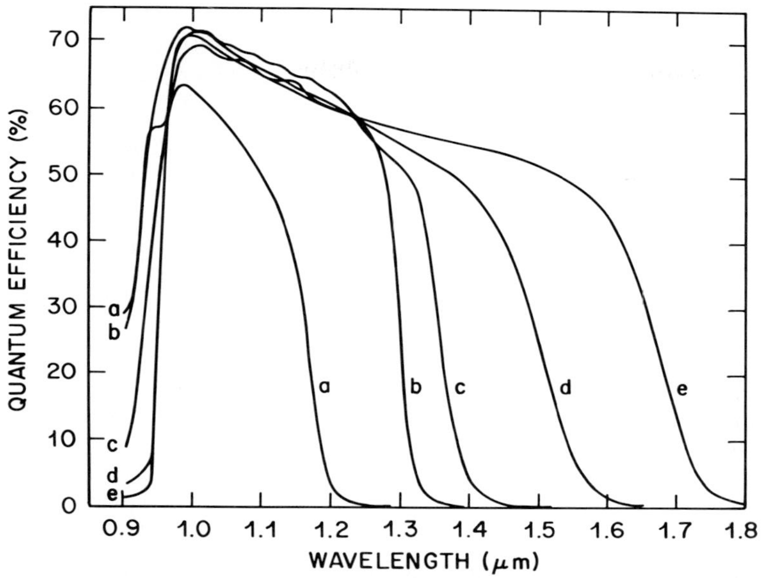

FIG. 14. Variation of quantum efficiency η_Q with wavelength for heterojunction photodiodes prepared from various $Ga_xIn_{1-x}As_yP_{1-y}$ alloy compositions on *n*-InP substrates. Curves a–c are for *p*-InP/*n*-GaInAsP devices top illuminated through the *p*-InP window. Curves c and d are for *p*-GaInAsP/*n*-InP devices rear illuminated through the *n*-InP substrate. [From Washington *et al.* 1978).]

junction devices in the mesa geometry, with short-wavelength response cut off by an InP substrate or InP window layer in each case. The long-wavelength response is controlled by the band gap of the quaternary absorbing layer, dropping by about a factor of 2 at the band-gap wavelength. The quantum efficiency [Eq. (9)] is a function of the absorption coefficient and the surface reflectivity. For the ~2-μm-thick absorption layers used in these photodiodes, the internal quantum efficiency increases toward unity with increasing absorption coefficient at the shorter wavelengths. The reflection loss at the InP–air interface limits η_Q to about 70%. The use of a silicon nitride dielectric antireflection layer would increase the maximum value of η_Q to 90% or more.

Because of its ability to respond over the entire 1–1.6-μm wavelength band and the relative ease with which high-purity material can be prepared, the simple and versatile GaInAs homojunction photodiode has received the greatest attention. Mesa geometries (Fig. 15a) have been described with front illumination through a thin *p* region (Pearsall and Hopson, 1978; Pearsall and Papuchon, 1978; Leheny *et al.*, 1979), window layer (Olsen, 1981), or

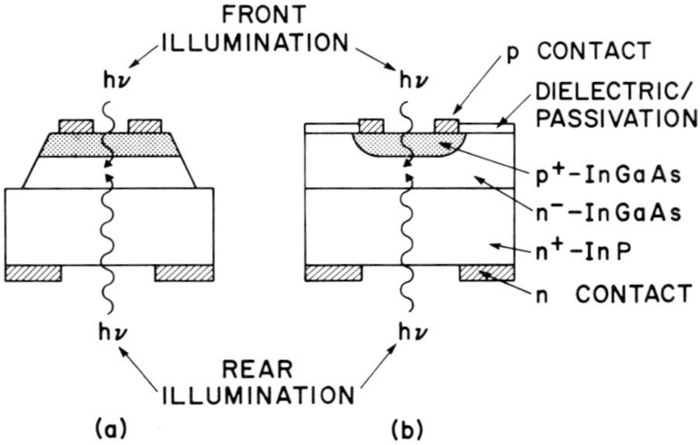

FIG. 15. (a) Mesa and (b) planar geometries of $Ga_{0.47}In_{0.53}As$ $p-i-n$ photodiodes. Illumination may be either through a thin p region or through the transparent InP substrate (compare Fig. 10).

with rear illumination through the InP substrate (Pearsall and Papuchon, 1978; Burrus *et al.*, 1981). Planar geometry devices (Fig. 15b) with front or rear illumination also have been reported (Susa *et al.*, 1979, 1980b; Forrest *et al.*, 1984a). The use of rear illumination through an InP wide-band-gap window layer offers a key advantage. All the photons will be absorbed within the n-type depletion region if it is made sufficiently wide. This eliminates the surface recombination that occurs when some of the light is absorbed in the p-GaInAs of a top-illuminated structure (Fig. 15a), and increases the quantum efficiency close to the band gap. Rear-illuminated devices may be faster, since carriers are generated directly in the depletion region and are not required to diffuse there from the p layer (Lee *et al.*, 1981b).

Restriction of the junction region to the smallest possible area minimizes both photodiode capacitance and dark current. In the case of the planar device, the junction is formed by diffusion through a mask, which if prepared of a proper dielectric material, can also serve as a surface passivation layer. The junction of a mesa device may be formed by diffusion, or may be formed directly during epitaxial growth by suitable doping. Diffusion can be replaced or supplemented by Be-ion implantation in some device designs. The mesa itself is generally formed by chemical etching. The practical range of photodiode-junction regions extends from diameters of about 25 μm for applications involving high speeds or photoreceivers for single-mode fiber systems (Lee *et al.*, 1981b) to diameters up to 1 mm for laser-power monitoring applications (Burrus *et al.*, 1981).

The design parameters for a GaInAs $p-i-n$ photodiode can be determined from the nomograph of Fig. 16 (Forrest, 1981). Shown are the deple-

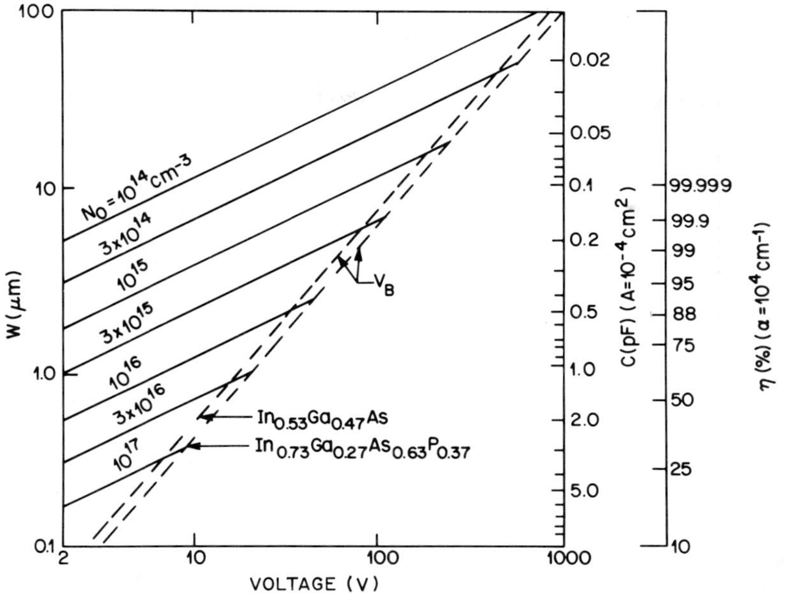

FIG. 16. Depletion-layer width W, junction capacitance C_j, and internal quantum efficiency $(1 - e^{-\alpha W})$ as functions of reverse-bias voltage V_b and net carrier concentration $N_D - N_A$ for a junction area of 10^{-4} cm^2. The dashed lines indicate the breakdown voltage for GaInAs and a GaInAsP composition. [From Forrest (1981).]

tion-layer width W, the junction capacitance $C = \varepsilon A/W$, scaled to a junction area $A = 10^{-4}$ cm^2, and the internal quantum efficiency $(1 - e^{-\alpha W})$, assuming an absorption coefficient $\alpha = 10^4$ cm^{-1}. All are functions of the reverse-bias voltage as well as the depletion-region net donor concentration N_D. The dashed lines indicate the breakdown voltage for GaInAs, as well as for a wider band-gap GaInAsP composition. From Fig. 16, design of a $p-i-n$ photodiode with greater than 90% internal quantum efficiency and capacitance less than 0.5 pF requires a depletion-layer width greater than about 2 μm. To achieve these characteristics with an operating bias below 10 V demands a net donor concentration below 3×10^{15} cm^{-3}. Devices with these characteristics have been reported by several authors (Leheny *et al.*, 1981; Lee *et al.*, 1981a).

The maximum speed of the $p-i-n$ photodiode depends only on junction capacitance and depletion layer width, as long as carrier diffusion effects are avoided, as in a properly designed rear-illuminated structure. By constructing a photodiode with a junction diameter of only 25 μm, a junction capacitance of 0.1 pF can be achieved, which in a high-speed 50-Ω circuit will give a

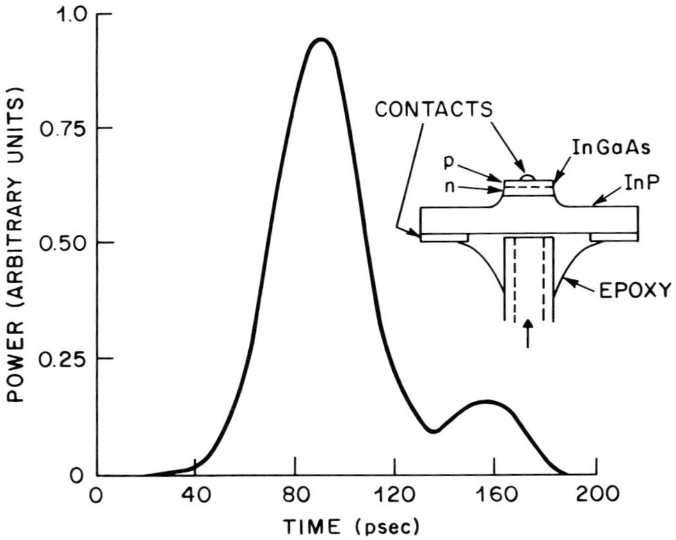

FIG. 17. Measured response (40-psec FWHM) of a high-speed 25-μm diameter GaInAs detector to an optical excitation pulse from a thin-film laser. The inset shows the device geometry. [From Stone and Cohen (1982).]

circuit-limited *RC* response-time contribution of only 5 psec. The transit-time contribution for a $W = 2$-μm-wide depletion region is about 20 psec (Section 6). Such ultrahigh-speed mesa-geometry GaInAs $p-i-n$ photodiodes have been demonstrated (Lee *et al.*, 1981b). The temporal response of one such device (Stone and Cohen, 1982) is shown in Fig. 17, with the device geometry shown in the inset. For the planar geometry of Fig. 15b, care must be taken to prevent illumination of the undepleted *n* region surrounding the depletion region. If this should occur, significantly slower response due to hole diffusion to the depletion region will result.

The dark-current characteristics of GaInAs $p-i-n$ photodiodes have been examined by a number of workers in some detail (e.g., Forrest *et al.*, 1980b). At low voltages, and near room temperature, the dark current is dominated by generation–recombination via traps in the depletion region. At higher voltages, but still below the region of avalanche breakdown, tunneling of carriers across the band gap increases the dark current significantly (Section 7). Figure 18 shows the current–voltage characteristics of a typical GaInAs $p-i-n$ photodiode (sample A) with $N_D = 1.1 \times 10^{15}$ cm^{-3} and area $A = 1.4 \times 10^{-5}$ cm^{-2}. The $I \propto V^{1/2}$ dependence characteristic of generation–recombination [Eq. (14)] is observed up to about 100 V, above which tunneling [Eq. (19)] is dominant, and I varies almost exponentially with bias.

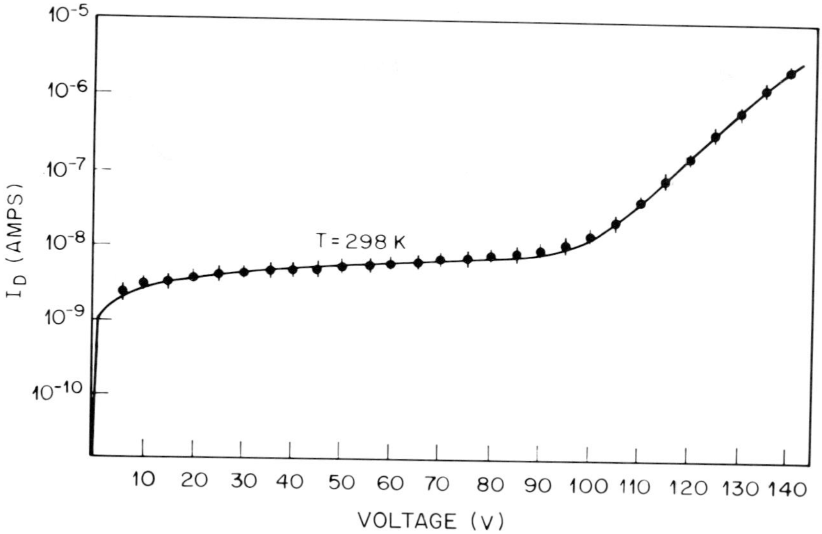

FIG. 18. Measured (points) and calculated (line) room-temperature current–voltage characteristic of a GaInAs $p-i-n$ photodiode with $N_D - N_A = 1.1 \times 10^{15}$ cm^{-3} and area $A = 1.4 \times 10^{-5}$ cm^2 ($\tau_{eff} = 340 \pm 30$ nsec). [From Forrest *et al.* 1980b).]

Avalanche breakdown for this carrier concentration takes place above $V \approx 175$ V. Tunneling currents are generally unimportant for the typical low-voltage operation of $p-i-n$ photodiodes. The effects of tunneling on APDs are discussed later.

Figure 19 shows the $I-V$ characteristics of a second diode (sample B) with larger area, $A = 2.9 \times 10^{-4}$ cm^2, and higher carrier concentration, $N_D = 1.5 \times 10^{16}$ cm^{-3}, for several temperatures. Again, the regions of generation–recombination and tunneling dominance of the dark current can be clearly discerned. Tunneling is significant even at 25 V in this case because with the higher carrier concentration, the required high field is reached at a lower applied voltage. Avalanche breakdown occurs just beyond 30 V bias.

In the low-voltage regime, $I \propto \exp(-E_g/2kT)$, indicative of generation–recombination via midgap traps, up to the highest temperature indicated [Eqs. (14) and (15)]. This can be clearly seen in Fig. 20, which shows the dark current for the diodes of Figs. 18 and 19 at 5 V bias as a function of temperature. In the vicinity of room temperature, I increases by an order of magnitude for each increase of 50°. Above about 100°C (373 K), the contribution of diffusion current, proportional to $\exp(-E_g/kT)$, is expected to be significant.

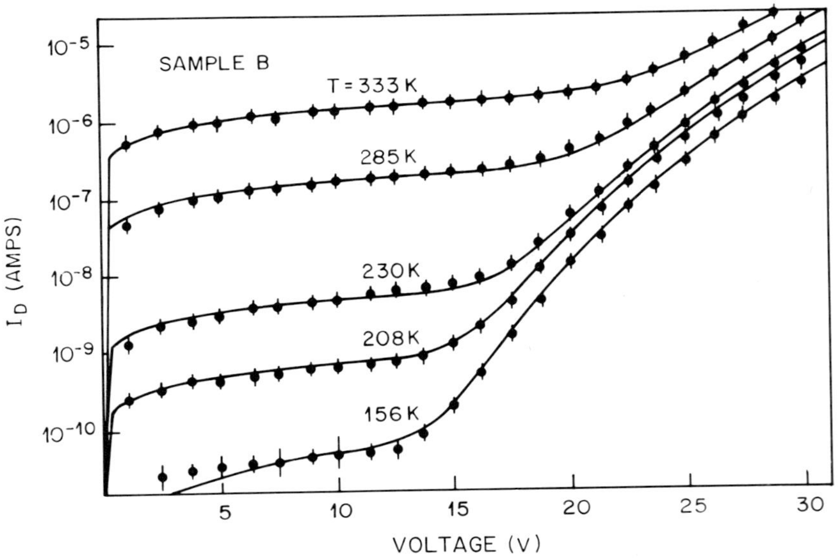

FIG. 19. Measured (points) and calculated (lines) current–voltage characteristics of a GaInAs *p–i–n* photodiode with $N_D - N_A = 1.5 \times 10^{16}$ cm^{-3} and area $A = 2.9 \times 10^{-4}$ cm^2 at five different temperatures ($\tau_{eff} = 15 \pm 2$ nsec). [From Forrest *et al.* (1980b).]

For system speeds below about 45 Mbit/sec, even the relatively small level of generation–recombination current in bulk GaInAs may pose a serious problem for receiver sensitivity (Fig. 5). This contribution, given by Eq. (14), is fundamental in nature, so that perfection in material preparation beyond maximizing τ_e cannot be expected to lead to a significant reduction in dark current. Nishida *et al.* (1979) were the first to show that special device structures employing wider band-gap junctions and that using a narrow-band-gap alloy only where necessary (e.g., to absorb the incident photons) could be used to significantly reduce the dark-current level. The major application of this concept is for APDs, which operate at large reverse voltages near breakdown, such as the SAM-APD discussed in Subsection c. Another adaptation of this idea, however, is to *p–i–n* photodiodes designed to be operated at low bias voltages (0–10 V) (Capasso *et al.,* 1980b; Pearsall *et al.,* 1981).

One such heterostructure (Pearsall *et al.,* 1981) consists of a p^+ layer of photon-absorbing GaInAs grown on low-doped *n*-type InP. Dopant diffusion leads to a junction in the InP layer close to the heterointerface. A 50-fold reduction in dark current over a homostructure has been demonstrated, in good agreement with theory. A small applied bias is required in order to

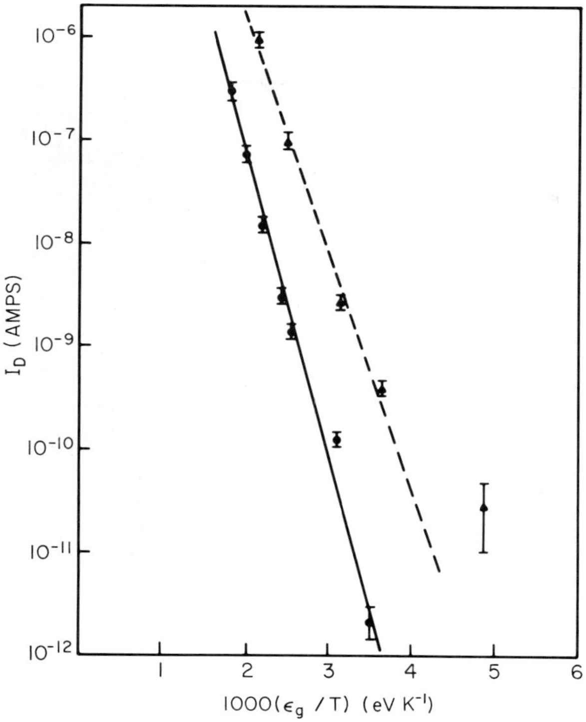

FIG. 20. Dark current versus E_g/T at 5 V for the samples of Fig. 18. The curves follow $I \propto \exp(-E_g/nkT)$. [From Forrest *et al.* (1980b).]

reduce the barrier at the heterointerface. In Fig. 21, the time-resolved response of the photodiode is shown at various levels of bias for optical excitation with a 1.3-μm laser source (Pearsall *et al.,* 1983b). The speed of the diode, (180 psec FWHM) which is diffusion limited, is unchanged by the bias under these conditions; only the quantum efficiency is affected. At 14 V, nearly the maximum quantum efficiency (60%) is obtained. Using this structure, the room-temperature dark current for a photodiode of 200-μm diameter is 10^{-10} A, and at 60°C, the current is 4×10^{-9} A, under an operating bias of 1 V (Pearsall *et al.,* 1981).

c. APD Performance

Early work on APDs in the GaInAsP- and GaInAs-material systems was based exclusively on homojunction structures (e.g., Hurwitz and Hsieh, 1978; Olsen and Kressel, 1979; Lee *et al.,* 1979b; Takanashi and Horikoshi,

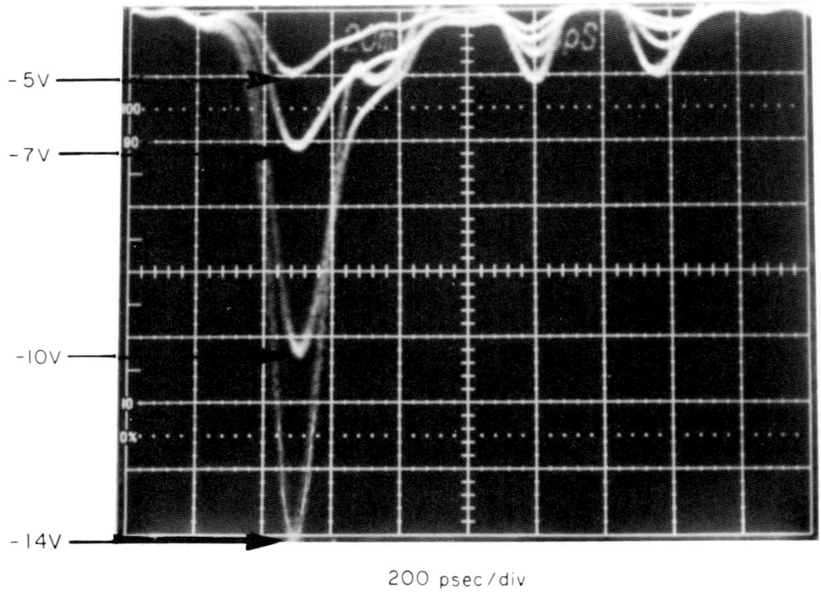

FIG. 21. Time-resolved diffusion-limited photoresponse of a heterostructure *p–i–n* photodiode showing 160-psec FWHM. Excitation is from a pulsed 1.3-μm diode laser with 38-psec FWHM. [From Pearsall *et al.* (1983b).]

1979; Yeats and Chiao, 1980). It soon became apparent, however, that "soft" breakdowns were typical and that large tunneling contributions to the dark current were the case at the high fields required for avalanche gain (Forrest *et al.,* 1980a; Ando *et al.,* 1980).

In Section 7, the tunneling current I_{tun} was given as a function of maximum junction field. Tunneling-current calculations assuming a uniform field have been reported by Ito *et al.* (1981) and by Takanashi *et al.* (1980), but these tend to overestimate the tunneling contribution. A more accurate approach is to treat the differential contribution to the tunneling current from each small segment of the high-field region and to integrate across this region, substituting at each point the actual value for the electric field. When the tunneling currents calculated in this way are compared to measured GaInAs diode currents, good agreement between theory and measurement is obtained (Pearsall, 1980). By using this procedure, one can find the bias voltage at which the tunneling-current density exceeds a given value, taken here as 10^{-1} A cm^{-2}. This result is shown in Fig. 22 as a function of carrier concentration in the high-field region. The avalanche breakdown voltage for

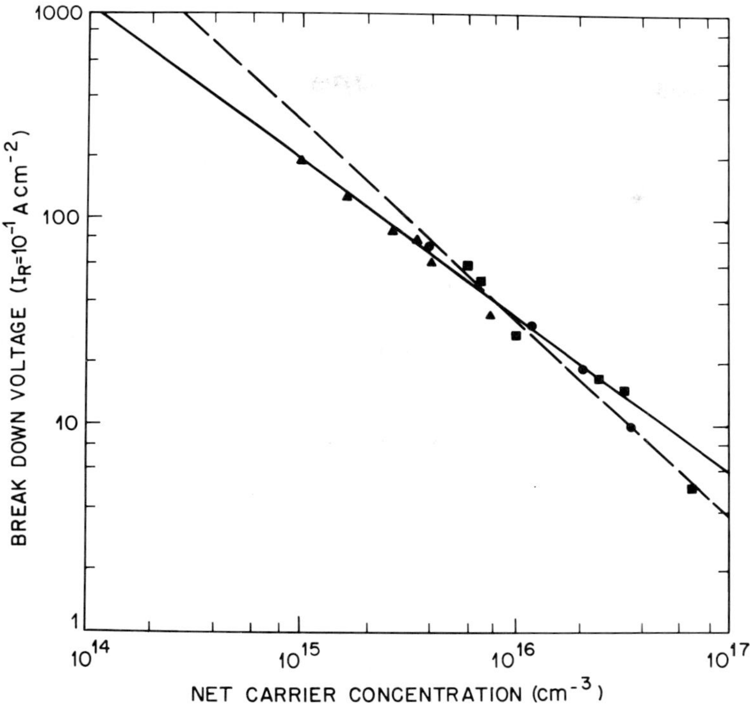

FIG. 22. Bias voltage for a tunneling current density (––) of 10^{-1} A cm^{-2} and for avalanche breakdown (—), as a function of carrier concentration for a $Ga_{0.47}In_{0.53}As$ p^+-n abrupt homojunction at $T = 295$ K.

impact ionization is also shown in Fig. 22. A comparison of these curves shows that tunneling dominates over avalanche breakdown for carrier concentrations greater than about 10^{16} cm^{-3} in GaInAs, making the observation of avalanche breakdown and avalanche gain difficult. In order to achieve substantial avalanche gain in GaInAs, material with carrier concentration below 10^{15} cm^{-3} must be used.

The application of heterostructures to eliminate the large tunneling currents found in homojunction APDs has permitted significant progress to be made toward practical avalanche devices. As mentioned briefly in the previous discussion on $p-i-n$s, such structures, first demonstrated by Nishida *et al.* (1979), make use of a narrow-band-gap region to absorb the incident photons and a separate wide-band-gap region to support avalanche multiplication. Any APD structure having these separated regions has become known generally as a separate absorption and multiplication APD (SAM-

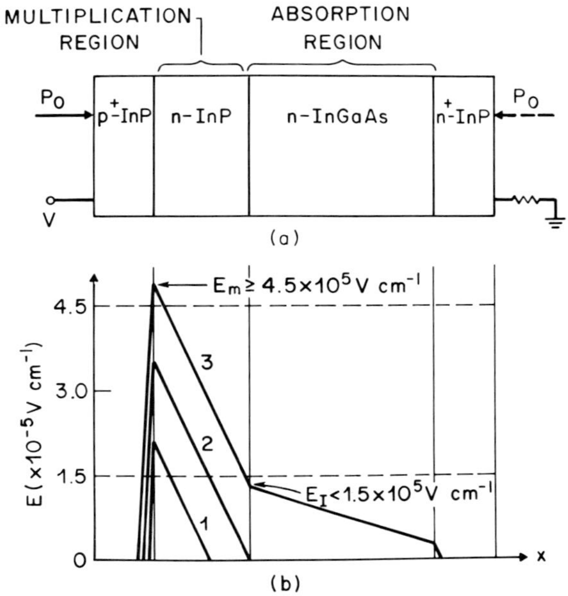

FIG. 23. (a) Schematic of a SAM-APD with *n*-GaInAs absorption region and *n*-InP multiplication region. Optical power may be incident from either side. (b) Electric field distribution with increasing bias from 1 to 3. Proper operation requires biasing for field configuration 3, $E_I < 1.5 \times 10^5$ V cm^{-1}, to avoid tunneling in *n* GaInAs, and $E_m > 4.5 \times 10^5$ V cm^{-1}, for multiplication in *n* InP.

APD). Figure 23 represents such a structure with a GaInAs absorption region and an InP multiplication region, as well as the corresponding field distribution required for proper operation. As in the simple $p-i-n$ structure, the narrow-band-gap absorbing region must be sufficiently thick to absorb all of the incident photons in order to yield a high quantum efficiency and must be sufficiently low in carrier concentration so that it can be fully depleted with a modest applied field. The SAM-APD is designed so that the field at the heterointerface remains sufficiently small to avoid significant tunneling currents when avalanche breakdown is reached in the multiplication region. For a GaInAs absorber, the field appearing at the heterointerface E_I must be kept below approximately 1.5×10^5 V cm^{-1} (Cook *et al.*, 1981).

The tight requirements on the heterointerface field—large enough to deplete the absorbing region yet low enough to avoid significant tunneling—in turn require precise control of the characteristics of the *n*-InP multiplication region. The maximum field in this region must be sufficiently high to ensure avalanche multiplication, or $E_m \geq 4.5 \times 10^5$ V cm^{-1} (Armiento *et*

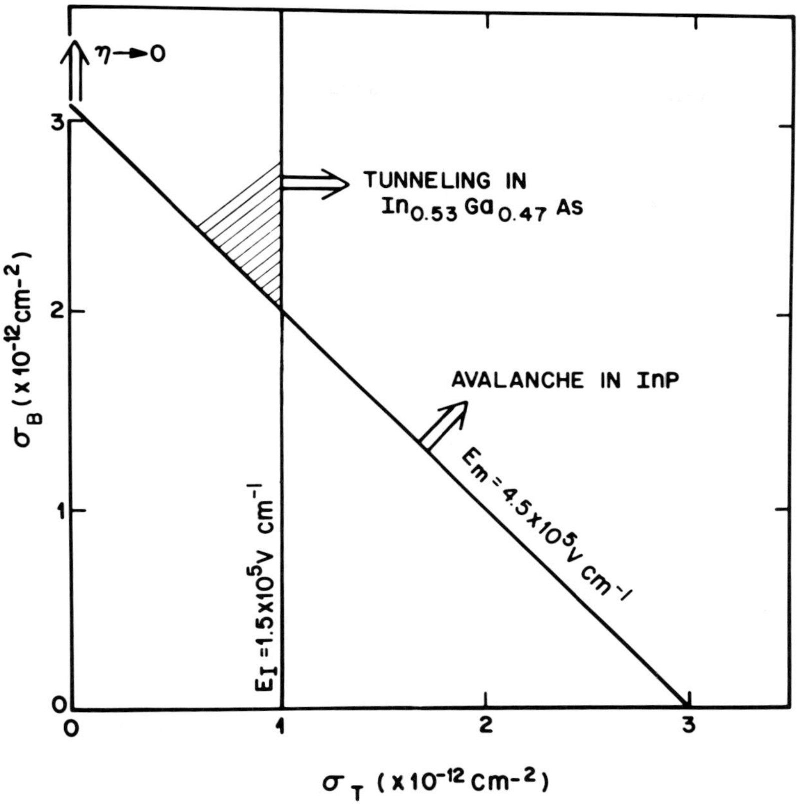

FIG. 24. Design range of a GaInAs/InP SAM-APD. The axes are the total swept-out charge in the ternary absorption region σ_T and in the binary multiplication region σ_B. [From Kim *et al.* 1981).]

al., 1979; Umebu *et al.*, 1980). Kim *et al.* (1981) have shown that this corresponds to a requirement on the total charge swept out from the binary *n*-InP layer σ_B and from the ternary *n*-GaInAs layer σ_T such that $\sigma_B + \sigma_T \geq 3 \times 10^{12}$ C cm^{-2}. The design range of the structure for multiplication without tunneling is shown in Fig. 24. The requirement that the heterointerface field $E_I \leq 1.5 \times 10^5$ V cm^{-1} corresponds to $\sigma_T \leq 1.0 \times 10^{12}$ C cm^{-2} and is given by the vertical line in Fig. 24. Tunneling in the absorption region is avoided by a design to the left of this line. The requirement on total charge $\sigma_B + \sigma_T \geq 3 \times 10^{12}$ C cm^{-2} to ensure multiplication in the InP region corresponds to a design above the diagonal line. When $\sigma_B > 3 \times 10^2$ C cm^{-2}, breakdown in the *n*-InP region occurs before punchthrough of the field into

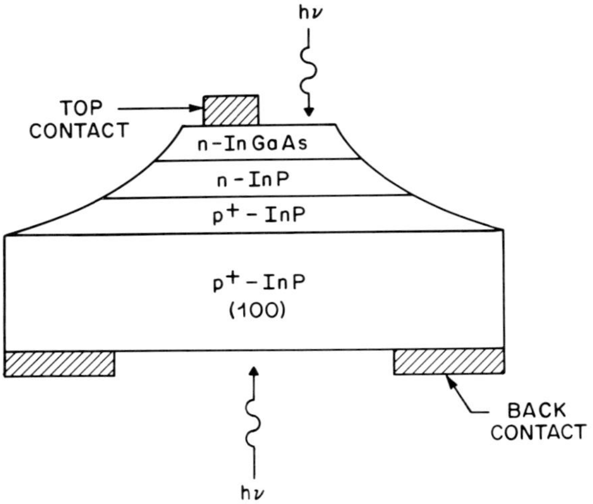

FIG. 25. GaInAs/InP mesa-type SAM-APD, grown on a p^+-InP substrate.

the absorbing region, and the quantum efficiency is negligible. For these reasons, a design falling in the shaded region of Fig. 24 gives the highest performance APD. This design can be achieved, for example, with an n-InP layer thickness of 2 μm and carrier concentration of $1.0-1.5 \times 10^{16}$ cm^{-3}. Appreciable tunneling currents may arise in the InP homojunction itself if the carrier concentration N_B is made greater than 5×10^{16} cm^{-3} (Ando *et al.*, 1980). Such tolerances on layer thickness and/or carrier concentrations stretch the limits of the current state of the LPE growth art. The development of vapor-phase-growth techniques may be required to achieve the necessary reproducibility.

Mesa-type GaInAs/InP SAM-APDs using the basic structure of Fig. 25 have been reported by a number of workers (Kanbe *et al.*, 1980; Susa *et al.*, 1980a, 1981b; Kim *et al.*, 1981). The dark-current and photocurrent multiplication of an experimental high-gain low-tunneling-leakage-current device constructed using the previous design criteria are shown in Fig. 26 (Forrest *et al.*, 1982a, 1983). In this device, the n-InP multiplication region and n-type GaInAs absorbing region have been grown sequentially on a (100) p^+-InP substrate–buffer-layer combination by LPE. The dark current and photocurrent are both negligible for low applied bias voltages. Under these conditions, the depletion region does not extend beyond the wide-band-gap InP layer, so the generation–recombination current is low. Photogenerated holes are prevented from reaching the junction by the valence-band heterobarrier. This condition corresponds to the field condition 1 in Fig. 23b.

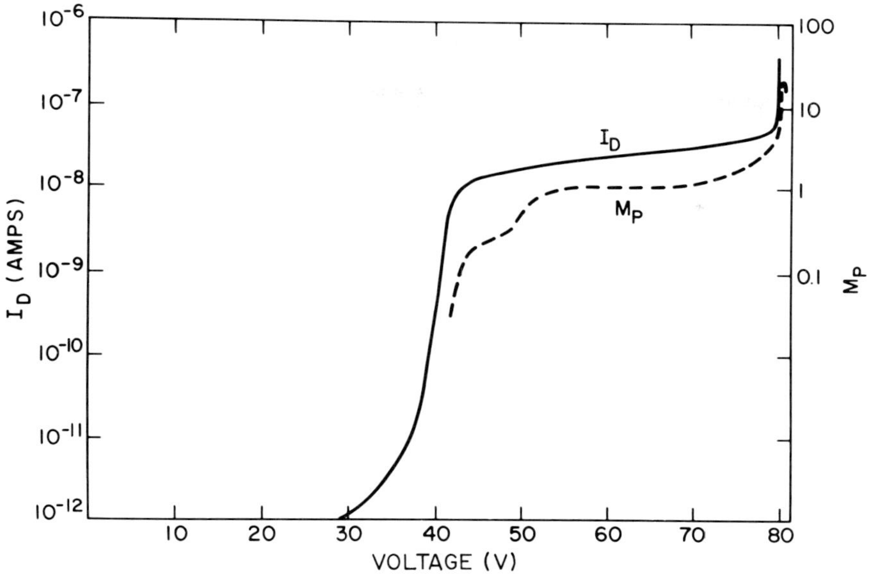

FIG. 26. Dark current I_D and photocurrent multiplication M_0 of the GaInAs/InP mesa-type SAM-APD of Fig. 25 as functions of the reverse-bias voltage. [From Forrest *et al.* (1983).]

At a sufficiently high reverse bias, about 40 V for this device, the depletion region reaches through to the InP/GaInAs interface (condition 2, Fig. 23b). The photocurrent increases rapidly with voltage as all photogenerated holes are collected, and the internal quantum efficiency approaches 100%. At this point, the dark current increases as well, for now the generation–recombination volume includes a portion of the narrow-band-gap GaInAs region. At still higher reverse bias, near 79 V, the onset of avalanche gain in the InP multiplication region is reached, and both photocurrent and dark current increase rapidly (condition 3, Fig. 23b). Spatially uniform gains of 20–30 were obtained in this particular device, with gains as high as 100 observed for other photodiodes (Forrest *et al.*, 1983). The rather small increase in dark current from "reach-through" to breakdown demonstrates that the tunneling-current contribution is negligible, because of the proper choice of layer thicknesses and carrier concentrations. However, the absolute level of dark current is still significant. Further experimental efforts will be required to reduce the dark current through increased emphasis on material and interface quality (bulk and interface generation–recombination current), as well as processing related factors (surface- or defect-generated leakage current).

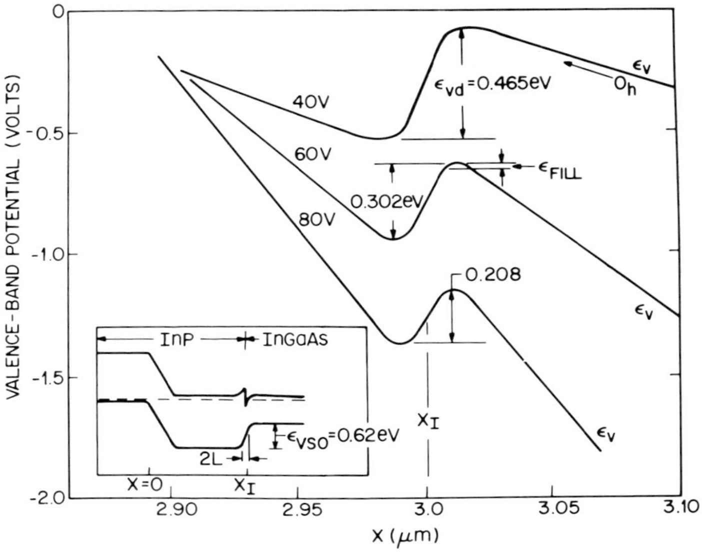

FIG. 27. Band structure of the GaInAs/InP SAM-APD at zero bias (inset) and details of the valence-band-edge potential (with voltage offset) for three values of bias voltage for a typical diode with grading length $2L = 100$ Å. Photogenerated holes are prevented from reaching the junction by the valence-band interface barrier at low bias voltage. [From Forrest *et al.* (1982b).]

In addition to affecting the magnitude of the photoresponse and dark current of the SAM-APD, the heterobarrier between the InP multiplication region and GaInAs absorption region also affects the photoresponse speed. A slow component in the photoresponse has been observed in some of the devices of the type described earlier and attributed to a pile-up of holes at the heterointerface valence-band discontinuity (Forrest *et al.*, 1982a,b). The inset to Fig. 27 shows the valence-band discontinuity of 0.62 eV at zero bias. Measurements of photoresponse speed as a function of bias voltage, which controls interface electric field, and temperature are in good agreement with a model describing the thermal emission of holes over the interface barrier. The reduction of the barrier with applied bias is shown in Fig. 27. Compositional grading of the interface, either continuously in a vapor-phase growth method or by the addition of one or more stepwise buffer layers during LPE growth, has been proposed as a solution to the carrier pile-up problem. Calculations show that a grading layer of only about $2L = 600$ Å is sufficient to make the effect of the heterobarrier hole trap negligible (Forrest *et al.*, 1982b). Such devices, with GaInAs absorption regions and a single GaInAsP buffer layer, have been reported by Mikawa *et al.* (1983) and by Campbell *et al.* (1983). In these devices, the 3-dB cutoff frequency at a gain of 10 was extended from about 100 MHz to over 1 GHz by the inclusion of a single buffer layer between the GaInAs absorption and InP multiplication regions. Devices with multiple buffer layers have also been reported (Matsushima *et al.*, 1981, 1982).

The complete frequency response of a SAM-APD with a single buffer layer has been examined in detail by Campbell *et al.* (1985). At low values of multiplication ($M \lesssim 5$) the bandwidth is typically limited to 1.2–1.8 GHz by the effective hole transit time as well as by the *RC* time constant of the circuit. At higher values of multiplication, a constant gain–bandwidth product of 10–18 GHz is observed. This arises both from the avalanche process itself, as discussed in Section 8 and shown in Fig. 12, and from some residual trapping of holes at the heterointerfaces. Similar results have also been observed by Yasuda *et al.* (1984).

The planar geometry SAM-APD, like the planar geometry p–i–n, has the potential advantages of a fully passivated junction-surface region and reproducible manufacture. However, it is a more difficult design from the viewpoint of fabrication technology. In particular, the quality of the dielectric–semiconductor interface associated with the passivation layer is of critical importance to the performance and reliability of the device. Poor interface quality can lead to excessive leakage current and ultimately to device failure. The very first SAM-APD reported (Nishida *et al.*, 1979) was in fact a planar GaInAsP/InP device. Since then, planar structures with improved performance have been reported by many workers (Ando *et al.*, 1981; Shirai *et al.*, 1981a,b,c, 1982a,b; Matsushima *et al.*, 1982; Ikeda *et al.*, 1983; and Matsushima *et al.*, 1984). The high-speed performance of planar geometry SAM-APDs has also been confirmed by Kobayashi *et al.* (1984), Sugimoto *et al.* (1984), and Yasuda *et al.* (1984).

Figure 28 shows the cross section of a planar SAM-APD with GaInAs absorbing layer and two GaInAsP buffer layers (Matsushima *et al.*, 1982). Multiplications of 15 with dark currents of ~ 100 nA have been reported for these devices. The epitaxial layers of the device are all grown n type, with the p-InP region formed by Zn diffusion. The p-region formation in all planar devices must be carefully controlled so that the remaining n-InP region meets the requirement on total charge given in Fig. 24. Control of the periphery of the junction region and the intersection of the p–n junction with the surface are even more demanding. Several guard-ring structures have been proposed and demonstrated, as in Si or Ge APD designs. These have made use of Zn or Cd diffusion (e.g., Ando *et al.*, 1982) or Be ion implantation (Shirai *et al.*, 1981c) to form a peripheral region with higher breakdown voltage than that required for the central avalanche gain region.

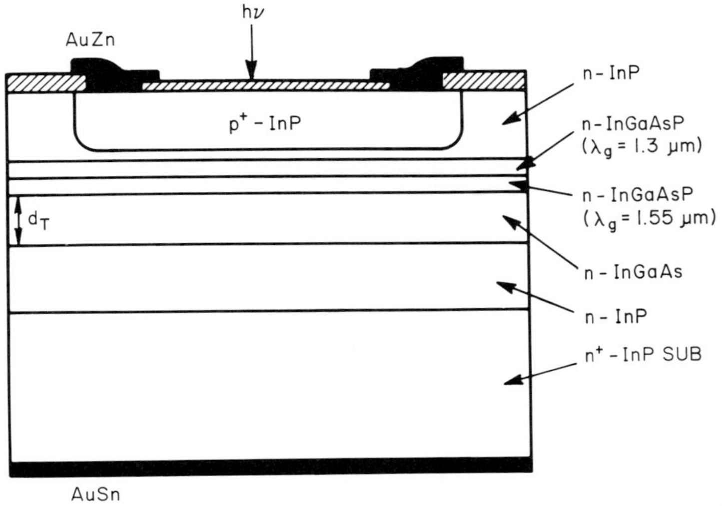

FIG. 28. Typical planar version of the GaInAs/InP SAM-APD with two GaInAsP buffer layers. [From Matsushima *et al.* (1982).]

Many of these structures incorporate double *n*-InP regions, with lighter doping near the surface, to form the guard-ring structure. Uniform avalanche multiplication in the central region without undesirable edge breakdown is the goal of all of these designs. Whether a guard-ring structure can be implemented that does not conflict with the charge requirements of Fig. 24 remains a serious question for high-performance planar GaInAs/InP SAM-APDs.

In addition to its dark current, capacitance, speed, and average avalanche multiplication properties, the performance of the GaInAs/InP SAM-APD in an optical receiver depends upon the excess noise it contributes. As discussed in Section 9, the excess-noise factor $F(M_0)$ is a function of the average multiplication and of the ionization rate ratio k. For the case of pure hole injection into InP ($\beta > \alpha$), which corresponds closely to the operating condition of the GaInAs/InP SAM-APD, Eq. (24) applies with $k = \alpha/\beta$, or more approximately, $k \sim F(M_0)/M_0$ from Eq. (4). A measurement of the excess-noise factor by Forrest *et al.* (1981b) for the mesa-geometry device discussed earlier (Fig. 25) is shown in Fig. 29. The data can be represented by $F(M_0) \approx 0.5M_0$ for $M_0 < 25$ and is close to the theoretical curves for $k \approx 0.5$ given in Fig. 13. This value is in good agreement with the $F(M_0)/M_0$ value of 0.42 reported by Yeats and Von Dessonneck (1981) for a similar structure with a GaInAsP-absorbing layer. Diadiuk *et al.* (1980, 1981) have reported much lower effective k values at low multiplications in GaInAsP/InP SAM-

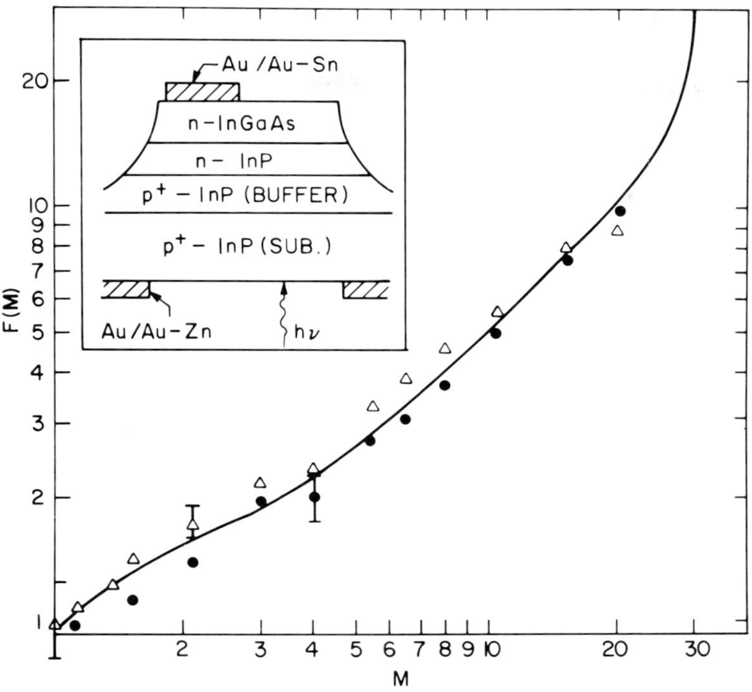

FIG. 29. Excess noise factor $F(M_0)$ as a function of average multiplication M_0 measured at 10 MHz for the mesa-type GaInAs/InP SAM-APD of Fig. 25 at two primary photocurrent levels: (△) 168 nA; (●) 16 nA. [From Forrest *et al.* (1981b).]

APDs, although these results have not been observed by others. Measurements on similar devices have been fit to Eq. (24) with comparable k values (Shirai *et al.*, 1981a,b) ranging from $\alpha/\beta = 0.52$ to 0.59.

The determination of k for InP from direct photocurrent measurements of the ionization rates α and β has also been reported (Cook *et al.*, 1982a). Careful measurements are required to obtain consistent results for these quantities (Bulman *et al.*, 1981). When properly applied, these k-value measurements have been able to predict the noise properties of SAM-APDs reasonably well (Stillman *et al.*, 1982). The fact that k depends on the electric field and possibly on the crystal orientation of the device structure (Mikawa *et al.*, 1983) should lead to further optimization of the GaInAs/InP SAM-APD.

The SAM-APD has been demonstrated in a number of different versions to date, and further improvements may be anticipated before an optimum structure evolves. For applications in which the spectral response need not extend fully to 1.65 μm, a GaInAsP composition with band-gap wavelength shorter than that of GaInAs may be used for the absorbing layer. The advantages of using an absorber with a wider energy gap are reduced generation–recombination dark current and reduced susceptibility to tunneling processes (Taguchi *et al.,* 1979; Diadiuk *et al.,* 1981). In addition, the smaller valence-band discontinuity potentially results in a less severe hole-trapping problem and greater photoresponse speed. Improvements may also be obtained by using GaInAsP to optimize the multiplication-region composition (Takanashi and Horikoshi, 1981).

11. AlGaAsSb Photodiodes

a. Material and Processing Technology

The epitaxial growth of $Al_xGa_{1-x}Sb$ on GaSb substrates is analogous in many ways to the more well-developed epitaxy of $Al_xGa_{1-x}As$ on GaAs. However, the lattice mismatch between the binary AlSb and GaSb endpoints, 0.65%, is significantly greater than the 0.13% mismatch between GaAs and AlAs. Compositions incorporating appreciable amounts ($x \gtrsim 0.2$) of aluminum therefore must contain small amounts of arsenic ($0 < y \lesssim 0.02$) to avoid serious lattice mismatch. In this case, the resulting epitaxial layer is the quaternary $Al_xGa_{1-x}As_ySb_{1-y}$, hereafter referred to simply as AlGaAsSb. The band gap in this lattice-matched alloy system can range from ~0.75 μm to the long-wavelength limit of GaSb itself, at 1.7 μm.

Although some work has been reported on the MBE (Yano *et al.,* 1979) and OM-CVD (Hess *et al.,* 1982) growth of these alloys, the primary growth method has been LPE. The substrate material GaSb is readily available commercially. It is less difficult to prepare than bulk InP and so ultimately may be less costly. Also, it can be prepared with very low defect density (~10^3 cm^{-2}). Growth by LPE usually takes place at lower temperatures, 350–500°C, than for the GaInAsP alloys (Law *et al.,* 1981). Material grown at the higher end of this temperature range tends to be *p* type, perhaps due to antimony vacancies. At the lower growth temperatures, unintentionally doped material is normally *n* type.

The LPE growth of low-background-carrier-concentration GaSb and AlGaSb layers has been achieved at the higher growth temperatures by compensating with Te, or at the lower growth temperatures by careful preparation of the starting materials. Carrier concentrations in the 5×10^{14}–5×10^{15}-cm^{-3} range have been obtained by prebaking the Ga solution in flowing H_2 for 24–48 hr at 700°C and then growing the epilayer starting at a temperature of ~350°C (Capasso *et al.,* 1981a,b). The growth of the perfectly lattice-matched quaternary alloy AlGaAsSb requires higher temperatures in order to incorporate adequate amounts of arsenic into the solid.

Although reports of extremely low carrier concentration have so far been limited to GaSb and AlGaSb, they suggest that such high-purity AlGaAsSb can also be grown for many photodiode application. The AlGaAsSb alloys can be readily prepared with a variety of dopants; tellurium and germanium have been found to be convenient donor and acceptor impurities, respectively.

b. p–i–n Photodiode Performance

The use of the AlGaAsSb alloy for *p–i–n* photodiodes permits optimization of the spectral response for the particular application. Photodiodes containing *p–n* homojunctions have been reported with quaternary absorbing layers having band-gap wavelengths in the 1.3–1.4-μm-wavelength range (Capasso *et al.,* 1981a) or with GaSb absorbing layers (Capasso *et al.,* 1980a). Heterostructure devices have also been reported (Chin and Hill, 1982). Figure 30 shows the quantum efficiency of a homojunction device in the mesa geometry. It consists of an n^+-GaSb substrate and buffer layer, an

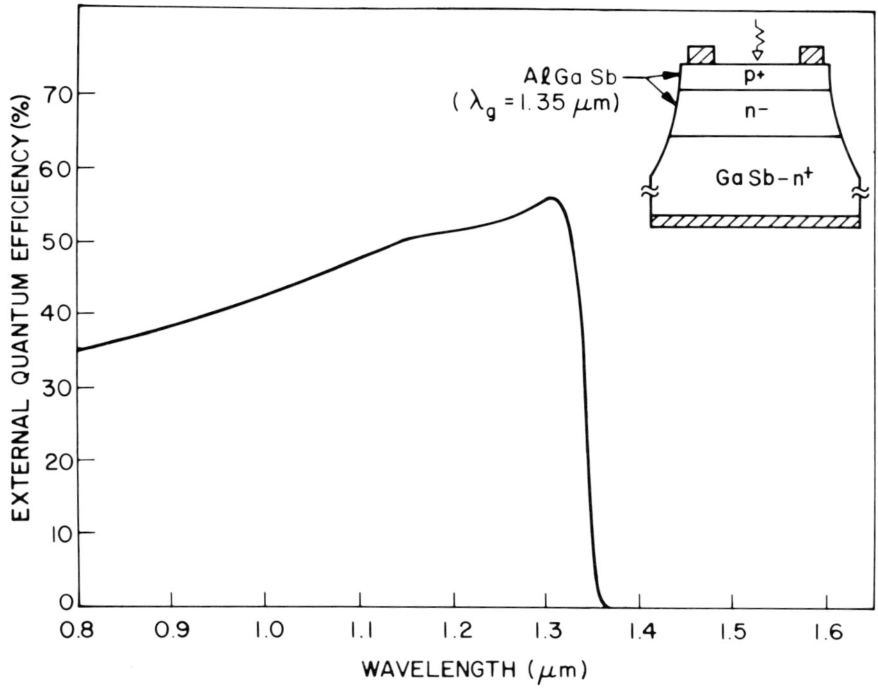

FIG. 30. Quantum efficiency as a function of wavelength for a homojunction AlGaSb mesa-type *p–i–n* photodiode (shown in inset). [From Capasso *et al.* (1981a).]

n^--$Al_xGa_{1-x}Sb$ absorption layer, and a thin ($\leq 1\ \mu m$) p^+-$Al_xGa_{1-x}Sb$ top layer. In this case, $x = 0.2$ for a band-gap wavelength of 1.35 μm. The 55% reported quantum efficiency (Capasso *et al.*, 1981a) is less than the maximum possible (~70%) because of surface recombination. Unlike the GaInAs photodiode geometry with its transparent InP substrate, the narrow-band-gap GaSb substrate does not make the rear-illuminated geometry possible, although transparent window layers can be used. The very low carrier concentration of the 3-μm-thick absorption layer, $\sim 5 \times 10^{14}\ cm^{-3}$, permits it to be depleted fully at a bias as low as 1.5 V. For a mesa diameter of $10^{-4}\ cm^2$, a capacitance of 0.4 pF has been obtained at these low voltages, together with a dark current as low as 15 nA.

The source of the relatively large dark currents typical of AlGaAsSb devices, in comparison with GaInAsP devices, has not yet been identified. Surface-leakage currents appear to be more serious in this material system, and efforts at improved processing and surface treatment have led to somewhat better device performance (Law *et al.*, 1981). Detailed measurements of the temperature and voltage dependence of dark current in homojunctions formed by Be-ion implantation have been reported (Tabatabaie *et al.*, 1982) and show that tunneling processes become dominant at relatively low bias voltages, ~10 V. At still lower voltages the surface, rather than the bulk generation–recombination contribution to the dark current, appears to be dominant.

A planar heterostructure AlGaAsSb photodiode has been reported by Chin and Hill (1982) that makes use of a wide-band-gap window layer to reduce surface leakage and improve quantum efficiency (Fig. 31). As in the SAM structure described previously, the p–n junction is situated in a wide-band-gap layer—in this case, one with a band-gap wavelength of 0.95 μm—and absorption takes place in an adjacent layer with a smaller band-gap wavelength, here ~1.35 μm. The p^+ region and n^- guard-ring structure are formed by Be-ion implantation, with a combination of SiO_2 and SiN_x used as both the implantation mask and surface passivation layer.

For a device with a ~150-μm active-region diameter and an absorbing-layer carrier concentration of $3 \times 10^{15}\ cm^{-3}$, a capacitance of 1 pF was obtained at a bias of 20 V. At this bias, the dark current was 10 nA, and the quantum efficiency was more than 70%. It should be noted that, although these values are comparable to those of GaInAs photodiodes, the dark currents are still at least an order of magnitude larger than those of GaInAsP photodiodes of comparable spectral response cutoff. At present, photodetectors suitable for the wavelength region out to 1.6 μm, whether consisting of the appropriate AlGaAsSb composition or GaSb, have considerably higher leakage currents than their GaInAs counterparts.

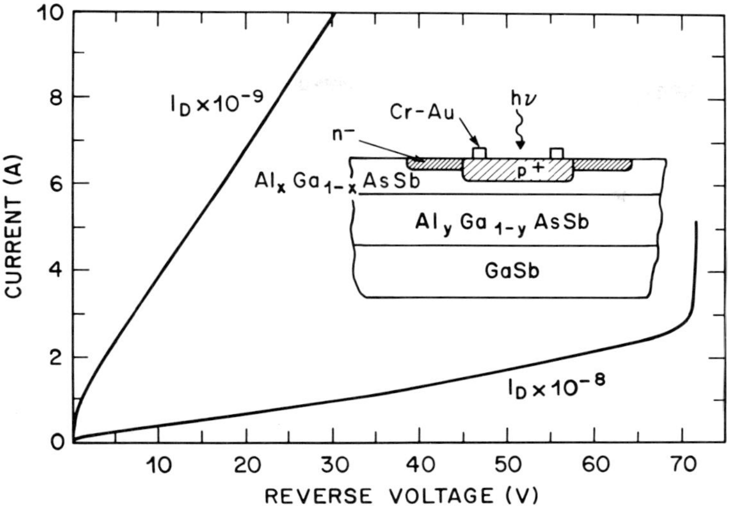

FIG. 31. Current–voltage characteristic of a planar-type AlGaAsSb $p-i-n$ photodiode (shown in inset). [From Chin and Hill (1982).]

c. *APD Performance*

In Subsection 11b it was shown that AlGaAsSb $p-i-n$ photodiodes may approach the performance of GaInAsP devices, but as yet show no advantages over them. Avalanche photodiodes fabricated from a particular $Al_xGa_{1-x}Sb$ composition, however, have the potential benefit of a very advantageous ionization rate ratio k. This is due to a feature of the $Al_xGa_{1-x}Sb$ valence band; the band-gap energy E_g and the spin–orbit splitting Δ become equal for $x \sim 0.065$, resulting in a large resonant enhancement of β, the ionization rate for hole-initiated multiplication (Hildebrand *et al.*, 1980). The band-gap wavelength for this material composition is near $\lambda = 1.55\ \mu$m, and the aluminum content is sufficiently small that it does not require the addition of As for close lattice matching.

The resonant enhancement of β arises from the reduction of the threshold energy to the band-gap energy for hole-initiated ionization (Pearsall *et al.*, 1977) and to the increase in impact-ionization probability to unity because no momentum transfer is necessary (Mikhailova *et al.*, 1976b). It has been shown by Pearsall *et al.* (1977) that for the approximation of parabolic energy bands with unequal effective masses, the ionization threshold energy E_{th} can be expressed for electrons as

$$E_{th,e} = E_g\{1 + [m_e/(m_e + m_{hh})]\}, \tag{25}$$

and for holes as

$$E_{\rm th,h} = E_g\left(1 + \frac{m_{\rm lh}(1 - \Delta/E_g)}{2m_{\rm hh} + m_e - m_{\rm lh}}\right) \quad \text{for} \quad \Delta \le E_g \qquad (26a)$$

$$E_{\rm th,h} = \Delta \quad \text{for} \quad \Delta > E_g, \qquad (26b)$$

where m_e, $m_{\rm lh}$, and $m_{\rm hh}$ are the electron, light-hole, and heavy-hole effective masses, respectively. The threshold for electron-initiated ionization is independent of Δ/E_g, whereas that for hole-initiated ionization has a minimum at $\Delta/E_g = 1$.

The compositional dependence of β/α has been measured in AlGaSb using electron-beam excitation to achieve pure hole and pure electron injection into a series of mesa homojunction diodes. The results of these measurements are shown in Fig. 32 in terms of Δ/E_g (Hildebrand *et al.*, 1980, 1981). An enhancement in β/α by almost a factor of 20 has been demonstrated, in comparison with the value of $\beta/\alpha \sim 1$ measured for GaSb alone. The small value of $k = \alpha/\beta \sim 0.05$ resulting from this enhancement would lead to an extremely low excess-noise factor. If dark currents could be reduced significantly in this material system, truly low-noise APDs for the long-wavelength region would become a distinct possibility.

12. HgCdTe Photodiodes

a. Material and Processing Technology

The material synthesis and processing technology of the II–VI alloy $Hg_{1-x}Cd_xTe$ has been developed at least as fully as the technology of the III–V alloys discussed previously. Most of this work has centered, however, on detectors and detector arrays for the 3–5-μm- and 8–14-μm-wavelength ranges (e.g., Vèrié and Ayas, 1967; Long and Schmit, 1970). The alloy $Hg_{1-x}Cd_xTe$ can be formed for any value of x, and the lattice parameter varies by only 0.3% between the HgTe and CdTe binary endpoints. The

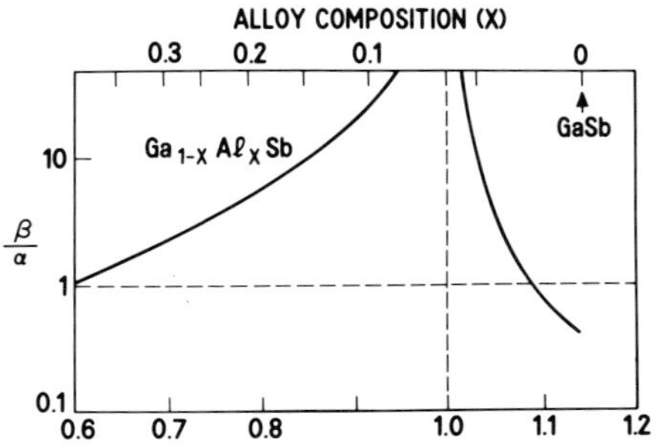

FIG. 32. Experimental dependence of the ratio β/α on composition (upper scale) and on Δ/E_g (lower scale) for impact ionization in AlGaSb. [From Hildebrand *et al.* (1981).]

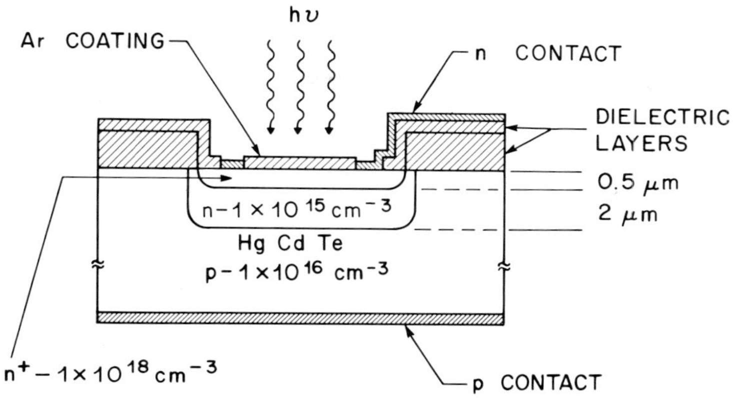

FIG. 33. Structure of a planar $Hg_{1-x}Cd_xTe$ $p-i-n$ photodiode. [After Pichard *et al.* (1982).]

direct band-gap energy can be adjusted from $E_g = 1.529$ eV (0.8 μm) for $x = 1$ (CdTe) to $E_g = 0$ for $x \approx 0.1$. The HgTe endpoint is a semimetal, with $E_g = -0.115$ eV. For photodiode applications in the 1 – 1.6-μm region, the composition can be tuned for peak photoresponse at 1.3 μm with $x = 0.7$ or at 1.6 μm with $x = 0.62$. In principle, either of these compositions can be grown epitaxially on CdTe substrates with lattice mismatches of less than ~0.1% (Chu *et al.*, 1980). However, the most advanced material technology now involves the growth of bulk crystal (Nguyen-Duy *et al.*, 1980).

The usual solution-growth methods practiced for III – V bulk materials are difficult to apply to $Hg_{0.3}Cd_{0.7}Te$, primarily due to the high vapor pressure of mercury, ~ 100 atm at the melting temperature of ~ 1000°C. The traveling-heater method permits the growth to take place from a tellurium-rich solution at reduced temperature and pressure, much like the solution growth method of LPE (Pichard *et al.*, 1982). The crystals grown in this way are *p* type, with free-carrier concentrations $N_A - N_D$ in the neighborhood of 10^{15} cm^{-3}. For a band-gap wavelength of 1.3 μm, a uniformity of ± 0.03 μm along the length of a 10-cm ingot, and of ± 0.01 μm across the 2-cm diameter, have been reported by Pichard *et al.* (1982). Junction formation in $Hg_{1-x}Cd_xTe$ has been based on Al-ion implantation followed by annealing in a Hg atmosphere.

b. p – i – n Photodiode Performance

Photodiodes of $Hg_{1-x}Cd_xTe$ can be optimized for any wavelength longer than ~0.8 μm; devices peaking at 1.3 μm receive the most attention for lightwave communications applications. Although mesa photodiodes can be fabricated readily from these materials, the processing technology is also well developed for planar structures. Such a structure is shown in Fig. 33 (Pichard

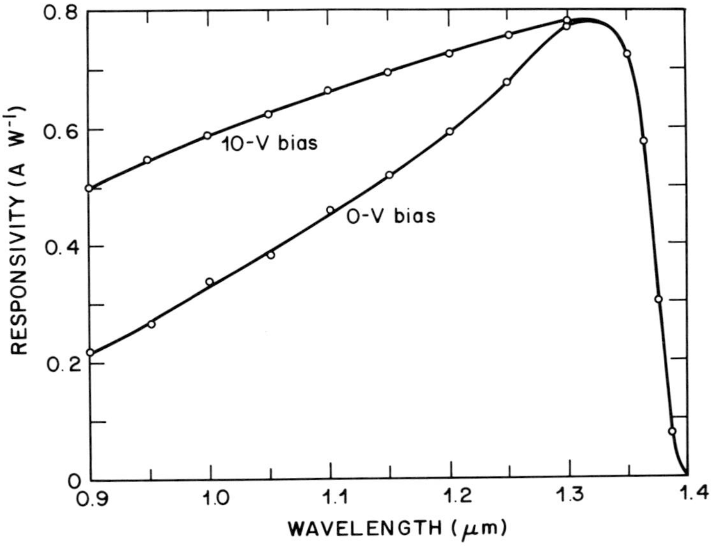

FIG. 34. Spectral responsivity of the $Hg_{1-x}Cd_xTe$ $p-i-n$ photodiode of Fig. 33 at 0 and 10 V reverse bias. [From Pichard *et al.* 1982).]

et al., 1982). It consists of a 2-μm wide Al-implanted i region ($n = 1 \times 10^{15}$ cm^{-3}) for absorption and a 0.5-μm wide implanted n region ($n = 1 \times 10^{16}$ cm^{-3}).

The spectral responsivity of this device is shown in Fig. 34 for a reverse bias of 10 V. The quantum efficiency exceeds 75% at 1.3 μm because of the large absorption coefficient ($\gtrsim 10^4$ cm^{-1}) and the use of a dielectric antireflection coating. At this bias, a receiver-type photodiode with an 80-μm-diameter active region has a capacitance of 1 pF and a typical dark current of 1–5 nA. The dependence of dark current on temperature suggests that the generation–recombination process is dominant. The dark current increases by a factor of 1.8 for each increase of 10°C. Large-area photodiodes suitable for monitor applications have also been fabricated in this material system. Devices with areas of 3×10^{-2} cm^2 have dark currents of ~100 nA at a reverse bias of 5 V (Pichard *et al.*, 1982).

e. APD Performance

The $Hg_{1-x}Cd_xTe$-alloy system offers the same potential advantage for low-noise APDs as does the $Al_{1-x}Ga_xSb$ system—a resonantly enhanced ionization rate ratio. Estimates of the spin–orbit splitting Δ, combined with measurements of the variation of band-gap energy E_g with x, lead to the conclusion that $\Delta/E_g = 1$ for $x \approx 0.73$ (Vèrié *et al.*, 1982). This alloy composition corresponds to $E_g = 0.97$ eV, or $\lambda_g = 1.28\ \mu m$. As discussed in Section 11 for $Al_{1-x}Ga_xSb$, the $E_g = \Delta$ resonance in HgCdTe results in a lowering of the threshold ionization energy and an increase in the ionization probability for holes. The ionization rate for electrons is unaffected, so that $k = \alpha/\beta$ should be reduced significantly.

Preliminary results have been reported which suggest that the expected reduction in α/β actually takes place. The electron- and hole-initiated multiplication in a planar photodiode with a spectral response cutoff at 1.41 μm were measured separately and an ionization rate ratio $k = \alpha/\beta = \frac{1}{9}$ deduced from the data (Vèrié *et al.*, 1982). Additional measurements on a range of alloy compositions are required to explore the resonant enhancement of β, but the results so far are very encouraging. Finally, it is possible that the band-gap wavelength corresponding to the resonance condition can be increased to permit optimization at 1.6 μm or beyond by making use of the quaternary alloy $Hg_{1-x}Cd_xSe_yTe_{1-y}$.

IV. Integrated Photodiode Devices

13. General Considerations for Integrated Devices

The alloys GaInAsP and HgCdTe share the common feature that they can be grown epitaxially over a wide range of compositions that can be lattice matched to commercially available single-crystal substrate material. This exceptional situation opens the way for a single wafer to contain a host of optical and electronic devices, whose detailed structure and composition can be independently optimized to implement the desired optoelectronic function. In principle, it would be possible to build an entire photoreceiver or regenerator on a single chip. However, in practice, the niche for integrated optoelectronics for communications applications is not developing precisely along these lines. In considering any scheme to integrate optoelectronic devices exploiting the common growth technology, important criteria have to be met which will justify the development expense:

1. *Functionality.* Does the integrated structure permit new kinds of information processing not possible using discrete components?
2. *Cost.* Does the integrated structure substantially reduce the system cost by eliminating labor-intensive processing and wiring operations?
3. *Performance.* Does the integrated device improve system performance by increasing system capacity or by reducing the noise?
4. *Reliability.* Does the integrated structure lead to a lower system failure rate?

5. *Process yield.* Does the integrated wafer have a higher cost-effective production yield than the individual components?

6. *Device size.* Can the integrated structure lead to critical reduction in component size?

Since light waves and electrical impulses travel at roughly the same speed, optical information processing on a chip does not necessarily lead to system speed reductions, compared to electronic processing. Guided-wave interactions require device lengths in millimeters, whereas transistor dimensions are many orders of magnitude smaller. Most important is the astounding capacity of the silicon technology to produce high-reliability low-cost integrated circuits for information-processing applications. These considerations indicate a more limited area for integrated optoelectronic devices than may have been envisioned originally.

One of the more costly components in present-day optical-fiber systems is the optical fiber connector. In the wiring of each new optical local office or industrial park where hundreds of thousands of demountable optical connections have to be made, the cost of the fiber connectors (at present about $100 each) may dominate the entire system cost. Figure 35 shows an optical

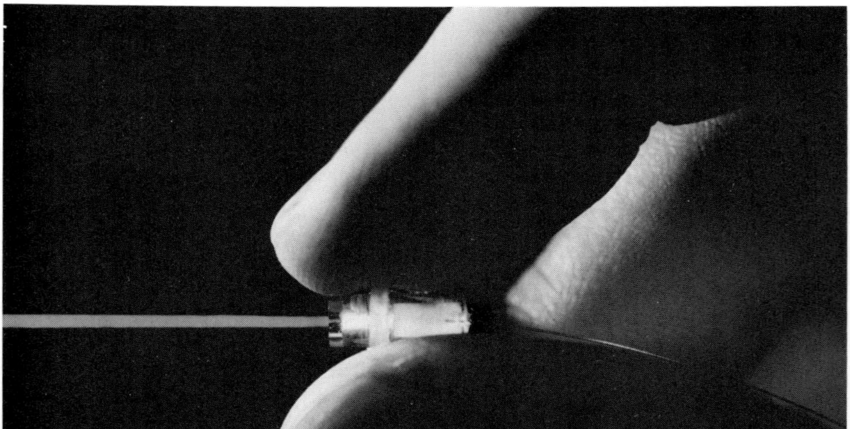

FIG. 35. An expanded beam lens core for fiber-optic connectors made by Polaroid for a major manufacturer in the fiber-optics industry. The specifications called for an aspheric lens with a diameter of less than 2 mm in a precision-molded core of an overall size less than 10 mm with mechanical dimenions and alignments to optical tolerances. The core is used to align precisely optical fibers and to register the signal transmitted between the fibers.

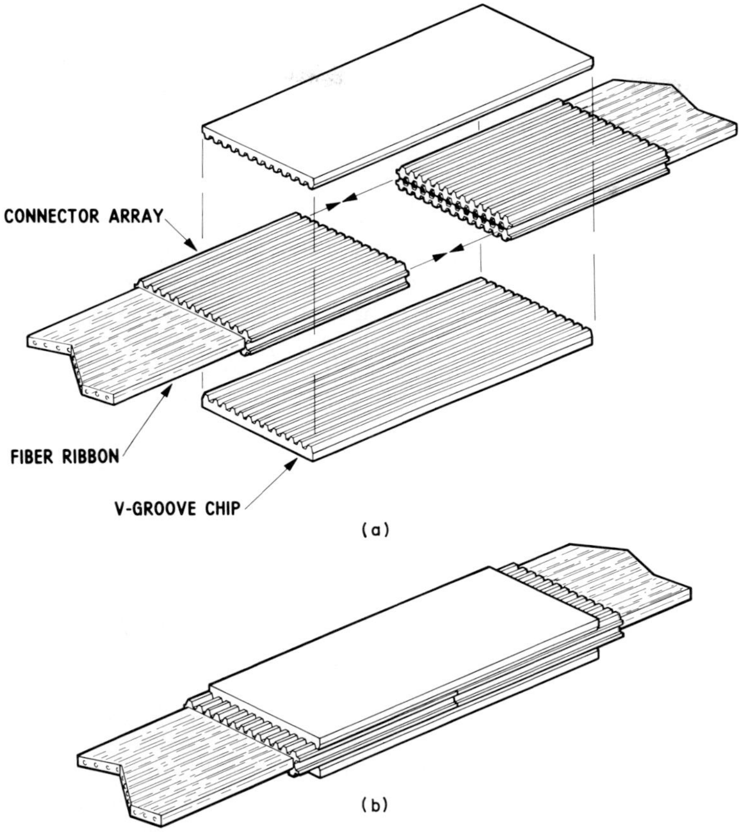

FIG. 36. Multifiber cable connector: (a) pre-assembly; [from Schroeder (1978)] (b) finished connector [from Miller (1978)]. Mechanical alignment of fibers is achieved by precision grooves etched chemically in silicon wafers. The alignment of these grooves can be controlled to fractions of a micrometer. [Reprinted with permission from the Bell System Technical Journal. Copyright 1978, AT&T.]

approach to this problem. By using a mechanical approach (Schroeder, 1978; Miller, 1978), Western Electric Co. has developed a single connector capable of aligning 12 fibers at once, as shown in Fig. 36. Substantial and immediate cost savings can be realized by integrating detectors (and sources) to form arrays of devices, all of which may be joined to the appropriate fiber at once by a single connector.

Stabilization of the output characteristics of lasers and LEDs requires a photodetector as part of the feedback loop. Integration of this photodetector immediately adjacent to the emitting device appears to be clearly advantageous with respect to cost, size, and reliability by comparison to the same function achieved by discrete devices. The principal requirements for feedback stabilization are the correct spectral response and response time. High quantum efficiency and low noise are of secondary importance because of the proximity of source and detector. Since photodetector yield of manufacture is usually determined by dark-current-related noise, the integrated source monitor should be a device function that can be incorporated with little additional penalty in the wafer yield.

In the introduction to this chapter (Part I), we discussed the effect of the receiver capacitance on minimum detectable optical power. At high bit rates, generally above 200 Mbit/sec, the photodetector capacitance needs to be below 0.3 pF so as not to contribute to the receiver noise. Reducing the capacitance by only 0.2 pF at this level will increase sensitivity by about 1 dB and the transmission distance by as much as 5 km. The interconnection or lead capacitance between a hybrid photodetector and preamplifier in the same package is typically about 0.15 pF, which could be reduced substantially if these two devices could be integrated on the same chip. The improvement in sensitivity gained by the integrated $p-i-n$/FET becomes more significant for higher bit rates, and the $p-i-n$/FET may be an alternative to the APD as a means of increasing receiver performance.

Because of the enormous bandwidth of optical fibers, wavelength-division multiplexing can be used to dramatically increase the capacity of an optical fiber by transmitting in parallel, to deliver many independent channels of information. Practical technical solutions to the multiplexing of sources are in the realm of present-day capabilities. Similarly elegant solutions for demultiplexing are not quite so evident, because of the fundamentally more difficult problem of unmixing as compared to mixing. Demultiplexing is a problem well suited to integrated optoelectronic solutions. The principal improvement to be gained is a savings of space. The nonintegrated approach requires precision optics with dimensions on the order of a centimeter, whereas wavelength-discriminatory absorption takes place on the order of 10^{-4} cm. In the rest of this part, we discuss the current state of the art in each of the four principal areas for integrated photodetector development mentioned previously: arrays, emitter-feedback monitors, $p-i-n$/FETs, and wavelength demultiplexing.

14. Detector Arrays

A detector array in fiber-optic communications can be used as an integrated element to be matched to an array of fibers or as the detector element in a wavelength-division demultiplexer relying on diffraction to achieve spatial separation of different wavelengths of light. This latter application is discussed later. The matching of detector arrays to a bundle of fibers can, in principle, be accomplished in a straightforward manner. This kind of device is likely to be used whenever groups of fibers need to be plugged in or unplugged, especially if space is at a premium. Examples of such applications are computer mainframe-to-mainframe links and at switching and distribution points in local area networks, which will be found in the "wired" office building or complex.

There are some important performance requirements, unique to arrays that further define the device technology beyond that defining the single device:

1. *Crosstalk.* What are the interdevice spacing, capacitance, and resistance?
2. *Fiber alignment.* How is the alignment of the fiber array and detector array achieved? Can fibers be plugged into the array?
3. *Packaging.* The geometry of the integrated device poses a particular problem because the fiber is naturally coupled perpendicular to the plane of the device wafer, whereas the circuitboard environment may dictate that the fiber bundle lie in the plane of the wafer.
4. *Reliability and redundancy.* How does one deal with the problem of a ten-element array with only nine working detectors: (i) before packaging? (ii) after installation?

Device crosstalk arises from three sources: optical signal, resistance, and capacitance. Although it is not shown here, contributions to signal crosstalk from optical leakage or interdevice resistance are immeasurably small compared to background noise. Interdevice capacitance, however, may have an important effect. The effect of interdevice capacitance on crosstalk becomes more important as the bit rate increases, for reasons similar to those shown earlier in the consideration of the effect of capacitance on receiver noise power.

The fabrication of arrays is a process with a finite yield. There are two basic approaches to selecting the parts of the wafer with, for example, 10 working devices in a row. One is to sequentially test every device on a wafer. With automated probing, this strategy may prove to be the simplest, especially if the individual device yield y is low. Low device yield is estimated from the expression

$$y \lesssim 1 - (1/n), \tag{27}$$

where n is the number of devices in the array.

An alternative strategy may be first to separate the wafer, without testing, into $1 \times m$ arrays, where $m \geq n$. The probability that the resulting chip will contain at least one working $1 \times n$ array can be approximated from statistical considerations and can be expressed by a recurrence relation

$$P(n, m, y) = y^n + \sum_{i=0}^{m-n-1} P(n, m-i-1)y^i(1-y). \qquad (28)$$

Equation (28) is based on the premise that device failures or defects occur randomly. In fact, this is not the case. Defect locations are strongly correlated. Hence Eq. (28) will give a somewhat pessimistic view of the situation. If the bar length does not contain more than twice the number of devices in the finished array (i.e., $m \leq 2n$), Eq. (28) may be approximated by

$$P(n, m, y) = y^n[1 + (m-n)(1-y)]. \qquad (29)$$

With a high individual device yield (e.g., for a 10-element array, $y \geq 0.9$), the wafer is first cut into $1 \times m$ bars with the probability that at least one good array is found to be given by Eq. (29). This result is shown graphically in Fig. 37. The probability of getting a good array depends dramatically on the

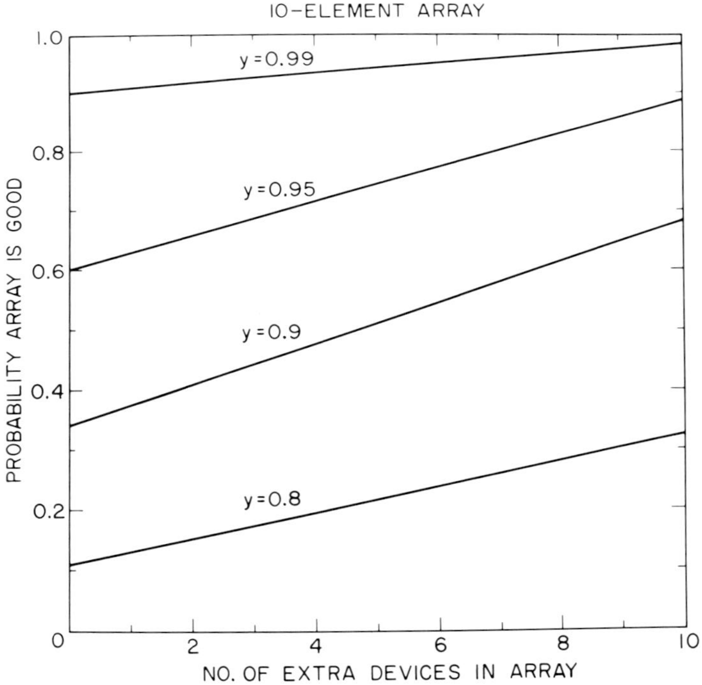

FIG. 37. Array yield as a function of extra devices in the array bar for a 10-element array. Results are shown for various values of y, the individual device yield, assuming defects, are randomly distributed. Incorporating this kind of redundancy may prove useful if the individual device yield is low.

individual device yield when the yield falls below the value given by Eq. (27). In Fig. 37 it can be seen that the strategy of increasing the number of devices in the bar improves the probability linearly. At the same time, increasing the number of elements in the bar means that an increasing fraction of devices will be wasted even if a good array is found. To obtain the yield fraction for the wafer as a whole, Eq. 29 must be multiplied by n/m. The result shows that redundancy improves the total wafer yield only if the individual device yield is low. If the individual device probability is 0.7, a 20-element bar will double the wafer yield, compared to cutting the wafer into 10-element arrays. If the wafer yield is high, cutting the wafer into 20-element bars will actually decrease the total wafer yield.

This discussion of device yield and redundancy, although elementary, serves to underscore the importance of incorporating only forgiving and reliable technology in integrated optoelectronics. Arrays must be composed of devices whose individual yields are high and whose reliability in service is dependable. Hence, the array technology may be better suited for LEDs than for lasers, and for $p-i-n$ photodiodes rather than for APDs. The use of LEDs as sources limits the system performance so that only shorter links are feasible ($L < 10$ km) at modest bit rates (<100 Mbit/sec). For these uses, the array will have to satisfy the criterion of reduced cost per circuit over that of using discrete devices.

15. Monitors for Emitter Stabilization

Optical emitter characteristics need to be controlled in three important areas: intensity, wavelength, and mode profile. Integrated detector–emitter pairs offer a convenient means of achieving the necessary monitoring for feedback stabilization. In Fig. 38, we show an integrated laser monitor (Iga and Miller, 1980) developed for output-intensity control. Because of the proximity of the source to the detector, sensitivity and noise are not such important considerations as in the case of high-bit-rate signal detection in an optical-fiber link. Deviations from the desired emitter characteristics can be controlled conveniently by means of the temperature and/or the injected current density.

It has been shown (Coldren *et al.*, 1981; Tsang *et al.*, 1983) that close-coupled resonator cavities can be used to achieve both rapid and precise mode control for semiconductor lasers. In addition, one wishes to achieve stability against long-term degradation in intensity or drift in output wavelength. For this application, the monitoring detector functions as an optical power meter and need not have a rapid response. Short-term fluctuations are those whose duration is a small fraction of the output pulse length. In order to obtain effective short-term control over variations in output wavelength or intensity, the bandwidth of the feedback circuit would have to be an order of

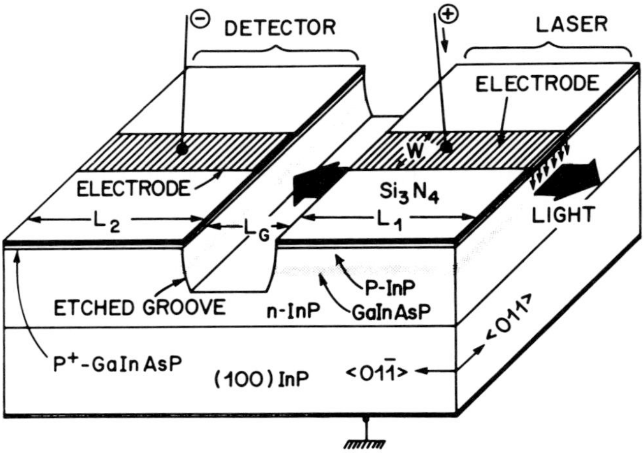

FIG. 38. Integrated laser with monitoring detector. [From Iga and Miller (1980b).]

magnitude higher than the bit rate of the signal stream. This condition implies a modulation rate greater than can be easily achieved with currently known techniques. Instead, for stabilization using integrated detectors and electronic feedback, we must be concerned with variations or drift in output characteristics that occur over many bit periods, i.e., long-term fluctuations. Rapid stabilization is more appropriately achieved by direct optical feedback, as in the case of the coupled-cavity laser (Coldren *et al.,* 1983).

Stabilization of the total optical power is achieved by detecting the near-field emission from the laser or LED. Increasing output power by increasing the drive current may additionally shift the wavelength. For some single-mode communications, this shift may not be desirable, and additional adjustment of the laser heat-sink temperature may be necessary to maintain both emission wavelength and intensity. Such a feedback scheme has been demonstrated (Brown and Smith, 1983) that results in a stability in the output laser frequency, which is better than 50 kHz. This system requires an external monochromator to achieve wavelength discrimination.

As shown in Fig. 39, the dual-wavelength photodetector (Campbell *et al.,* 1979, 1980; Ogawa *et al.,* 1981) is a promising solution to the problem of obtaining wavelength resolution sufficiently well defined to achieve and control mode stability in single-mode laser systems using an integrated device as the sensing and discriminating element. The basic properties of the dual-wavelength photodiode are shown in Fig. 39. Although it may be seen that the spectral response is quite wide, the ratio of the photocurrents detected by each of the two photodiodes integrated in this structure can be used to discriminate the output wavelength to within 20 Å (Campbell *et al.,*

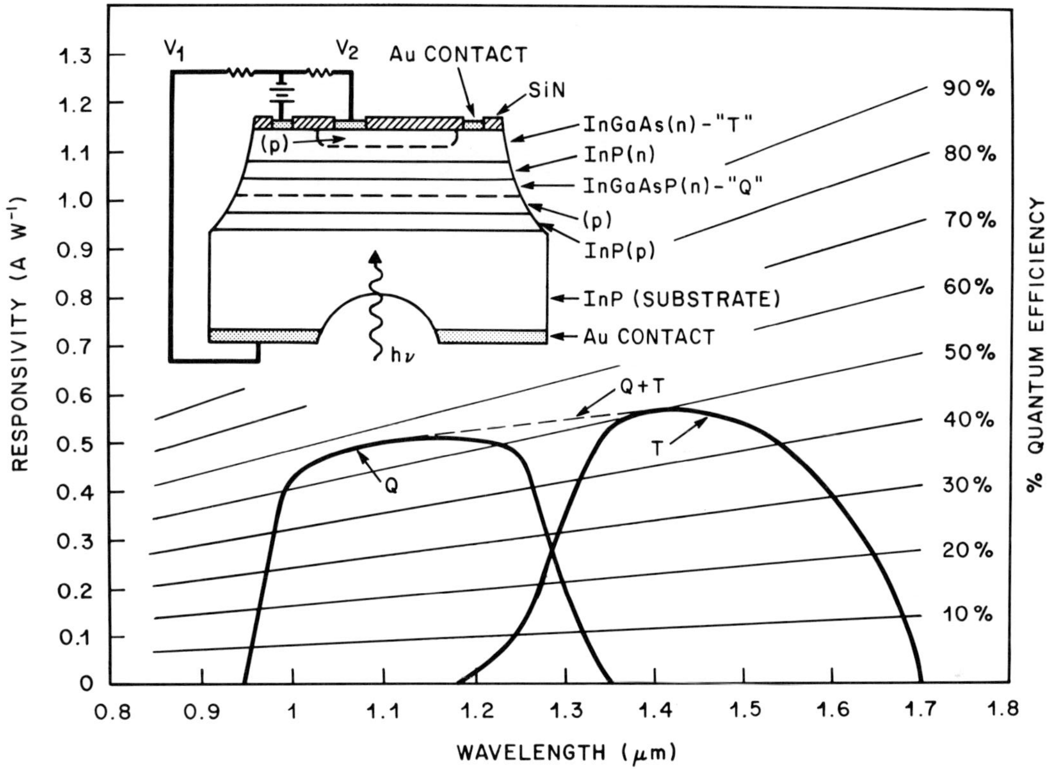

FIG. 39. Dual-wavelength photodiode. Two photodiodes of differing absorption edge are stacked one on top of the other. For emitter stabilization purposes, the absorption edge of the quaternary (Q) higher-band-gap diode should coincide with the wavelength region of interest. [From Ogawa *et al.* (1981).]

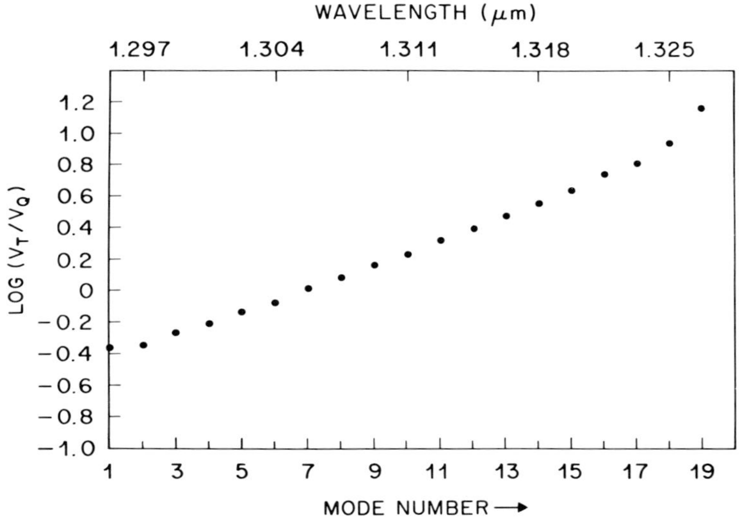

FIG. 40. Ratio of the photocurrents detected at two different wavelengths for the dual-wavelength photodetector. The results show a wavelength discrimination of 20 Å, which is sufficient to detect mode jumping in a single-mode laser. Thus, the device acts like an integrated monochromator/detector over a wavelength range of 250 Å. [From Campbell *et al.* (1983a).]

1983a). This resolution is sufficient to stabilize the mode of a single-mode laser, as shown in Fig. 40 where the log ratio of the two photodetector signals is displayed as a function of laser mode and wavelength. The appropriate stabilizing circuit is shown in Fig. 41. Similar feedback circuitry can be used to control the output wavelength and power of LEDs. Although the half-width of emission is quite large ($\Delta E \simeq 60$ meV), in this case the range shown in Fig. 40 is sufficiently large to detect drift of the peak position of the LED output. Hence, the dual-wavelength photodetector seems to be well suited for emitter stabilization against long-term drift in wavelength or operating power for both lasers and LEDs in optical-fiber telecommunications.

16. *p–i–n*/FET RECEIVERS

In Part I of this chapter it was shown that receiver noise power depends both on shot noise from the current carried by the photodiode and on the total receiver capacitance. At high bit rates (> 100 Mbit/sec), the noise associated with capacitance, which increases as the bit rate cubed [Eq. (2)], becomes the dominant component of total receiver noise. The interconnection capacitance between a discrete photodetector and amplifier contributes an important fraction of the total receiver capacitance. Integration of the

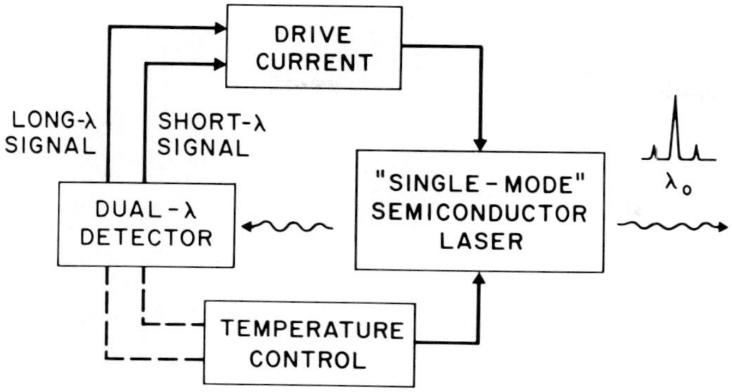

FIG. 41. Stabilizing and feedback circuit used with the dual-wavelength detector for emitter stabilization. Temperature and current-injection levels are used to control the wavelength and intensity. [From Campbell (1983).]

detector monolithically with the receiver amplifier is one way to reduce this capacitance. From Fig. 6, it can be seen that the benefits of this reduction occur chiefly above 100 Mbit/sec. For this reason, the $p-i-n$/FET can be thought of as an alternative to an APD, which also offers some performance improvement at very high bit rates.

However, at lower bit rates, the $p-i-n$/FET, unlike the APD, remains an attractive choice because it also represents a lower cost component while maintaining adequate performance. For many systems operating at these lower bit rates, component cost is a very important part of the product specification, because literally millions of these components will be needed.

There are two logical choices for the FET in $p-i-n$/FETs: GaInAs and InP. Field-effect transistors showing excellent performance characteristics have been made from each of these materials. In addition to being the same material from which photodetectors are made, electrons in GaInAs have the highest peak velocity of any known semiconductor suitable for room-temperature operation (Cappy *et al.,* 1980). Although InP does not perform as well in this regard as GaInAs for FET applications, it has the advantage of a large band gap, which leads to more stable, lower noise operation. Many of the integrated-optics applications occur at moderate bit rates, (10–100 Mbit/sec), where low noise is more important than speed, and InP seems to be a more logical choice for integrated-circuit development for these applications.

In order to obtain the benefit of reduced noise power made possible by integration, the amplifier chip must contain all the elements of the first stage. Further integration in order to fabricate the entire receiver on a chip may well bring added benefits in cost and reliability; however, the basic improve-

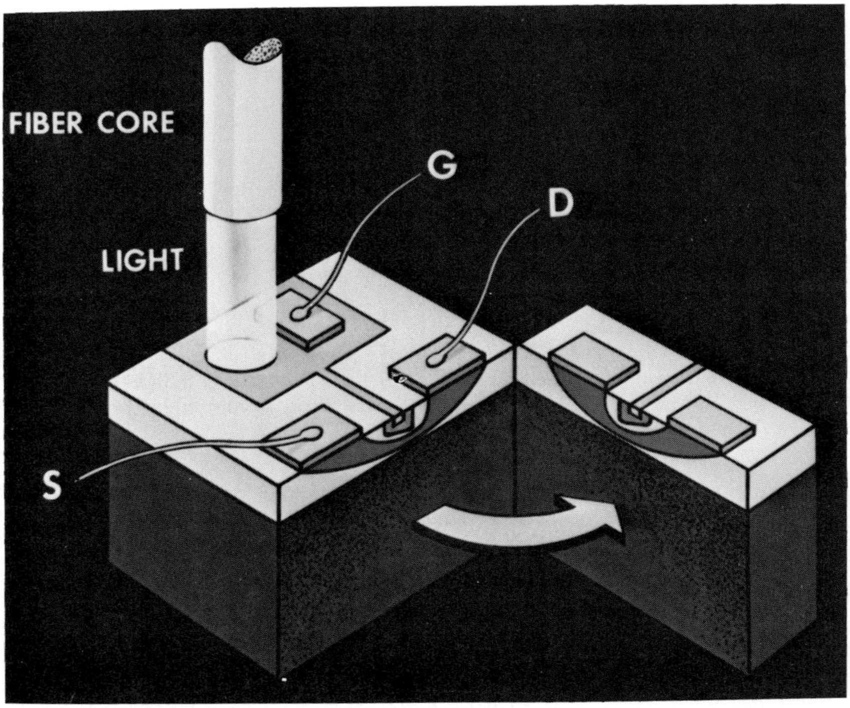

FIG. 42. Schematic diagram of photodiode integrated with the first-stage amplifier. [From R. F. Leheny (1980, unpublished.)]

ments in noise performance are all obtained in the integration of the first stage.

The photocurrent typically is amplified in an FET channel about 2000–5000 Å in thickness and containing about 10^{17} cm^{-3} free electrons. The photocurrent is generated in the depletion region of the detector, which is 2–4 μm thick and formed in material with about 10^{15} cm^{-3} free electrons. There is therefore a basic incompatibility between structural requirements for these two devices. One straightforward solution is to develop a nonplanar design incorporating the FET amplifier circuit on one level and the $p-i-n$ device on another. A schematic digram showing the nonplanar integration of a $p-i-n$ and FET is shown in Fig. 42. An alternative is an all-planar structure in which the FET channel depth and carrier concentration are defined electronically rather than structurally. This kind of FET can be made using a selectively doped heterostructure (Pearsall *et al.*, 1983a) or an insulated-gate inversion-mode transistor (Shinoda and Kobayashi, 1982).

The first integrated $p-i-n$/FET suitable for $1-1.6$-μm telecommunications was fabricated using a nonplanar structure and a junction FET (Leheny *et al.*, 1980a). A schematic of this device, which contained only one transistor, is shown in Fig. 43. Although not optimized for noise performance, as discussed earlier, this device showed useful gain and served to stimulate much work toward the development of a useful FET technology that could be integrated with GaInAs photodetectors.

Inversion-mode transistors have been made on both InP (Wieder *et al.*, 1981; Shinoda and Kobayashi, 1982) and GaInAs (Liao *et al.*, 1982). These devices present relatively minor problems from the point of view of fabrication technology. On the other hand, performance achieved using the inversion-mode structure has been uniformly disappointing. The best results achieved so far show gains less than 10% of that achieved using the same materials in device structures that are not so selective to the quality of the semiconductor surface.

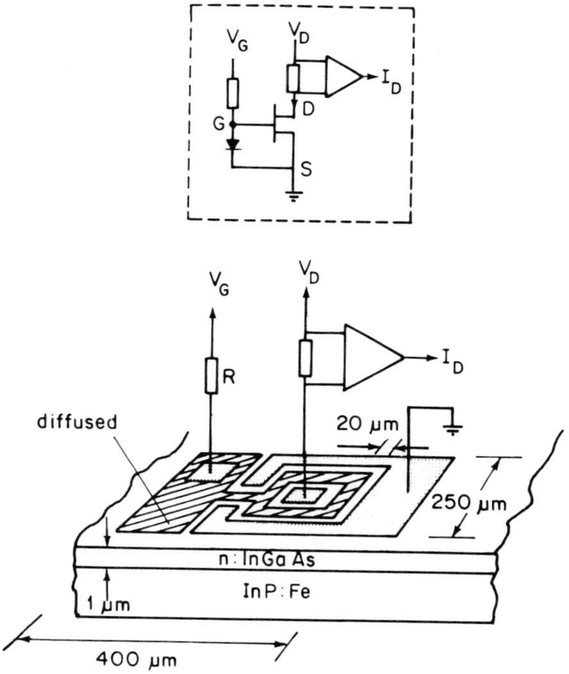

FIG. 43. Schematic diagram of the first integrated $p-i-n$ and junction field-effect transistor (JFET) structure. This device is constructed from one layer of active material and does not permit optimization of FET channel doping and $p-i-n$ depletion-layer capacitance. Despite this drawback, the successful demonstration of this device indicated the potential of integrated-circuit technology for lightwave communications. [From Leheny *et al.* (1980b). © 1980 IEEE.]

Selectively doped heterostructure transistors offer the same nearly two-dimensional confinement of the channel current as the insulated-gate inversion-mode transistor. At the same time, the heterointerface state density remains low, so that the transport properties are quite favorable for high-performance FETs. Room-temperature transconductances of 140 msec mm^{-1} (Pearsall *et al.,* 1983a) have been measured in these devices.

Depletion-mode FETs for integrated $p-i-n$/FETs have been made using two different gate technologies. The simple MESFET Schottky-barrier gate cannot be made on either InP or GaInAs without introducing excessively high gate-leakage current. Grown or diffused $p-n$ junctions can be used to form the gate and represent the most satisfactory gate technology known in terms of electronic properties. Field-effect transistors made using this approach show good performance (Leheny *et al.,* 1980b). The junction gate is, however, difficult to handle during processing. There are no currently known etching techniques that can be used to form the short gate lengths ($\leq 1\ \mu$m) needed to achieve high-speed high-gain FETs. The resulting structure contains several nonplanar levels, and this type of three-dimensional geometry leads to degradation problems.

The second design technology used for depletion-mode FETs is an enhanced Schottky-barrier gate (Morgan and Frey, 1978). The leakage current associated with the low Schottky barrier of metals on InP or GaInAs can be dramatically reduced by placing a thin layer of insulating material (O'Connor *et al.,* 1982) or a higher-gap semiconductor (Ohno *et al.,* 1981) between the gate metal and the FET channel. This latter structure has been successfully integrated (Barnard *et al.,* 1981) with a photodetector, and this dual-gate OP-FET is shown in Fig. 44.

17. Wavelength-Division Demultiplexing

The bandwidth available in optical fibers is truly extraordinary. Reserving only one optical wavelength for transmission over a fiber represents a staggering waste of the potential transmission capabilities of a single strand of fiber. It is easy to imagine how many independent communication channels could be set up between 1.25 and 1.35 μm with each channel spaced comfortably 50 Å apart. Yet, wavelength-division multiplexing has been stymied by a difficulty in coupling these various wavelengths into a single optical fiber (multiplexing) and of separating them spatially (demultiplexing) at the receiver.

Practical solutions to this problem can be divided into two areas, depending basically on whether the sources are lasers (and are thus closely spaced in wavelength) or LEDs (and spaced farther apart). For the purposes of this problem, close together versus far apart is measured relative to the absorption edge of a photodiode, which is about 400 – 500 Å wide at this range of

FIG. 44. Dual-gate OP-FET. This transistor/detector pair is a "one-level" design incorporating bias control for optimization of the operating point of the transistor independently of the bias on the photodetector. [From Barnard *et al.* (1981).]

band-gap energies. If the wavelength-division spacing is made less than this, the approach to demultiplexing tends toward the use of gratings or interference filters to separate the light spatially onto an array of detectors all made from the same semiconducting material.

If the wavelengths are spaced far apart, as is likely to be the case when LEDs are used, then the detection can be made in a single device consisting of photodiodes integrated one on top of the other. The dual-wavelength photodetector (Campbell *et al.*, 1979, 1980) described above for wavelength monitoring, is such a component. This integrated photodiode is capable of simultaneously detecting and demultiplexing optical signals from two wavelength bands without the need for additional optical components, such as gratings, dielectric filters, or prisms. Wavelength discrimination is accomplished in this multilayer structure by a GaInAsP quaternary layer (labeled Q in Fig. 39) and a GaInAs ternary layer (T). Photons with energy less than the band gap of the Q layer $[E_g(Q) > \hbar\omega > E_g(T)]$ are absorbed in the ternary

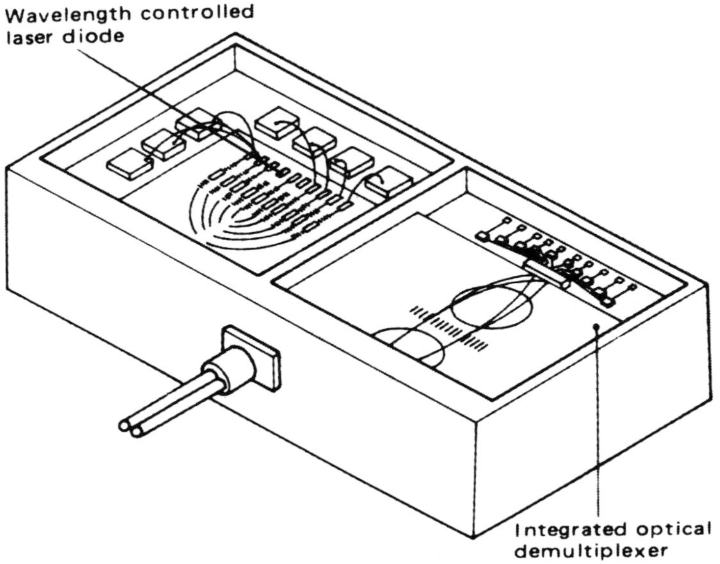

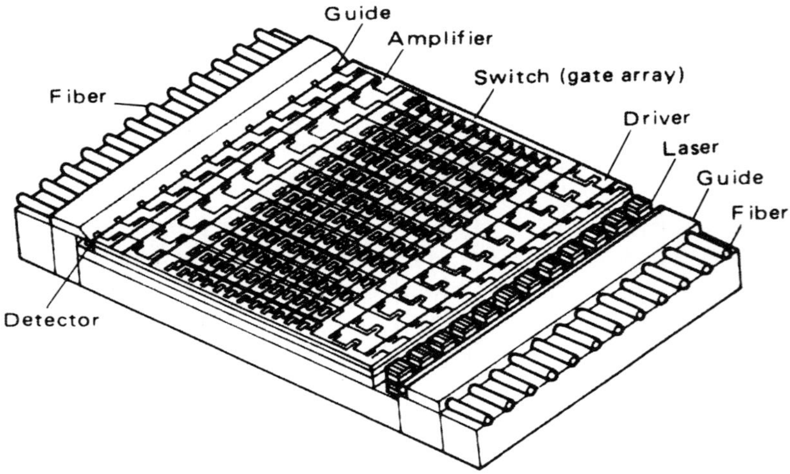

FIG. 45. Both figures illustrate proposed optical-fiver telecommunication devices from Japan's Optoelectronics Applied Measurement and Control System. (a) Demultiplexing is accomplished using a holographic grating and an integrated photodetector array. (b) Proposed integrated $n \times n$ optoelectronic switch. Incoming signals carried on optical fibers are coupled into three fiber waveguides and detected by an integrated detector detector–receiver–amplifier array. (Courtesy of the Japanese Ministry of International Trade and Industry.)

layer. Higher-energy photons are absorbed in the Q layer. Crosstalk levels as low as −19 and −30 dB have been measured in the long- and short-wavelength regions, respectively. With a few modifications in the doping and layer thicknesses, the structure shown in Fig. 39 may also be used as an LED. A wavelength-division-multiplexing system based on these modifications has been built (Ogawa *et al.*, 1981) and operated at 33 Mbit/sec, with a selectivity of about −39 dBm at a 10^{-9} bit error rate (BER). Although this performance is about 15 dB below the best sensitivity achieved at this bit rate, it is still quite adequate for many applications and shows the potential of wavelength-division multiplexing for reducing costs in optical-fiber systems by expanding the transmission rate without increasing the bit rate.

Some of the likely future applications of integrated optoelectronics using arrays of photodiodes and FETs are shown in Figs. 45 and 46. These figures show project goals for the Optical Joint Research Laboratory organized by the Japanese Ministry of International Trade and Industry (M.I.T.I.) to develop new device technologies for optical communications. These two projects illustrate the commitment to wavelength-division multiplexing and high-bit-rate electronic switching. Both of these projects underscore the importance of developing an optical device array technology, particularly, photodiode and $p-i-n$/FET arrays. In many applications, such as computer data links, interoffice trunking, or local area networks, optical fibers are naturally used in bundles or ribbons of 4–12 fibers, and array technology is a natural component of the emerging optical fiber telecommunications industry.

REFERENCES

Abott, S. M., and Muska, W. M. (1979). *Electron Lett.* **15,** 250.

Amano, T., Takahei, K., and Nagai, H. (1981). *Jpn. J. Appl. Phys.* **20,** 2015.

Ando, H., Kanbe, H., Ito, M., and Kaneda, T. (1980). *Jpn. J. Appl. Phys.* **19,** L277.

Ando, H., Yamauchi, Y., Nakagome, H., Susa, N., and Kanbe, H. (1981). *IEEE J. Quantum Electron.* **QE-17,** 250.

Ando, H., Susa, N., and Kanbe, H. (1982). *IEEE Trans. Electron Devices* **ED-29,** 1408.

Armiento, C. A., Groves, S. H., and Hurwitz, C. E. (1979). *Appl. Phys. Lett.* **35,** 333.

Barnard, J., Ohno, H., Wood, C. E. C., and Eastman, L. F. (1981). *IEEE Electron Devices Lett.* **EDL-2,** 7.

Baumeister, P., and Picus, G. (1969). *Sci. Amer.* **223,** 59.

Bell Telephone Magazine (1979). Vol. 58, No. 5.

Brown, M. G., and Smith, R. G. (1983). *Proc. Globecom* **83,** 21.4.

Bulman, G. E., Cook, L. W., and Stillman, G. E. (1981). *Appl. Phys. Lett.* **39,** 813.

Burrus, C. A., Dentai, A. G., Lee, T. P. (1979). *Electron. Lett.* **15,** 655.

Burrus, C. A., Dentai, A. G., and Lee, T. P. (1981). *Op. Commun.* **38,** 124.

Campbell, J. C. (1983). Private communication (Bell Laboratories).

Campbell, J. C., Lee, T. P., Dentai, A. G., Burrus, C. A. (1979). *Appl. Phys. Lett.* **34,** 401.

Campbell, J. C., Dentai, A. G., Lee, T. P., Burrus, C. A. (1980). *IEEE J. Quantum Electron.* **QE-16,** 601.

Campbell, J. C., Burrus, C. A., Copeland, J. A., Dentai, A. G. (1983a). *Electron. Lett.* **19,** 672.
Campbell, J. C., Dentai, A. G., Holden, W. S., Kasper, B. L. (1983b). *Electron. Lett.* **19,** 818.
Campbell, J. C., Holden, W. S., Qua, G. J., Dentai, A. G. (1985). *IEEE J. Quantum Electron.* (To be published.)
Capasso, F., Panish, M. B., Sumski, S., Foy, P. W. (1980a). *Appl. Phys. Lett.* **36,** 165.
Capasso, F., Logan, R. A., Hutchinson, A., and Manchon, D. D. (1980b). *Electron. Lett.* **16,** 893.
Capasso, F., Hutchinson, A. L., Foy, P. W., Bethea, C., and Bonner, W. A. (1981a). *Appl. Phys. Lett.* **39,** 736.
Capasso, F., Panish, M. B., Sumski, S. (1981b). *IEEE J. Quantum Electron.* **QE-17,** 273.
Capasso, F., Tsang, W. T., Hutchinson, A. L., and Williams, G. F. (1982). *Appl. Phys. Lett.* **40,** 38.
Cappy, A., Carnez, B., Fauquembergues, R., Salmer, G., and Constant, E. (1980). *IEEE Trans. Electron Devices* **ED-27,** 158.
Chin, R., and Hill, C. M. (1982). *Appl. Phys. Lett* **40,** 332.
Chin, R., Holonyak, N., Jr., Stillman, G. E., Tang, J. Y., and Hess, K. (1980). *Electron. Lett.* **16,** 467.
Chu, M., Shin, S. H., Law, H. D., and Cheung, D. T. (1980). *Appl. Phys. Lett.* **37,** 318.
Coldren, L. A., Miller, B. I., Iga, K., and Rentschler, J. A. (1981). *Appl. Phys. Lett.* **38,** 315.
Coldren, L. A., Ebeling, K. J., Miller, B. I., and Rentschler, J. A. (1983). *IEEE J. Quantum Electron.* **QE-19,** 1057.
Cook, L. W., Tabatabaie, N., Tashima, M. M., Windhorn, T. W., Bulman, G. E., and Stillman, G. E. (1981). *In* "GaAs and Related Compounds, 1980 (8th Int. Conf.)," p. 361. Inst. Phys., Bristol.
Cook, L. W., Bulman, G. E., and Stillman, G. E. (1982a). *Appl. Phys. Lett.* **40,** 589.
Cook, L. W., Tashima, M. M., Tabatabaie, N., Low, T. S., and Stillman, G. E. (1982b). *J. Cryst. Growth* **56,** 475.
Diadiuk, V., Groves, S. H., Hurwitz, C. E. (1980). *Appl. Phys. Lett.* **37,** 807.
Diadiuk, V., Groves, H., Hurwitz, C. E., Iseler, G. W. (1981). *IEEE J. Quantum Electron.* **QE-17,** 260.
Emmons, R. B. (1967). *J. Appl. Phys.* **38,** 1705.
Forrest, S. R. (1981). *IEEE J. Quantum Electron.* **QE-17,** 217.
Forrest, S. R., DiDomenico, M., Jr., Smith, R. E., and Stocker, H. J. (1980a). *Appl. Phys. Lett.* **36,** 580.
Forrest, S. R., Leheny, R. F., Nahory, R. E., and Pollack, M. A. (1980b). *Appl. Phys. Lett.* **37,** 322.
Forrest, S. R., Camlibel, I., Kim, O. K., Stocker, H. J., Zuber, J. R. (1981a). *IEEE Electron. Lett.* **EDL-2,** 283.
Forrest, S. R., Williams, G. F., Kim, O. K., Smith, R. G. (1981b). *Electron. Lett.* **17,** 917.
Forrest, S. R., Smith, R. G., and Kim, O. K. (1982a). *IEEE J. Quantm Electron.* **QE-18,** 2040.
Forrest, S. R., Kim, O. K., and Smith, R. G. (1982b). *Appl. Phys. Lett.* **41,** 95.
Forrest, S. R., Kim, O. K., and Smith, R. G. (1983). *Solid-State Electron.* **26,** 951.
Grove, A. S. (1967). "Physics and Technology of Semiconductor Devices." Wiley, New York.
Hess, K. L., Chin, R., and Dapkus, P. D. (1982). *Dig. TMS/AIME Electron. Mater. Conf., Fort Collins, Colo., 1982.*
Hildebrand, O., Kuebart, W., and Pilkuhn, M. H. (1980). *Appl. Phys. Lett.* **39,** 801.
Hildebrand, O., Kuebart, W., Benz, K. W., Pilkuhn, M. H. (1981). *IEEE J. Quantum Electron.* **QE-17,** 284.
Hurwitz, C. E., and Hsieh, J. J. (1978). *Appl. Phys. Lett.* **32,** 487.
Iga, K., and Miller, B. I. (1980a). *Electron. Lett.* **16,** 342.

Iga, K., and Miller, B. I. (1980b). *Appl. Phys. Lett.* **37,** 339–341.
Ikeda, M., Wakita, K., Hata, S., Kondo, S., and Kanbe, H. (1983). *Electron. Lett.* **19,** 61.
Ishida, K., Imoto, K., and Suganuma, T. (1982). *Fiber Integr. Opt.* **4,** 191.
Ito, M., Kaneda, T., Nakajima, K., Toyama, Y., and Ando, H. (1981). *Solid-State Electron.* **24,** 421.
Kanbe, H., Susa, N., Nakagome, H., Ando, H. (1980). *Electron. Lett.* **16,** 163.
Kane, E. D. (1959). *J. Phys. Chem. Solids* **12,** 181.
Keck, D. B. (1982). *Proc. Soc. Photo-Opt. Instrum. Eng.* **340,** 2.
Kim, O. K., Forrest, S. R., Bonner, W. A., Smith, R. G. (1981). *Appl. Phys. Lett.* **39,** 402.
Kobayashi, M., Noda, Y., Kushiro, Y., Seki, N., Akiba, S. (1984). *Appl. Phys. Lett.* **45,** 759.
Kuphal, E., and Pocker, A. (1981). *J. Cryst. Growth* **58,** 133.
Law, H. B., Chin, R., Nakano, K., Milano, R. A. (1981). *IEEE J. Quantum Electron.* **QE-17,** 275.
Lee, T. P. (1983). Photodetectors, *in* "Fiber Optics" (J. C. Daley, ed.). CRC Press, Cleveland.
Lee, T. P., and Li, Tingye (1979a). Photodetectors, *in* "Optical Fiber Communications" (S. E. Miller and A. G. Chynoweth, eds.), Chap. 18. Academic Press, New York.
Lee, T. P., Burrus, C. A., and Dentai, A. G. (1979b). *IEEE J. Quantum Electron.* **QE-15,** 30.
Lee, T. P., Burrus, C. A., Dentai, A. G., (1981a). *IEEE J. Quantum Electron.* **QE-17,** 232.
Lee, T. P., Burrus, C. A., Ogawa, K., and Dentai, A. G. (1981b). *Electron. Lett.* **17,** 431.
Leheny, R. F., Nahory, R. E., and Pollack, M. A. (1979). *Electron. Lett.* **15,** 713.
Leheny, R. F., Nahory, R. E., Pollack, M. A., Ballman, A. A., Beebe, E. D., DeWinter, J. C., and Martin, R. J. (1980a). *Electron. Lett.* **16,** 353.
Leheny, R. F., Nahory, R. E., Pollack, M. A., Ballman, A. A., Beebe, E. D., DeWinter, J. C., and Martin, R. J. (1980b). *IEEE Electron Device Lett.* **EDL-1,** 110.
Leheny, R. F., Nahory, R. E., Pollack, M. A., Beebe, E. D., and DeWinter, J. C. (1981). *J.Quantum Electron.* **QE-17,** 227.
Liao, A. S. H., Tell, B., Leheny, R. F., Nahory, R. E., De Winter, J. C., and Martin, R. J. (1982). *Electron Device Lett.* **EDL-3,** 158.
Long, D., and Schmit, J. L. (1970). *In* "Semiconductors and Semimetals" (R. K. Willardson and A. C. Beer, eds.) Vol. 5, p. 75. Academic Press, New York.
Matsushima, Y., and Sakai, K. (1982). *In* 'GaInAsP Alloy Semiconductors," (T. P. Pearsall, ed.), Chap. 16. Wiley, New York.
Matsushima, Y., Sakai, K., Noda, Y. (1981). *IEEE Electron. Device Lett.* **EDL-2,** 179.
Matsushima, Y., Akiba, S., Sakai, K., Kushiro, Y., Noda, Y., and Utaka, K. (1982). *Electron. Lett.* **18,** 945.
Matsushima, Y., Noda, Y., Kushiro, Y., Seki, N., Akiba, S. (1984). *Electron. Lett.* **20,** 235.
McIntyre, R. J. (1966). *IEEE Trans. Electron Devices* **ED-13,** 164.
McIntyre, R. J. (1972). *IEEE Trans. Electron Devices* **ED-19,** 703.
Mikawa, T., Shirai, T., Nakajima, K., and Kaneda, T. (1983). *Tech. Dig. Lasers and Electro-Opt. Conf., Baltimore, 1983.* 220.
Mikhailova, M. P., Rogachev, A. A., and Yassievich, I. N. (1976a). *Sov. Phys.—Semicond. (Engl. Trans.)* **10,** 866.
Mikhailova, M. P., Smirnova, N. N., and Slobodchikov, S. V. (1976b). *Sov. Phys.—Semicond. (Engl. Trans.)* **10,** 509.
Miller, C. M. (1978). *Bell Syst. Tech. J.* **57,** 75.
Moll, J. L. (1964). *In* "Physics of Semiconductors." McGraw Hill, New York.
Morgan, D. V., and Frey, J. (1978). *Electron. Lett.* **14,** 737.
Nguyen-Duy, T., Pichard, G., and Raymond, F. (1980). *Proc. Fiber Opt. and Commun., 80, San Francisco,* 177. Information Gatekeepers, Brookline.
Nishida, K. (1977). *Electron. Lett.* **13,** 419.

Nishida, K., Taguchi, K., and Matsumoto, Y. (1979). *Appl. Phys. Lett.* **35,** 251.
O'Connor, P., Pearsall, T. P., Cheng, K. Y., Cho, A. Y., Hwang, J. C. M., Alavi, K. (1982). *IEEE Electron Device Lett.* **EDL-3,** 64.
Ogawa, K., Lee, T. P., Burrus, C. A., Campbell, J. C., Dentai, A. G. (1981). *Electron. Lett.* **17,** 857.
Ohno, H., Barnard, J., Rathbun, L., Wood, C. E. C., and Eastman, L. F. (1981). *In* "GaAs and Related Compounds, 1980," p. 465. Inst. Phys., Bristol.
Oliver, J. D., Jr., and Eastman, L. F. (1980). *J. Electron. Mater.* **9,** 693.
Olsen, G. H. (1981). *IEEE Electron. Device Lett.* **EDL-2,** 217.
Olsen, G. H., and Kressel, H. (1979). *Electron. Lett.* **15,** 141.
Pearsall, T. P. (1980). *Electron. Lett.* **16,** 771.
Pearsall, T. P. (1981). *J. Opt. Commun.* **2,** 42.
Pearsall, T. P. (1982). *Fiber Integr. Opt.* **4,** 107.
Pearsall, T. P., Nahory, R. E., and Chelikowsky, J. R. (1977). *In* "GaAs and Related Compounds, 1976," (L. I. Eastman, ed.), p. 331. Inst. of Phys., Bristol.
Pearsall, T. P., and Hopson, R. W. (1978). *J. Electron. Mater.* **7,** 133.
Pearsall, T. P., and Papuchon, M. (1978). *Appl. Phys. Lett.* **33,** 640.
Pearsall, T. P., Piskorski, M., Brochet, A., and Chevier, J. (1981). *IEEE J. Quantum Electron.* **QE-17,** 255.
Pearsall, T. P., Hendel, R., O'Connor, P., Alavi, K., Cho, A. Y. (1983a). *IEEE Electron Device Lett.* **EDL-4,** 5.
Pearsall, T. P., Logan, R. A., and Bethea, G. G. (1983b). *Electron. Lett.* **19,** 611.
Pichard, G., Meslage, J., Fragnon, P., and Raymond, F. (1982). *Proc. Eur. Conf. Opt. Commun., Cannes, Fr. 1982,* 389.
Pollack, M. A., Nahory, R. E., DeWinter, J. C., and Ballman, A. A. (1978). *Appl. Phys. Lett.* **33,** 314.
Schroeder, C. M. (1978). *Bell Syst. Tech. J.* **57,** 91.
Shinoda, Y., and Kobayashi, T. (1982). *In* "GaAs and Related Compounds, Oise, 1981", p. 347. *Inst. Phys.,* Bristol.
Shirai, T., Osaka, F., Yamazaki, S., Kaneda, T. and Susa, N. (1981a). *Appl. Phys. Lett.* **39,** 168.
Shirai, T., Yamazaki, S., Osaka, F., Nakajima, K., and Kaneda, T. (1981b). *Appl. Phys. Lett.* **40,** 532.
Shirai, T., Oska, F., Yamazaki, S., Nakajima, K., Kaneda, T. (1981c). *Electron. Lett.* **17,** 826.
Shirai, T., Yamazaki, S., Kawata, H., Nakajima, K., and Kaneda, T. (1982a). *IEEE Trans. Electron Device* **ED-20,** 1404.
Shirai, T., Yamazaki, S., Yasuda, K., Mikawa, T., Nakajima, K., and Kaneda, T. (1982b). *Electron. Lett.* **18,** 576.
Smith, D. R., Hooper, R. C., and Webb, R. P. (1979). *Proc. Int. Symp. Circuits and Syst. Tokyo, 1979,* 791.
Stillman, G. E., and Wolfe, C. M. (1977). Avalanche photodiodes, *in* "Semiconductors and Semimetals" (Willardson, R. K. and Beers, A. C., eds.) Vol. 12, p. 291. Academic Press, New York.
Stillman, G. E., Cook, L. W., Roth, T. J., Low, T. S., and Skromme, B. J. (1981). *In* "GaInAsP Alloy Semiconductors" (T. P. Pearsall, ed.) Chap. 6. Wiley, New York.
Stillman, G. E., Cook, L. W., Bulman, G. E., Tabatabaie, N., Chin, R., and Dapkus, P. D. (1982). *IEEE Trans. Electron Devices* **ED-29,** 1355.
Stone, J. and Cohen, L. G. (1982). *IEEE J. Quantum Electron.* **QE-18,** 511.
Sugimoto, Y., Torikai, T., Makita, K., Ishihara, H., Minemura, K., Taguchi, K., Iwakami, T. (1984). *Electron. Lett.* **20,** 653.
Susa, N. (1982). Private Communication (NTT, Japan).

Susa, N., and Okamoto, H. (1984). *Jpn. J. Appl. Phys.* **23,** 317.
Susa, N., Yamauchi, Y., Kanbe, H. (1979). *Electron. Lett.* **15,** 238.
Susa, N., Nakagome, H., Mikami, O., Ando, H., Kanbe, H., (1980a). *IEEE J. Quantum Electron.* **QE-16,** 864.
Susa, N., Yamauchi, Y., Ando, H., Kanbe, H. (1980b). *IEEE Electron Device Lett.* **EDL-1,** 55.
Susa, N. and Yamauchi, Y. (1981a). *J. Cryst. Growth* **51,** 518.
Susa, N., Nakagome, H., Ando, H., Kanbe, H. (1981b). *IEEE J. Quantum Electron.* **QE-17,** 243.
Sze, S. M. (1981). *In* "Physics of Semiconductor Devices." Wiley, New York.
Tabatabaie, N., Stillman, G. E., Chin, R., and Dapkus, P. D. (1982). *Appl. Phys. Lett.* **40,** 415.
Taguchi, K., Matsumoto, Y., and Nishida, K. (1979). *Electron. Lett.* **15,** 453.
Takanashi, Y. and Horikoshi, Y. (1979). *Jpn. J. Appl. Phys.* **18,** 1615.
Takanashi, Y., Kawashima, M. and Horikoshi, Y. (1980). *Jpn. J. Appl. Phys.* **19,** 693.
Takanashi, Y., and Horikoshi, Y. (1981). *Jpn. J. Appl. Phys.* **20,** 1271.
Tsang, W. T., Olsson, N. A., and Logan, R. A. (1983). *Appl. Phys. Lett.* **42,** 650.
Umebu, I., Choudhury, A. N. M. M., and Robson, P. N. (1980). *Appl. Phys. Lett.* **36,** 302.
Vèrié, C. and Ayas, J. (1967). *Appl. Phys. Lett.* **10,** 241.
Vèrié, C., Raymond, F., Besson, J., and Nguyen-Duy, T. (1982). *J. Cryst. Growth* **59,** 342.
Washington, M. A., Nahory, R. E., Pollack, M. A., and Beebe, E. D. (1978). *Appl. Phys. Lett* **33,** 854.
Webb, P. P., McIntyre, R. J., and Conradi, J. (1974). *RCA Rev.* **35,** 234.
Wemple, S. H. (1983). Private Communication. (Bell Laboratories, Murray Hill, N.J.)
White, K. I., and Nelson, B. P. (1979). *Electron. Lett.* **15,** 396.
Wieder, H. H., Clawson, A. R. Elder, D. L., and Collins, D. A. (1981). *IEEE Electron Device Lett.* **EDL-2,** 73.
Williams, G. F., and Leblanc, H. P. (1983). Private communication (Bell Laboratories, Murray Hill, New Jersey).
Yamada, J-I, Saruwatari, M., Asatami, K. Tsuchiya, H Sugiyama, K., and Kimuro, T. (1978). *IEEE J. Quantum Electron.* **QE-14,** 791.
Yamada, J., Kawana, A., Nagai, H., and Kimura, T. (1982). *Electron Lett.* **18,** 98.
Yamauchi, Y., Susa, N., and Kanbe, H. (1982). *J. Cryst. Growth* **56,** 402.
Yano, M., Suzuki, Y., Ishii, T., Matsushima, Y., and Kimata, M. (1979). *Jpn J. Appl. Phys.* **17,** 2091.
Yasuda, K., Mikawa, T., Kishi, Y., Kaneda, T. (1984). *Electron. Lett* **20,** 373.
Yeats, R. and Chiao, S. H. (1980). *Appl. Phys. Lett.* **36,** 167.
Yeats, R., and Von Dessonneck, K. (1981). *IEEE Electron. Device Lett.* **EDL-2,** 268.

CHAPTER 3

Silicon and Germanium Avalanche Photodiodes

Takao Kaneda

OPTICAL SEMICONDUCTOR DEVICES LABORATORY
FUJITSU LABORATORIES, LTD.
ATSUGI, JAPAN

I. Introduction

Avalanche photodiodes are unrivaled for optical detector applications from the viewpoint of detectable power in optical-fiber communications. There are two attractive wavelength regions in this communication system: 0.8–0.9 and 1.0–1.6 μm. In the 0.8–0.9-μm range, silicon avalanche photodiodes are the most suitable detectors because of their high performance, high reliability, and inexpensive manufacture. In the 1.0–1.6-μm range, germanium avalanche photodiodes are used in practice. Most systems working in this region have been carried out using germanium photodiodes (Ka-

ISBN 0-12-752153-4

gawa *et al.*, 1982), whereas III–V alloys, such as InP/GaInAs avalanche photodiodes (Susa *et al.*, 1981; Kim *et al.*, 1981; Matsushima *et al.*, 1981; Shirai *et al.*, 1982; Kobayashi *et al.*, 1984) and InGaAs $p-i-n$/FETs (Leheny *et al.*, 1980; Smith *et al.*, 1982), are being investigated with the aim of obtaining a photodetector with lower noise and dark current.

This chapter reviews the state of the art of silicon and germanium avalanche photodiodes for optical-fiber communications. Indirect band-gap semiconductors, such as silicon and germanium, are preferable materials for forming avalanche photodiodes because the band-to-band tunneling current is much lower than that of direct band-gap semiconductors, and the avalanche process dominates the tunneling process at high electric fields, where carrier multiplication takes place. This leads to avalanche photodiodes with a low dark current. In germanium, despite the narrow band gap (0.67 eV), high avalanche gains are obtained with relatively low dark currents of $(6-10) \times 10^{-4}$ A cm^{-2} at 90% of a breakdown voltage (Kagawa *et al.*, 1981), whereas for a direct semiconductor of $In_{0.53}Ga_{0.47}As$ (0.75 eV), the tunneling process dominates the avalanche process in the carrier concentration range of greater than 1×10^{15} cm^{-3}, and high avalanche gains have not been obtained (Takanashi *et al.*, 1980; Ando *et al.*, 1980; Forrest *et al.*, 1980; Ito *et al.*, 1981).

Avalanche photodiodes are considered to be those having a carrier-multiplication operation added to an ordinary photodiode operation. Therefore, the properties can be understood as the sum of two contributions: one from the carrier-multiplication process and the other from the photodiode. There are three principal features for photodetectors, namely, quantum efficiency, response speed, and noise. Quantum efficiencies for avalanche photodiodes can be studied in the same way as those for ordinary photodiodes. Response speeds are determined by the sum of the buildup time for the avalanche-multiplication process and the usual frequency limits for photodiodes, such as the *RC* time constant, the carrier transit time in the depletion layer, and the carrier diffusion time in the undepleted layer. Noise in avalanche photodiodes is governed by multiplication noise, which comes from the statistical fluctuation in the multiplication process, whereas ordinary photodiodes have no multiplication noise.

The multiplication process distinguishes avalanche photodiodes from ordinary and $p-i-n$ photodiodes, which have a nearly intrinsic region between the p and n layer. Multiplication noise and avalanche buildup time are specific features of avalanche photodiodes. Especially, multiplication noise is the key property in the practical use of avalanche photodiodes because multiplication noise is the main noise source of optical receiver systems as well as amplifier noise following the avalanche photodiode, whereas the buildup time influences the response speed only in the high-frequency region of approximately 1 GHz (Kaneda and Takanashi, 1973a; Kaneda *et al.*, 1976c). Therefore, multiplication noise is the most important property in designing and characterizing avalanche photodiodes.

II. Design Considerations

Part II discusses various factors that determine the signal-to-noise ratio of avalanche photodiodes. The discussion includes multiplication noise, quantum efficiency, and response speed. These are the main features of avalanche photodiodes. Photodiode housing related to quantum efficiency and reliability of devices is also discussed.

1. Multiplication Noise

In the high-electric-field region ($>1 \times 10^5$ V cm^{-1}) of a highly reverse-biased junction, carriers drift in a saturation velocity and gain sufficiently high energy from the electric field to release new electron–hole pairs through impact ionization (Chynoweth, 1968). A chain of these impact ionizations leads to carrier multiplication. The average number of electron–hole pairs created by a carrier per unit distance traveled is denoted as an impact-ionization coefficient. The measured values of the impact-ionization coefficient for silicon and germanium are shown in Fig. 1 as a function of the electric

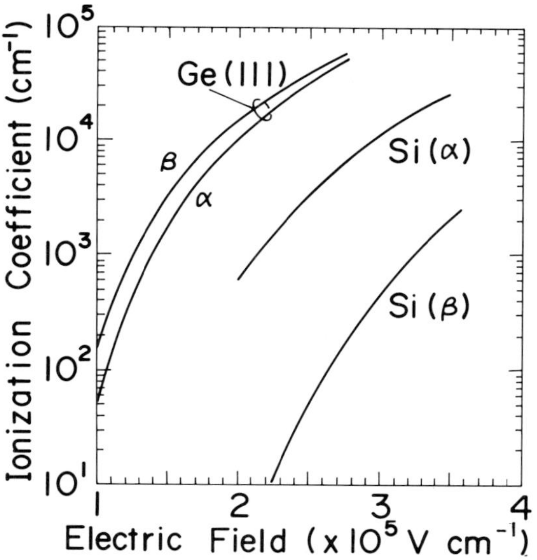

FIG. 1. Carrier ionization coefficients at 300 K for silicon and germanium. α, electrons; β, holes. (Data from Lee *et al.*, 1964; Mikawa *et al.*, 1980).

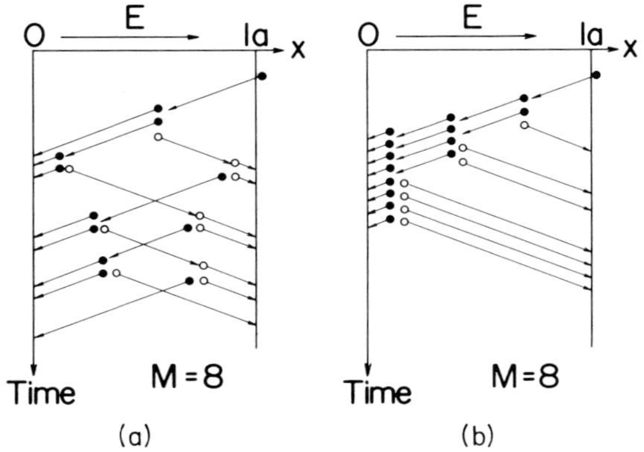

FIG. 2. Schematic representation of the carrier-multiplication process, where the multiplication factor is 8: (a) carrier impact ionization takes place for both carriers; (b) only electron impact ionization takes place.

field (Lee *et al.,* 1964; Mikawa *et al.,* 1980). In silicon, there is a great difference between the ionization coefficients of electrons α and holes β, especially in the low-electric-field region; however, in germanium, ionization coefficients are nearly equal.

The schematic representation of the carrier-multiplication process is shown in Fig. 2 (Ruegg, 1967), where the multiplication factor is 8, for example. Figure 2 shows a case of an ideal semiconductor; that is, Fig. 2a shows an equal ionization coefficient between electrons and holes ($\alpha = \beta$), and Fig. 2b shows the impact ionization occurring solely by electrons. As seen in Fig. 2a, the multiplication-process buildup is aided by the hole-feedback mechanism. (The buildup time is much longer than that in Fig. 2b, where no hole feedback exists, as discussed later.) The feedback greatly depends on the impact-ionization coefficient ratio between electrons and holes and is more pronounced as the symmetry in the carrier ionization coefficients becomes larger. Moreover, not every carrier-injected avalanche region undergoes the same feedback. In fact, there is quite a wide probability distribution for the feedback mechanism, so the possible multiplication factor for injected carriers is also widely distributed. Therefore, the mean square of the multiplication factor $\langle M^2 \rangle$ becomes larger than the square of the average multiplication factor $\langle M \rangle^2$ where $\langle\ \rangle$ denotes an ensemble average. Multiplication noise is commonly characterized by an excess-noise factor F defined as $F = \langle M^2 \rangle / \langle M \rangle^2$ (Webb *et al.,* 1974). Expressions for the excess-noise factor have been derived by McIntyre (1966, 1972) for the various ionization coefficients between electrons and holes. An expression for F can be written as

$$F = \phi/2qI_0\langle M\rangle^2, \tag{1}$$

where q is the electronic charge, I_0 the total injected currents, and ϕ the noise spectral density of multiplication noise given by

$$\phi = 2q\left[2I_nM^2(0) + 2I_pM^2(1_a) + 2\int_0^{l_a} G(x)M^2(x)\,dx + I\left\{2\int_0^{l_a} \beta M^2(x)\,dx - M^2(l_a)\right\}\right], \tag{2}$$

where I_p and I_n are the injected hole and electron currents respectively, l_a the length of the avalanche region, $G(x)$ the generation rate of electron–hole pairs [Eq. (14)], I the total photocurrent after multiplication, and $M(x)$ the position-dependent multiplication factor given by

$$M(x) = \frac{\exp\{-\int_0^x(\alpha-\beta)\,dx'\}}{1-\int_0^{l_a}\alpha\exp\{-\int_0^x(\alpha-\beta)\,dx''\}\,dx} \tag{3}$$

According to relative values of the electron- and hole-ionization coefficients, $M(x)$ has different dependences of position x, as shown in Fig. 3.

The schematic representation of hole and electron currents in the ava-

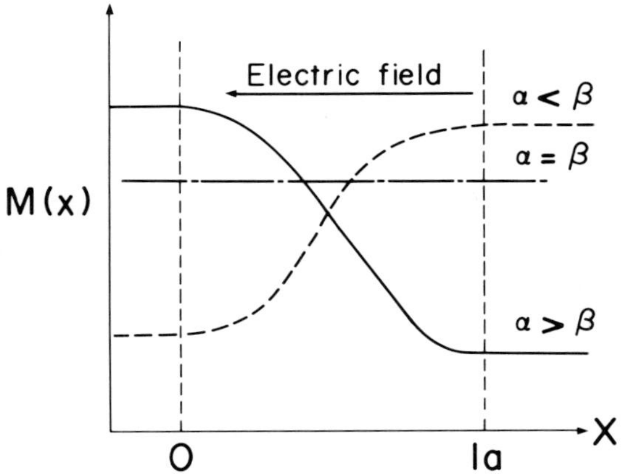

FIG. 3. Position-dependent multiplication factors for various relative values of (α) electron- and (β) hole-ionization coefficients.

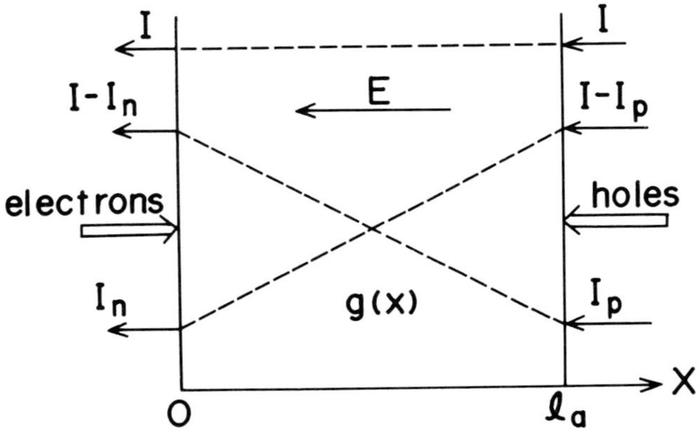

FIG. 4. Schematic representation of hole and electron currents in the avalanche region.

lanche region is shown in Fig. 4. The total injected currents I_0 is given by

$$I_0 = I_n + I_p + \int_0^{l_a} G(x)\,dx. \tag{4}$$

Then, the total photocurrent after multiplication I is given by

$$I = I_n M(0) + I_p M(l_a) + \int_0^{l_a} G(x)M(x)\,dx. \tag{5}$$

The average multiplication factor of avalanche photodiodes is expressed as

$$\langle M \rangle = I/I_0 \tag{6}$$

and is obtained from Eqs. (4) and (5).

Here, to consider the outline of the multiplication noise, we assume that the avalanche region is thin where the carrier generation can be neglected and the ratio of the hole to electron ionization coefficients k has a constant value. In this case, Eq. (2) reduced to (Kaneda *et al.*, 1976b)

$$\phi = 2qI_p M_p^3\left[1 + \frac{1-k}{k}\left(\frac{M_p - 1}{M_p}\right)^2\right] + 2qI_n M_n^3\left[1 - (1-k)\left(\frac{M_n - 1}{M_n}\right)^2\right], \tag{7}$$

where M_p and M_n are the multiplication factors for injected holes and electrons, respectively, and k is given by

$$k = (M_p - 1)/(M_n - 1), \tag{8}$$

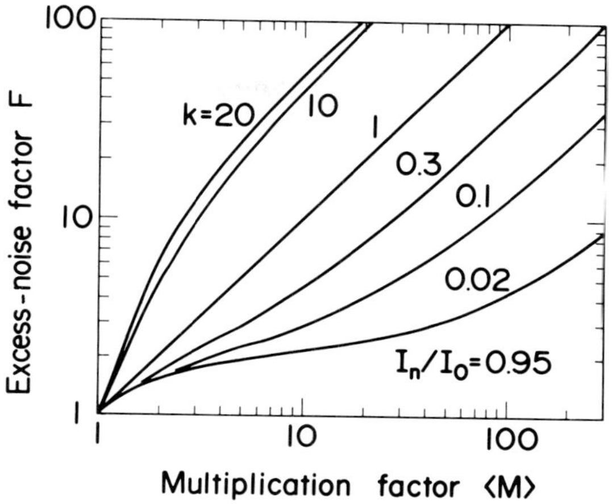

FIG. 5. Excess-noise factor F versus multiplication factor $\langle M \rangle$ with k as a parameter, where the electron-injection ratio $I_n/I_0 = 0.95$.

and $\langle M \rangle$ is given by

$$\langle M \rangle = (I_p M_p + I_n M_n)/I_0. \tag{9}$$

From Eqs. (8) and (9), M_p and M_n are given by

$$M_p = \frac{k(\langle M \rangle - 1)}{1 - (I_p/I_0)(1 - k)} + 1, \tag{10}$$

$$M_n = \frac{\langle M \rangle - (I_p/I_0)(1 - k)}{1 - (I_p/I_0)(1 - k)}. \tag{11}$$

By using Eqs. (7)–(11), we can study the excess-noise factor as a function of various parameters of interest, such as k, $\langle M \rangle$, and the carrier-injection ratio (I_n/I_0 or I_p/I_0), which depends on the incident wavelength. Figure 5 shows excess-noise factors for a diode having an electron-injection ratio I_n/I_0 of 0.95 as a function of $\langle M \rangle$ with k as a parameter. Because an electron injection is higher than a hole injection ($I_n > I_p$), a small k value is desirable to minimize excess noise, as shown in Fig. 5. Conversely, when $I_p > I_n$, a larger k value is desirable. Thus, a large asymmetry in the carrier-ionization coefficients is required to obtain low multiplication noise by decreasing the carrier-feedback effect. Excess-noise factors also depend strongly on carrier injection (Naqvi, 1972). Figure 6 shows excess-noise factors for a diode having a k value of 0.02 as a function of $\langle M \rangle$ with an electron-injection ratio as a parameter. To attain low multiplication noise, a carrier having higher

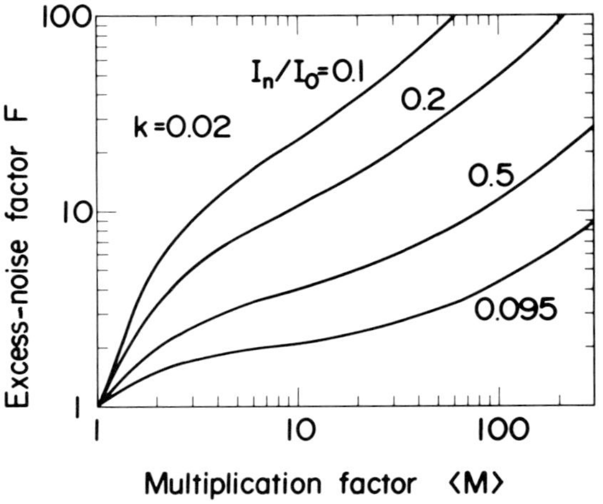

FIG. 6. Electron-injection-ratio dependence of excess-noise factor F as a function of multiplication factor $\langle M \rangle$, where $k = 0.02$.

ionization coefficients should be injected mainly into the avalanche region to minimize the carrier-feedback effect.

Summarizing the previous discussion, it is important for low-noise avalanche photodiodes that the carrier feedback in the multiplication process be minimized; that is, the ratio of electron and hole impact-ionization coefficients should be significantly different from unity, and the avalanche process should be initiated by the carrier species with the higher ionization coefficient.

The dependence of signal and noise powers on multiplication factors is shown graphically in Fig. 7. In the low-multiplication region, the multiplication noise of avalanche photodiodes is lower than the amplifier noise following the avalanche photodiode in optical receiver systems. The noise power of the amplifier is considered to remain unchanged for increasing multiplication factors, whereas the signal power is amplified by the square of the multiplication factor and the multiplication noise power increases in proportion to the 2 – 3 power of the multiplication factor. There is therefore an optimum multiplication factor that maximizes the signal-to-noise ratio. The optimum value usually lies at around the intersection point of the multiplication noise and the amplifier noise. Thus, significant improvement in overall sensitivity for optical receiver circuits is obtained from the carrier-multiplication process, even at microwave frequencies (Anderson *et al.*, 1965; Melchior and Lynch, 1966) because the buildup time in the avalanche process is considerably rapid (Kaneda and Takanashi, 1973a; Kaneda *et al.*,

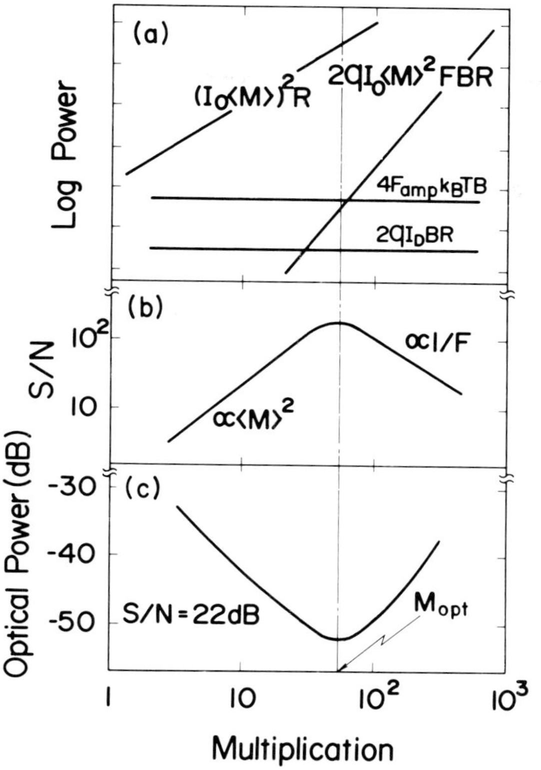

FIG. 7. Graphic representation of (a) signal and noise powers, (b) S/N ratio, and (c) detectable power as a function of multiplication factors: signal power, $I_0\langle M\rangle^2R$; multiplication-noise power, $2qI_0\langle M\rangle^2FBR$; amplifier-noise power, $4F_{\mathrm{amp}}k_{\mathrm{B}}TB$; and dark-current-noise power, $2qI_{\mathrm{D}}BR$.

1976c). The avalanche photodiode is therefore the most suitable detector in view of the detectable signal power.

2. QUANTUM EFFICIENCY

When the energy of the photon that falls on the optically active area of avalanche photodiodes is greater than or equal to the band gap of the semiconductor material, photocarriers are generated by the transition of an electron from the valence band into the conduction band through intrinsic excitation; thus, electron–hole pairs are generated optically, as illustrated in Fig. 8. The light-absorption coefficients of silicon and germanium are shown in Fig. 9 (Dash and Newman, 1955; Braunstein *et al.*, 1958; Sze, 1969). Since

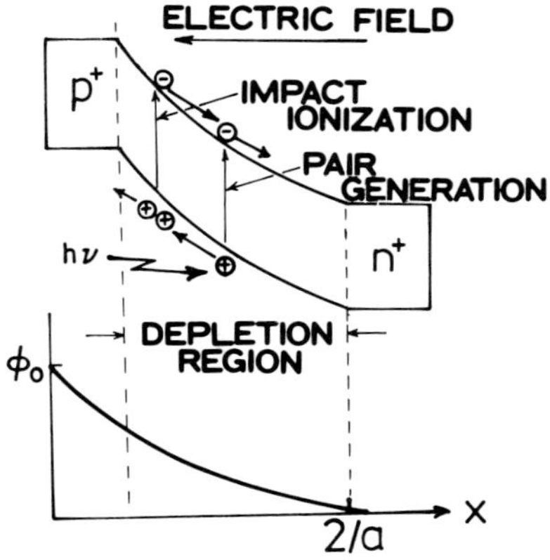

FIG. 8. Schematic representation of the principle of avalanche photodiode operation: energy-band diagram under reverse bias, carrier-pair generation, and light intensity in the photodiodes; ϕ_0, incident-photon flux density; a, light-absorption coefficient.

silicon and germanium are indirect semiconductors, the intrinsic carrier excitation takes place assisted by the phonon in the long-wavelength region, which is longer than 0.5 μm for silicon and longer than 1.55 μm for germanium. As the coincidence of photon and phonon absorption is much less probable, the light-absorption coefficient decreases in this long-wavelength

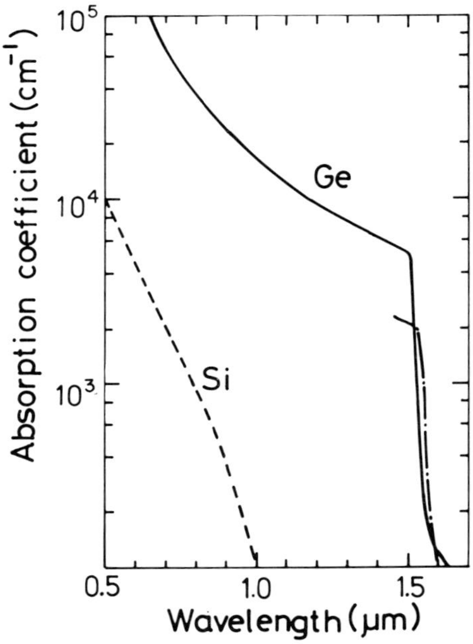

FIG. 9. Optical absorption coefficient versus wavelength for silicon and germanium. (Data from Dash and Newman, 1955; Braunstein *et al.*, 1958; Sze, 1969.)

region, as shown in Fig. 9. The long-wavelength cutoff λ_c is obtained by

$$\lambda_c = hc/E_G, \tag{12}$$

also commonly expressed by the following numerical equation:

$$\lambda_c = 1.241/E_G\,(\text{eV}) \quad (\mu\text{m}) \tag{13}$$

where h is Planck's constant ($= 6.624 \times 10^{-34}$ J sec), c the velocity of light ($= 2.998 \times 10^{10}$ cm sec^{-1}), and E_G the band-gap energy. The values of λ_c established by the band-gap energy of the materials are approximately 1.1 μm for silicon and 1.7 μm for germanium. To obtain a highly sensitive photodiode, the absorption coefficient of the semiconductor used must lie in the appropriate range of $5 \times 10^3 - 1 \times 10^4$ cm^{-1}. Silicon and germanium meet this requirement in the wavelength ranges 0.8–0.9 and 1–1.55 μm, respectively.

Carrier generation usually occurs both inside and outside the depletion region of the $p-n$ junction. Carriers are generated in the depletion region, and the electrons and holes are separated and drift in opposite directions because of the electric field there. Carriers generated outside, but within an average of a diffusion length on either side of the depletion region, will diffuse into the depletion region and drift there. While the carriers run across the depletion region, a photocurrent is induced in the outside circuit following the avalanche photodiode. The carrier generation rate G is given by

$$G(x) = (P/h\nu)a \exp(-ax), \tag{14}$$

where $P/h\nu$ is the incident-photon number, a the light-absorption coefficient, and x the depth from the photodiode surface. The incident light decays exponentially in the semiconductor material with a constant a. Not all of the generated carriers will contribute to the photocurrent. Carrier recombination takes place both at the surface and at the undepleted layer of photodiodes and decreases the conversion rate of an incident photon into a photocurrent. The conversion rate is denoted as quantum efficiency η:

$$\eta = \frac{I_{ph}/q}{P/h\nu}, \tag{15}$$

where I_{ph} is the photocurrent induced in the outside circuit. The quantum efficiency is generally calculated by using both the continuity and diffusion equations for electrons and holes. These are dealt with in Section 9. To obtain a high quantum efficiency, the depletion region needs to widen and extend from as near the surface of the photodiode materials as possible. In a well-designed avalanche photodiode, the depletion region has a length of about $2/a$ where the incident optical power falls to a fraction $1/e^2$ of its initial value. Because almost all carriers are generated in the high electric field of the depletion region, an avalanche photodiode shows a fast response as well as a high sensitivity. Moreover, for high quantum efficiency, it is important to make the fraction of incident photons reflected at the semiconductor surface as small as possible. The transparency at the air–semiconductor transition can be much improved by coating the surface with an appropriate antireflection film. For normal incidence, the fraction of transparent optical power T_E is given by

$$T_E = \frac{4n_0 n_1^2 n_2}{n_1^2(n_0+n_2)^2 - (n_0^2 - n_1^2)(n_1^2 - n_2^2)\sin^2(\delta/2)}, \tag{16}$$

where n_0 is the refractive index of the semiconductor, n_1 that of the antireflection coating films, n_2 that of air, and δ the so-called retardation given by

$$\delta = \pi n_1 t_1/\lambda, \tag{17}$$

where λ is the light wavelength and t_1 the thickness of the antireflection coating film. All photons incident on the photodiode surface enter into it when the refractive index of the film is

$$n_1 = (n_0 n_2)^{1/2}, \tag{18}$$

and the thickness is one-quarter wavelength given by

$$t_1 = \lambda/4n_1. \tag{19}$$

Silicon nitride or silicon dioxide is usually used as the antireflection film for practical silicon and germanium avalanche photodiodes. In particular, by using silicon nitride ($n_1 \sim 2$), the transparency is improved nearly 100% because the refractive index is 3.6 for silicon and 4.0 for germanium, and the square root of these values is nearly equal to that of silicon nitride. When there is no antireflection film, the transparency is only about 70%.

3. Response Speed

There are four time constants involved in determining the response speed of avalanche photodiodes: (1) the depletion-layer transit time t_{tr}, (2) the RC time constant t_{RC}, (3) the diffusion time in the undepleted layer t_D, (4) and the avalanche buildup time t_a. The first three time constants greatly depend on the length of the depletion region designed primarily from the light-absorption coefficient of semiconductor materials. Therefore the time constants limiting the response speed differ between silicon and germanium avalanche photodiodes, because the light-absorption coefficient in the wavelength region of interest differs by more than an order of magnitude between silicon ($\lambda = 0.8-0.9\ \mu$m) and germanium ($\lambda = 1-1.55\ \mu$m), as shown in Fig. 9. In silicon avalanche photodiodes, the depletion layer needs to be 30–50 μm long; thus, the transit time governs the response speed. In germanium, however, the transit time is not the main limiting factor of response speed, because the length of the depletion layer is only 2–3 μm, and the transit time across the depletion layer is a small value of approximately 5×10^{-11} sec, which is obtained from

$$t_{tr} = l_D / v_s, \tag{20}$$

where l_D is the length of the depletion layer and v_s the saturated drift velocity ($\sim 6 \times 10^6$ cm sec^{-1}). Both the *RC* time constant and the avalanche buildup time govern the response speed of germanium avalanche photodiodes.

The response speed of avalanche photodiodes can be obtained from the time-dependent continuity equations for electrons and holes, including the aforementioned time constants such as t_{tr}, t_{RC}, t_D, and t_a (Emmons and Lucovsky, 1966; Chang, 1967). However, the discussions are dealt with separately here to make the physical meanings clear. In a nonavalanche photodiode, for a worst case analysis, it is assumed that all the carriers are generated at one edge of the depletion layer. The transit-time cutoff frequency has been shown by Gärtner (1959) to be

$$f_{tr} = 0.446/t_{tr}. \tag{21}$$

For the case of avalanche photodiodes, the generated carriers transit across the depletion layer and then enter the multiplication region, where they are multiplied. After this, some of the carriers transit across the depletion layer again; that is, the carriers have to travel across the depletion layer twice for a worst case. The transit-time cutoff frequency is therefore given by

$$f_{tr} = 0.446/2t_{tr} = 0.223/t_{tr}. \tag{22}$$

This is a reasonable assumption for avalanche photodiodes because the avalanche region usually locates at one edge of the depletion layer and is thin enough compared to the thickness of the whole depletion layer.

The simplified equivalent circuit of an avalanche photodiode is shown in Fig. 10. The *RC* time constant is given by

$$t_{RC} = (R_s + R_L)C, \tag{23}$$

where R_s is the diode series resistance, R_L the load resistance, and C the diode capacitance, which is the sum of the junction and package capacitances. For an avalanche photodiode with a one-sided abrupt junction, R_s is governed primarily by the sheet resistance of a highly doped layer, and a small value of 10–20 Ω is typically obtained for a diode made with reasonable care. The value of R_L is determined from the requirements of the receiver system (noise, speed, etc.). The diode capacitance depends on the diode area, the length of the depletion layer, and the package used. The diode area should be as small as possible for an *RC* time constant that is short compared to the

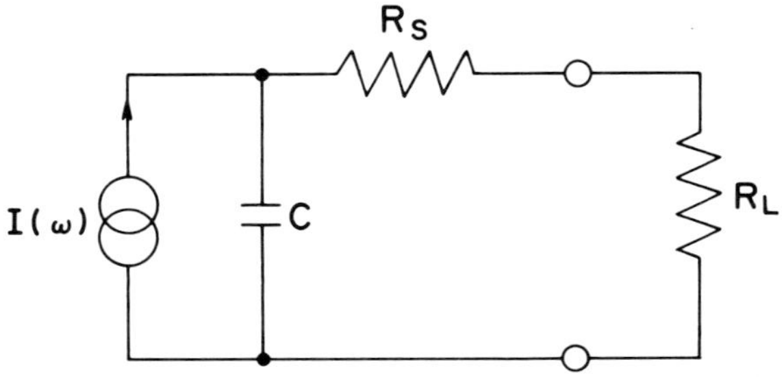

FIG. 10. Simplified equivalent circuit of an avalanche photodiode.

diameter of the incident optical signal. The junction capacitance is in inverse proportion to l_D, which varies with the applied voltage V and is expressed for a one-sided abrupt junction as

$$C = \frac{\varepsilon A}{l_D} = A\left(\frac{q\varepsilon N_B}{2(V + \phi_B)}\right), \tag{24}$$

where ε is the dielectric constant of the semiconductor material, N_B the carrier concentration of bulk materials, and ϕ_B the builtin voltage. Since the transit time is proportional to the length of the depletion layer, a trade-off exists between the RC time constant and the transit time that is related to quantum efficiency. This suggests that an optimization is necessary in the carrier concentration of wafers used for fabrication.

The diffusion time in the undepleted layer t_D has been shown by Sawyer and Rediker (1958) to be

$$t_D = l_0^2/2.4D, \tag{25}$$

where l_0 is the length of the carriers diffused in the undepleted region and D the minority-carrier diffusion constant. In well-designed avalanche photodiodes, the depletion length is designed to be about $2/a$, where a is the light-absorption coefficient. Carrier diffusion from the bulk-material region can therefore be considered to be negligibly small, since almost all carriers are generated in the depletion region. In a highly doped (undepleted) surface layer, however, photogenerated carriers diffuse into the depletion region. The surface layer is typically thin (0.2–0.3 μm) owing to the requirements of quantum efficiency and response speed. The minority-carrier diffusion length in the surface layer is nearly equal to or often longer than the thickness of this layer. In this case, since the diffusion length is limited by the thin surface layer, the carrier diffusion becomes sufficiently fast—below 0.1 nsec, as calculated from Eq. (25). Carrier diffusion in the thin surface layer is affected by surface recombination, especially in the wavelength region of high light-absorption coefficients; for example, in the 1.0–1.5-μm wavelength for germanium. The diffusion current decreases as the surface recombination velocity increases, but this leads to a decrease in quantum efficiency.

Carrier multiplication builds up by the contribution of carrier feedback, as shown in Fig. 2. The multiplication process is not instantaneous. The avalanche buildup time t_a depends on the number of feedback processes and is expressed proportionate to the multiplication factor as

$$t_a = \tau\langle M\rangle, \tag{26}$$

where τ is the intrinsic response time (Emmons, 1967). This value depends strongly on the ratio of the electron and hole impact-ionization coefficients. If there is a large asymmetry in the ionization coefficients of the two carriers, the avalanche buildup time becomes small, as shown in Fig. 2. This leads to small τ values. The relationship between τ and the ionization coefficient ratio of the two carriers has been shown by Emmons (1967) to be

$$\tau = N(\beta/\alpha)(l_a/v_s), \tag{27}$$

where N is a number varying slowly from $\frac{1}{3}$ to 2 as β/α varies from 1 to 10^{-3}; l_a is the length of the avalanche region, and l_a/v_s is the avalanche-region transit time. The intrinsic response time has been investigated by a multiplication-factor dependence of the shot-noise power in the gigahertz region (Kaneda and Takanashi, 1973a, 1975; Kaneda *et al.*, 1976c). These values have been determined, including their dependences on both l_a and the wavelengths exciting the avalanche process. In silicon, the τ values are approximately 5×10^{-13} sec, and the l_a dependence is in good agreement with the calculated results from Eq. (27). The τ values for germanium avalanche photodiodes are 5×10^{-12} sec and are an order of magnitude larger than those of silicon. The difference depends primarily on the β/α relation. Fortunately, the condition for shortening the avalanche buildup time also minimizes the multiplication noise, because these characteristics are governed by the carrier-feedback process in the multiplication process.

4. Photodiode Housing

The package of a detector must be designed so as to give a high overall quantum efficiency, a hermetic seal, and handling convenience. As an example, a package developed by Fujitsu Lab. and NTT is shown in Fig. 11 (Yamaoka *et al.*, 1976). To satisfy the previous requirements, this package has a 100-μm-thick sapphire window with a silicon dioxide (SiO_2) antireflection coating. The window is welded hermetically to the titanium (Ti) cap, which is also welded hermetically in a dry-nitrogen atmosphere to the gold-plated kovar stem on which is mounted a photodiode chip. For a reliable

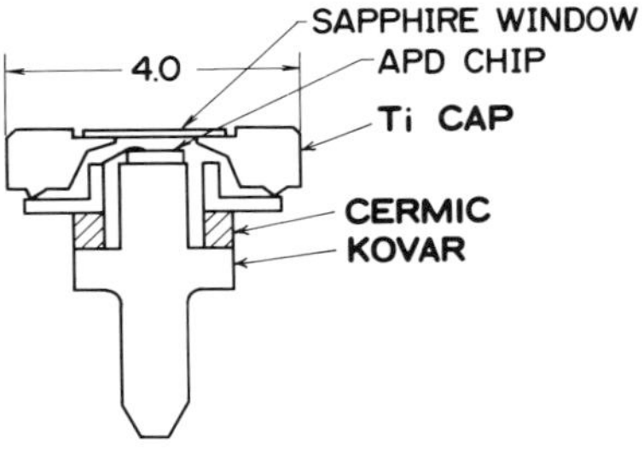

FIG. 11. A package of avalanche photodiodes for use in fiber-optic communication systems.

device, the leakage of the package is kept below 10^{-10} atm cm^3 sec^{-1}. To obtain a high optical coupling efficiency, the distance between the outside of the window and the photosensitive surface is held to about 200 μm, and the tolerance in the alignment between the center of the active area and the center of the package is kept as low as possible (typically less than 50 μm). As a result, incident light is easily focused onto the photosensitive area of the diode. Since the package has a pill structure, it allows easy mounting in coaxial and strip-line microwave circuits. This leads to an excellent wide-bandwidth performance of up to the gigahertz region. Figure 12 shows the pulse-response waveform for a mode-locked Nd:YAG laser ($\lambda = 1.06$ μm) obtained by using the package mounted on a germanium avalanche photodiode. A good waveform free from impedance mismatch was obtained. Pulse

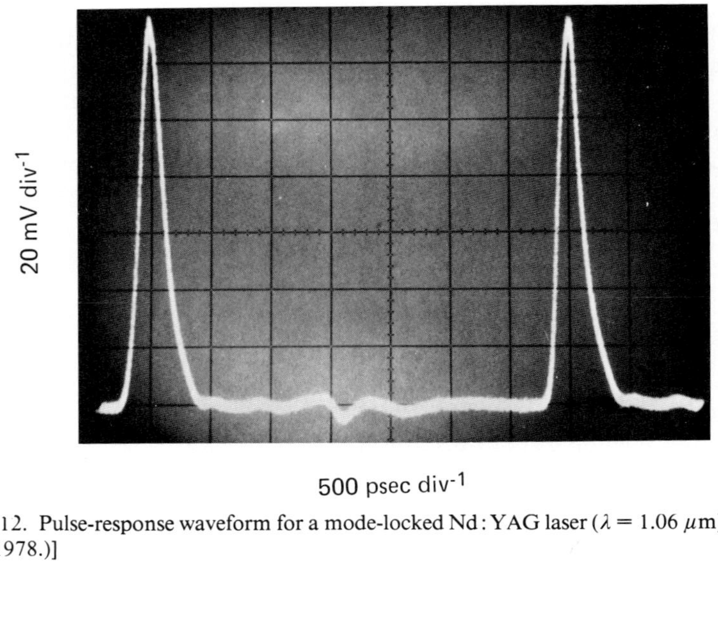

FIG. 12. Pulse-response waveform for a mode-locked Nd:YAG laser ($\lambda = 1.06$ μm). [Ando *et al.* (1978.)]

risetime and half-pulsewidth are 100 and 200 psec, respectively. This response time is restricted by the photodiode chip itself, because the junction capacitance is 1.8 pF, and the load resistance is 50 Ω; thus, the *RC* time constant is nearly 100 psec.

III. Silicon Avalanche Photodiodes

Because of a large asymmetry in the electron- and hole-impact-ionization coefficients and its appropriate value of light-absorption coefficients, silicon is the most suitable material for avalanche photodiodes in a low-transmission-loss region, $\lambda = 0.8-0.9\ \mu m$, of silica fiber. It is also easy to apply conventional processing technology to silicon. In Part III, we discuss the structures, design details, and performance of silicon avalanche photodiodes.

5. Reach-Through Structure

In optical-fiber communications, a detector is required to operate in low-bias voltages, usually less than 200 V or so. A lower operating voltage is of course more desirable for the volume and reliability requirements of a dc/dc converter. From the light-absorption coefficient of silicon in the wavelength range of $0.8-0.9\ \mu m$, a large depletion layer of $30-50\ \mu m$ is needed to obtain high quantum efficiency. In a one-sided abrupt junction, to deplete this long depletion layer and to obtain avalanche gains at the same time, it is necessary to apply an extremely high voltage of more than 500 V, where the applied voltage is obtained from $V_a = \int_0^{l_D} E\, dx$ and E is the electric field strength. The electric field profile is schematically shown in Fig. 13. Carrier multiplication takes place in the highest electric field region, and the layer (avalanche region) is typically very thin compared with the entire depletion length. In the remaining layer, carriers only drift in a saturation velocity, which is attained at an electric field greater than 1×10^4 V cm^{-1} for silicon and germanium (Sze, 1969). Therefore, for low operating voltages, the electric field of the carrier-drift region can be decreased to a low value, but it must be sufficiently high to attain the carrier-saturation velocity. To satisfy this requirement, silicon avalanche photodiodes for optical-fiber communications have the so-called reach-through structure having an $n^+-p-\pi-p^+$ construction (Ruegg, 1967; Conradi, 1974; Kaneda *et al.*, 1976a; Nishida *et al.*, 1977; Melchior *et al.*, 1978; Smeets and Politiek, 1979). The electric field of this structure reaches through from the n^+ to the p^+ layer, and only the avalanche region has such a high electric field that impact ionizations take place there, as shown by the dashed line in Fig. 13. An operating voltage of the reach-through photodiodes can decrease to as low as 200 V, and voltages of about 50 V are needed for the avalanche region and the

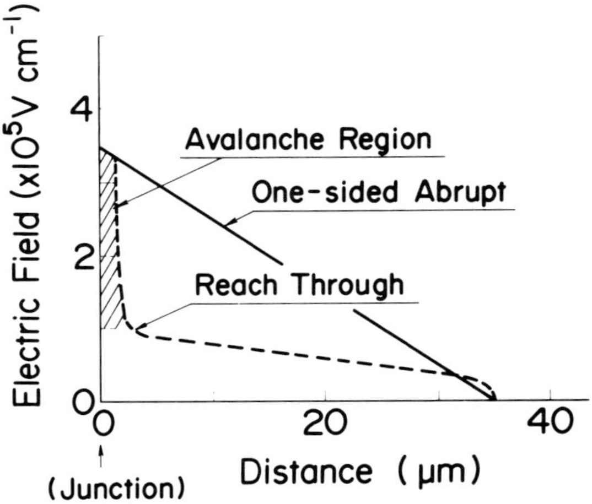

FIG. 13. Schematic profile of electric field for a one-sided abrupt diode (solid line) and a reach-through diode (dashed line).

remainder for the depletion layer. The device structure is shown in Fig. 14, where the light is incident on the n^+ surface. As seen in Fig. 14, planar diodes with a guard ring are easier to fabricate and are expected to show higher electrical stability and durability because the technologies of channel stop and/or field plate for suppressing surface-leakage currents are readily adaptable. The incident light is mainly absorbed in the π layer. Since photogener-

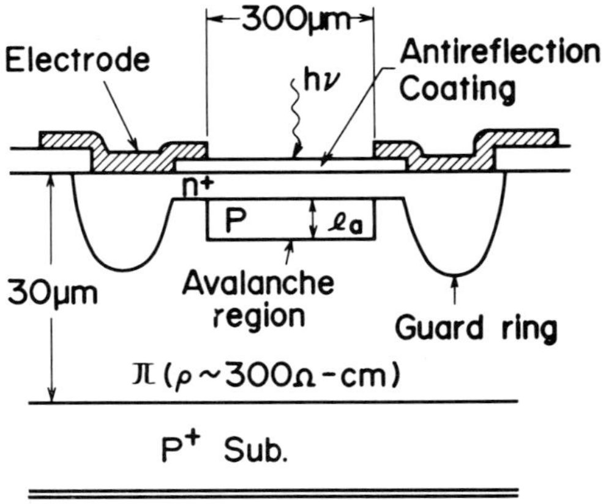

FIG. 14. Cross-sectional view of a silicon reach-through avalanche photodiode. [From Kaneda *et al.* (1978).]

ated electrons with higher ionization coefficients are injected into the avalanche region, this structure has low multiplication noise.

A purer electron injection is achieved by an inverted construction consisting of $p^+-\pi-p-n^+$ layers formed in the mesa configuration, in which the light is incident on the p^+ surface (Takamiya *et al.*, 1975). A lower multiplication noise than that of $n^+-p-\pi-p^+$ diodes has been reported, but the difference is very small. Regarding the reproducibility and reliability of these devices, planar $n^+-p-\pi-p^+$ diodes are more widely studied and fabricated (Conradi, 1974; Amouroux *et al.*, 1975; Berchtold *et al.*, 1975; Kaneda *et al.*, 1976a; Kanbe *et al.*, 1976a; Nishida *et al.*, 1977; Melchior *et al.*, 1978; Goedbloed and Smeets, 1978).

The reach-through avalanche photodiode shows low multiplication noise, high quantum efficiencies, and high speed with a low operating voltage. Thus, it is the most suitable structure for use in the 0.8–0.9-wavelength fiber-optic communication systems. However, understanding its characteristics and its fabrication are more difficult than a simple one-sided junction device.

6. Design Details

a. A Model for Reach-Through Diodes

The major design criteria for the avalanche photodiode is reduction of multiplication noise. The noise value can be calculated from Eqs. (1)–(3) if the electric field profile of the diode is known. The electric field of the reach-through diode, however, is rather complex, and also Eqs. (2) and (3) do not have immediately understandable physical meanings because they include double-integral terms. It is therefore important to derive an analytical expression for designing reach-through avalanche photodiodes. In a silicon reach-through diode, the p layer is the key region because it determines the most important characteristics of excess-noise factor F and an operating voltage, which lies in the vicinity of breakdown voltage V_B. The doping level of the p layer is relatively low, on the order of 10^{12} cm^{-2}. If both doping concentration and depth are not precisely controlled, the values of F and V_B will not lie in the predicted region. Ion implantation allows superior control of both the doping level and depth. We therefore discuss the diodes formed using the ion-implantation techniques for the p layers. A schematic representation of the concentration profile is shown in Fig. 15a. The peak concentration of the p layer is usually on the order of 10^{16} cm^{-3} and is two orders of magnitude higher than the π layer. The actual electric field is illustrated by the dashed line in Fig. 15b and is rather complex, depending on the distance. To simplify design of the parameters of interest (F and V_B), the electric field is modified as shown by the solid line in Fig. 15b. The magnitude of the

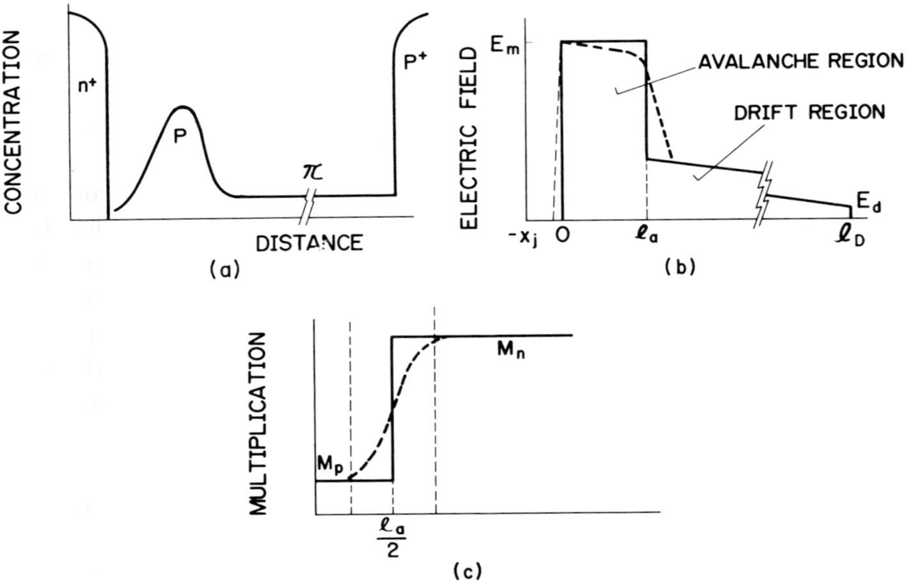

FIG. 15. Model for reach-through avalanche photodiodes. Schematic view of (a) the actual impurity-concentration profile; (b) the actual profile (dashed line) and the model (solid line) of the electric field; and (c) the actual distribution (dashed line) and the model (solid line) of the multiplication factor. [From Kaneda *et al.* (1976b).]

electric field in the avalanche region is assumed to be a constant value that varies according to the applied voltage. The relationship between E_m, the maximum electric field at V_B, and the length of the avalanche region l_a can be calculated from the avalanche-breakdown condition given by

$$\int_0^{l_a} \alpha \exp\left[\int_0^x (\beta - \alpha)\,dx'\right] dx = \int_0^{l_a} \beta \exp\left[\int_0^x (\alpha - \beta)\,dx'\right] dx = 1. \tag{28}$$

In the modified electric field shown by the solid line in Fig. 15b, Eq. (28) is reduced to

$$\ln k/(k-1) = \alpha l_a. \tag{29}$$

The calculated results of E_m as a function of l_a are shown in Fig. 16, where the ionization coefficients used are the data given by Lee *et al.* (1964) as

$$\alpha = 3.8 \times 10^6 \exp\left(-\frac{1.75 \times 10^6}{E}\right) \text{cm}^{-1}$$

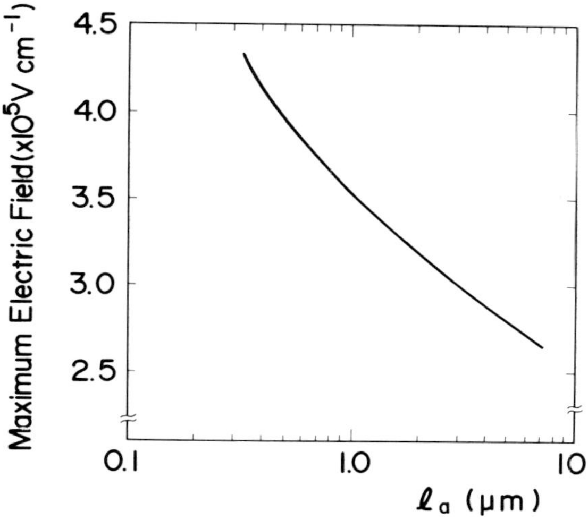

FIG. 16. Calculated maximum electric field at breakdown voltage as a function of the avalanche-region length, where carrier-ionization coefficients used are the data given by Lee *et al.* (1964).

for electrons and

$$\beta = 2.25 \times 10^7 \exp\left(-\frac{3.26 \times 10^6}{E}\right) \text{cm}^{-1}$$

for holes. The value of E_m decreases with increasing l_a. The relationship between l_a and k_m (k_m is the k value at E_m) can also be obtained from Eq. (29), as shown in Fig. 17. The value of k_m decreases as l_a increases. This indicates that multiplication noise decreases as l_a increases.

In the modified electric field shown by the solid line in Fig. 15b, the breakdown voltage is given by

$$V_B = E_m l_a + \{[qN_A(l_D - l_a)^2]/2\varepsilon\} + E_D(l_D - l_a), \qquad (30)$$

where N_A is the carrier concentration of the π layer and E_D the electric field strength at the deep edge of the depletion layer. For the response-time requirement, E_D must have a high value so carriers can drift by a saturation velocity.

Another modification is used for the distribution of carrier multiplication, as illustrated by the solid line in Fig. 15c, whereas the actual profile is the shape shown by the dotted line. Carrier injections are supposed to be divided into pure electron I_n and hole I_p injections at the center of the avalanche

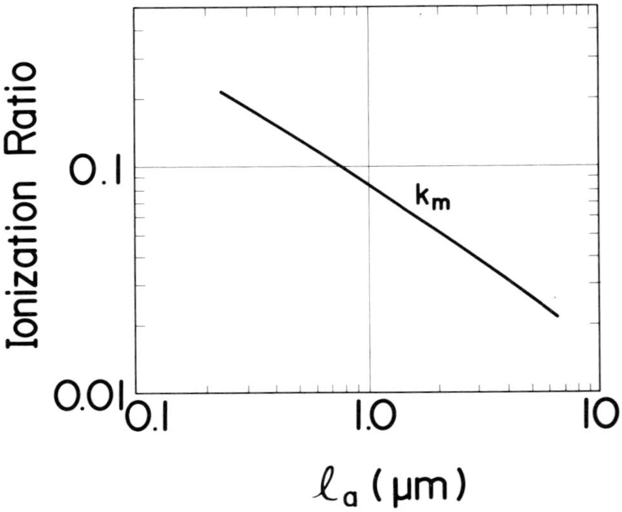

FIG. 17. The calculated ratio of hole–electron-ionization coefficients at breakdown voltage as a function of the avalanche-region length.

region, $l_a/2$. The values of I_n and I_p are given by

$$I_n = \exp\{-a(x_j + \tfrac{1}{2}l_a\}/I_0 \tag{31}$$

and

$$I_p = 1 - I_n, \tag{32}$$

where x_j is the junction depth and I_0 the initial photocurrent obtained from Eqs. (4) and (14) as

$$I_0 = P/hv\{1 - \exp(-al_D)\}, \tag{33}$$

where the photogenerated carriers in the n^+ and p^+ layers are assumed to recombine there and do not contribute to photocurrents. If the k value is known, we can study easily the characteristics of the excess-noise factor by using Eqs. (7)–(11) and (31)–(33). In Subsection 6b, we obtain the effective value of k.

b. Effective Value of k

The hole–electron ionization-coefficient ratio (k value) depends on the strength of the electric field (Fig. 1), and the strength varies according to multiplication factors. Because the electric field varies according to distance, as shown by the dashed line in Fig. 15b, the k value also depends on that distance. That is, the k value varies both according to multiplication factors and the position of the avalanche region. Therefore the constant-k approxi-

TABLE I

PROPERTIES OF SAMPLES USED FOR MULTIPLICATION-NOISE MEASUREMENTS

Sample	Dose ($\times 10^{12}$ cm^{-2})	l_a (μm)	E_m ($\times 10^5$ V cm^{-1})
1	2.7	0.4	5.1
2	2.0	0.8	4.1
3	1.8	1.2	3.7
4	1.7	1.9	3.4
5	1.4	2.5	3.2

mation is valid for simplifying the design and characterization of avalanche photodiodes, where k_{eff}, the effective value of k, is used. The k_{eff} values are determined from actual multiplication-noise measurements.

Five diodes in Table I were prepared for measurements by using boron (B^{2+}) implantation with an acceleration energy of 300 keV. To obtain different lengths of the avalanche region, diodes 1, 2, and 3 had different implanting directions; that is, random, ⟨111⟩, and ⟨110⟩, respectively. Diodes 4 and 5 were made by a subsequent π-layer epitaxial growth after the random implantation. The carrier profiles of diodes 4 and 5 are therefore broader than the others because of additional heat processes, as shown in Fig. 18.

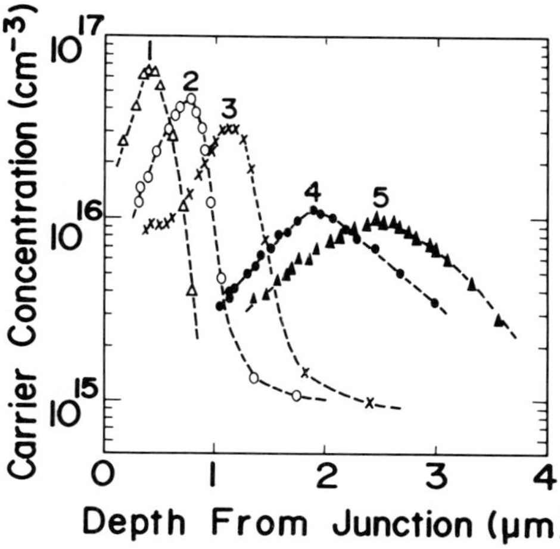

FIG. 18. Carrier-concentration profiles of the p layer. Boron-ion-implantation directions of diodes 1, 2, and 3 were random, ⟨111⟩, and ⟨110⟩, respectively. Diodes 4 and 5 were made by a subsequent π-layer epitaxial growth after random implantations (300 keV). [From Kaneda *et al.* (1976b).]

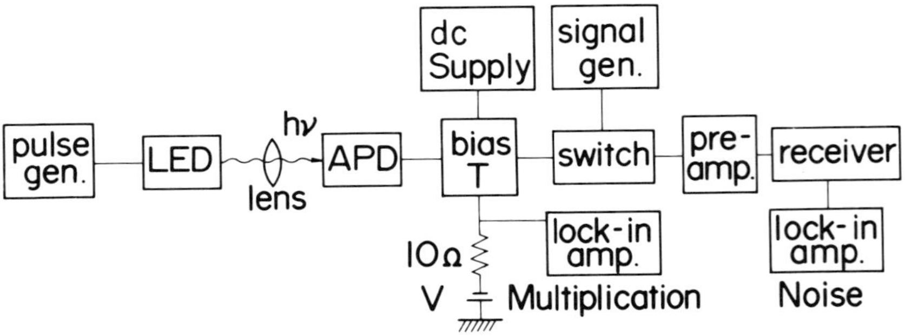

FIG. 19. Apparatus used to measure multiplication noise.

Multiplication-noise measurements were performed by using the apparatus shown in Fig. 19. Light-emitting diodes ($\lambda = 0.83\ \mu$m) were used for exciting the avalanche process. Light modulated in a square waveform at a frequency of 1 kHz was focused onto the diode. The multiplication noise was amplified and detected by a preamplifier and precision test receiver. The receiver output, synchronized with the incident light pulse, was measured with a lock-in amplifier. In this way, multiplication noise excited by photocurrents could be measured, and the influence of thermal and dark-current noise was eliminated because these components were not synchronized with the incident light pulse. The absolute noise power was obtained by comparing the multiplication noise to the reference rf power. By using this measuring apparatus, we can obtain an absolute noise power as low as about -130 dBm (10^{-13} mW).

Figure 20 shows an example of the calculated results of F versus $\langle M \rangle$ for various values of k_{eff} and the experimental values for sample 3. As seen in Fig. 20, the experimental points follow the calculated curves where $k_{\text{eff}} = 0.037 \pm 0.005$ quite well. The absolute level of the measured points is estimated as accurate to about ± 0.5 dB, which corresponds to the ambiguity of the k_{eff} value.

The k_{eff} values thus found are shown in Fig. 21 and relate to k_{m} as follows:

$$k_{\text{eff}} \approx 0.52 k_{\text{m}}. \tag{34}$$

The optimum gain of silicon avalanche photodiodes in fiber transmission systems is generally in the range of $\langle M \rangle$ 50–100, so that the relationship between F and l_{a} at $\langle M \rangle = 100$ is calculated by using Eq. (34) at a light-absorption coefficient of 700 cm ($\lambda = 0.83\ \mu$m for silicon). The results are shown by the solid line in Fig. 22. Experimental results are also shown in Fig. 22 and are in broad agreement with the calculations. The F value at

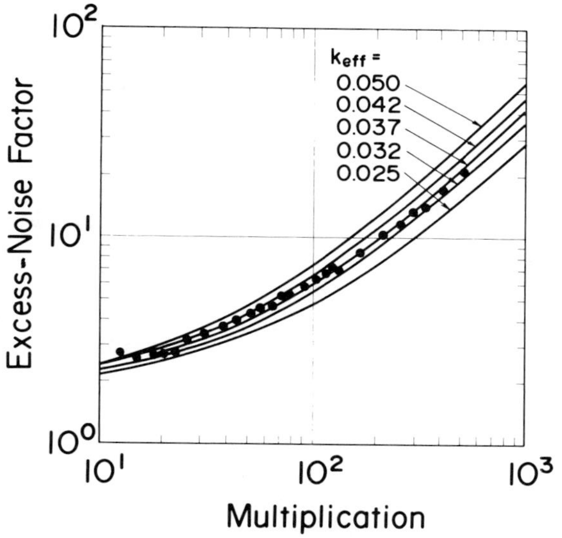

FIG. 20. Comparison of experimental and calculated results of the excess-noise factor as a function of the multiplication factor for sample 3. $\lambda = 830$ nm, $I_n \sim 0.94$, $I_p \sim 0.061$. [From Kaneda *et al.* (1976b).]

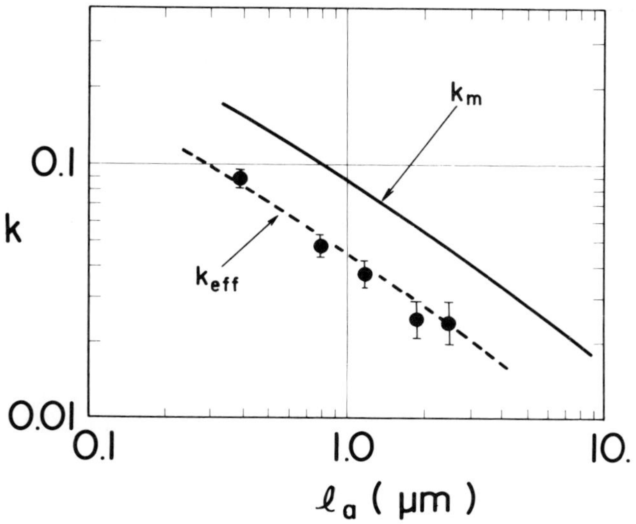

FIG. 21. Effective hole–electron-ionization coefficients ratio as a function of the avalanche-region length. The solid line shows the calculated hole–electron-ionization coefficients ratio at breakdown voltage, where the ionization coefficients used are the data given by Lee *et al.* (1964). Solid circles show the experimental results.

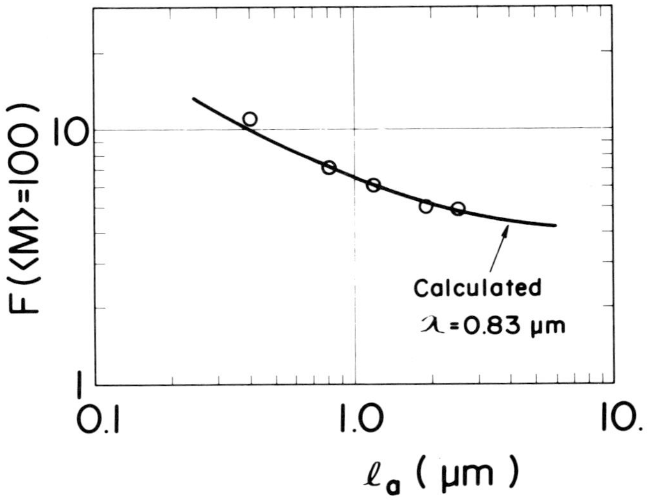

FIG. 22. Excess-noise factor at multiplication factor of 100 as a function of the avalanche-region length. [From Kaneda *et al.* (1976b).]

$\langle M \rangle = 100$ decreases steadily as l_a increases. In the larger l_a region, however, the curve of F shows a saturation tendency because of the increasing holes injection on l_a.

c. Optimum Length of Avalanche Region

Because k_{eff} values are determined as a function of l_a from Eqs. (29) and (34), we can calculate the various characteristics of multiplication noise as a function of $\langle M \rangle$, l_a, and λ. Figure 23 shows the F-versus-λ characteristics as a parameter of $\langle M \rangle$, where the values of $x_j = 0.3\ \mu m$, $l_a = 2.0\ \mu m$, $k_{eff} = 0.027$, and $l_D = 30\ \mu m$ were selected. The light-absorption coefficients used are the data shown in Fig. 9. The F value increases noticeably as the wavelength decreases in a range smaller than 0.7 μm, because the hole-injection ratio becomes larger due to increased light-absorption coefficients. Figure 24 also shows the F-versus-l_a characteristics with a parameter of $\langle M \rangle$ at a wavelength of 0.83 μm. From the results shown in Figs. 23 and 24, we find that the optimum value for l_a lies in the range from 2 to 3 μm for the multiplication region of 50–100 and wavelength of 0.8–0.9 μm. At this time, the breakdown voltage calculated from Eq. (30) is in the range of 150–200 V, where the values of $l_D = 30\ \mu m$, $N_A = 1 \times 10^{14}\ cm^{-3}$ and $E_D = 1 \times 10^4\ V\ cm^{-1}$ were selected for the calculation. The voltage range is adequate for practical use in fiber-optic communications systems.

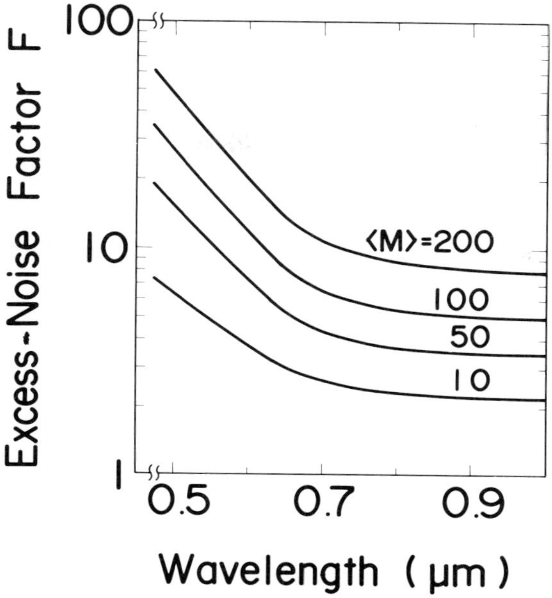

FIG. 23. Calculated results of excess-noise factor versus wavelength characteristics with the parameter of multiplication factors. $l_a = 2.0\ \mu m$, $k_{eff} = 0.027$.

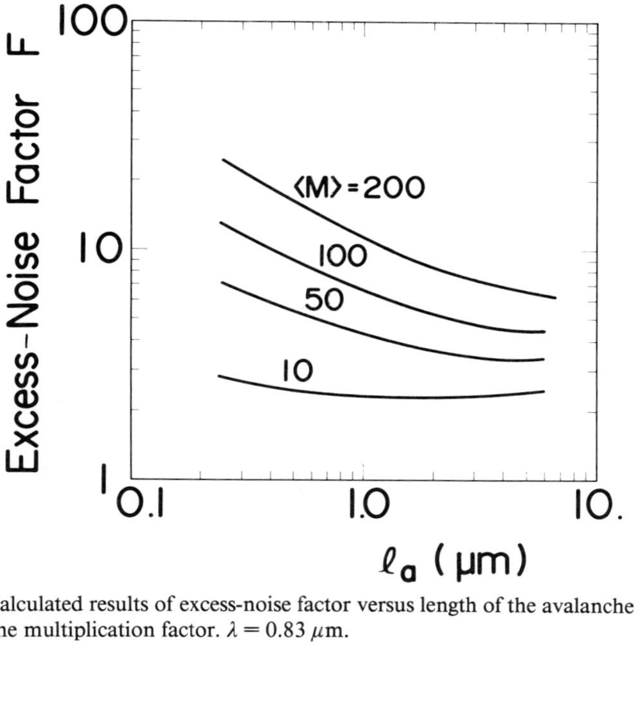

FIG. 24. Calculated results of excess-noise factor versus length of the avalanche region as a function of the multiplication factor. $\lambda = 0.83\ \mu m$.

7. *p*-Layer Formation

There are several methods for forming a long avalanche region, such as ion implantation, epitaxy, diffusion, and a combination of these technologies. For obtaining a deep *p* layer, which must be formed with a precise dosage and depth, the channeled implantation is a successful method for its reproducibility. Figure 25 shows the concentration of boron obtained from a channeled implantation with acceleration energy of 800 keV compared to that of random implantations. The ⟨110⟩ axis is used for a deep *p* layer because the axis is the most open channel into silicon (Lecrosnier *et al.*, 1977). The peak depth of the *p* layer from the junction depth, which is almost equal to the length of the avalanche region, is larger than the 2 μm for the ⟨110⟩ channeled implantation. By using the parallel-scanning method for channeled-ion implantations, a uniform *p* layer can be obtained for the whole area of a wafer (Nishi *et al.*, 1978). Uniformity can be measured from the breakdown

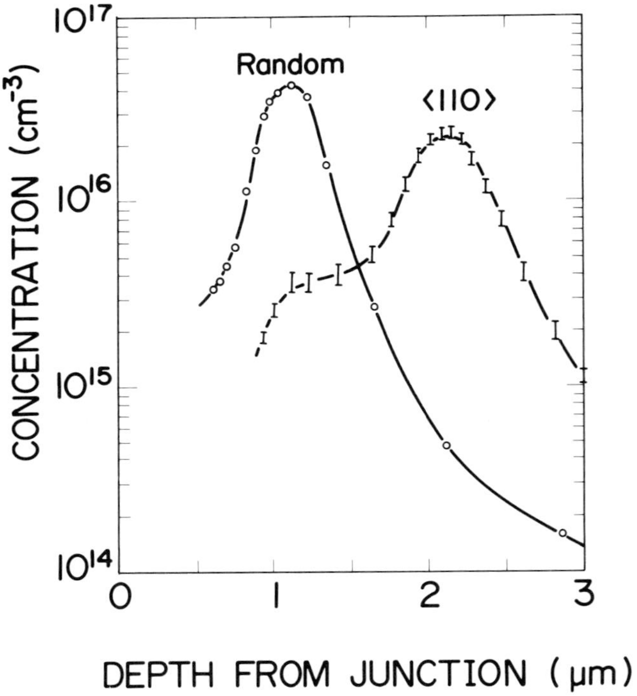

Fig. 25. Concentration profiles of boron-ion-implanted layers: random implantations, ion dose = 2×10^{12} cm^{-2}; ⟨110⟩ channeled implantations, ion dose = 1.9×10^{12} cm^{-2}; the acceleration energy of 800 keV. [From Kaneda *et al.* (1978).]

Numeral	V_B (volts)
0	140-149
1	150-159
2	160-169
3	170-179
4	180-189
5	190-199
6	200-219
7	220-239
8	240-259
9	260-279
X	≳280

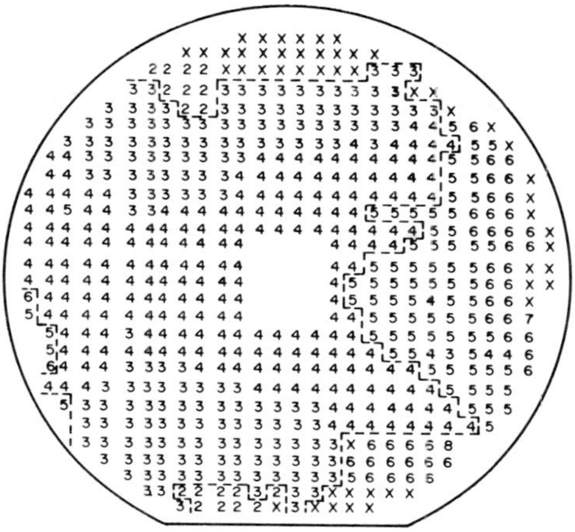

FIG. 26. Spacial distribution of breakdown voltages obtained by the parallel-scanning system. [From Nishi *et al.* (1978).]

voltage of reach-through diodes, because the breakdown voltage depends strongly on the *p*-layer profile. Figure 26 shows the distribution of the breakdown voltage of the diodes on one wafer. Breakdown voltages lie in the range 170–200 V, except for the periphery of the wafer where the wafer quality is lacking and/or is affected by fabrications. The uniformity is comparable to that obtained by random implantations. The parallel-scanning method for channeled-ion implantations is extremely suitable for a *p*-layer formation in silicon reach-through photodiodes because it provides deep and uniform *p*-type doping that is reproducible.

8. CHARACTERISTICS

By using the reach-through structure, silicon avalanche photodiodes show low multiplication noise, high quantum efficiencies, and high speed with a low operating voltage in the 0.8–0.9-μm-wavelength region. The other specific features appear in the characteristics of capacitance, multiplication, and temperature dependence of breakdown voltage. In Section 8, optical and electrical characteristics of the reach-through photodiode formed by the $\langle 110 \rangle$ channeled-ion implantation are discussed. Some of them are compared to a diode made by random implantations (Kaneda *et al.*, 1978). The properties of the diodes discussed are summarized in Table II. Figure 27 shows the capacitance-versus-voltage characteristics of diode chips at a frequency of 1 MHz. The capacitance falls steeply when the voltage of the depletion layer extends to the π layer through the p layer. The voltage is

TABLE II

CHARACTERISTICS OF DIODES

Sample	Boron implant	Dose ($\times 10^{12}$ cm^{-2})	π-Layer concentration ($\times 10^{13}$ cm^{-3})	V_B (V)	F ($M = 100$)
1	110	1.9	5	159	4–5
2	Random	2.0	6	168	6–7

denoted as reach-through voltage V_{rt} and is given by

$$V_{rt} = (qQ_p/\varepsilon)l_a, \tag{35}$$

where Q_p is the number of uncompensated charges per unit area contained in the depleted p region and is almost equal to an implanted dosage of boron. As seen in Fig. 27, V_{rt} for the diode by $\langle 110 \rangle$ channeled-ion implantation is twice as large as the random implantations. This is mainly due to the difference of l_a. The capacitance shows a constant value in the higher voltage region where the depletion region reaches the p^+ substrate. A low value of ≈ 0.6 pF was obtained for the diode with an active-area diameter of 300 μm and a 30-μm depletion length. The electric field profile can be obtained from the capacitance–voltage characteristics. The results at the breakdown volt-

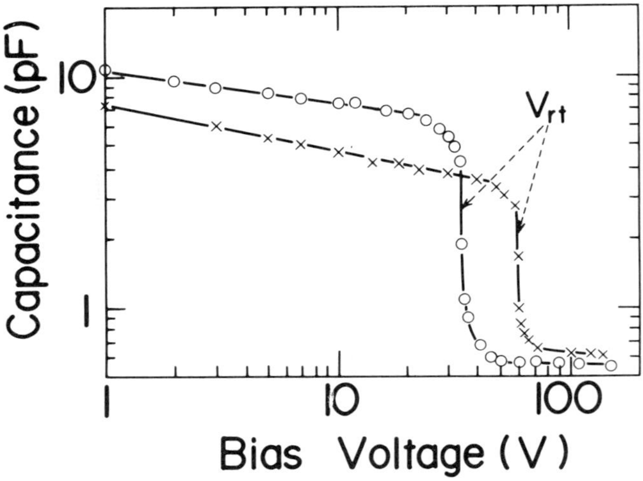

FIG. 27. Capacitance versus bias-voltage characteristics obtained at 1 MHz. ×, 800 keV, $\langle 110 \rangle$; ○, 800 keV, random.

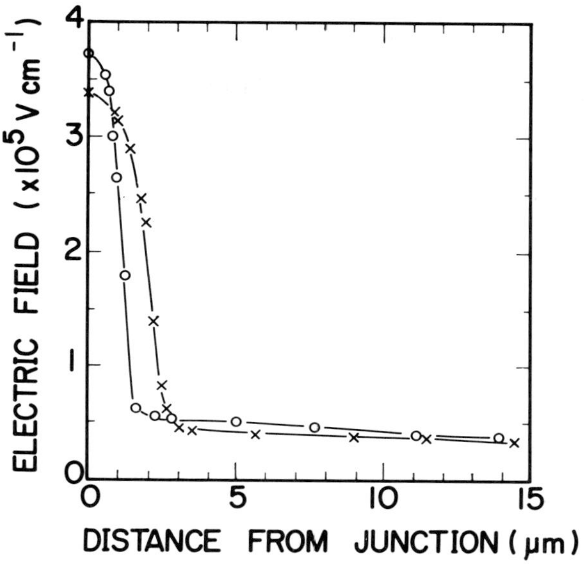

FIG. 28. Electric field profiles at breakdown voltage as a function of position obtained from capacitance measurements. ○, 800 keV, random; ×, 800 keV, ⟨110⟩.

age are shown in Fig. 28. There are extremely high electric field regions in the junction region, which is a specific feature of reach-through avalanche photodiodes. For the ⟨110⟩ channel-implanted diode, the maximum value of the electric field is lower, and the high-field region is longer than that of the random one. In the lower electric field, we obtain a larger asymmetry in the electron- and hole-impact-ionization coefficients. The electric field in the π layer is approximately 5×10^4 V cm^{-1} and is sufficiently high to obtain a carrier-saturation velocity.

Avalanche-multiplication characteristics are shown in Fig. 29. Multiplication factors rise steeply up to about 5. In this multiplication range, the depletion layer has not yet extended into the π layer and applied bias voltage effectively increases the electric field strength, thus causing a steep increase in multiplication factors. When the depletion region extends to the π layer, applied bias voltage is shared in the whole depletion region. This lead to a relatively slow rise in multiplication-voltage characteristics. A knee in the multiplication characteristics is one of the notable features of reach-through avalanche photodiodes. A uniform multiplication factor of more than 1000 was usually obtained. Multiplication factors at a bias voltage vary with an incident wavelength. Because electrons have higher ionization coefficients than holes in silicon, a larger multiplication factor is obtained by increasing

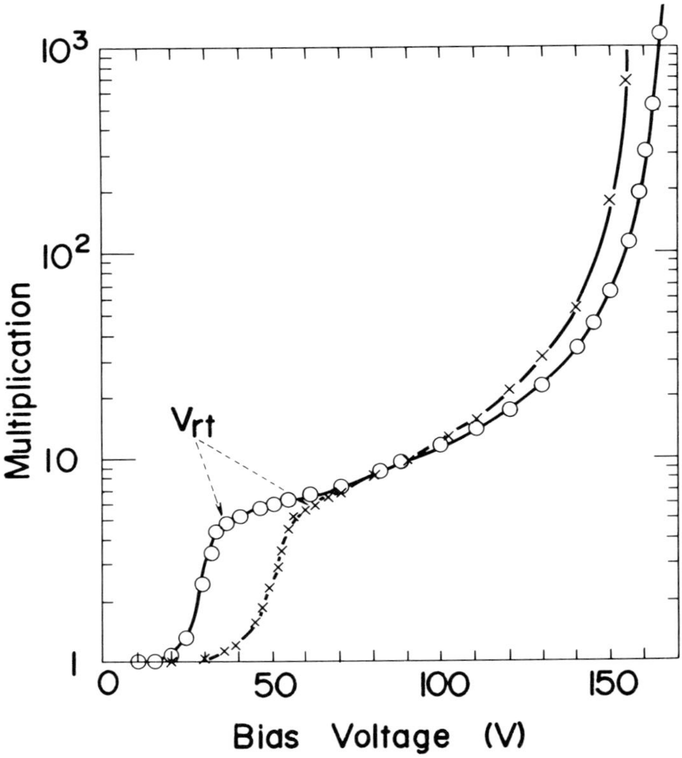

FIG. 29. Multiplication versus bias-voltage characteristics at 1 kHz. ○, 800 keV, random; ×, 800 keV, ⟨110⟩; $\lambda = 830$ nm; $I_0 = 0.1\ \mu$A.

an electron injection; namely, the multiplication factor shows a high value as the incident wavelength increases in the diodes discussed.

Quantum efficiency of avalanche photodiodes is often difficult to determine definitely because carrier multiplication takes place at operating voltages. If quantum efficiency depends on bias voltage, it is hard to distinguish the variation of quantum efficiency from carrier multiplication. Then, the $p-i-n$ photodiodes, which merely lack the p layer of the $n^+-p-\pi-p$ construction, were made in the same wafer. The quantum efficiency of the $p-i-n$ photodiodes showed a constant value in the bias-voltage range 1–200 V and was equal to that of avalanche photodiodes measured at approximately 1 V, where no carrier multiplication takes place. We therefore can conclude that quantum efficiency of the silicon reach-through structure is unchanged by bias voltages and that the $p-\pi$ junction is not affected for hole

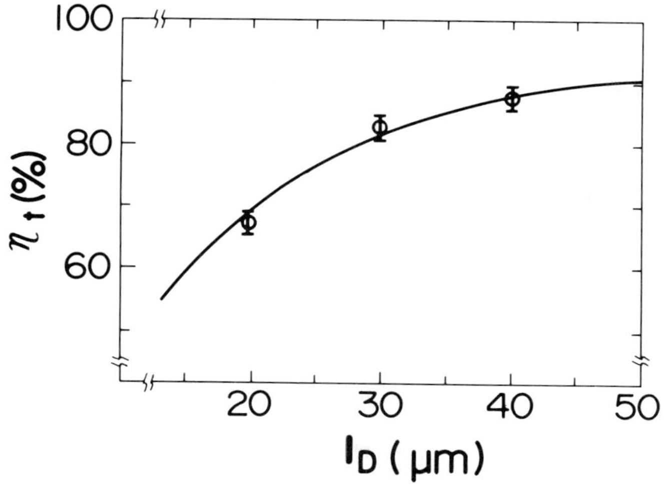

FIG. 30. Depletion-layer-length dependence of quantum efficiencies at a wavelength of $\lambda = 830$ nm. The thickness of AR films (Si_3N_4) is 105 nm. The solid line gives the calculated values.

transportations even at the low-bias-voltage region. Thus, quantum efficiency was studied at a bias voltage of 1 V.

Quantum efficiency depends largely on the length of the depletion layer l_D. Figure 30 shows l_D dependence of quantum efficiency at a wavelength of 830 nm. The diodes used had a chemical vapor deposition (CVD) Si_3N_4 film in the active region for an antireflection coating. The results calculated were obtained from Eqs. (16) and (33) and agree well with the experimental ones. High quantum efficiencies of greater than 80% were obtained by using more than 30 μm of l_D. Figure 31 shows the spectral response for the diode with l_D of 30 μm. The response shows the maximum value at around 800 nm. Recombination in the thin n^+ layer becomes substantial for wavelengths of less than 700 nm. Quantum efficiency decreases rapidly beyond 900 nm as the incident light penetrates into the p^+ layer, where the generated carriers recombine.

The frequency response of the diode was studied as a function of multiplication factors by using the apparatus shown in Fig. 32. The results obtained at a wavelength 830 nm (GaAlAs laser) are shown in Fig. 33. In the low-multiplication region of about 3, the diode shows a slow response because the photogenerated carriers in the π layer diffuse into the junction region because the depletion layer has not yet extended into the π layer in this multiplication range. The diode shows a fast response of -3 dB cutoff fre-

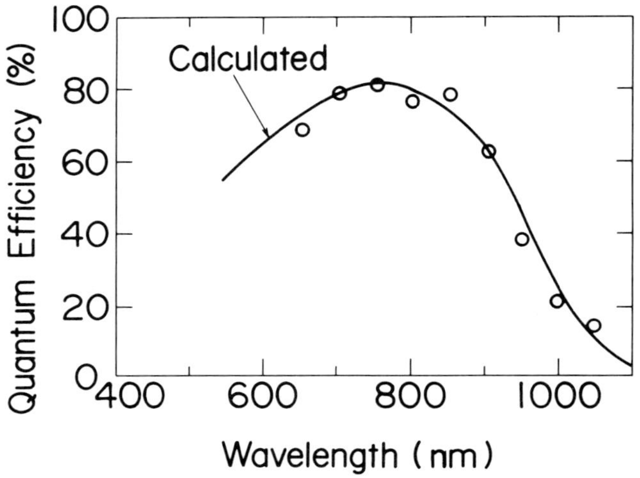

FIG. 31. Quantum efficiency as a function of wavelength; where the depletion-layer-length $l_D = 30\ \mu$m.

quency of above 500 MHz in the multiplication range greater than 10, which is of interest for practical use, where the depletion region reaches the p^+ substrate and carriers run in the saturation velocity in the π layer. The multiplication dependence of frequency response was not observed in the range of $\langle M \rangle = 10-100$ and at a frequency up to 500 MHz because of the short avalanche buildup time in silicon. Since the intrinsic response time τ is approximately 5×10^{-13} sec (Kaneda *et al.,* 1976c), the maximum gain–bandwidth product limiting the avalanche buildup time is calculated from $1/(2\pi\tau)$ to be about 320 GHz. This value is larger by more than an order of

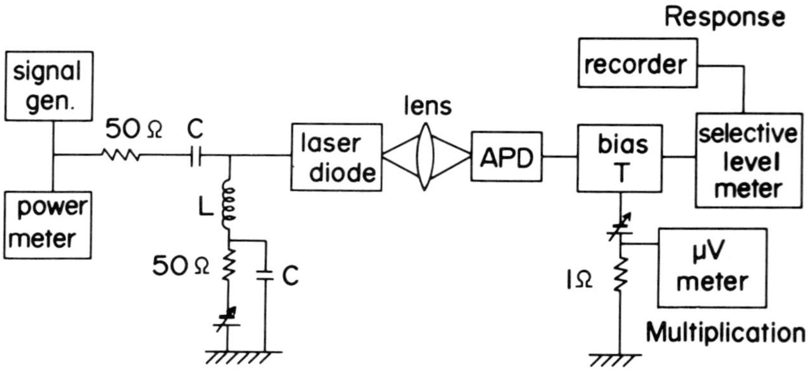

FIG. 32. Apparatus used to measure frequency response.

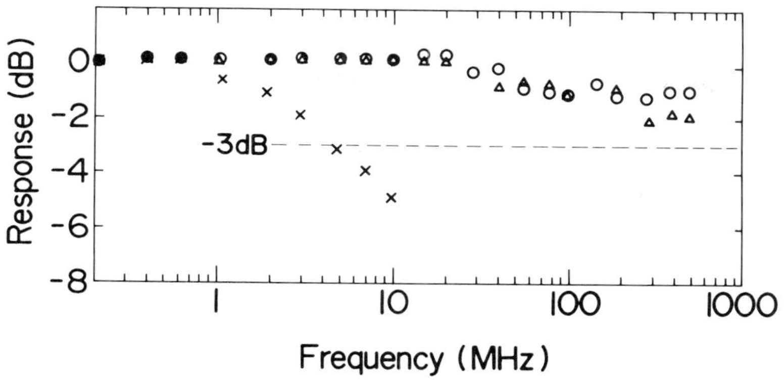

FIG. 33. Frequency response of silicon avalanche photodiodes as a parameter of multiplication factors obtained at 830 nm. ○, $\langle M \rangle \sim 100$; △, $\langle M \rangle \sim 10$; ×, $\langle M \rangle \sim 3$.

magnitude than those of Ge (~30 GHz) and InP/GaInAs (10–20 GHz) APDs (Kaneda and Takanashi, 1973a; Campbell *et al.*, 1983; Yasuda *et al.*, 1984; Kobayashi *et al.*, 1984).

The frequency response shown in Fig. 33 deteriorated about 1 dB at around 50 MHz and showed a rather flat configuration above this frequency. Since for even a worst-case analysis, the transit-time cutoff frequency is 520 MHz, which was evaluated from Eq. (22) by selecting $l_D = 30\ \mu m$ and $v_s = 7 \times 10^6$ cm sec^{-1}, the deterioration at around 50 MHz is due to neither the transit-time effect nor the *RC* time constant and carrier diffusion time. Kanbe *et al.* (1976b) have found that the built-in field at the $\pi - p^+$ junction influences response speed in reach-through avalanche photodiodes. The built-in field E_{bi} is given by (Shive, 1959) as

$$E_{bi} = \frac{k_B T}{q} \frac{1}{N_A(x)} \frac{dN_A(x)}{dx}, \tag{36}$$

where $N_A(x)$ is the carrier concentration of the $\pi - p^+$-junction region, T the absolute temperature, and k_B Boltzmann's constant. The electric field has a constant strength of $(-k_B T\theta/q)$, where the concentration profile is assumed to be $N_A(x) = N_s \exp(-\theta x)$, θ is a constant, and N_s is the carrier concentration of the p layer. The strength is evaluated at about 600 V cm^{-1} by selecting the values of $N_s = 1 \times 10^{19}$ cm^{-3}, the π-layer concentration (1×10^{14} cm^{-3}) and the transition region of the $\pi - p^+$ junction (5 μm). The carrier-drift velocities at this electric field are rather low values of 1×10^6 cm sec^{-1} for electrons and 3×10^5 cm sec^{-1} for holes (Sze, 1969). Carriers with these drift velocities cannot respond above several tens of megahertz and lead to a

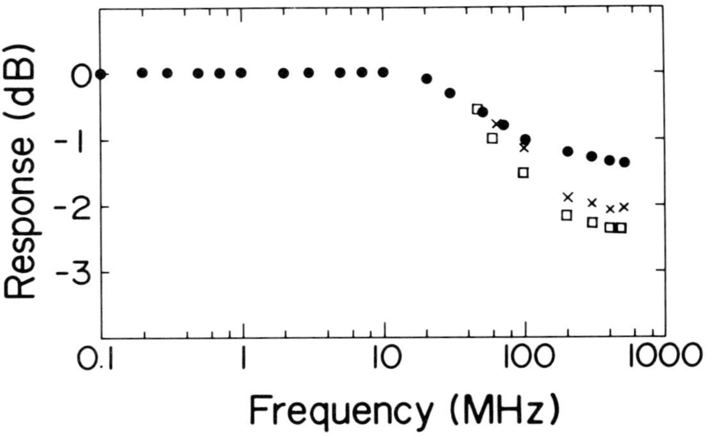

FIG. 34. Depletion-layer-length dependence of frequency response for silicon avalanche photodiodes ($\lambda = 830$ nm). $\langle M \rangle = 10$; ●, $l_D = 30\ \mu$m; ×, $l_D = 35\ \mu$m; □, $l_D = 40\ \mu$m.

reduction of response. This would cause a 1-dB deterioration above 50 MHz.

Figure 34 shows the frequency response as a function of the length of the depletion region at a multiplication factor of 10. The transit time in the depletion region becomes substantial in the frequency range above 100 MHz. The -3-dB cutoff frequency is obtained to be greater than 500 MHz, even for a diode with $l_D = 40\ \mu$m.

Multiplication-noise characteristics for the diodes in Table II are shown in Fig. 35 and were obtained at a frequency of 30 MHz using the apparatus shown in Fig. 19. The primary photocurrents were kept at 2.0 μA for the incident wavelength of 830 nm. The recorder traces fluctuate in accordance with varying bias voltages, which were changed in increments of 0.1 V. A noise reduction of 2 dB was realized with the $\langle 110 \rangle$ channeled implantations compared with the random implantations. By using the expression of excess noise factor F, the low values of $F = 4-5$ at a multiplication factor of 100 were obtained for the channel-implanted diode, as shown in Fig. 36.

a. Temperature Characteristics

Figure 37 shows the dark current versus bias voltage characteristics as a function of temperature in the range from -25 to 100°C. Dark currents increase as temperature increases. The current consists of three components: the generation–recombination current in the depletion layer I_{DB}, the generation–recombination current in the surface depletion layer I_{DS}, and

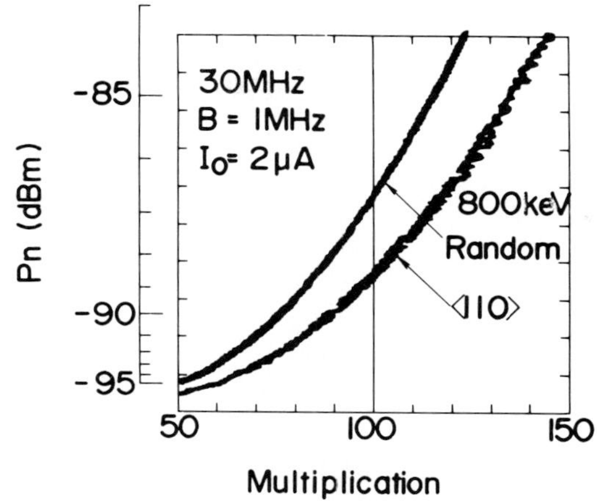

FIG. 35. Experimental results of avalanche noise as a function of multiplication factors. [From Kaneda *et al.* (1978).]

the diffusion current from the neutral region I_{DD}. They are given by

$$I_{DB} = \tfrac{1}{2}q(n_i/\tau_0)A_j l_D, \qquad I_{DS} = \tfrac{1}{2}qS_0 n_i A_s,$$
$$I_{DD} = q(\tau_n^{-1})(n_i^2/N_A)A_j L_n, \tag{37}$$

where n_i is the intrinsic carrier density, A_j the junction area, A_s the area of the surface-depletion layer, S_0 the surface recombination velocity, τ_0 the electron lifetime in the layer, τ_n the electron lifetime in the p layer, and L_n the

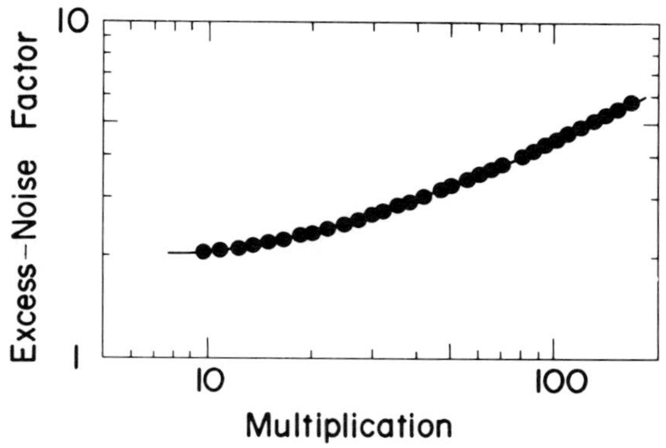

FIG. 36. Excess-noise factors as a function of multiplication factors. $f = 30$ MHz, $B = 1$ MHz, $\lambda = 830$ nm, $I_0 = 1.0$ μA.

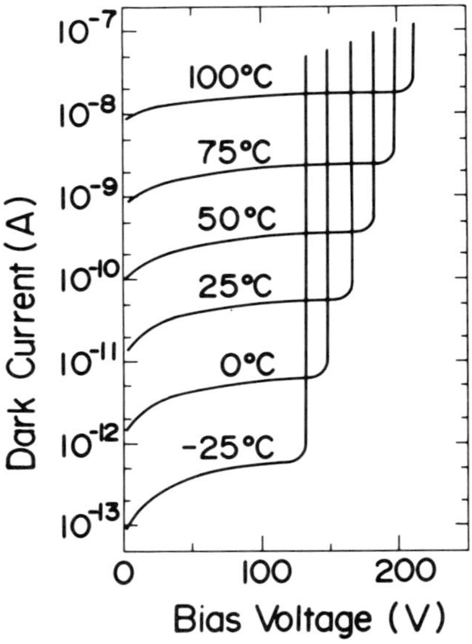

FIG. 37. Temperature dependence of dark current versus bias-voltage characteristics for a Si APD.

diffusion length of electron. In these values, n_i is the most sensitive to temperature and is given as

$$n_i \propto \exp(-E_G/2k_B T). \tag{38}$$

Then, Eq. (35) can be expressed in the following two ways:

$$I_{DB}, I_{DS} \propto \exp(-E_G/2k_B T), \qquad I_{DD} \propto \exp(-E_G/k_B T). \tag{39}$$

In the high-temperature range, dark currents in avalanche photodiodes are determined by I_{DD}. In the low-temperature range, however, the currents consist of I_{DB} and I_{DS}; these cannot be distinguished only by the temperature dependence of dark currents. Dark currents at 10 V are shown as a function of inverse temperature in Fig. 38. Above 70°C, dark currents increase in proportion to $\exp(-E_G/k_B T)$. In the range of less than 70°C, which is practically used in optical-fiber communication systems, dark currents vary by temperature as $\exp(-E_G/2k_B T)$ and show a low value of less than 1 nA.

As shown in Fig. 37, breakdown voltage increases with temperature. Since phonon scattering becomes significant as temperature increases, impact-ionization coefficients of carriers decrease as temperature increases. This leads to the above-mentioned temperature dependence on breakdown volt-

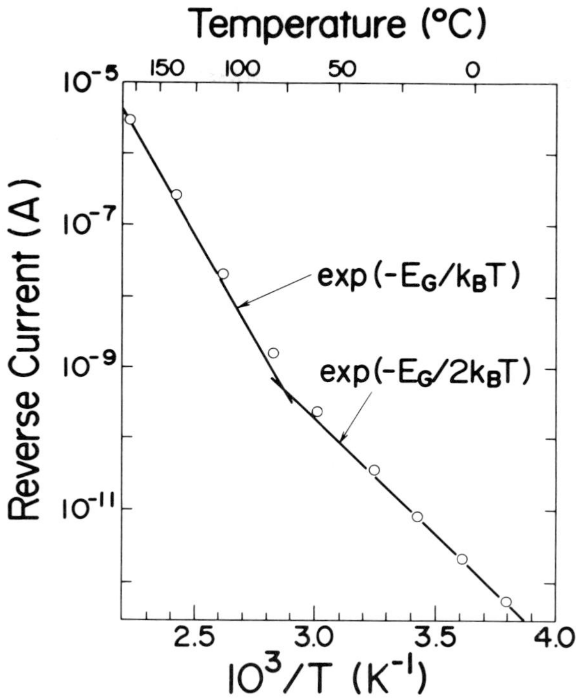

FIG. 38. Reverse dark current as a function of reciprocal temperature for a Si APD with $V_R = 10$ V, $E_G = 1.17$ eV.

age. When the temperature increases (dT), the required change of the electric field distribution is illustrated in Fig. 39, where dE_m is the necessary change in the electric field for maintaining the breakdown condition. The necessary change of breakdown voltage dV is given by

$$dV = dE_m l_D. \tag{40}$$

The temperature dependences of the ionization coefficients are given by Grant (1973) as

$$\alpha = 6.2 \times 10^5 \exp[-(1.05 \times 10^6 + 1.3 \times 10^3 T)/E]$$

and (41)

$$\beta = 2.0 \times 10^6 \exp[-(1.95 \times 10^6 + 1.1 \times 10^3 T)/E],$$

where T is in degrees celsius and E is the electric field. From Eqs. (29) and (41), dE_m can be calculated as a function of l_a. Using Eq. (40), the tempera-

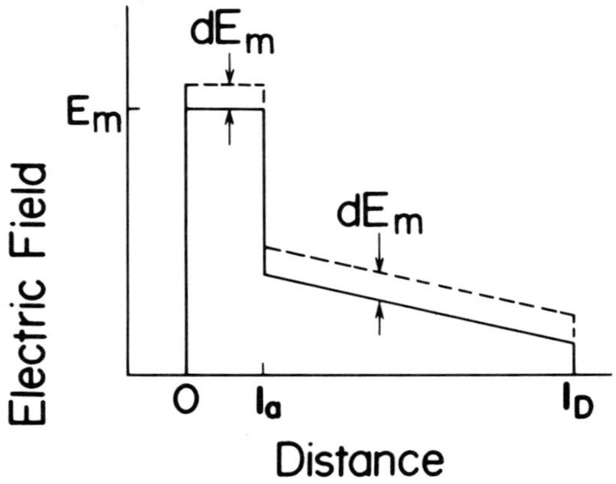

FIG. 39. Schematic representation of the electric field for temperature variations. [From Kaneda *et al.* (1976b).]

ture dependence of the breakdown voltage is expressed as

$$\gamma = (dV/V_0)(1/dT) = (l_D/V_0)[dE_m(l_a)/dT], \tag{42}$$

where V_0 is the breakdown voltage at 25°C. Measurements of γ were carried out for several diodes, each having a different l_a. The V_0 values of these diodes were 100 ± 10 V, and l_D was about 10 μm. The experimental and calculated results of γ versus l_a are shown in Fig. 40 and show sufficiently good agreement. Figure 41 shows the l_D dependence of breakdown voltage as

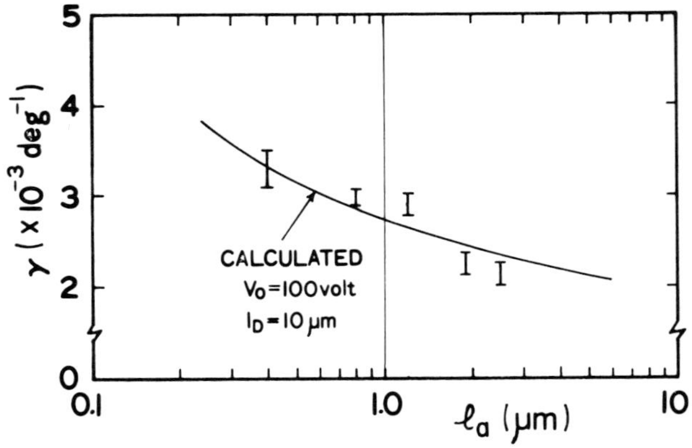

FIG. 40. Temperature coefficient of breakdown voltage as a function of avalanche-region length for $V_0 = 100$ V, $l_D = 10$ μm. [From Kaneda *et al.* (1976b).]

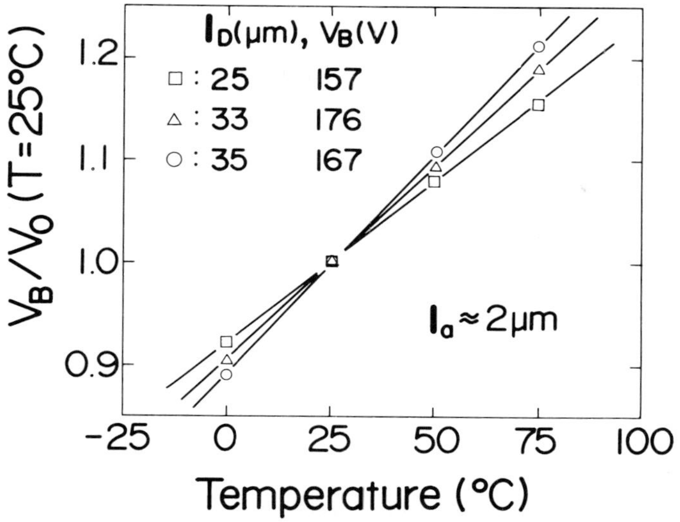

FIG. 41. Temperature dependence of breakdown voltage as a parameter of the depletion-layer length.

a function of temperature. The values of V_B/V_0 increase with increasing temperature in proportion to l_D, as expected from Eq. (41). The value in silicon reach-through avalanche photodiodes lies in $(3-5) \times 10^{-3}\,\text{deg}^{-1}$ and is larger than that of a one-sided junction diode whose γ value is around $1 \times 10^{-3}\,\text{deg}^{-1}$. This depends on the factor (l_D/V_0) shown in Eq. (42). In reach-through diode, the depletion-layer length is designed to be long for its breakdown voltage.

Multiplication-versus-bias-voltage characteristics also vary with temperature depending on breakdown voltage, as shown in Fig. 42. Multiplication curves have a similar bias-voltage dependence for temperature variation. By using Eqs. (40) and (42), the temperature variation of bias voltages for sustaining a multiplication factor is given by

$$\left.\frac{dV}{dT}\right|_{\langle M\rangle=M_0} = \left.\frac{dV}{dE_m}\frac{dE_m}{dT}\right|_{\langle M\rangle=M_0} = \gamma V_0. \tag{43}$$

Thus, the voltage for sustaining a multiplication has the same temperature dependence as the breakdown voltage. This leads to similar multiplication characteristics for temperature variation.

The temperature dependence of multiplication noise was studied at 30 MHz in the range 25–75°C. Figure 43 shows the noise power as a function of multiplication factor. Multiplication noise exhibits a constant value

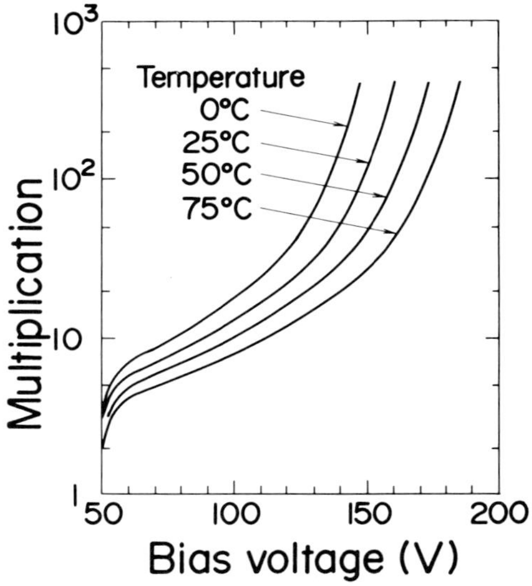

FIG. 42. Temperature dependence of multiplication versus bias-voltage characteristics for $\lambda = 830$ nm, $I_0 = 0.1\ \mu$A.

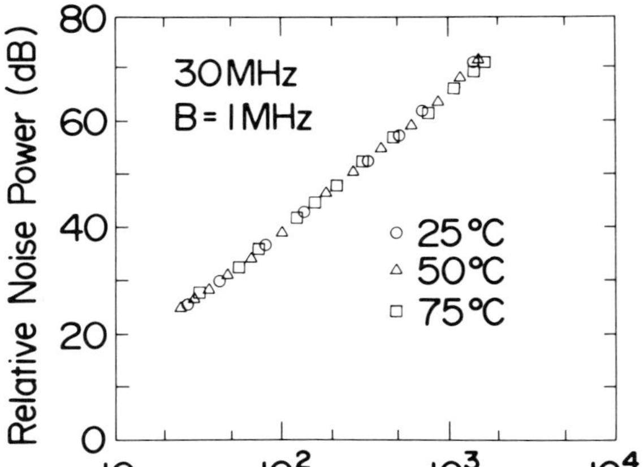

FIG. 43. Temperature dependence of multiplication noise measured at 30 MHz with $B = 1$ MHz. ○, 25°C; △, 50°C; □, 75°C.

in this temperature range. This shows that the hole–electron ionization-coefficient ratio has a constant value by changes in temperature, although the electron- and hole-ionization coefficients depend largely on temperature [Eq. (41)].

IV. Germanium Avalanche Photodiodes

Germanium has a suitable range of light-absorption coefficients ($\sim 10^4\ cm^{-1}$) in the wavelength region of 1–1.55 μm (Fig. 9). Therefore the light-absorption length of 3 μm is sufficiently long for detecting optical signals. In this short absorption length, an electric field sufficiently high to cause an avalanche multiplication can be obtained while avoiding the tunneling effect by using wafers having a carrier concentration of around $10^{16}\ cm^{-3}$. This leads to an avalanche photodiode with a low operating voltage, fast response, and high quantum efficiency. Naturally, germanium is a useful material for avalanche photodiodes in the 1-μm-wavelength region, where transmission loss in optical fiber is lowest. However, germanium is not the ideal material for avalanche photodiodes because it has nearly equal ionization coefficients and a high intrinsic carrier concentration. Therefore germanium avalanche photodiodes must be designed and fabricated having the highest performance for use in optical-fiber communications systems. Next, in Section 9, design details, structure, and performance of germanium avalanche photodiodes are discussed.

9. Design Details

a. Ionization Coefficients

To construct a low-multiplication-noise avalanche photodiode, studies on ionization coefficients are crucial. The ionization coefficients are determined from the statistical interrelation between the threshold energy for ionization and phonon-scattering mechanisms that depend on crystal orientation (Anderson and Crowell, 1972). Several workers have reported on ionization coefficients in germanium (Miller, 1955; Decker and Dunn, 1970; Dai and Chang, 1971). The experimental results, however, are not necessarily in agreement, and crystal orientation of germanium has not been specified. Mikawa *et al.* (1980) have studied the crystal-orientation dependence of ionization coefficients in germanium from the analysis of photocurrent-multiplication data, which is the most reliable procedure for determining the ionization coefficients. A special diode was designed and fabricated to obtain pure electron- and hole-injection currents for orientations of $\langle 111 \rangle$ and $\langle 100 \rangle$ as shown in the inset of Fig. 44. An n^+ layer with a thickness of 800 nm was formed by arsenic diffusion. A guard ring was

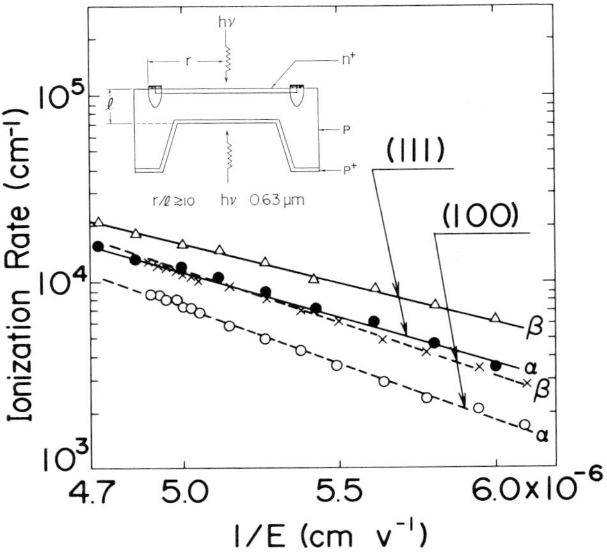

FIG. 44. Ionization coefficients in ⟨111⟩ and ⟨100⟩ germanium as a function of reciprocal electric field at room temperature: (△, ●) ionization coefficients for holes and electrons, respectively, in ⟨111⟩ germanium; (×, ○) for holes (β) and electrons (α), respectively, in ⟨100⟩ germanium. The inset shows a cross-sectional view of an n^+-p Ge APD used in the ionization measurements. [From Mikawa *et al.* (1980).]

formed by antimony diffusion. A p^+ layer was fabricated by boron-ion implantation. Impurity concentrations of the substrate were 1×10^{16} and 4×10^{15} cm^{-3} for ⟨111⟩ and ⟨100⟩ orientations, respectively. Diameters of the light-sensitive area were 300 and 100 μm for ⟨111⟩- and ⟨100⟩-oriented Ge APDs, respectively. A local cavity was fabricated at the bottom of the substrate so that light incident on the p^+ layer surface could be absorbed near the depletion layer. The value of the ratio r/l was chosen to be more than 10 (r is the radius of the guard ring and l is the distance from the n^+ layer surface to the bottom of the cavity). He–Ne laser light of wavelength 630 nm was shined alternately on the n^+ and p^+ layer surfaces. When the light is incident on the n^+ layer surface, almost all the light is absorbed in the n^+ layer at this wavelength. This realizes pure hole injection into the avalanche region. When the light is incident on the p^+ layer surface, photogenerated electrons in the p^+ layer diffuse into both the depletion layer and the guard ring. The current diffused into the guard ring causes experimental error in the value of the multiplication factor. Because the dimension of the diodes was chosen to be $r/l > 10$, the diffusion current to the guard ring can be estimated to be an order of magnitude less than the current to the depletion layer. Thus the photocurrent observed becomes almost pure electron injection only to the depletion layer. Then, hole- (M_p) and electron (M_n) initiated multiplication can be obtained. The ionization coefficients were calculated from the multiplication characteristics by using the following equations applicable to an n^+-p one-sided abrupt junction (Law and Lee, 1978):

$$\alpha(E_{\mathrm{m}}) = \frac{E_{\mathrm{m}}}{M_p M_n} \frac{dM_n}{dV},$$

$$\beta(E_{\mathrm{m}}) = \frac{E_{\mathrm{m}}}{M_p} \frac{dM_p}{dV} + (1 - M_p)\alpha(E_{\mathrm{m}}).$$

The maximum electric field E_{m} in the junction is calculated from $E_{\mathrm{m}} = [2qN(V + V_{\mathrm{bi}})/\varepsilon]^{1/2}$, where the impurity concentration N of the substrate and built-in voltage V_{bi} are obtained from $C-V$ measurements. Figure 44 shows the experimental results of ionization coefficients as a function of reciprocal of electric field. The solid and dashed lines represent the results for $\langle 111 \rangle$ and $\langle 100 \rangle$, respectively. A least squares fit gives an electric field range $1.67 \times 10^5 \leqq E \leqq 2.13 \times 10^5$ V cm^{-1} as

$$\alpha = 2.72 \times 10^6 \exp(-1.1 \times 10^6/E) \quad \mathrm{cm}^{-1}$$

and (44)

$$\beta = 1.72 \times 10^6 \exp(-9.37 \times 10^5/E) \quad \mathrm{cm}^{-1},$$

for the $\langle 111 \rangle$ orientation, and

$$\alpha = 8.04 \times 10^6 \exp(-1.4 \times 10^6/E) \quad \mathrm{cm}^{-1}$$

and (45)

$$\beta = 6.39 \times 10^6 \exp(-1.27 \times 10^6/E) \quad \mathrm{cm}^{-1},$$

for the $\langle 100 \rangle$ orientation. Holes have higher ionization coefficients than electrons for both $\langle 111 \rangle$ and $\langle 100 \rangle$ germanium, and the hole–electron ionization-coefficient ratios of $\langle 100 \rangle$ are greater than those of $\langle 111 \rangle$.

Crystal-orientation dependence of multiplication noise in germanium avalanche photodiodes was studied for all three crystal directions (Kaneda *et al.*, 1979a). Pearsall *et al.* (1978) found that impact ionization of hot carriers in GaAs depends on crystal orientations. A similar argument can be made for germanium because the threshold energies for impact ionization are different for the three crystal directions (Anderson and Crowell, 1972). If impact ionization coefficients have different values by crystal orientations, the magnitudes of multiplication noise in germanium avalanche photodiodes would vary according to crystal directions. The diodes used in this study have an n^+-p structure fabricated using $\langle 111 \rangle$-, $\langle 110 \rangle$-, and $\langle 100 \rangle$-

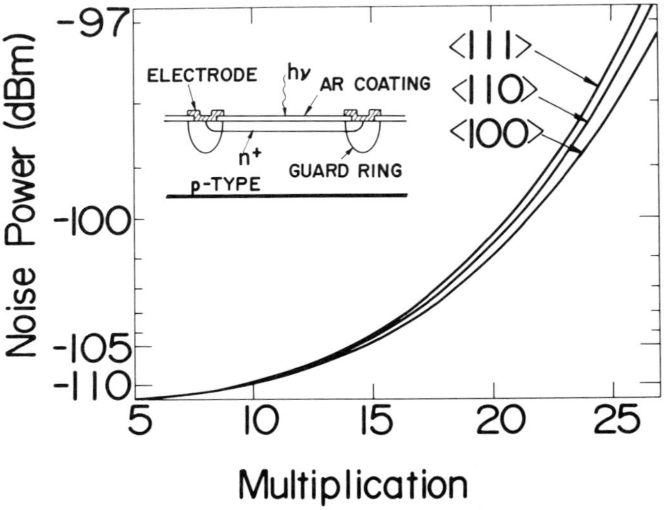

FIG. 45. Experimental results of avalanche noise as a function of multiplication factor for three orientations of electric field for $f = 30$ MHz, $B = 1$ MHz, $\lambda = 630$ nm, $I_0 = 1.0\ \mu$A. The inset shows a cross-sectional view of an n^+-p Ge APD used in noise measurements.

oriented p-type wafers. A cross-sectional view is shown in the inset of Fig. 45, and the characteristics of the diodes are summarized in Table III. The guard ring layer was made by Sb-diffusion $\approx 6\ \mu$m in depth and the n^+ layer by shallow As-diffusion $\approx 0.5\ \mu$m in depth. An antireflection coating was prepared by CVD of silicon dioxide (SiO_2).

Noise measurements were carried out at a frequency of 30 MHz by illuminating the diodes with a He–Ne laser (633 nm). The primary photocurrents I_{p0} were kept at 1.0 μA. The diodes studied are the selected ones, which show the largest multiplication factor among several diodes tested for each direction, since, if there are microplasmas and/or field nonuniformities, the multiplication factor will deteriorate. The maximum multiplication factors $\langle M \rangle_{max}$ obtained at $I_{p0} = 1.0\ \mu$A are listed in the fifth column of Table III.

TABLE III

CHARACTERISTICS OF SAMPLES

Sample	Orientation	Bulk conc. ($\times 10^{15}$ cm^{-3})	V_B (at 100 μA) (V)	I_D (at 0.9V_B) (μA)	$\langle M \rangle_{max}$ (at $I_{p0} = 1.0\ \mu$A)
1	⟨111⟩	8.5	27.3	0.40	80
2	⟨110⟩	7.6	29.7	0.45	80
3	⟨100⟩	3.5	55.8	7.0	40

The larger value of $\langle M\rangle_{max} = 80$ is obtained for the $\langle 111\rangle$- and $\langle 110\rangle$-oriented diodes, and a low value of $\langle M\rangle_{max} = 40$ for the $\langle 100\rangle$-oriented diode is due to the higher dark current. The size of the light spot was about 30 μm and was small in comparison with the diameter of the light-sensitive area of the diodes (100 μm). The multiplication dependence of avalanche noise was investigated by using a recorder tracing method in order to improve the accuracy of measurements and clarify the noise differences among different diodes. Almost all photocurrents are generated in the n^+ layer since the skin depth of germanium at $\lambda = 633$ nm is evaluated to be ≤ 0.1 μm, whereas the junction depth is ≈ 0.5 μm. Pure hole injection, therefore, is achieved in this case, and multiplication noise power dependent only on the ratio of hole to electron ionization coefficients, because the other parameters determining noise power are kept constant.

The experimental results are shown in Fig. 45. The results show that the noise power of the $\langle 100\rangle$-oriented diode is the lowest and the $\langle 111\rangle$-oriented diode is the highest. This difference of 0.5 dB between the $\langle 100\rangle$- and $\langle 111\rangle$-oriented diodes agrees with the calculated results of Eqs. (7), (44), and (45). However, considering the estimated experimental errors (± 0.3 dB) limited by the calibration of noise power and the fact that the measured noise dispersion for diodes in the same wafer is ± 0.3 dB, we can say that a distinct difference of multiplication noise connected with crystal orientations is not obtained from the results of Fig. 45.

Ionization coefficients of hot carriers are determined theoretically from the statistical interrelation between the threshold energy for ionization and the phonon-scattering mechanism. In germanium, equivalent intervalley scattering among the $\langle 100\rangle$ valleys is the dominant scattering mechanism (Zulliger, 1971). From the results of Fig. 45, this intervalley scattering is considered to be more significant in the ionization process of electrons than threshold energies.

Ionization coefficients depend also on the magnitudes of the electric field. The values of the maximum electric field at breakdown voltage are evaluated to be 2.3×10^5 V cm^{-1} for the $\langle 111\rangle$- and $\langle 110\rangle$-oriented diodes and 2.1×10^5 V cm^{-1} for the $\langle 100\rangle$-oriented diode. Values of $K(=\beta/\alpha)$ are considered to increase with decreasing electric field strength (Miller, 1955; Mikawa *et al.*, 1980). Therefore, the results of Fig. 45 may include the effect of the magnitude of electric field. This requires a further investigation for multiplication noise dependent both on crystal orientations and on the magnitude of the electric field.

b. Multiplication Noise

There are three different structures for use in wavelengths below 1.5 μm where the absorption coefficient of germanium is considerably high. They are a shallow junction of n^+–p (Melchior and Lynch, 1966; Shibata *et al.*,

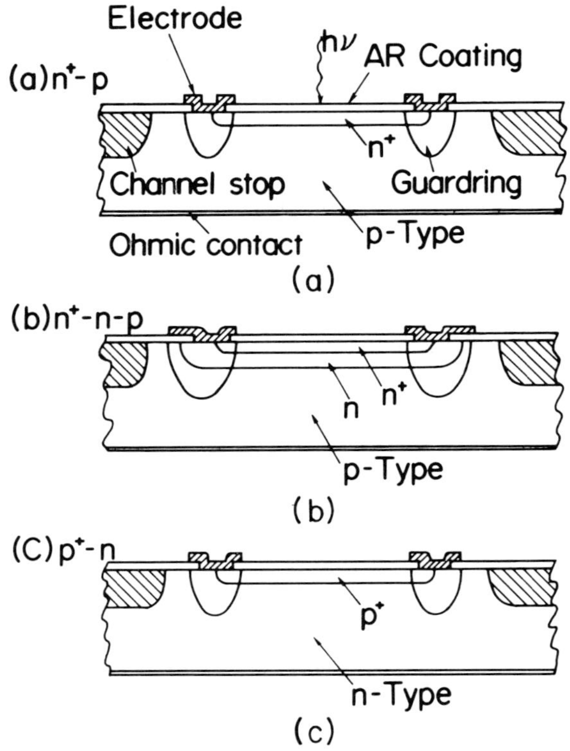

FIG. 46. Cross-sectional view of germanium avalanche photodiodes for the wavelength 1.3 μm. [From Kaneda and Kanbe (1983).]

1971; Kaneda and Takanashi, 1973b); a shallow junction of p^+-n (Kaneda *et al.,* 1979b; Kagawa *et al.,* 1981); and a deep junction of an n^+-n-p diode (Kaneda *et al.,* 1980; Mikami *et al.,* 1980; Mikawa *et al.,* 1981). Their structures are shown in Fig. 46 (Kaneda and Kanbe, 1983). Excess-noise factors were calculated by using Eqs. (1), (2), and (44) at a multiplication factor of 10 because the optimum multiplication factor is approximately this value in communication systems. The results obtained are shown in Fig. 47. In the p^+-n device, excess-noise factors decrease as wavelengths increase because the hole–electron-injection ratio increases with increasing wavelengths, corresponding to the monotonical decrease of the absorption coefficient. In the n^+-p device, however, excess-noise factors increase with increasing wavelength because the injection ratio decreases with increasing wavelengths. In the wavelength above 1.2 μm, the low-loss region of optical

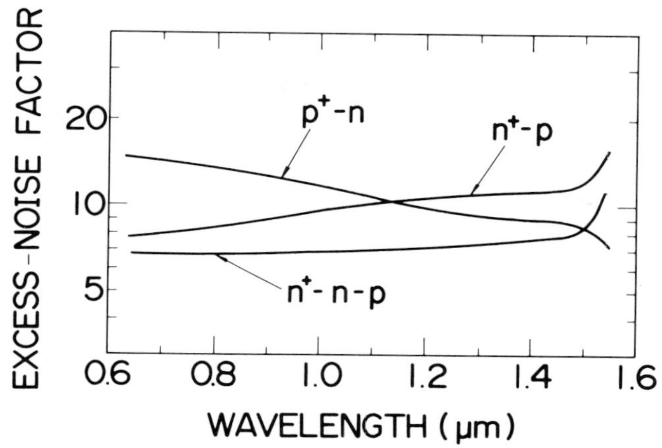

FIG. 47. Calculated excess-noise factors for p^+-n, n^+-p, and n^+-n-p germanium avalanche photodiodes as a function of wavelengths at multiplication factor of 10. [From Kaneda and Kanbe (1983).]

fibers, p^+-n diodes show lower excess-noise factors than do n^+-p diodes. In the n^+-n-p device, almost all incident light is absorbed in the thick n region, and pure hole injections are achieved. Thus, the lowest excess-noise factors are expected in wavelengths below 1.5 μm. In the wavelengths above 1.5 μm, however, because most incident light enters the p layer through the n layer, excess-noise factors show high values because of increased electron-injection currents.

c. *Quantum Efficiency*

Because germanium has high absorption coefficients of the order of 10^4 cm^{-1}, incident photons are mainly absorbed in the surface region. The quantum efficiency of germanium avalanche photodiodes is therefore strongly affected by diffusion currents injected from the highly doped surface region. Using a p^+-n diode as an example, the electron diffusion currents I_{n0} injected into the junction from the p layer are given by

$$I_{n0} = \frac{q\phi_0 aL}{a^2L_n^2 - 1}\left[\frac{aL_n + C - B\exp(-ax_j)}{A} - aL_n\exp(-ax_j)\right], \quad (46)$$

where

$$\begin{aligned} A &= C\sinh(x_j/L_n) + \cosh(x_j/L_n), \\ B &= C\cosh(x_j/L_n) + \sinh(x_j/L_n), \qquad C = S_0L/D_n, \end{aligned} \quad (47)$$

and S_0 is the surface recombination velocity, ϕ_0 the incident photon flux density, D_n the diffusion coefficient of electrons in the p layer, and L_n the diffusion length of electrons in the p layer.

On the other hand, hole-injection currents I_{p0} from the n layer are given by

$$I_{p0} = q\phi_0 \exp(-ax_j), \tag{48}$$

where almost all holes generated in the n layer are considered to diffuse into the junction because in germanium the diffusion length is much longer than the light-absorption length for wavelengths below 1.5 μm. The quantum efficiency η is then given by

$$\eta = T_E(I_{n0} + I_{p0})/q\phi_0, \tag{49}$$

where T_E is the fraction of transparent optical power given by Eq. (16). The quantum efficiency for $n^+ - p$ diodes can be evaluated in the same way. Figure 48 shows the calculated and experimental results of the quantum efficiency of diodes having various n^+-layer thicknesses as a function of wavelength. The diodes have 200-nm-thick SiO_2 film on the active region as an antireflection coating. Agreement between the calculated and experimental results is fairly good. The quantum efficiency decreases by increasing the n^+-layer width. This is because a carrier recombination becomes substantial in the wider n^+ layer, thus decreasing quantum efficiency.

Because the diffusion length of electrons in the p^+ layer is greater than that of the holes in the n^+ layer, quantum efficiency of the $p^+ - n$ structure is expected to be greater than the $n^+ - p$ structure. Figure 49 shows the calcu-

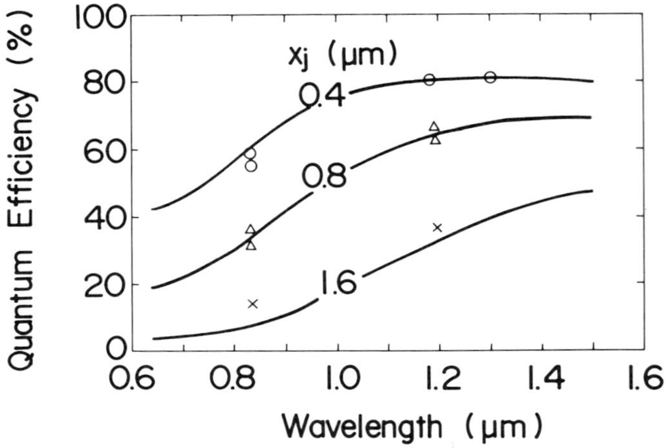

FIG. 48. Quantum efficiency versus wavelength characteristics as a parameter of junction depth. The diodes have 200-nm-thick antireflection-coating films (SiO_2) on the active region.

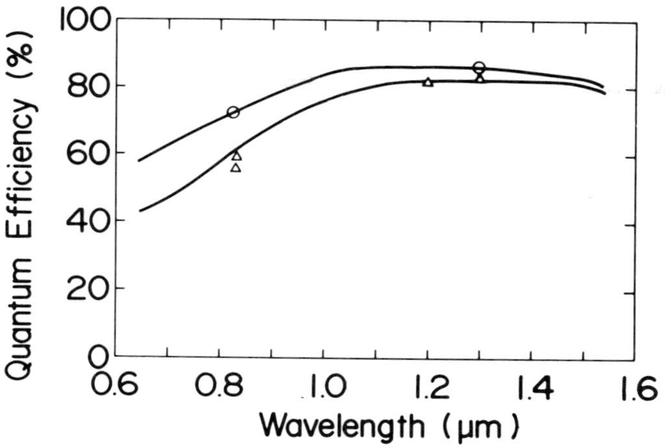

FIG. 49. Quantum efficiency for p^+-n (○) and n^+-p (△) diodes as a function of wavelength. The diodes have 200-nm-thick antireflection-coating films (SiO_2) on the active region. [From Kaneda and Kanbe (1983).]

lated results as a function of wavelength by selecting the following parameters: $D_p = 6.5$ cm² sec^{-1}, $L_p = 0.44$ μm, $D_n = 40$ cm² sec^{-1}, $L_n = 1.1$ μm, $S = 10^4$ sec^{-1} $x_j(p^+-n) = 0.2$ μm, and $x_j(n^+-p) = 0.4$ μm. The difference in quantum efficiency between n^+-p and p^+-n diodes is remarkable in the short-wavelength-region below 1 μm, where incident light is mainly absorbed in the highly doped surface region. These results are in good agreement with the experimental ones shown in Fig. 49.

In the n^+-n-p diode, quantum efficiency is expected to be nearly equal to the n^+-p diode because the hole diffusion length in the n-region is much greater than the thickness (~ 2 μm).

d. Frequency Response

In germanium avalanche photodiodes, the depletion-layer transit time and the *RC* time constants are not main elements governing the response speed in the frequency range below 1 GHz, since they are evaluated from Eqs. (20) and (23) to be about 5×10^{-11} and 1×10^{-10} sec, respectively, where the length of the depletion layer $l_D = 3$ μm, the carrier-saturation velocity $v_s = 6 \times 10^6$ cm sec^{-1}, the load resistance $R_L = 50$ Ω, the series resistance $R_s = 20$ Ω, and the diode capacitance 1 pF are selected for the evaluations.

The influence of carrier diffusion on response speed is expected to be very small for p^+-n and n^+-p devices because the highly doped surface region is very thin (0.2–0.4 μm); thus, carrier diffusion time becomes very short. In

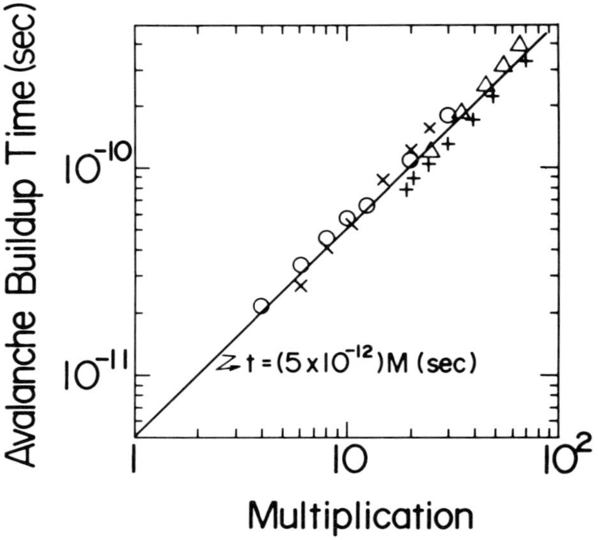

FIG. 50. Avalanche buildup time versus multiplication factor obtained for n^+-p diodes. $\lambda = 1.06\ \mu$m: △, 1.2 GHz; +, 1.9 GHz. $\lambda = 6328$ Å: ×, 3.0 GHz; ○, 4.0 GHz. [From Kaneda and Takanashi (1973a).]

the n^+-n-p device, however, the carrier must diffuse in the thick n region. This would cause a poor frequency response compared to the p^+-n and n^+-p diodes.

Avalanche buildup time can be very long in germanium avalanche photodiodes because ionization coefficients between holes and electrons are nearly equal (they are greatly different in silicon). Avalanche buildup time is obtained from a shot-noise measurement in the gigahertz region. Figure 50 shows the results for n^+-p diodes. The intrinsic response time is 5×10^{-12} sec and is an order of magnitude larger than that of silicon. The influence of avalanche buildup time is remarkable above several hundred megahertz and in the higher multiplication region above 10 (Kaneda and Takanashi, 1973b).

10. Surface Passivation

Surface-leakage current and electrical stability of devices are determined mainly by surface passivation. This requires passivation film having both a low interface state density and a suitable interface state charge that does not generate a surface inversion layer. Because the native oxides GeO and GeO_2 are unstable and cannot be used alone as passivation film, surface passivation is a serious problem in achieving a germanium avalanche photodiode

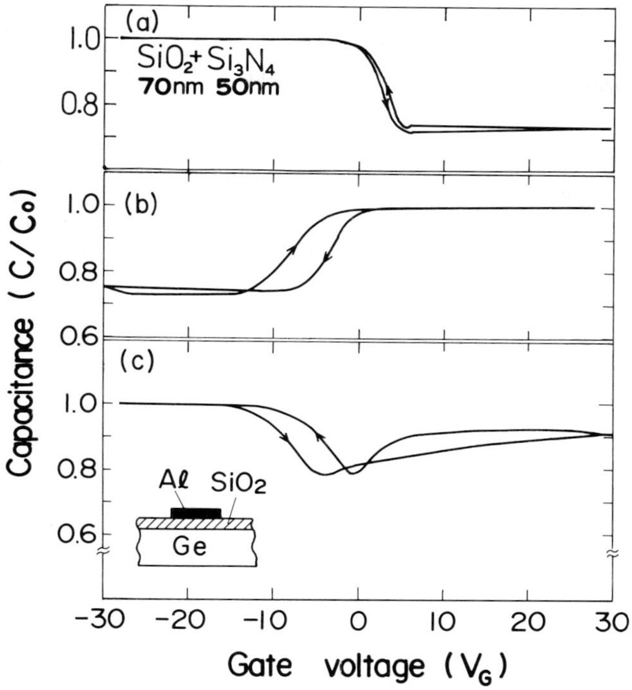

FIG. 51. Capacitance-versus voltage-characteristics of germanium MOS diodes: (a) double layers of SiO_2 (70 nm) and Si_3N_4 (50 nm) on *p*-type germanium ($\sim 1 \times 10^{16}$ cm^{-3}) with $C_0 = 36.6$ pF; (b) SiO_2 (200 nm) on *n*-type germanium with $C_0 = 22.5$ pF ($\sim 1 \times 10^{16}$ cm^{-3}); (c) SiO_2 (200 nm) on *p*-type germanium with $C_0 = 21.7$ pF ($\sim 1 \times 10^{16}$ cm^{-3}).

with stabilized low dark currents. This situation is different from that of silicon devices, which have stable and effective natural SiO_2 passivation film. Several kinds of films have been studied for photodiodes, FETs, and CCDs by several workers (Kuisl, 1972; Wang and Gray, 1976; Wang and Storms, 1976; Kagawa *et al.*, 1979; Hino *et al.*, 1982). Interface state densities as low as 1×10^{12} cm^{-2} eV^{-1} are obtained by CVD SiO_2 on germanium, and lower densities of around 10^{-11} cm^{-2} eV^{-1} are achieved by the double layer of CVD SiO_2 and Si_3N_4 film and by a CVD SiO_2 film annealed in oxygen. Figure 51 shows the results of the $C-V$ measurement of the MOS structure having an SiO_2 film and a double layer of CVD SiO_2 and Si_3N_4 films. The SiO_2 film is suitable for p^+-n diodes because the film has positive charges, and electrons are attracted to the surface, hence suppressing generation of a surface inversion layer. A double layer of CVD SiO_2 and Si_3N_4 film is utilized for n^+-p and n^+-n-p diodes because the flat-band voltage is close to zero; thus, a surface inversion layer is harder to generate than SiO_2 film with negative flat-band voltages.

11. Junction Formation

Doping techniques for both *n*- and *p*-type impurities are indispensible for fabricating a well-designed avalanche photodiode. In germanium, diffusion of *n*-type impurities, such as antimony (Sb) and arsenic (As), are better developed than for *p*-type impurities. Both the vapor pressure and the diffusion constant of *p*-type impurities, such as zinc (Zn), boron (B), indium (In), and gallium (Ga), are rather low; thus, high temperatures and long diffusion times are needed to form the *p* layer. This often leads to poor device properties and has prevented the study of *p*-type impurities until recently. Zinc diffusion for forming a guard ring has been studied because zinc has a higher diffusion constant and vapor pressure than other *p*-type impurities. The diffusion was carried out in an argon atmosphere at 830°C. The results obtained from the four-point probe technique are shown in Fig. 52 and show a linear graded junction with a higher breakdown voltage of 90 V for the wafer with a concentration of 4×10^{15} cm^{-3}. The diodes were formed by using Zn diffusion, with CVD SiO_2 used as a diffusion mask. The diodes showed a rather high dark current of about 1 μA at $0.9\,V_B$, where the diode diameter was 140 μm. This dark current is attributed to crystal defects

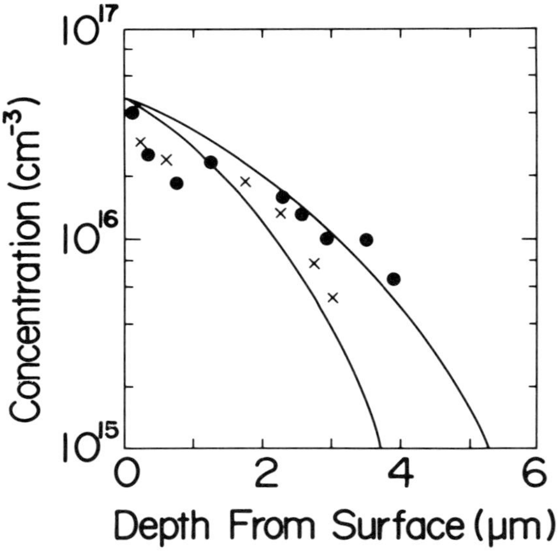

FIG. 52. Doping profiles of zinc diffusion at 830°C. ●, 240 min; ×, 120 min. *D* is calculated to be 2.3×10^{-12} cm^2 sec^{-1}.

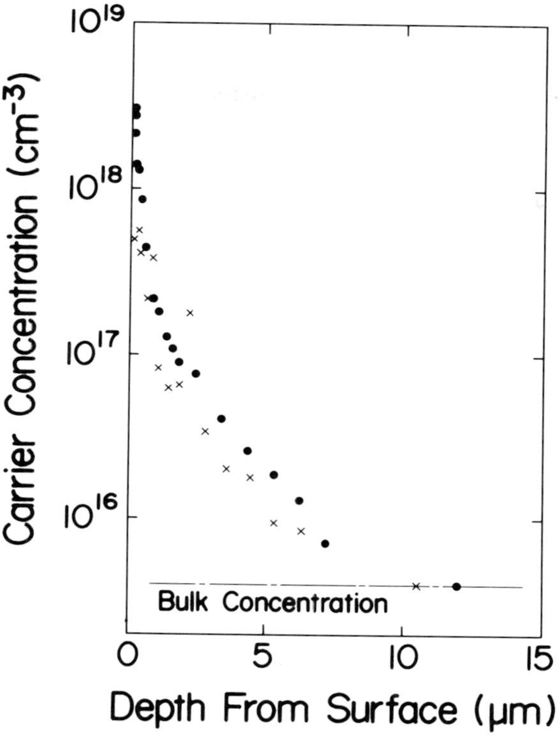

FIG. 53. Doping profiles of beryllium-ion-implanted layers (100 keV) annealed at 650°C for 1 hr. Dose: ●, 5×10^{14} cm^{-2}; ×, 2×10^{14} cm^{-2}. [From Kagawa *et al.* (1982a).]

induced at the surface layer of germanium by thermal stress because the coefficients of thermal expansion between SiO_2 and germanium are very different.

Beryllium-ion implantation is promising for forming a guard-ring layer. As shown in Fig. 53, the carrier profiles of beryllium annealed at 650°C for 1 hr show a long tail in the low-concentration region, resulting in a linearly graded junction. Because the annealing temperature was rather low, 650°C, defect formations can be decreased. This leads to a reproducible fabrication of diodes with low dark currents, as described in Section 12.

Boron- and indium-ion implantations were studied for forming the active region of photodiodes where a one-sided abrupt junction is required to attain a lower avalanche breakdown than in a guard-ring region. Figure 54 shows the doping profiles of In-ion implantation annealed at 650°C for 1 hr. Shallow and abrupt junctions suitable for p^+-n diodes were obtained from doses

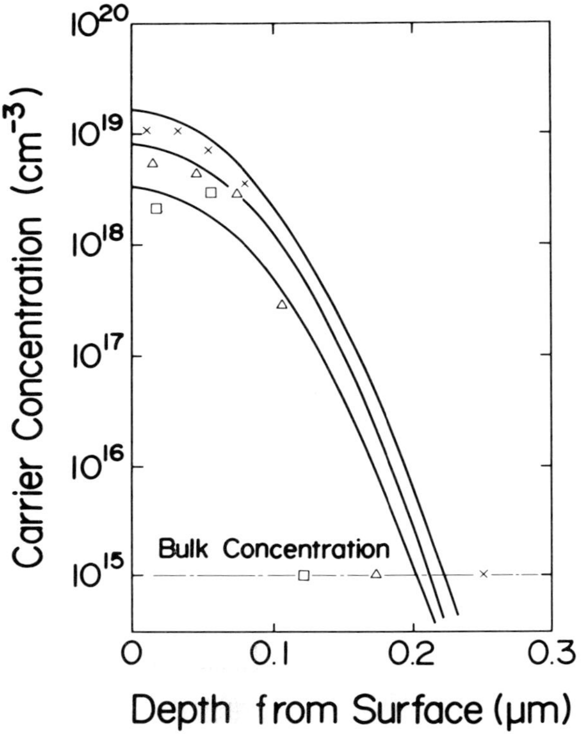

FIG. 54. Doping profiles of indium-ion-implanted layers (90 keV) annealed at 650°C for 1 hr. Dose: ×, 1×10^{14} cm^{-2}; △, 5×10^{13} cm^{-2}; □, 2×10^{13} cm^{-2}. [From Kagawa *et al.* (1982a).]

2×10^{13} to 1×10^{14} cm^{-2}. Similar junction profiles were also obtained by B-ion implantations. Thus, in the case of germanium, by using ion implantation for *p*-type impurities both linearly graded and one-sided abrupt junctions can be formed at rather low temperature of 650°C, whereas to use diffusion techniques, high temperatures above 800°C are needed.

Formation of *n*-type layers are made by both diffusion and ion implantation. The doping profiles of Sb diffusion are shown in Fig. 55 as an example. Diffusion and Sb-source temperatures were 730 and 420°C, respectively. Argon was used as the carrier gas. The junction shows a linear graded profile and is applicable for a guard-ring layer in $n^+ - p$ diodes. Arsenic ion implantation is a reproducible technique for forming *n* layers, such as the guard-ring and active and channel-stop layers, as described next.

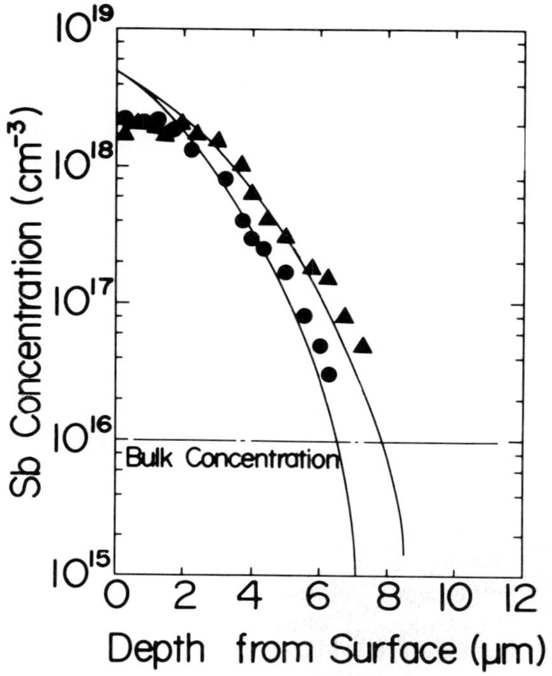

FIG. 55. Doping profiles of antimony diffusion for $T = 730°C$, $D = 3.3 \times 10^{-12}$ cm^2 sec^{-1}. For ▲, $t = 180$ min; ●, $t = 125$ min.

12. Device Fabrication and Characteristics

As shown in Fig. 46, there are three different structures for use in wavelengths below 1.5 μm. Diode formations and their characteristics are discussed here.

The first studies of germanium avalanche photodiodes were carried out using a shallow-junction n^+-p structure, adapted from the facility of fabrication, because diffusion techniques for n-type impurities such as antimony (Sb) and arsenic (As) were better developed for p-type impurities (Melchior and Lynch, 1966). The planar device with guard ring shown in Fig. 46 was studied because it was expected to display greater durability and electrical stability. For the basic device requirements of avalanche photodiodes, such as high quantum efficiency, first response speed and then avalanche breakdown occurred ahead of tunnel breakdown. The carrier concentration of wafers used were selected at around 1×10^{16} cm^{-3}. Surface etching determines the leakage currents of germanium diodes. Etching solutions for germanium have been reported by several workers (Camp, 1955; Wallis and Wang, 1959; Irving, 1962). The etching rates of most of these solutions, however, are too high, and it is difficult to obtain a smooth and uniform surface and to control the etching depth. Moreover, these solutions contain fluoric acid (HF), so SiO_2 films and photoresist cannot be used as a preferential etching mask. Two kinds of new etching solution consisting of H_3PO_4, H_2O_2, and H_2O, and HCl, H_2O_2 and H_2O were studied for use in forming germanium photodiodes (Kagawa *et al.*, 1982b). By using these solutions, a smoothly etched germanium surface was obtained at a controlled etching rate of 0.02–0.4 μm min^{-1}. Also, SiO_2 films and photoresist can be used as preferential etching masks without dissolution and separation. This allows highly reproducible fabrication. The n^+ layer was made by As diffusion at 620°C for 10 min. The junction depth was 0.4 μm, which was optimized by a signal-to-noise ratio at wavelengths of 1 μm. The guard-ring layer was made by Sb diffusion at 730°C for 3 hr. The channel stopper, which cuts off surface-inversion channels, was made by Be-ion implantation. Diffusion and annealing were carried out in an Ar atmosphere. Doping masks were CVD SiO_2 deposited at 450°C. Typical dark-current dependence on temperature is shown in Fig. 56 as a function of applied reverse-bias voltage for a diode having an active-region diameter of 100 μm. Breakdown voltage was around 30 V and increased with increasing temperature. The temperature coefficient of breakdown voltage was a small value of about 1.0×10^{-3} deg^{-1} because the electric field has a one-sided abrupt profile, different from the reach-through type adopted for silicon avalanche photodiodes. Reverse-bias-voltage dependence of dark currents becomes weak when temperature increases. This shows that dark currents are determined by diffusion currents in temperatures above 24°C. In the low-temperature range, however, dark currents are governed by generation–recombination currents, which are in proportion to the depletion-layer length varied by bias voltage. In fact, activation energies of dark currents at a reverse-bias voltage are equal to the energy gap of germanium ($E_G = 0.67$ eV) in the high-temperature region and equal to $E_G/2$ in the low-temperature region.

Junction-diameter dependence of dark currents was studied by forming diodes with various diameters on the same substrate (Ando *et al.*, 1978). Dark currents for diodes after Sb diffusion for the guard ring are proportional to the guard-ring diameter and are thought to be leakage currents flowing through the inside and outside of the guard-ring peripheries. Dark currents for diodes after As diffusion for the active region are proportional to the square of the diameter when it is larger than 200 μm and consists mainly of reverse currents flowing through the n^+-p junction.

In another classification, dark current can be divided into two components: dark current that is multiplied, I_{DM}, by the avalanche effect and that

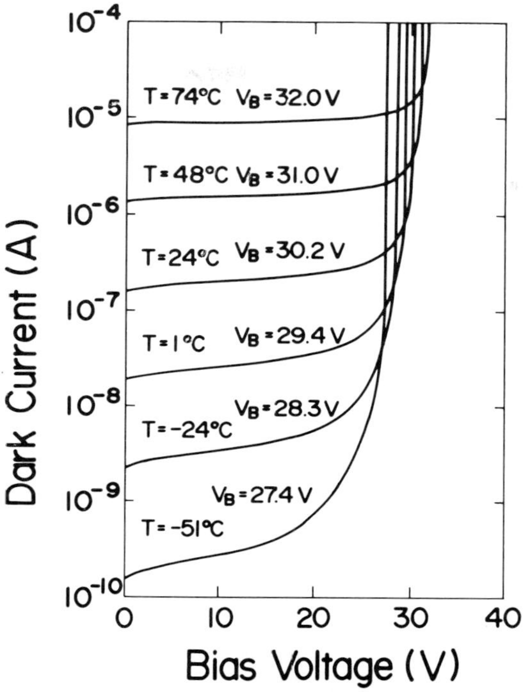

FIG. 56. Temperature dependence of dark current versus bias-voltage characteristics for n^+–p Ge APDs.

which is not multiplied, I_{D0}, flowing through the periphery of the guard ring. The multiplied dark current is generally most important in germanium avalanche photodiodes since this current governs the shot noise as well as the signal current. The shot noise i_n^2 is given by

$$i_n^2 = 2qB\{F\langle M\rangle^2(I_{P0} + I_{DM}) + I_{D0}\}, \tag{50}$$

where B is the bandwidth and I_{P0} the signal current. Because the optimum multiplication factors in communication systems are approximately 10 and the excess-noise factor F is nearly equal to $\langle M\rangle$, the value of $F\langle M\rangle^2$ is on the order of 10^3. Because I_{D0} is only larger than I_{DM} by a factor of 2–3 in 100-μm-diameter devices, unmultiplied dark current can be neglected with regard to the shot-noise level.

The values of I_{DM} are usually obtained from the dark-current-versus-multiplication characteristics (Ando *et al.,* 1978). In the higher multiplication region, where dark current is governed by the multiplied dark current, a linear relationship is obtained between dark current and multiplication factors, and I_{DM} is determined by extrapolating the linear relationship to the unity multiplication.

Another means of obtaining the values of I_{DM} that is convenient for germanium avalanche photodiodes is now described. As shown in Fig. 56, dark current consists of diffusion current for temperatures above 24°C. Carrier-diffusion length is evaluated to be about 400 μm by selecting values of $\tau_n = 30\ \mu$sec and $D_n = 50\ \text{cm}^2\ \text{sec}^{-1}$ and is two orders of magnitude longer than the depletion region, which is 3 μm at breakdown voltage. That is, diffusion current is independent of bias voltage. Increased dark current can be considered to be due to the avalanche multiplication. Consequently, I_{DM} is given from two diffusion currents at two different bias voltages by

$$I_{DM} = [I_{DD}(V_1) - I_{DD}(V_2)]/[M(V_1) - M(V_2)]. \tag{51}$$

The value of I_{DM} obtained from this method is in the range of 20–30 nA for diodes having an active-area diameter of 100 μm at 25°C at unity multiplication and is in good agreement with the value determined from the dark-current-versus-multiplication characteristics.

Carrier multiplication was measured at a frequency of 1 kHz. Maximum multiplication as high as 100 was obtained at the primary photocurrent before multiplication ($M = 1$) $I_{p0} = 1\ \mu$A. Noise measurements were carried out at a center frequency of 30 MHz with a bandwidth of 1 MHz. At a multiplication factor of 10, excess-noise factors of $F \approx 9$ and $F \approx 11$ were experimentally obtained at $\lambda = 0.83$ and 1.3 μm, respectively. These F values are in good agreement with the calculated results shown in Fig. 47. Quantum efficiency is 80% at $\lambda = 1.3\ \mu$m (Fig. 48).

The n^+-n-p structure shown in Fig. 46b was made by using p-type wafers with a carrier concentration of $1.3 \times 10^{16}\ \text{cm}^{-3}$. Sketches of a doping and an electric field profile are shown in Fig. 57. A guard-ring layer was made by arsenic deposition (620°C for 10 min) and subsequent drive-in diffusion at 780°C for 3 hr. By using this method, a well-graded junction with a high breakdown voltage was fabricated. The breakdown voltage of the guard-ring layer was 38 V. The n layer, which is the key layer in this diode, was formed by arsenic implantation at 150 keV and subsequent drive-in diffusion at 730°C for 1 hr. The implanting dose was $2 \times 10^{13}\ \text{cm}^{-2}$. The junction depth was estimated to be about 2.5 μm from the diffusion constant of arsenic, $1.7 \times 10^{-12}\ \text{cm}^2\ \text{sec}^{-1}$ at 730°C, and from the drive-in diffusion time, assuming the doping profile of a Gaussian distribution. The n^+ layer was made by arsenic diffusion at 620°C for 10 min. The layer thickness was 0.3 μm. Avalanche breakdown voltage was about 31 V and was about 7 V higher than the n^+-p diode made by the same diffusion on the same substrate. The silicon dioxide deposited by CVD was used for passivation and the antireflection coating film. Dark current was about 1 μA at 90% of the breakdown

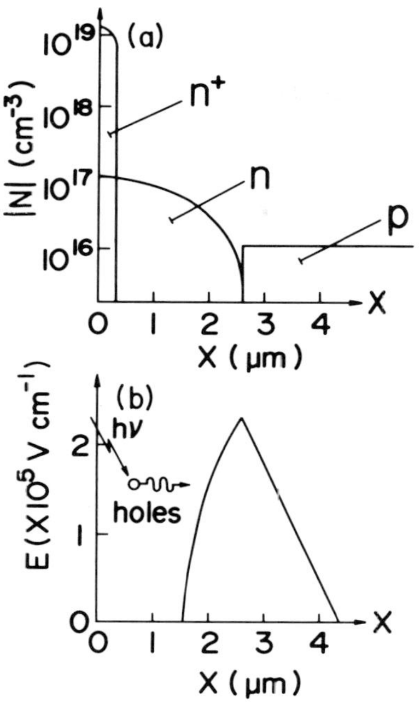

FIG. 57. Sketches of (a) doping and (b) electric field profiles for n^+-n-p diodes. [From Kaneda *et al.* (1980).]

voltage $0.9\,V_B$. Maximum multiplication of 100 was obtained at $I_{P0} = 1\ \mu A$. Figure 58 shows the avalanche gain uniformity of the diode at $\langle M \rangle = 10$. The diameter of the light spot (He–Ne laser, $\lambda = 633$ nm) was about 10 μm. A fairly uniform multiplication was obtained in the entire sensitive area. Low-excess-noise factors $F \approx 6.5$ and $F \approx 7$ were obtained at $\langle M \rangle = 10$ and at $\lambda = 0.83$ and 1.3 μm, respectively, and were in good agreement with the calculated one shown in Fig. 47. A quantum efficiency as high as 70–80% was obtained at $\lambda = 1.15$ and 1.3 μm. Distinct deterioration of quantum efficiency was not observed with the introduction of the n layer. This is due to the longer diffusion length of holes $L_p = 150\ \mu$m compared with the width of the n-layer, assuming $D_p = 22.8$ cm sec^{-1} and $\tau_p = 10\ \mu$sec, where D_p and τ_p are the diffusion constant and the hole lifetime in the n layer, respectively. The response speed of n^+-n-p diodes was studied for the frequencies up to 500 MHz using a sinusoidally modulated light from a InGaAsP laser ($\lambda = 1.3\ \mu$m). The results obtained for a multiplication factor of 10 are shown in Fig. 59, where the results for n^+-p diodes are also shown. The response deterioration at 500 MHz was about 3 dB for n^+-n-p diodes and about 1 dB for n^+-p diodes. This difference is due to the hole-diffusion effect in the n layer. The diffusion-limited cutoff frequency f_c (-3 dB) is

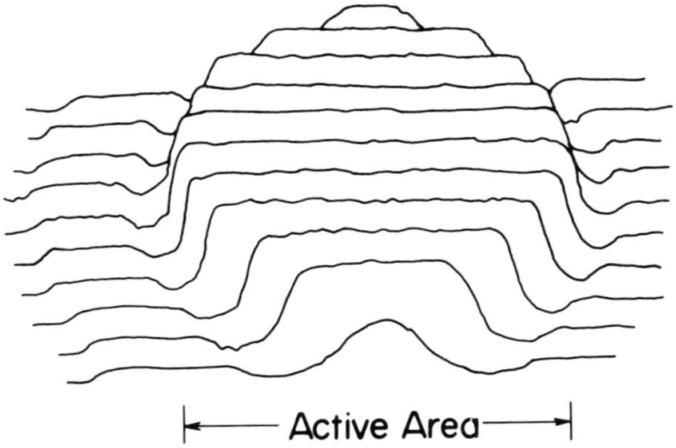

FIG. 58. Photoresponse scans at 633 nm of an n^+-n-p diode at a multiplication factor of 10. The diameter of the light spot is about 10 μm. [From Kaneda *et al.* (1980).]

given from Eq. (25) by $f_c = 2.4D_p/2\pi l_0$, where l_0 is the length of the carrier diffused in the undepleted region. The calculated result is $f_c = 440$ MHz for $l_0 = 1.4$ μm and $D_p = 22.8$ cm sec^{-1} and is in good agreement with the experimental results.

The p^+-n germanium avalanche photodiodes were studied for the first time by Kaneda *et al.* (1979) using $\langle 111 \rangle$-oriented n-type wafers (4×10^{15} cm^{-3}). The p layers in these diodes were made by zinc diffusion at 830°C for 3 hr for the guard ring and by boron implantation at 40 keV for the p^+ active region. The junction depth was about 0.3 μm, and the surface concentration was 1×10^{18} cm^{-3}, obtained by the four-point probe tech-

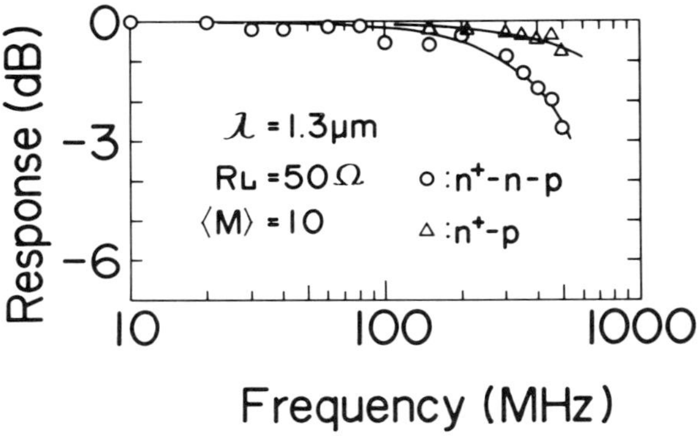

FIG. 59. Frequency response of n^+-n-p and n^+-p diodes at a multiplication factor of 10 for $\lambda = 1.3$ μm, $R_L = 50$ Ω, $\langle M \rangle = 10$.

nique. In these diodes, dark current was a rather high value of 0.7 μA at 0.9 V_B, which is considered to be due to the effects of surface-layer defects caused by high temperature and the long diffusion process for the guard ring.

To decrease the dark current, Kagawa *et al.* (1981) have fabricated a p^+-n diode using full ion implantation. The diode has a planar structure with a guard ring and a channel stop. A cross-sectional view is shown in Fig. 46c. The diameter of the active area is 100 μm. A ⟨111⟩-oriented n-type wafer with a carrier concentration of about 8×10^{15} cm^{-3} was used. The guard-ring layer was formed by implantation of beryllium ions at an implanting energy of 100 keV and a dose of 1×10^{14} cm^{-2}. An implantation of indium ions at an energy of 90 keV and a dose of 2×10^{13} cm^{-2} was used to form the p^+ layer. Annealing both implantation layers was then carried out in a single stage at the low temperature of 650°C for 1 hr. The deep-graded junction was made at a depth of 4 μm in the guard-ring layer, and a shallow junction of about 0.2 μm was obtained in the active region. The channel stop was formed by arsenic ion implantation (130 keV, dose 1×10^{14} cm^{-2}), and the surface-passivation film was coated by CVD SiO_2 at 450°C for 10 min. Arsenic atoms were electrically activated during this deposition process. The p^+ electrode was formed by aluminum evaporation. The breakdown voltage of the diode was about 32 V. The dark current at 0.9 V_B was as low as 150 nA. By using the low-temperature single-stage annealing process, a low dark-current diode was made reproducibly. Quantum efficiency as high as $\eta = 84-90\%$ at $\lambda = 1.3$ μm was obtained by using the antireflection coating of plasma CVD of silicon nitride (Si_3N_4), whereas $\eta = 75-80\%$ for the antireflection coating of CVD SiO_2. This is because the reflective index of Si_3N_4, which was determined by ellipsometry to be 1.82, is larger than that of SiO_2 (1.45). The Si_3N_4 film is a better antireflection coating for germanium having a reflective index of 4.0. A multiplication factor of more than 100 was typically obtained at 1 kHz and at $I_{p0} = 1.0$ μA. Gain uniformity was flat throughout the active region. Multiplication noise was studied at a frequency of 30 MHz. The results obtained are shown in Fig. 60. Wavelengths of $\lambda = 1.3$ and 1.55 μm were used to excite the avalanche process. At a multiplication factor of 10, the excess-noise factor was 8–9 at $\lambda = 1.3$ μm and about 6.5 at $\lambda = 1.55$ μm. These F values are in good agreement with the calculated results (Fig. 47). The p^+-n diode shows a low excess-noise factor in the wavelength range 1.3–1.55 μm. The frequency response was measured as a function of multiplication factor at $\lambda = 1.3$ μm (InGaAsP laser). At a multiplication of 10, where the optimum gain is approximately this value in communication systems, the response deterioration was 0.5–1.0 dB at a frequency of 500 MHz, as shown in Fig. 61. The deterioration is mainly due to the RC time constant. The -3-dB cutoff frequency is considered to be more than 1 GHz. At a higher multiplication region, the response

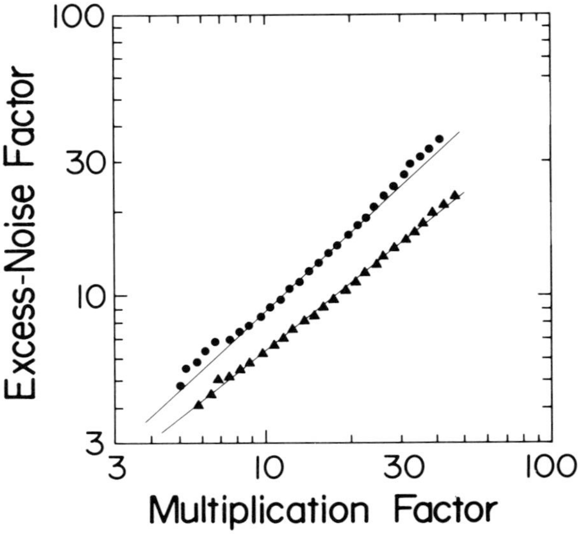

FIG. 60. Excess-noise factor of p^+-n diodes as a function of multiplication factor for wavelengths 1.3 (●) and 1.55 (▲) μm for a p^+-n Ge APD with $f = 30$ MHz, $B = 1$ MHz, $I_{p0} = 1.0\ \mu$A.

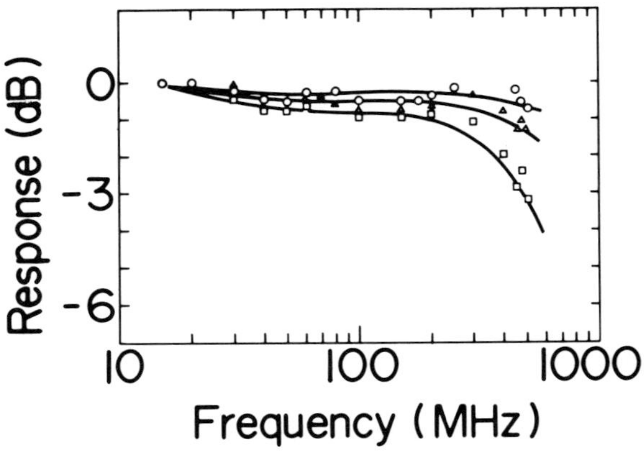

FIG. 61. Frequency response of p^+-n Ge APDs as a parameter of the multiplication factor at $\lambda = 1.3\ \mu$m. $M = 10$ (○), 30 (△), 60 (□). [From Kagawa *et al.* (1982a).]

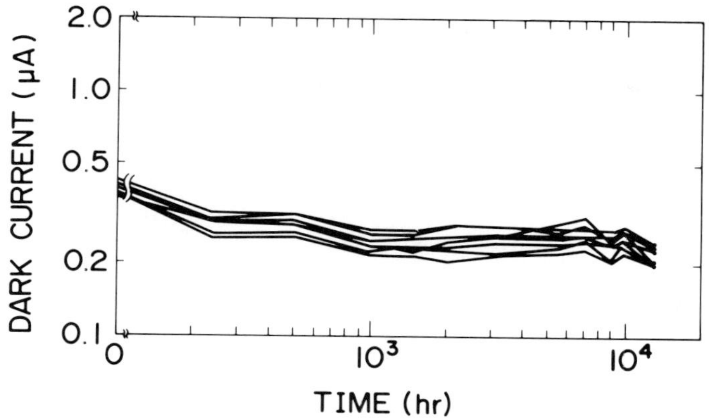

FIG. 62. Aging characteristics of dark current as a function of time for p^+-n Ge APDs with $T_a = 125°C$, $V_R = 25$ V. [From Kagawa *et al.* (1982a).]

decreases with increasing multiplication factor because of the avalanche buildup time.

a. Reliability

Bias-temperature treatment was carried out for testing the reliability of the p^+-n diodes that have CVD SiO_2 film for surface passivation. As an example, Fig. 62 shows the results of dark currents at 0.9 V_B as a function of time at an ambient temperature of 125°C and a bias voltage of −25 V. The dark currents were measured at 25°C by lowering the temperature. No degradation was observed after operating more than 10,000 hr. Other main characteristics of excess-noise factor versus multiplication factor, quantum efficiency at 1.3 μm, and breakdown voltage, were also tested, and we have observed no degradation in these. The diodes are recognized to be highly reliable for practical use in fiber-optic communications systems.

Fabrications and device properties are now discussed for three types of device structure: n^+-p, n^+-n-p, and p^+-n diodes. As a result, in the 1-μm-wavelength region, the p^+-n diodes were found to meet the all detector requirements for use in communication systems. The characteristics of p^+-n diodes are summarized in Table IV. The diodes operate at low bias voltage (~30 V) and show low multiplication noise, high quantum efficiency (~90%), high-speed response ($f_c \geq 1$ GHz), and high reliability. Moreover, their fabrications are reproducible and inexpensive because the conventional processing technology is applicable to germanium.

TABLE IV

THE CHARACTERISTICS OF p^+-n GERMANIUM AVALANCHE PHOTODIODES HAVING A 100-μm ACTIVE-AREA DIAMETER

Parameter	Value	Unit	Test condition
Breakdown voltage (V_B)	30	V	
Temperature Coefficient of V_B	0.12	$\times 10^{-3}$ deg^{-1}	$\gamma = \frac{V_B(25°C + \Delta T°C) - V_B(25°C)}{V_B(25°C)\,\Delta T(°C)}$
Dark current	0.15–0.3	μA	$V_a = 0.9\ V_B$
Capacitance	1.3	pF	1 MHz, $V_a = 0.9 V_B$
Quantum efficiency	84–90	%	$\lambda = 1.3\ \mu$m, dc
Excess-noise factor	8–9		$\lambda = 1.3\ \mu$m, $\langle M \rangle = 10$
Cutoff frequency	1	GHz	$\lambda = 1.3\ \mu$m, $\langle M \rangle = 10$, $R_L = 50$

13. REACH-THROUGH STRUCTURE FOR 1.55 μm

As shown in Fig. 9, the light-absorption coefficients of germanium decrease rapidly in the 1.55-μm-wavelength region, where silica fiber shows the lowest transmission loss of 0.2 dB km^{-1}. It is important therefore to find the long cutoff wavelength where germanium avalanche photodiodes can be used in optical-fiber communication systems. Abrupt junction-type germanium photodiodes, such as n^+-p and p^+-n diodes developed for the 1.3-μm region, have a narrow depletion-layer length of 2–3 μm. If they receive 1.55 μm of light, most of the light is absorbed outside the depletion layer. Then, the photogenerated carriers will travel in the neutral region by diffusion. This leads to a photodiode with a degraded response speed. For response speed and high quantum efficiency, the depletion-layer length should be 10 μm or more. In the abrupt-junction photodiodes, however, it is hard to obtain the long depletion layer, as described later. In the abrupt-junction diode with a wafer concentration of 1×10^{16} cm^{-3}, the breakdown voltage is around 30 V, and the difference in breakdown voltages between the active and the guard-ring layer is about 15 V. The breakdown-voltage ratio between the active and the guard-ring layers V_A/V_G is about 0.7 or less, which is required empirically for obtaining a uniform and high gain in the active region. Here, if we fabricate the diode with a longer depletion layer of more than 10 μm by using a low-concentration wafer, the breakdown voltage will approach 100 V. However, the breakdown-voltage difference between the active and guard-ring layers will still remain around 15 V. This is because the maximum electric field at the breakdown voltage and the depletion length extended into the guard-ring layers are almost the same as that for the diode with the wafer concentration of 1×10^{16} cm^{-3}. Thus, V_A/V_G

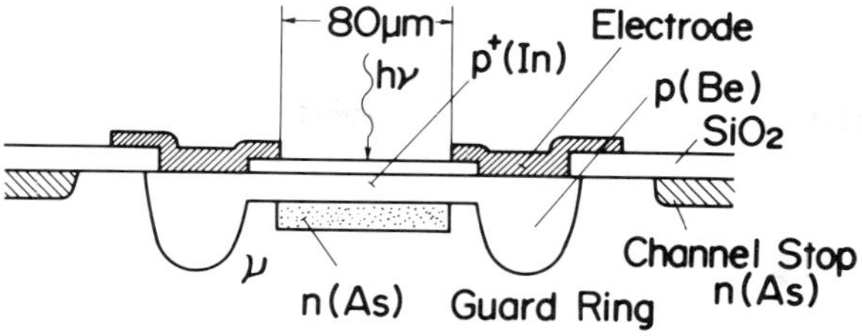

FIG. 63. Cross-sectional view of germanium reach-through avalanche photodiodes; $\nu \sim 3 \times 10^{14}$ cm^{-3}. [From Mikawa *et al.* (1983b).]

attains a large value of 0.8–0.9. This often leads to a local enlargement of avalanche gain and/or microplasma breakdown in the periphery of the guard ring.

By using the reach-through structure, as described in Section 5, we can obtain the long depletion layer and V_A/V_G of 0.7 or less. Figure 63 shows the cross-sectional view of the reach-through germanium avalanche photodiode composed of $p^+-n-\nu$ layers for use in wavelengths of 1.55 μm (Mikawa *et al.*, 1983b, 1984). The concentration of the ν layer was selected to be 3×10^{14} cm^{-3} because of the requirement that the depletion layer length be more than 10 μm at breakdown voltage of ~ 100 V. The guard-ring layer was formed by beryllium (Be)-ion implantation with an accelerating energy of 100 keV and a dose of 1×10^{14} cm^{-2}. The breakdown voltage of Be implanted layer is 90–140 V for wafers with a concentration of 3×10^{14} cm^{-3}. The n-type active and channel-stop layers were made by arsenic (As)-ion implantation with an acceleration energy of 140 keV. The dose for the channel-stop layer was 1×10^{14} cm^{-2}, and the surface concentration obtained was near 1×10^{18} cm^{-3}, which is sufficiently high to cut the surface-inversion channel. The dose for the active layer influences strongly the breakdown voltage of the diode. The reason is that in germanium, the maximum electric field at breakdown voltage has a lower value of about $E_m = 2.5 \times 10^5$ V cm^{-1} compared with that of silicon, because impact-ionization coefficients in germanium are much higher than that in silicon, as shown in Fig. 1. By using E_m, the number of space charges per unit area Q_n contained in the depleted n and ν regions is calculated from

$$Q_n = \varepsilon E_m / q. \tag{52}$$

In germanium, Q_n is evaluated as being about 2.2×10^{12} cm^{-2}, and most of Q_n comes from the n layer, because the ν-layer concentration has a low value of 3×10^{14} cm^{-3} and a thickness of 10 μm or more. The total number of Q_n is therefore governed by the dose of the n layer, and thus the breakdown voltage depends strongly on the implanted dose. Figure 64 shows the As dose

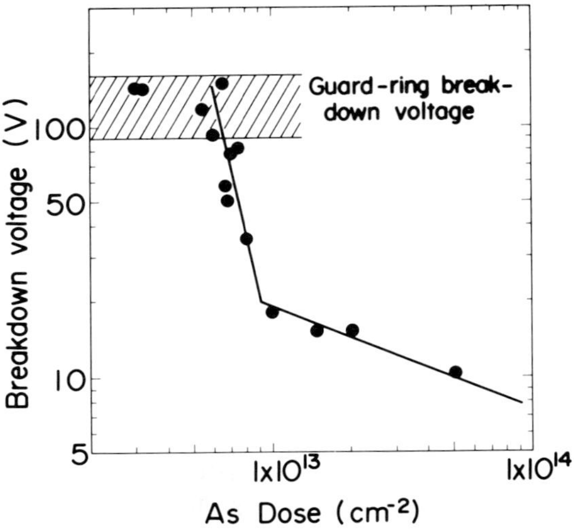

FIG. 64. Breakdown voltage as a function of arsenic dose in germanium reach-through avalanche photodiodes. The hatched region indicates the breakdown voltage of the guard-ring layer. [From Mikawa *et al.* (1983b).]

dependence on the breakdown voltage. The hatched region in Fig. 64 indicates the breakdown voltage in the guard-ring layer. When the dose is higher than 1×10^{13} cm^{-2}, the breakdown voltage shows a low value of less than 20 V, because the concentration in the n layer becomes higher than 1.5×10^{16} cm^{-3}. The breakdown occurs in the n layer, and the depletion layer does not extend into the ν layer. If the dose is under 5×10^{12} cm^{-2}, on the other hand, the breakdown voltage lies in the same voltage range as the guard-ring layer. The breakdown is considered likely to occur at the periphery of the guard-ring junction. Therefore the As dose is suitable in the range of $5 \times 10^{12}-1 \times 10^{13}$ cm^{-2}. To obtain the longer depletion length when V_A/V_G of less than 0.7, the dose of 7×10^{12} cm^{-2} was selected for photodiode fabrications.

Typical characteristics of the dark current and photomultiplication at $\lambda = 1.55$ μm are shown in Fig. 65. The breakdown voltage V_B was about 70 V, and the dark current at 0.9 V_B was 0.8 μA. A maximum multiplication of near 100 was obtained at the initial photocurrent of $I_{p0} = \mu$A. From this multiplication characteristic, the reach-through voltage, where the depletion layer extends into the ν layer, was found to be about 20 V. The voltage was confirmed from the $C-V$ characteristics; that is, the capacitance decreases rapidly at around 20 V. Because the diode has a long depletion layer, which was determined to be about 20 μm from the $C-V$ measurements, a low

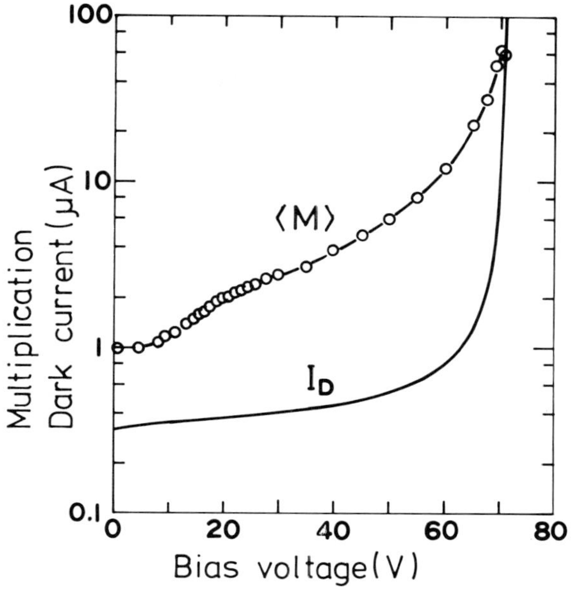

FIG. 65. Multiplication (○) and dark-current (——) characteristics as a function of bias voltage. Multiplication measured at $\lambda = 1.55\ \mu$m, $I_{p0} = 2\ \mu$A. [From Mikawa *et al.* (1984).]

capacitance of 0.5–0.6 pF was obtained at $0.9\,V_B$, including the package capacitance (0.3 pF). The quantum efficiency was 85% at 1.55 μm for the diode with an SiO_2 antireflection coating. Gain uniformity in the active region was studied by spot-scanned photoresponse at 633 nm (He–Ne laser light). A flat response was obtained in the entire active region with a diameter of 80 μm. Multiplication noise was measured at a frequency of 30 MHz and at 1.55 μm (InGaAsP LED light). The excess-noise factor F was 6.1 at a multiplication factor of 10. The F value is expressed as $\langle M \rangle^{0.79}$ in the multiplication range 5–50 and is the lowest ever reported for germanium avalanche photodiodes.

Frequency response was studied at 1.55 μm by using sinusoidally modulated InGaAsP laser. The results obtained at a multiplication factor of 10 are shown in Fig. 66, where the results for p^+-n diodes are also shown. A flat response up to around 300 MHz was obtained for the reach-through diode. A response degradation was 2 dB or less at 500 MHz, and the degradation was due to the carrier transit time. The calculated transit-time cutoff frequency becomes about 900 MHz from Eq. (22), where the saturation drift velocity of 8×10^6 cm sec^{-1} and the length of the depletion layer of 20 μm are selected. In the p^+-n diode, however, the cutoff frequency was only

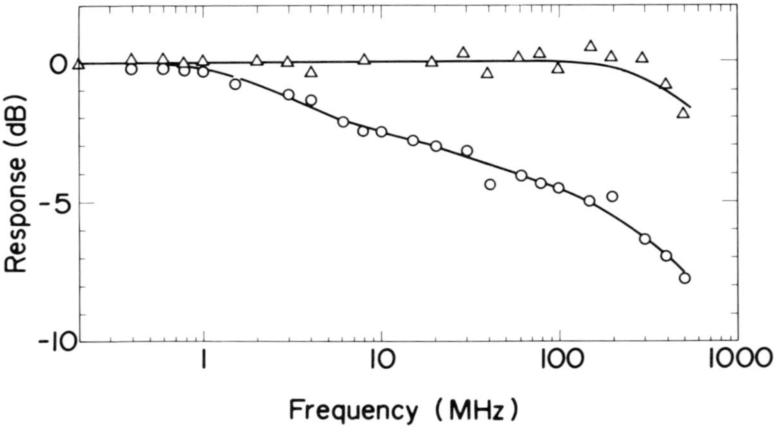

FIG. 66. Frequency response of Ge reach-through APDs (△) obtained at a wavelength of 1.55 μm and at a multiplication factor of 10. The results for p^+-n avalanche photodiodes are also shown (○).

20 MHz, which is limited by the carrier diffusion time in the undepleted layer [Eq. (25)], because the light-absorption length is much longer than the depletion length of 3 μm or less. Figure 67 shows the pulse-response characteristics of both the reach-through and p^+-n diodes for the pulse rate 100 Mbit/sec. The reach-through diode shows first rise and fall characteristics free from interference between the received pulse forms, whereas the p^+-n diode shows a slow component in rise characteristics and a long tail in fall characteristics. As a result, the reach-through germanium avalanche photodiodes show a first-response speed ($f_c \approx 900$ MHz), high quantum efficiency of 85%, and low noise ($F = \langle M \rangle^{0.79}$) and meet the detector requirements for fiber-optic communication systems operating in the 1.55-μm region.

14. Small Active-Area Diodes for Single-Mode Fibers

The germanium avalanche photodiodes thus far discussed have an 80–100-μm active-area diameter and are utilized for single- and multimode optical fibers. With the increasing need for long-distance and high-bit-rate optical-fiber communication systems, one has to deal with single-mode fibers, and further improvements in photodetector sensitivity are required. However, the values of quantum efficiency, response speed, and multiplication noise have almost approached the theoretical limits, as described in Section 12. There is little room for improving dark-current characteristics. As seen in Eq. (47), the multiplied dark current I_{DM} strongly affects the shot-noise level of photodiodes. Because I_{DM} flows through the active region

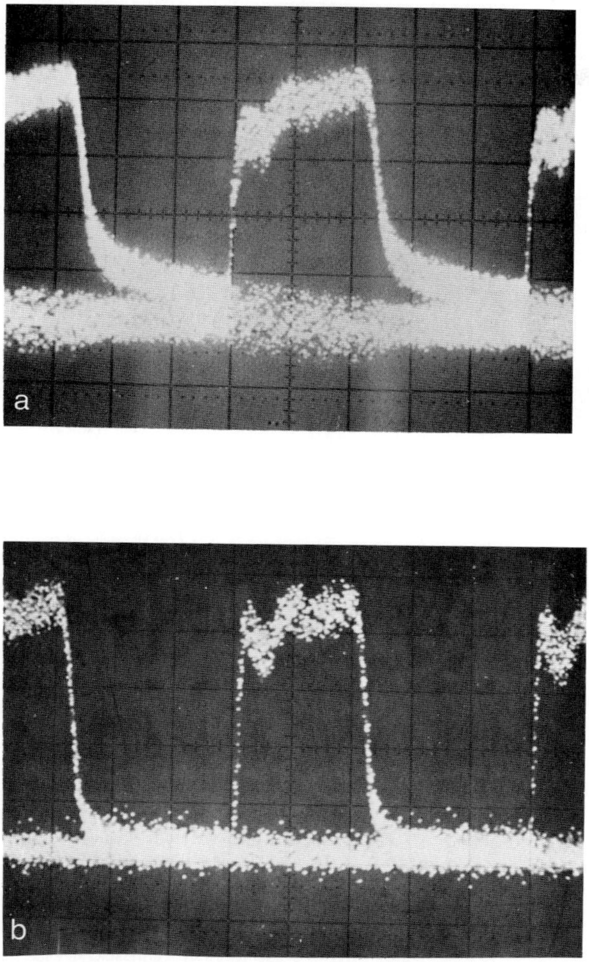

FIG. 67. Pulse-response waveform for (a) p^+-n Ge APDs (density ~ 1×10^{16} cm^{-3}) and (b) reach-through APDs (density ~ 3×10^{14} cm^{-3}). The pulse rate is 100 Mbit/sec; $\lambda = 1.55$ μm. [From Mikawa *et al.* (1983b).]

of the diode, the current is considered to decrease by decreasing the active-area diameter. This can be done for single-mode fibers without degrading the coupling efficiency between an optical fiber and a photodetector because the core diameter of the fiber is small, around 10 μm. For this purpose, p^+-n diodes having various diameters, such as 20, 30, 50, 80, and 100 μm, were made on the same wafer (Mikawa *et al.*, 1984b). A schematic cross section of

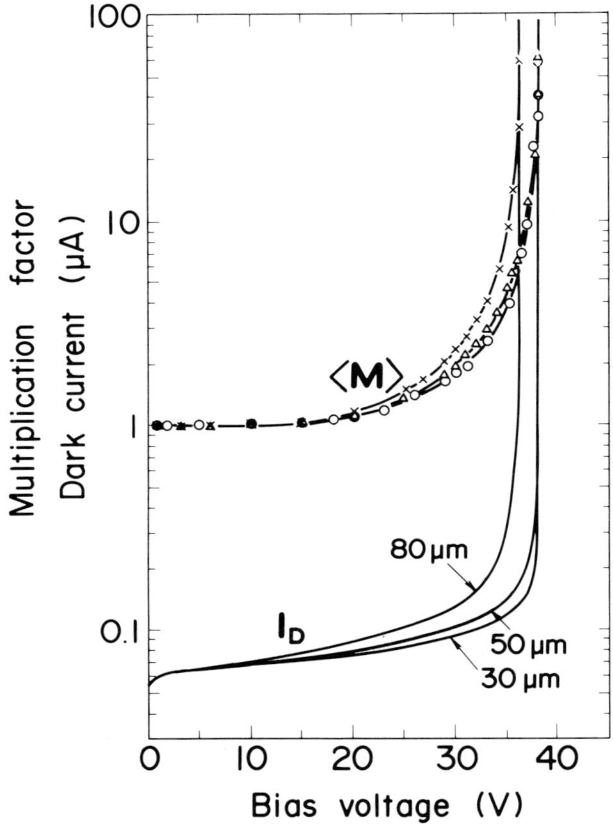

FIG. 68. Multiplication and dark current versus bias-voltage characteristics with the parameter of an active-area diameter of 30 (○), 50 (△), and 80 (×) μm and with $\lambda = 1.3\ \mu$m and $I_{p0} = 2\ \mu$A.

the diode is shown in Fig. 46c. The p^+ layer was formed by B-ion implantation, the guard ring by Be, and the channel stop by As, respectively. The carrier concentration of the wafer used was about 6×10^{15} cm^{-3}. The surface of the diode was passivated by CVD SiO_2 films. A plasma CVD SiN was used for antireflection coating. Figure 68 shows the multiplication factor and dark-current-versus-bias-voltage characteristics of the diodes having 30–80 μm active-area diameters. Multiplication factors as high as 100 for the initial photocurrent of 2 μA were obtained at 1.3 μm for these diodes. Dark current near breakdown voltage, where the multiplied dark current governs the total current, decreases with decreasing the active-area diameter, as seen in Fig. 68. Multiplied dark currents were obtained from dark-current and

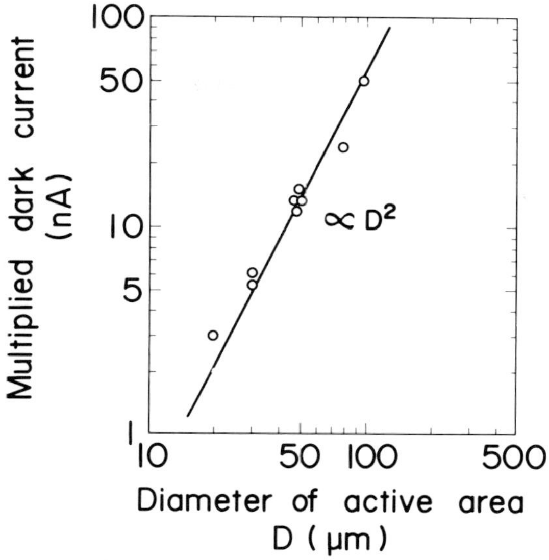

FIG. 69. Multiplied dark current versus active-area diameter at 25°C.

multiplication characteristics in the vicinity of the breakdown voltage using Eq. (48). The results obtained are shown in Fig. 69, and it is found that I_{DM} is in proportion to the square of the active-area diameter D as expected. In the 30-μm-diameter diode, for example, I_{DM} is 5–6 nA at 25°C and is an order of magnitude lower than that of the conventional 100-μm diameter. Because I_{DM} is governed by diffusion current, I_{DM} increases with temperature in proportion to $\exp(-E_G/kT)$. A low dark current of $I_{DM} = 30$ nA can be obtained for the 30-μm-diameter diode even at 50°C, which is normally the upper limit for communication systems.

The other characteristics were studied using the 30-μm-diameter diode. Diode capacitance, of course, decreases with decreasing the active-area diameter. A value of 0.5 pF was obtained at 30 V, including the package capacitance (0.3 pF), whereas about 1.2 pF was obtained for the 100-μm-diameter diode. A flat response up to 1 GHz, which is limited by the measuring apparatus (Fig. 33), was obtained at a multiplication factor of 10 for the load resistance of 50 Ω. A module was assembled connecting the diode with a single-mode fiber through a spherical lens with antireflection-coating film. The quantum efficiency of the module was 90% at $\lambda = 1.3$ μm and did not differ from results using the diode alone. This shows that the coupling efficiency between the photodiode and the fiber is nearly 100%. Temperature variation of the module quantum efficiency was within 3% in the temperature range of 5–50°C, and it was found that the module can be used in communication systems. A discussion of detectable power for this diode next follows in Part V.

V. Minimum Detectable Power

The overall performance of avalanche photodiodes for use in fiber-optic communication systems can be evaluated from the minimum detectable power, considering the excess-noise factor, quantum efficiency, and dark current. In digital communication systems, a simplified expression of the signal-to-noise (S/N) ratio, including the following amplifier noise, is given by

$$\frac{S}{N}=\frac{(I_0\langle M\rangle)^2R}{2q\{I_{p0}+I_{DM})\langle M\rangle^2F+I_{D0}\}BR+4F_{amp}k_BTB,} \tag{53}$$

where F_{amp} is the noise factor of the following amplifier, R the equivalent input resistance of the following amplifier, T the absolute temperature, and k_B Boltzmann constant; the S/N value depends strongly on the multiplication factor, as shown in Fig. 7. The optimum multiplication factor, where S/N has the maximum value, can be obtained from

$$\frac{d(S/N)}{dM}=0. \tag{54}$$

By using Eq. (15), the average detectable power P is given by

$$P=\frac{1}{4}\frac{BFh\nu}{\eta}\left(\frac{S}{N}\right)\left\{1+\left[1+\frac{\chi}{F^2B(S/N)}\right]^{1/2}\right\} \tag{55}$$

and

$$\chi=\frac{I_{DM}F}{q}+\frac{2k_B}{q^2\langle M\rangle^2R}\left(TF_{amp}+\frac{2qI_{D0}R}{4k_B}\right), \tag{56}$$

where the modulating signal is assumed to be a return-to-zero (RZ) pulse with 50% mark density and 50% duty factor. This corresponds to the factor of $\frac{1}{4}$ in Eq. (55). By using Eqs. (55) and (56), we can calculate detectable power versus multiplication characteristics as a function of various diode properties, such as quantum efficiency, dark current, and excess-noise factors. Figure 70 shows the calculated results for quantum efficiency in case of silicon avalanche photodiodes, as an example, where the data rate is 100 Mbit/sec, and the S/N is assumed to be 22 dB for a 10^{-9} error rate. Detectable power is the minimum when the multiplication factor is around 60 in this case. The minimum detectable power decreases as quantum efficiency

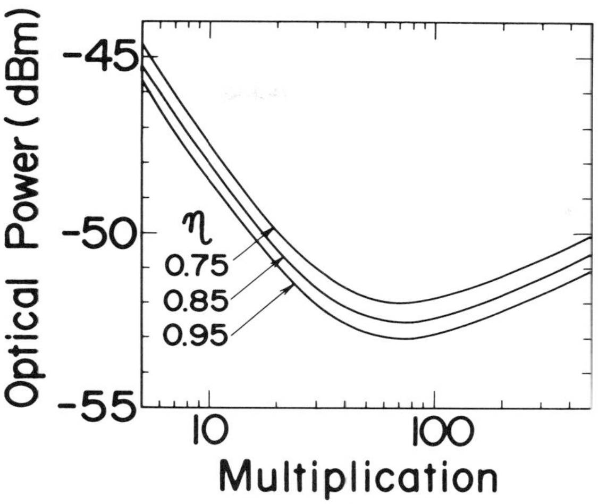

FIG. 70. Calculated optical power versus multiplication characteristics for Si APD with the parameter of quantum efficiency at 100 Mbit/sec data rate.

increases with a rate of about 0.5 dB for a quantum efficiency variation of 10%.

Studies on the effect of multiplied and unmultiplied dark current on detectable power are important for the design and fabrication of photodiodes, especially germanium diodes, because germanium has a large intrinsic carrier concentration and a small band-gap energy. In Fig. 71, the effect of a multiplied dark current I_{DM} is calculated at a data rate of 800 Mbit/sec for germanium avalanche photodiodes. With the current less than 20 nA, the I_{DM} effects on detectable power are negligible, whereas detectable power decreases as I_{DM} increases in the range of $I_{DM} \geq 20$ nA. The minimum detectable power becomes worse than 1 dB at $I_{DM} = 200$ nA compared with $I_{DM} = 20$ nA at a data rate of 800 Mbit/sec. This takes place in germanium avalanche photodiodes. In germanium, I_{DM} varies with temperature in proportion to $\exp(-E_G/k_B T)$, so that I_{DM} increases by a factor of about 10 with a temperature rise of 25°C above room temperature. The I_{DM} effect varies by a data rate and becomes greater in the lower data-rate region.

The influence of unmultiplied dark current I_{D0} on detectable power is very small. Figure 72 shows the calculated results at a data rate of 800 Mbit/sec. As shown in Fig. 72, the difference in minimum detectable power between $I_{D0} = 0.2\ \mu$A and $I_{D0} = 20\ \mu$A is negligibly small. This indicates that avalanche photodiodes must be designed and fabricated with low multiplied dark currents rather than with low, unmultiplied ones.

Detectable power in practical receiver systems is plotted in Fig. 73 and

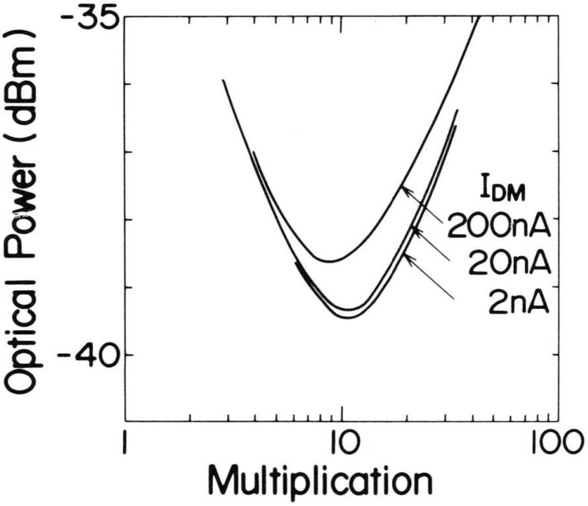

FIG. 71. Calculated optical power versus multiplication characteristics for a Ge APD with the parameter of multiplied dark current: data rate, 800 Mbit/sec; excess-noise factor, $\langle M \rangle^{0.9}$; unmultiplied dark current, 200 nA; quantum efficiency, 0.8; input impedance, 1 Ω.

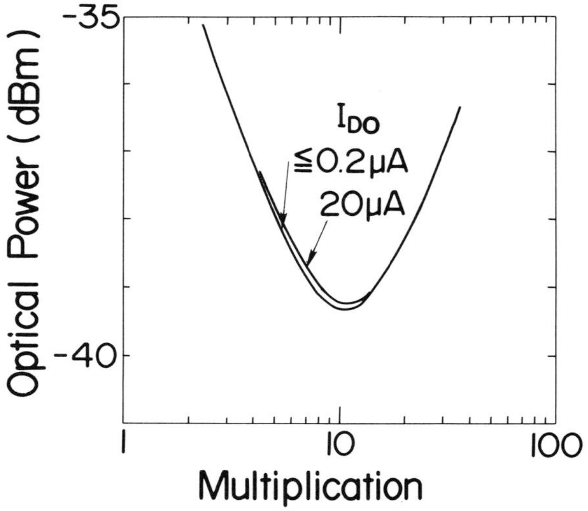

FIG. 72. Calculated optical power with the parameter of unmultiplied dark current. Boundary conditions are the same as Fig. 71, except for multiplied dark current, 20 nA.

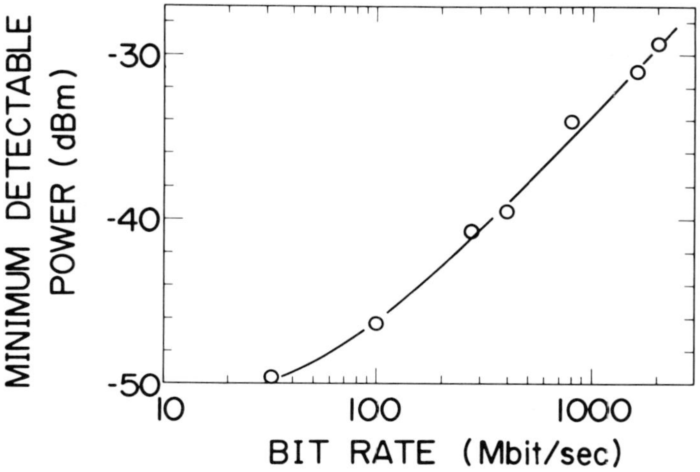

FIG. 73. Minimum detectable power as a function of bit rate in practical fiber-optic systems for a Ge APD with $P_e = 10^{-9}$, $\lambda = 1.3\ \mu m$ in an RZ data format. [From Kaneda and Kanbe (1983).]

were reported using germanium avalanche photodiodes at the wavelength of 1.3 μm. The photodiodes are now used in the entire range of data rates. It can be seen that the detectable power is low enough for practical use (Ito *et al.*, 1980; Yamada *et al.*, 1981; Ohta *et al.*, 1982; Hakamada *et al.*, 1982).

The detectable power for small active-area diodes was studied at 45 Mbit/sec (RZ-data format) and at 1.3 μm for operating temperatures of 25 and 50°C. The results are shown in Fig. 74, where the results for the 100-μm-diameter diode are also shown for comparison. The minimum detectable powers obtained for the 10^{-11} error rate were -51.9 and -50.7 dBm at 25 and 50°C, respectively, for the 30-μm-diameter diode. These values were better by 2.5 and 4.5 dB than those for the 100-μm diode. A temperature variation of the minimum detectable power between 25 and 50°C was 1.2 dB for the 30-μm diode but 3 dB for the 100-μm diode. This arises from the difference in the multiplied dark current between two types of diode. A small active-area-diameter diode greatly contributes to the enhancement of photodiode sensitivity.

At the wavelength of 1.55 μm, the detectable power for the p^+-n diode worsens because of the degraded frequency response. For example, the minimum detectable power at 100 Mbit/sec was about -39 dBm for the 10^{-11} error rate and was about 6 dB inferior to that at 1.3 μm. By using the reach-through diode, the detectable power has been improved remarkably. The minimum detectable power of -44.3 dBm was obtained for the 10^{-11}

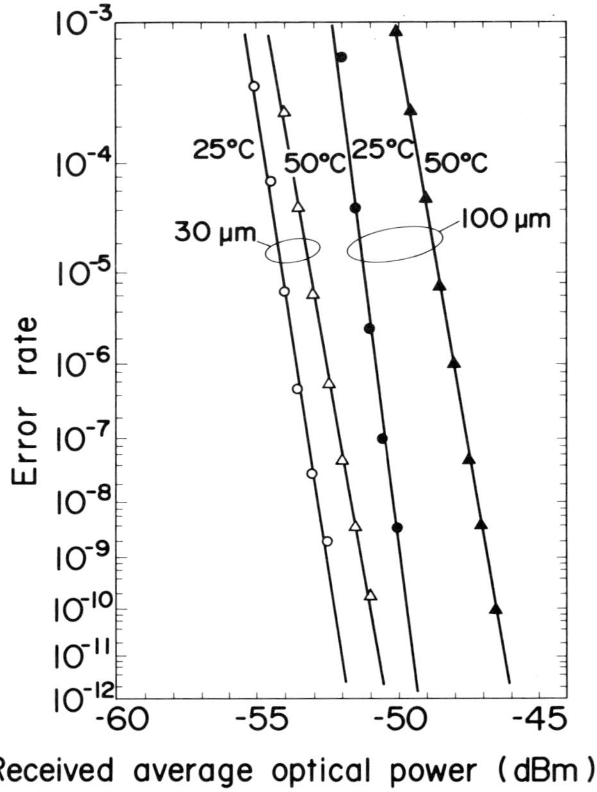

FIG. 74. Bit-error-rate characteristics at 45 Mbit/sec and at 1.3 μm for the 30-μm-diameter diode [(○) at 25°C, (△) at 50°C] and the 100-μm-diameter diode [(●) at 25°C, (▲) at 50°C] used in an RZ format. [From Mikawa *et al.* (1983a).]

error rate. This is equivalent to a receiving level obtained in the 1.3-μm region for the p^+-n diode.

VI. Concluding Comments

Silicon and germanium avalanche photodiodes for use in fiber-optic communication systems have been discussed in this chapter. The most important feature of avalanche photodiodes is multiplication noise. The characteristics are in good agreement with the calculated results from carrier-ionization coefficients and carrier injections. Quantum efficiency and response speed also meet with the predicted values from photodiode design. Multiplication noise, quantum efficiency, and response speed are the principal features of avalanche photodiodes. These characteristics in silicon and germanium photodiodes have almost attained the theoretical limits. There is little room left for improvement.

In the wavelength 0.8–0.9 μm, silicon is the most suitable material because of its large asymmetry in electron- and hole-ionization coefficients. Although GaAs is a potential material for use around 0.8 μm, its carrier-ionization coefficients are not so promising as silicon (Stillman *et al.,* 1974; Pearsall *et al.,* 1977; Law and Lee, 1978; Ito *et al.,* 1978; Ando and Kanbe, 1981), and the long cutoff wavelength falls rather short of 0.84 μm. GaAs photodiodes are therefore a poor choice for practical use, except in a monolithically integrated photoreceiver.

In the 1.3-μm wavelength, a shallow junction of p^+-n germanium avalanche photodiodes is the best device, as shown by its overall performance with regard to multiplication noise, quantum efficiency, response speed, and reliability. Although a deep junction of n^+-n-p diodes shows the lowest multiplication noise, its frequency response is inferior to other diodes.

In the 1.55-μm diode, where the lowest transmission-loss region of optical fiber occurs, the reach-through type of germanium photodiode is suitable. Multiplication noise of germanium avalanche photodiodes exhibits the lowest value in this wavelength region, and the excess noise-factor is expressed as $F = \langle M \rangle^{0.79}$. This value is almost the same to that of III–V detectors, which are under investigation (Shirai *et al.,* 1982).

Avalanche photodiodes have been fabricated using planar structures with a guard ring and a channel stop for greater durability and electrical stability. The guard ring is the key layer in obtaining high avalanche gain and good reproducibility of device. For this purpose, a higher breakdown voltage is more desirable. The breakdown-voltage ratio between the active and guard ring regions must be 0.7 or less, from empirical considerations. Because conventional processing technology is applicable to silicon and germanium, the photodiode fabrication is reproducible, reliable, and inexpensive.

Ion implantation plays an important role in fabricating avalanche photodiodes. It allows superior control of dosage and depth. This makes it easy to achieve reach-through-type silicon and germanium photodiodes whose active layers are made by boron- and arsenic-ion implantation, respectively. Doping techniques for both *n*- and *p*-type impurities are indispensible for forming well-designed avalanche photodiodes because a $p-n$ junction and a channel stop are formed by the different types of impurities. Ion implantation in the case of germanium is useful for making *p* layers, whereas *p*-type diffusion has not developed well.

In silicon avalanche photodiodes, multiplication noise decreases as the avalanche-region length increases because the hole-to-electron-ionization-coefficient ratio decreases according to it; quantum efficiency is also improved as the depletion length increases. These procedures lead to an increase of the breakdown voltage. Hence, the sensitivity of silicon avalanche photodiodes is improved as the breakdown voltage increases. A similar argument can be made for germanium because the asymmetry of carrier-ionization coefficients increases as the electric field strength decreases.

The minimum detectable power is strongly affected by the multiplied dark current, which flows through the avalanche region when it is greater than several nanoamperes. Because germanium has a large intrinsic carrier density and a long diffusion length, germanium photodiodes suffer from high multiplied dark current. Small-active-area germanium detectors are effective for improving sensitivity. With a 30-μm active-area-diameter diode, the multiplied dark current decreases as much as 5 nA at 25°C and is an order of magnitude less than the 100-μm-diameter diode. Temperature variations of detectable power are remarkably improved using this small-active-area device. Photodetectors must be designed to have low multiplied dark current rather than low unmultiplied dark current for use in fiber-optic communication systems.

In the 0.8–0.9-μm wavelength, silicon avalanche photodiodes exhibit extremely high performance, whereas optical fibers have relatively high transmission losses. In the 1-μm region, although germanium avalanche photodiodes exhibit rather high dark current and avalanche noise, the transmission loss of optical fiber is extremely low. This combination of fibers and detectors surely makes good pairs, which indicates that silicon and germanium avalanche photodiodes will hereafter play significant roles in fiber-optic communication systems.

References

Amouroux, C., Brilman, M., and Ripoche, G. (1975). *Tech. Dig.—Int. Electron Devices Meet.* p. 595.

Anderson, C. L., and Crowell, C. R. (1972). *Phys. Rev. B* **5,** 2267.

Anderson, L. K., McMullin, P. G., D'Asaro, L. A., and Goetzberger, A. (1965). *Appl. Phys. Lett.* **6,** 62.

Ando, H., and Kanbe, H. (1981). *Solid-State Electron.* **24,** 629.

Ando, H., Kanbe, H., Kimura, T., Yamaoka, T., and Kaneda, T. (1978). *IEEE J. Quantum Electron.* **QE-14,** 804.

Ando, H., Kanbe, H., Ito, M., and Kaneda, T. (1980). *Jpn. J. Appl. Phys.* **19,** L277.

Berchtold, K., Krumpholz, O., and Suri, J. (1975). *Appl. Phys. Lett.* **26,** 585.

Braunstein, R., Moore, A. R., and Herman, F. (1958). *Phys. Rev.* **109,** 695.

Camp, P. R. (1955). *J. Electrochem. Soc.* **102,** 586.

Campbell, J. C., Dentai, A. G., Holden, W. S., and Kasper, B. L. (1983). *Electron. Lett.* **19,** 818.

Chang, J. J. (1967). *IEEE Trans. Electron Devices* **ED-14,** 139.

Chynoweth, A. G. (1968). *In* "Semiconductors and Semimetals" (R. K. Willardson and A. C. Beer, eds.), Vol. 4, Chap. 4. Academic Press, New York.

Conradi, J. (1974). *Solid-State Electron.* **17,** 99.
Dai, B. T., and Chang, C. Y. (1971). *J. Appl. Phys.* **42,** 5198.
Dash, W. C., and Newman, R. (1955). *Phys. Rev.* **99,** 1151.
Decker, D. R., and Dunn, C. N. (1970). *IEEE Trans. Electron Devices* **ED-17,** 290.
Emmons, R. B. (1967). *J. Appl. Phys.* **38,** 3705.
Emmons, R. B., and Lucovsky, G. (1966). *IEEE Trans. Electron Devices* **ED-13,** 297.
Forrest, S. R., DiDomenico, M., Jr., Smith, R. G., and Stocker, H. J. (1980). *Appl. Phys. Lett.* **36,** 580.
Gärtner, W. W. (1959). *Phys. Rev.* **116,** 84.
Goedbloed, J. J., and Smeets, E. T. J. M. (1978). *Electron. Lett.* **14,** 67.
Grant, W. N. (1973). *Solid-State Electron.* **16,** 1189.
Hakamada, K., Nakagawa, K., Yamaguchi, N., Toge, T., and Mori, H. (1982). *Annu. Meet. Inst. Electron. Commun. Eng. Jpn.* p. 2210.
Hino, I., Hata, S., Torigai, T., Iwasaki, H., Minemura, K., Katayama, S., and Nishida, K. (1982). *Tech. Rep. IECE Jpn.* **OQE82-5,** p. 33.
Irving, B. A. (1962). *J. Electrochem. Soc.* **109,** 120.
Ito, M., Kagawa, S., Kaneda, T., and Yamaoka, T. (1978). *J. Appl. Phys.* **49,** 5607.
Ito, M., Kaneda, T., Nakajima, K., Toyama, Y., and Ando, H. (1981). *Solid-State Electron.* **24,** 421.
Ito, T., Nakagawa, K., Motegi, M., Minami, T., Higo, Y., and Suyama, S. (1980). *ICC'80 Conf. Rec.* **1,** 28.3.1.
Kagawa, S., Mikawa, T., Kaneda, T., Toyama, Y., and Ando, H. (1979). *Tech. Rep. IECE Jpn.* **ED79-73,** p. 17.
Kagawa, S., Kaneda, T., Mikawa, T., Banba, Y., and Toyama, Y. (1981). *Appl. Phys. Lett.* **38,** 429.
Kagawa, S., Mikawa, T., and Kaneda, T. (1982a). *Fujitsu Sci. Tech. J.* **18,** 397.
Kagawa, S., Mikawa, T., and Kaneda, T. (1982b). *Jpn. J. Appl. Phys.* **21,** 1616.
Kanbe, H., Kimura, T., Mizushima, Y., and Kajiyama, K. (1976a). *IEEE Trans. Electron Devices* **ED-23,** 1337.
Kanbe, H., Mizushima, Y., Kimura, T., and Kajiyama, K. (1976b). *J. Appl. Phys.* **47,** 3749.
Kaneda, T., and Kanbe, H. (1983). "Optical Devices and Fibers" (Y. Suematsu, ed.) Vol. 5, p. 48. Ohmsha, Tokyo.
Kaneda, T., and Takanashi, H. (1973a). *Jpn. J. Appl. Phys.* **12,** 1091.
Kaneda, T., and Takanashi, H. (1973b). *Jpn. J. Appl. Phys.* **12,** 1652.
Kaneda, T., and Takanashi, H. (1975). *Appl. Phys. Lett.* **26,** 642.
Kaneda, T., Matsumoto, H., Sakurai, T., and Yamaoka, T. (1976a). *J. Appl. Phys.* **47,** 1605.
Kaneda, T., Matsumoto, H., and Yamaoka, T. (1976b). *J. Appl. Phys.* **47,** 3135.
Kaneda, T., Takanashi, H., Matsumoto, H., and Yamaoka, T. (1976c). *J. Appl. Phys.* **47,** 4960.
Kaneda, T., Kagawa, S., Yamaoka, T., Nishi, H., and Inada, T. (1978). *J. Appl. Phys.* **49,** 6199.
Kaneda, T., Mikawa, T., Toyama, Y., and Ando, H. (1979a). *Appl. Phys. Lett.* **34,** 692.
Kaneda, T., Fukuda, H., Mikawa, T., Banba, Y., Toyama, Y., and Ando, H. (1979b). *Appl. Phys. Lett.* **34,** 866.
Kaneda, T., Kagawa, S., Mikawa, T., Toyama, Y., and Ando, H. (1980). *Appl. Phys. Lett.* **36,** 572.
Kim, O. K., Forrest, S. R., Bonner, W. A., and Smith, R. G. (1981). *Appl. Phys. Lett.* **39,** 402.
Kobayashi, M., Yamazaki, S., and Kaneda, T. (1984). *Appl. Phys. Lett.* **45,** 759.
Kuisl, M. (1972). *Solid-State Electron.* **15,** 595.
Law, H. D., and Lee, C. A. (1978). *Solid-State Electron.* **21,** 331.
Lecrosnier, D., Paugam, J., and Gallou, J. (1977). *Appl. Phys. Lett.* **30,** 323.
Lee, C. A., Logan, R. A., Batdorf, R. J., Kleimack, J. J., and Wiegmann, W. (1964). *Phys. Rev.* **134,** A761.

Leheny, R. F., Nahory, R. E., Pollack, M. A., Ballman, A. A., Beebe, E. D., DeWinter, J. C., and Martin, R. J. (1980). *Electron Lett.* **16,** 353.
McIntyre, R. J. (1966). *IEEE Trans. Electron Devices* **ED-13,** 164.
McIntyre, R. J. (1972). *IEEE Trans. Electron Devices* **ED-19,** 703.
Matsushima, Y., Sakai, K., and Noda, Y. (1981). *IEEE Electron Device Lett.* **EDL-2,** 179.
Melchior, H., and Lynch, W. T. (1966). *IEEE Trans. Electron Devices* **ED-13,** 829.
Melchior, H., Hartman, A. R., Schinke, D. P., and Seidel, T. E. (1978). *Bell Syst. Tech. J.* **57,** 1791.
Mikami, O., Ando, H., Kanbe, H., Mikawa, T., Kaneda, T., and Toyama, Y. (1980). *IEEE J. Quantum Electron.* **QE-16,** 1002.
Mikawa, T., Kagawa, S., Kaneda, T., Toyama, Y., and Mikami, O. (1980). *Appl. Phys. Lett.* **37,** 387.
Mikawa, T., Kagawa, S., Kaneda, T., Sakurai, T., Ando, H., and Mikami, O. (1981). *IEEE J. Quantum Electron.* **QE-17,** 210.
Mikawa, T., Kaneda, T., Nishimoto, H., Motegi, M., and Okushima, H. (1983a). *Electron. Lett.* **19,** 452.
Mikawa, T., Kagawa, S., and Kaneda, T. (1983b). *Fujitsu Sci. Tech. J.* **19,** 497.
Mikawa, T., Kagawa, S., and Kaneda, T. (1984). *IEEE Trans. Electron Devices* **ED-31,** 971.
Miller, S. L. (1955). *Phys. Rev.* **99,** 1234.
Naqvi, I. M. (1972). *Proc. IEEE* **60,** 1555.
Nishi, H., Inada, T., Sakurai, T., Kaneda, T., Hisatsugu, T., and Furuya, T. (1978). *J. Appl. Phys.* **49,** 608.
Nišhida, K., Ishii, K., Minemura, K., and Taguchi, K. (1977). *Electron. Lett.* **13,** 280.
Ohta, N., Hagimoto, K., Ohue, K., and Nakagawa, K. (1982). *Annu. Meet. Inst. Electron. Commun. Eng. Jpn.* p. 2219.
Pearsall, T. P., Nahory, R. E., and Chelikowsky, I. R. (1977). *Phys. Rev. Lett.* **39,** 295.
Pearsall, T. P., Capasso, F., Nahory, R. E., Pollack, M. A., and Chelikowsky, J. R. (1978). *Solid-State Electron.* **21,** 297.
Ruegg, H. W. (1967). *IEEE Trans. Electron Devices* **ED-14,** 239.
Sawyer, D. E., and Rediker, R. H. (1958). *Proc. IRE* **46,** 1122.
Shibata, T., Igarashi, Y., and Niimi, T. (1971). *Tech. Reps. IECE Jpn.* **SSD71-31.**
Shirai, T., Yamazaki, S., Yasuda, K., Mikawa, T., Nakajima, K., and Kaneda, T. (1982). *Electron. Lett.* **18,** 575.
Shive, J. N. (1959). "The Properties, Physics and Design of Semiconductor Devices." Van Nostrand, Princeton, New Jersey.
Smeets, E. T. J. M., and Politiek, J. (1979). *Appl. Phys. Lett.* **35,** 112.
Smith, D. R., Hooper, R. C., Smyth, P. P., and Wake, D. (1982). *Electron. Lett.* **18,** 453.
Stillman, G. E., Wolfe, C. M., Rossi, J. A., and Foyt, A. G. (1974). *Appl. Phys. Lett.* **24,** 471.
Susa, N., Nakagome, H., Ando, H., and Kanbe, H. (1981). *IEEE J. Quantum Electron.* **QE-17,** 243.
Sze, S. M. (1969). "In Physics of Semiconductor Devices," Chap. 2. Wiley, New York.
Takamiya, S., Kondo, A., and Shirahata, K. (1975). *Tech. Rep. IECE Jpn.* **SSD75-35,** p. 65.
Takanashi, Y., Kawashima, M., and Horikoshi, Y. (1980). *Jpn. J. Appl. Phys.* **19,** 693.
Wallis, G., and Wang, S. (1959). *J. Electrochem. Soc.* **106,** 231.
Wang, K. L., and Gray, R. V. (1976). *J. Electrochem. Soc.* **123,** 1392.
Wang, K. L., and Storms, H. A. (1976). *J. Appl. Phys.* **47,** 2539.
Webb, P. P., McIntyre, R. J., and Conradi, J. (1974). *RCA Rev.* **35,** 234.
Yamada, J., Machida, S., and Kimura, T. (1981). *Electron. Lett.* **17,** 479.
Yamaoka, T., Matsumoto, H., and Kaneda, T. (1976). *Fujitsu Sci. Tech. J.* **12,** 87.
Yasuda, K., Mikawa, T., Kishi, Y., and Kaneda, T. (1984). *Electron. Lett.* **20,** 373.
Zulliger, H. R. (1971). *J. Appl. Phys.* **42,** 5570.

SEMICONDUCTORS AND SEMIMETALS, VOL. 22, PART D

CHAPTER 4

Sensitivity of Avalanche Photodetector Receivers for High-Bit-Rate Long-Wavelength Optical Communication Systems

S. R. Forrest

AT&T BELL LABORATORIES
MURRAY HILL, NEW JERSEY

I. Introduction

Considerable effort has been invested world wide in developing very high-bit-rate, long-wavelength optical communication systems with ever-increasing distances between regenerators. Increasing the bit rate expands the capacity of the system, whereas increasing the distance between transmitter and receiver ensures higher reliability with a decrease in the number of components. In both cases, these efforts ultimately reduce the cost of the communication system.

Until recently, *p–i–n* photodiodes fabricated from InGaAsP alloys that

ISBN 0-12-752153-4

are grown epitaxially on high-quality InP substrates were regarded as the devices most suitable for use in long-wavelength communication systems operating at moderate bit rates of $\lesssim 300$ Mbit/sec (Gloge *et al.,* 1980; Smith *et al.,* 1980). In particular, low-capacitance, low dark-current $In_{0.53}Ga_{0.47}As$ $p-i-n$ detectors sensitive to wavelengths as long as $\lambda = 1.65\ \mu m$ have been demonstrated (Pearsall and Papuchon, 1978; Burrus *et al.,* 1979; Forrest *et al.,* 1980), thus making them suitable for use in low-noise optical receivers.

As receiver bandwidth is increased to accommodate high transmission bit rates, however, the sensitivity decreases, thereby reducing the allowable separation between regenerators. To increase sensitivity at high bit rates, considerable research has been done to develop detectors with internal gain. Among such detectors are avalanche photodiodes (APDs), phototransistors (Fritzche *et al.,* 1981; Campbell *et al.,* 1981), majority-carrier devices (Chen *et al.,* 1981), and high-speed photoconductors (Barnard *et al.,* 1981; Degani *et al.,* 1981). Of these devices, the APD is reverse biased and therefore can in principle exhibit high-speed response in the absence of large dark currents. On the other hand, phototransistors and photoconductive devices (Forrest, 1984b, 1985) can have gains equal to or higher than APDs, although the former devices must draw large dark currents in order to attain high-speed operation. Since these currents are a source of noise, the greatest success at improving receiver sensitivities over those achieved with $p-i-n$ detectors has been obtained using APDs.

Unfortunately, receiver sensitivity is strongly dependent on APD dark current due to the excess noise generated in the multiplication of these currents. Until 1980, efforts directed at fabricating APDs sensitive at long wavelengths with dark currents as small as those of $In_{0.53}Ga_{0.47}As$ $p-i-n$ detectors were largely unsuccessful. Thus, nearly all high-speed transmission experiments were carried out using Ge avalanche photodiodes, whose room-temperature primary dark currents typically exceeded 100 nA. This high value of dark current greatly limited the improvement in receiver sensitivity over that which could be obtained with a $p-i-n$ detector. In 1980, low dark-current InGaAsP/InP heterostructure APDs with sensitivity to $\lambda = 1.25\ \mu m$ were successfully fabricated (Diadiuk *et al.,* 1980). In this structure, the highest electric field and therefore carrier multiplication occurs at the $p-n$ junction located in the wide-band-gap InP layer. On the other hand, light absorption takes place in the narrow-band-gap InGaAsP layer, where the electric field is kept at a low level (Nishida *et al.,* 1979). In this way, carrier tunneling in the narrow-band-gap material is greatly reduced, resulting in low dark-current devices. Heterostructure APDs fabricated using $In_{0.53}Ga_{0.47}As$ as the light-absorbing material have also been shown to exhibit low dark currents (Kim *et al.,* 1981; Shirai *et al.,* 1981) and have been used to give improved receiver sensitivities at $B = 45$ Mbit/sec (Forrest *et al.,* 1981) and $B = 1$ Gbit/sec (Campbell *et al.,* 1983).

In this chapter, we discuss the sources of noise in receivers employing either $p-i-n$ or avalanche photodiodes and compare the sensitivities that can be obtained using these two device structures. In particular, the effects that the primary dark current and the ionization-rate ratio k have on the sensitivity of an APD receiver are considered in detail. Also, we compare the performance of Ge and $In_{0.53}Ga_{0.47}As/InP$ devices. We find that dark currents typical of the best $In_{0.53}Ga_{0.47}As/InP$-heterostructure APDs significantly degrade the receiver sensitivities from values achievable for devices with no dark current, especially at low bit rates. In addition, the ultimate sensitivity of a receiver can be increased by fabricating novel devices with reduced k values. Examples of such novel devices are superlattice and staircase APDs (Capasso *et al.*, 1983) and GaAlAsSb APDs with resonant ionization behavior (Hildebrand *et al.*, 1981). However, the improvement in sensitivity obtained using reduced k-value structures can be realized only if the dark current is small.

Next, we discuss several transmission experiments using APD and $p-i-n$ receivers. In these experiments, distances between regenerators as long as 161.5 km have been achieved, and systems employing bit rates higher than 2 Gbit/sec have demonstrated excellent performance.

We conclude the discussion by considering several mechanisms that limit the transmission-system sensitivity at the highest bit rates presently envisioned. The most important mechanisms for degrading receiver performance at high bit rates are the APD gain-bandwidth response-time limit, transmitter bandwidth, laser noise, and fiber dispersion.

II. Digital Receiver Sensitivity

In Part II we derive expressions for the sensitivity of digital receivers employing $p-i-n$ or avalanche photodetectors. The results obtained shall be used later to calculate the expected sensitivities by using component parameters typical of practical state-of-the-art receivers.

In analyzing the sensitivities of receivers employing avalanche or $p-i-n$ detectors, we follow the treatment of Personick and Smith (Personick, 1973; Smith and Personick, 1979), where the noise spectrum is assumed to be Gaussian. The probability of mistakenly identifying a mark as a space (or vice versa) during the detection interval is called the bit-error rate, given by:

$$\mathrm{BER} = \frac{1}{\sqrt{2\pi}} \int_Q^\infty \exp\left(-\frac{x^2}{2}\right) dx = \frac{1}{2}\,\mathrm{erfc}\left(\frac{Q}{\sqrt{2}}\right). \tag{1}$$

Here,

$$Q = (D - s_{0,1})/\sigma_{0,1}, \tag{2}$$

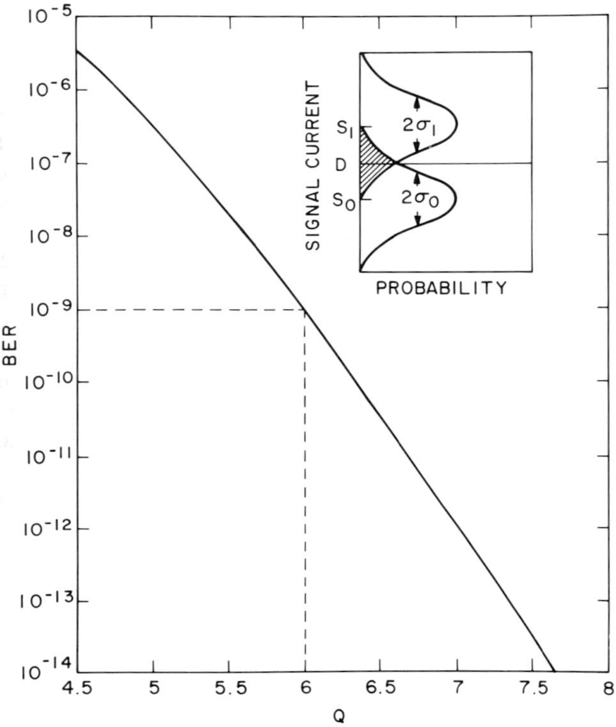

FIG. 1. Bit-error rate versus Q. Inset: Signal current s corresponding to a mark and a space indicating the variance σ and decision level D. A Gaussian probability distribution is assumed. The shaded area in the tails of the distributions overlapping D indicates the probability for detecting an error.

where s_1 (s_0) is the expected value of the signal associated with a mark (space), σ_1 (σ_0) is its variance, and D the decision level set by the optical receiver (see inset, Fig. 1). In Fig. 1 is plotted the bit-error rate as a function of the parameter Q. The allowable BER needed to achieve an acceptable level of error depends on the requirements of the particular transmission system, and typically varies between 10^{-7} and 10^{-10}. During the remainder of this treatment, we assume that BER $= 10^{-9}$, giving $Q = 6$.

1. *p–i–n* PHOTODETECTOR RECEIVER

The variance σ and signal strength s can be expressed explicitly in terms of current. Thus, $\sigma_0 = \sigma_1 = \sigma = \langle i_a^2 \rangle^{1/2}$, where $\langle i_a^2 \rangle^{1/2}$ is the combined rms noise current of the detector and receiver. Here we have assumed that the noise current for either a mark or a space is equivalent in a receiver employing a $p-i-n$ detector. For high light levels, however, this may not be the case, due to shot noise introduced by the photogenerated current. Nevertheless, in this treatment we are interested only in the ultimate sensitivity of digital optical receivers where the incident light levels are expected to be quite small. Under these conditions, the assumption of $\sigma_1 = \sigma_0 = \sigma$ introduces negligible error into the calculations. Now the signal current s can be written in terms of the detected optical power during the transmission of a "1" (p_1) or a "0" (p_0) such that Eq. (2) is given by

$$\begin{aligned} D - s_1 &= D - (q\lambda/hc)p_1 = -Q\langle i_a^2\rangle^{1/2}, \\ D - s_0 &= D - (q\lambda/hc)p_0 = Q\langle i_a^2\rangle^{1/2}. \end{aligned} \tag{3}$$

Here, q is the electronic charge, h is Planck's constant, c the velocity of light, and λ the wavelength of the light pulse.

From Eqs. (3a) and (3b),

$$p_1 - p_0 = 2Q(hc/q\lambda)\langle i_a^2\rangle^{1/2}, \tag{4}$$

$$D = \tfrac{1}{2}(q\lambda/hc)(p_1 + p_0). \tag{5}$$

Equation (5) implies that the optimum decision level is at a power midway between that received for a mark and that received for a space.

Now the received power p is related to the mean power incident on the receiver $\bar{P}$ via

$$\eta\bar{P} = p_1 f + p_0(1 - f), \tag{6}$$

where η is the efficiency of the receiver in converting incident optical power into signal current. Therefore η accounts for both the detector quantum efficiency and the optical coupling efficiency. Also, f is the probability that during a given time slot, a mark will be received. For many pulse codes $f \simeq \frac{1}{2}$, giving

$$\eta\bar{P} = \tfrac{1}{2}[p_1 + p_0]. \tag{7}$$

Next, the extinction ratio r is defined as the ratio of power received in the 0 state to that in the 1 state; thus,

$$r = p_0/p_1. \tag{8}$$

Combining Eqs. (4), (7), and (8), we obtain

$$\eta\bar{P}_{p-i-n} = \frac{1+r}{1-r}\,Q\left(\frac{hc}{q\lambda}\right)\langle i_a^2\rangle^{1/2}, \tag{9}$$

which is the minimum detectable power level for a digital optical receiver employing a $p-i-n$ photodetector. Here, the minimum power is a function of the BER as indicated by the presence of the factor Q in Eq. (9). For

BER $= 10^{-9}$, then $Q = 6$ (Fig. 1), implying that a signal-to-noise ratio of 6 is required.

For the initial part of our treatment, we assume that no power is incident during the transmission of a space. This corresponds to complete source extinction where $r = 0$. Thus,

$$\eta \bar{P}_{p\text{-}i\text{-}n} = A\langle i_{\mathrm{a}}^2\rangle^{1/2} \tag{10}$$

where $A = Q(hc/q\lambda)$ is proportional to the energy of the incident photon. At $\lambda = 1.3$ μm, we have $A = 5.7$ V, and at $\lambda = 1.55$ μm, then $A = 4.8$ V. Thus, the incident power required to achieve a given receiver sensitivity decreases with increasing wavelength. This power dependence results from the photodetection process whereby optical power is converted into electrical current via generation of pairs of charges whose *numbers* are linearly proportional to the number of incident photons.

2. APD Receiver

In deriving Eqs. (9) and (10), it was assumed that the noise currents associated with both the 1 and 0 states were equal. This implies that shot noise generated by the signal current is negligible. This assumption, however, cannot be applied to receivers employing APDs due to excess noise arising from the random nature of the multiplication process. Indeed, it can be shown that for an average gain M, the rms noise currents generated during the transmission of a mark or a space are given by

$$\begin{aligned} \langle i^2\rangle_1 &= \langle i_{\mathrm{a}}^2\rangle + 2q(q\lambda/hc)M^2F(M)I_1Bp_1, \\ \langle i^2\rangle_0 &= \langle i_{\mathrm{a}}^2\rangle + 2q(q\lambda/hc)M^2F(M)I_1Bp_0, \end{aligned} \tag{11}$$

respectively. Here B is the bit rate; I_1 is a Personick integral (Personick, 1973); and $F(M)$ is the excess multiplication-noise factor. The noise additional to $\langle i_{\mathrm{a}}^2\rangle$ is thus the shot noise generated by the multiplied signal current.

For the present analysis, we assume that the dark current of the APD is zero, and therefore $\langle i_{\mathrm{a}}^2\rangle$ represents only the noise current generated by the receiver itself. The case of nonzero APD dark current is treated in Section 6.

Combination of Eqs. (11) and (2) gives

$$\begin{aligned} D - (q\lambda/hc)Mp_1 &= -Q[\langle i_{\mathrm{a}}^2\rangle + 2q(q\lambda/hc)M^2F(M)I_1Bp_1]^{1/2}, \\ D - (q\lambda/hc)Mp_0 &= Q[\langle i_{\mathrm{a}}^2\rangle + 2q(q\lambda/hc)M^2F(M)I_1Bp_0]^{1/2}. \end{aligned}$$

Here, it has been assumed that the signal current is multiplied, on the average, by a factor M. Also, the effects of intersymbol interference have been neglected. That is, we assume that the signal-current tails of bits transmitted prior to the detection interval do not extend into that interval. Solving these equations in a manner similar to that used in Section 1, the sensitivity of a receiver employing an APD is given by

$$\eta \bar{P}_{\mathrm{APD}} = A\left(\frac{1+r}{1-r}\right)\left\{qQF(M)I_1 B\left(\frac{1+r}{1-r}\right) + \left[(2qQF(M)I_1 B)^2 \frac{r}{(1-r)^2} + \frac{\langle i_{\mathrm{a}}^2\rangle}{M^2}\right]^{1/2}\right\}. \tag{12}$$

In the limit of perfect source extinction during the transmission of a space ($r = 0$), Eq. (12) simplifies to

$$\eta \bar{P}_{\mathrm{APD}} = A[(\langle i_{\mathrm{a}}^2\rangle^{1/2}/M) + qQF(M)I_1 B]. \tag{13}$$

For the case of $M = 1$, the second term is negligible, and Eq. (13) reduces to the expression for $\eta \bar{P}_{p\text{-}i\text{-}n}$ [Eq. (10)], as expected.

To solve Eq. (13), we use the approximate expression of McIntyre (1966) for the excess-noise factor. Thus,

$$F(M) = M\{1 - (1-k)[(M-1)/M]^2\}, \tag{14}$$

where $k \leq 1$ is the ratio of ionization coefficients of the charge carriers, and it is assumed that multiplication is initiated by the most ionizing carrier. Since the excess-noise factor is an increasing function of the average multiplication M, we infer from Eq. (13) that the signal power required to achieve a given receiver sensitivity ($\eta \bar{P}_{\mathrm{APD}}$) decreases in proportion to the avalanche gain, provided that the gain is small. However, an increase in M results in an increase in $F(M)$, thereby increasing the second term in the expression for $\eta \bar{P}_{\mathrm{APD}}$. When $M = M_{\mathrm{opt}}$ (where M_{opt} is the optimum value of the avalanche gain), $\eta \bar{P}_{\mathrm{APD}}$ is minimized, thereby yielding the highest attainable receiver sensitivity for a given k value and amplifier noise current. As the gain is increased beyond M_{opt}, the second term in Eq. (13) becomes dominant, and $\eta \bar{P}_{\mathrm{APD}}$ begins to increase once again.

The dependence of $\eta \bar{P}_{\mathrm{APD}}$ on M for several values of k and for zero APD dark current ($I_{\mathrm{DM}} = 0$) is illustrated in Fig. 2. In calculating the curves, we have taken $\langle i_{\mathrm{a}}^2\rangle = 3 \times 10^{-15}\ \mathrm{A}^2$, $B = 500$ Mbit/sec, and $I_1 = 0.5$, which is characteristic of rectangular input pulses (Personick, 1973). Also, the values of k shown span the range typical of the most commonly employed photodetector materials, where $k \simeq 1$ for Ge, $k \simeq 0.5$ for InP and related compounds (Umebu *et al.*, 1980; Takanashi and Horikoshi, 1981), and $k \simeq 0.025$ for Si (Brain, 1978). As discussed earlier, $\eta \bar{P}_{\mathrm{APD}}$ has a minimum at $M = M_{\mathrm{opt}}$ (indicated in parentheses for the several k values shown) that increases with decreasing k. This in turn results in a decrease in the minimum detectable power of the receiver. Thus, for $I_{\mathrm{DM}} = 0$, $\eta \bar{P}_{\mathrm{APD}}$ is strongly dependent on the k value, implying that efforts at choosing material systems and device struc-

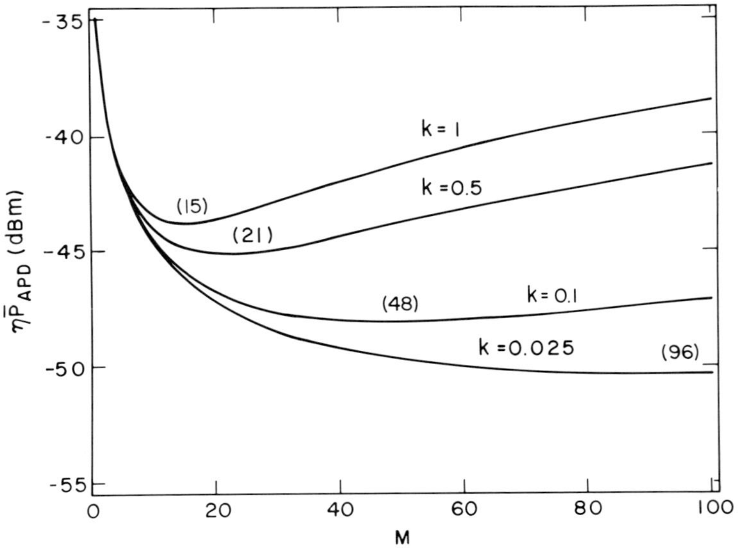

FIG. 2. APD receiver sensitivity versus gain at $B = 500$ Mbit/sec and $\langle i_a^2 \rangle = 3 \times 10^{-15}$ A^2; values of optimum gain are indicated in parentheses. Parameters are BER $= 10^{-9}$, $\lambda = 1.3$ μm, $I_{DM} = 0$.

tures that minimize k will lead to improved receiver performance. For practical detectors, however, the sensitivity improvement obtained by a decrease in k is rapidly lost with increasing APD dark current. The dependence of receiver sensitivity on I_{DM} is discussed further in Section 6.

III. Receiver Noise Current

In Part II, the sensitivity of digital receivers was derived in terms of the total noise current, $\langle i_a^2 \rangle^{1/2}$. In Part III, we briefly discuss several of the most important sources of noise in both $p-i-n$ and APD receivers. In later sections, the sensitivity obtained using an APD is normalized to that of a $p-i-n$ wherever practical, thereby avoiding a discussion of the details that govern receiver sensitivity. Nevertheless, the results obtained here will serve to establish a reference to the practical range of sensitivities achievable using state-of-the-art optical receivers in conjunction with low-noise long-wavelength photodiodes.

3. $p-i-n$ Noise Current

In Section 3, we consider the various noise-current sources in high-impedance and transimpedance amplifiers employing a $p-i-n$ photodiode in conjunction with an FET in the first stage of amplification. The discussion is confined to FET front ends since they are the most commonly used in low-noise optical receivers for moderate bit-rate applications. It has been shown (Ogawa and Campbell, 1982; Yamada and Kimura, 1982) that bipolar front ends afford greater sensitivity than FET front ends only at high bit rates (> 500 Mbit/sec), although the advantage is small even at the highest practical bit rates considered (> 1 Gbit/sec). In addition, when comparing high-frequency devices, bipolar transistor response is often slow compared to FETs, thereby placing further restrictions on their usefulness at high bit rates. Since we are interested here in estimating $\eta \overline{P}_{p-i-n}$ for later comparison with the sensitivity of APD receivers, the discussion is simplified by considering only FET amplifiers, following the treatment of Smith and Personick (1979).

In Fig. 3a is shown a functional block diagram of the photodiode and receiver circuit. The receiver amplifier consists of a preamplifier followed by an equalizer. The equalizer is used to reshape the input pulse, which is distorted by the transmitter, transmission medium, photodiode, and the preamplifer circuit. Following the equalizer is the postamplifier, and finally we show the filter, which is used to limit the total receiver bandwidth and therefore the noise. The transfer function of each circuit element is indicated within the functional blocks in Fig. 3a. In the treatment that follows, only those noise sources arising in the preamplifier are considered. In practice, however, noise generated in the postamplifier and equalizer can affect the sensitivity of low-noise circuits.

A schematic diagram of a high-impedance FET front end, along with its parallel equivalent circuit is shown in Fig. 3b. Here, R_L indicates the load resistor used in biasing the photodiode. The total receiver input resistance R_T therefore consists of the parallel combination of the amplifier resistance R_a plus R_L, and contributes an equivalent Johnson noise current of

$$\langle i_R^2 \rangle = (4kT/R_T)I_2B. \tag{15}$$

Here, kT is the Boltzmann energy at temperature T; B is the bit rate; and I_2 is an integral (Personick, 1973) that depends on the transfer function of the circuit. For rectangular input pulses that fill the bit time slot of duration $1/B$, and assuming a raised cosine output pulse, we have $I_2 \simeq 0.55$. Evaluation of the noise current for other pulse shapes is discussed in Section 12. In most high-impedance amplifiers, R_L is the dominant source of resistive noise such that we can make the approximation $R_T = R_L$.

The shot noise due to leakage current at the input is a second source of amplifier noise. Now the total input current can be written

$$I_T(t) = I_g + I_D + I_s(t), \tag{16}$$

where I_g is the sum of the FET gate leakage and other shunt-current sources;

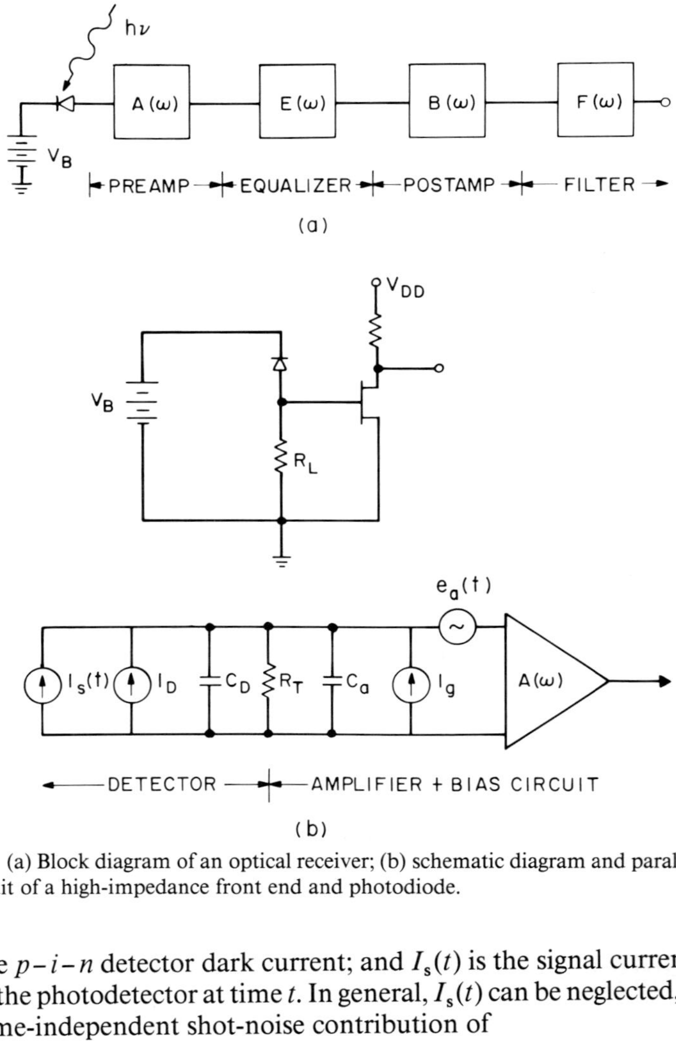

FIG. 3. (a) Block diagram of an optical receiver; (b) schematic diagram and parallel equivalent circuit of a high-impedance front end and photodiode.

I_D is the p–i–n detector dark current; and $I_s(t)$ is the signal current generated in the photodetector at time t. In general, $I_s(t)$ can be neglected, giving a total time-independent shot-noise contribution of

$$\langle i_T^2 \rangle = 2q(I_g + I_D)I_2 B. \tag{17}$$

At the output of the FET, the channel conductance gives rise to a thermally generated noise current. Referring this current to the input, we obtain a noise voltage per unit bandwidth of $\langle e_a^2 \rangle^{1/2}$, where

$$\langle e_a^2 \rangle = 4kT\Gamma/g_m. \tag{18}$$

Here, Γ is the excess-noise factor of the FET (Ogawa, 1981) and g_m the transconductance measured at the amplifier operating point. To write Eq. (18) in terms of current noise referred to the input, we use the input admittance $Y_{\text{in}} = (1/R_{\text{L}}) + j\omega C_{\text{T}}$. Here, C_{T} is the total capacitance of the front end given by

$$C_{\text{T}} = C_{\text{D}} + C_{\text{a}} = C_{\text{D}} + C_{\text{gs}} + C_{\text{gd}} + C_{\text{s}}, \tag{19}$$

where C_{D} is the photodiode capacitance and C_{a} the total preamplifier capacitance. Thus, C_{a} consists of the parallel combination of the stray capacitance due to packages and wire bonds to the devices C_{s} and the FET gate-source C_{gs} and gate-drain C_{gd} capacitances. It can be shown (Smith and Personick, 1979) that Eq. (18) leads to a channel-noise current of

$$\langle i_{\text{c}}^2 \rangle = \frac{4kT\Gamma}{g_{\text{m}}} \left[\frac{I_2 B}{R_{\text{L}}^2} + (2\pi C_{\text{T}})^2 I_3 B^3 \right]. \tag{20}$$

Here, I_3 is a Personick integral whose value depends on the frequency response of the linear channel. Typically, I_3 has a value of approximately 0.085, where once again we assume rectangular input pulses and a raised cosine output-pulse shape.

Finally, the contribution to the noise current due to $1/f$ noise in the front end FET can be approximately modeled using

$$\langle i_f^2 \rangle = (4kT\Gamma/g_{\text{m}})(2\pi C_{\text{T}})^2 f_{\text{c}} I_{\text{c}} B^2, \tag{21}$$

where f_{c} is the $1/f$ noise corner frequency, and $I_{\text{c}} \simeq 0.12$ (Chen, 1978; Forrest, 1985).

The total noise current $\langle i_{\text{a}}^2 \rangle$ of the amplifier is the sum of the currents previously given. Thus,

$$\langle i_{\text{a}}^2 \rangle_{p\text{-}i\text{-}n} = \langle i_{\text{R}}^2 \rangle + \langle i_{\text{T}}^2 \rangle + \langle i_{\text{c}}^2 \rangle + \langle i_f^2 \rangle + \langle i_{\text{p}}^2 \rangle, \tag{22}$$

where we have also added the noise-current contribution due to the postamplifier and other circuits $\langle i_{\text{p}}^2 \rangle$ following the first stage of amplification.

The previous discussion describes the noise of high-impedance (HZ) front ends. Another commonly used receiver configuration that is slightly less sensitive but has increased dynamic range is the transimpedance (TZ) amplifier. The transfer function of the TZ amplifier is simply

$$v_{\text{o}} = -Z_{\text{F}} i_{\text{in}}, \tag{23}$$

where v_{o} is the output voltage generated by the input current i_{in} and Z_{F} the feedback impedance. This impedance is usually due to the parallel combination of a feedback capacitance C_{F} and a resistor R_{F} through which the diode is biased. The noise current of a transimpedance amplifier can be calculated using the treatment given previously by replacing R_{L} with R_{F} in Eqs. (15) and (20). Note that the frequency response of a TZ amplifier is approximately

TABLE I

REPRESENTATIVE COMPONENT VALUES (AT 20°C) FOR LONG-WAVELENGTH $p-i-n$/GaAs FET TRANSIMPEDANCE AMPLIFIERS

Parameter	Symbol and unit	Value	Parameter	Symbol and unit	Value
Feedback resistance	R_F (Ω)	10^6-10^3	FET noise factor	Γ	1.5
FET gate leakage	I_g (nA)	15	Total preamplifier capacitance	$C_{gs} + C_{gd} + C_s$ (pF)	1.0
$p-i-n$ dark current	I_D (nA)	85	$p-i-n$ capacitance	C_D (pF)	0.5
FET transconductance	g_m (msec)	35	FET $1/f$ noise-corner frequency	f_c (MHz)	25

limited by $R_F C_F$, whereas $R_L C_T$ limits the HZ receiver bandwidth. Since C_F can be made small ($\lesssim$0.1 pF) compared with C_T (~2 pF), the frequency response, and therefore the dynamic range of the TZ amplifier, is improved. On the other hand, since the maximum closed-loop gain cannot exceed the open-loop gain of the amplifier, in practice R_F is smaller than R_L. This leads to an increase in the noise of the TZ over the HZ front end.

Table I gives several of the component values needed in calculating the noise terms at room temperature in Eqs. (15)–(21). The values given are typical of GaAs FETs and $In_{0.53}Ga_{0.47}As$ $p-i-n$ photodiodes used in sensitive TZ and HZ optical receivers in the bit-rate range 50 Mbit/sec $\leq B \leq$ 2 Gbit/sec. Note that a range of values is given for the feedback resistor. As the bit rate is increased, it is necessary to decrease R_F to obtain the required amplifier frequency response, which we recall is approximately determined by $R_F C_F$ for TZ amplifiers and by $R_L C_T$ for HZ circuits.

Figure 4 illustrates the contribution of each noise term in Eq. (22) (assuming $\langle i_p^2 \rangle = 0$) using the component values listed in Table I. We see that at low bit rates, $\langle i_a^2 \rangle_{p-i-n}$ is dominated by shot noise due to FET and $p-i-n$ leakage currents. However, over most of the bit-rate range of interest, the channel noise $\langle i_c^2 \rangle$ dominates. Since this term is determined by the rato C_T^2/g_m [Eq. (20)], it is extremely important in constructing low-noise high-bit-rate optical receivers to minimize the FET and $p-i-n$ capacitance while maximizing the transconductance of the FET. For these reasons, GaAs FETs with gate lengths of ~1 μm (and therefore high g_m) employed in conjunction with small-area (low C_D) $In_{0.53}Ga_{0.47}As$ $p-i-n$ photodiodes have been extensively used for optical receiver applications.

Also indicated in Fig. 4 is the noise contribution from the feedback resistor R_F. Once again this term is only appreciable at the low end of the frequency spectrum but becomes increasingly insignificant at higher bit rates in well-

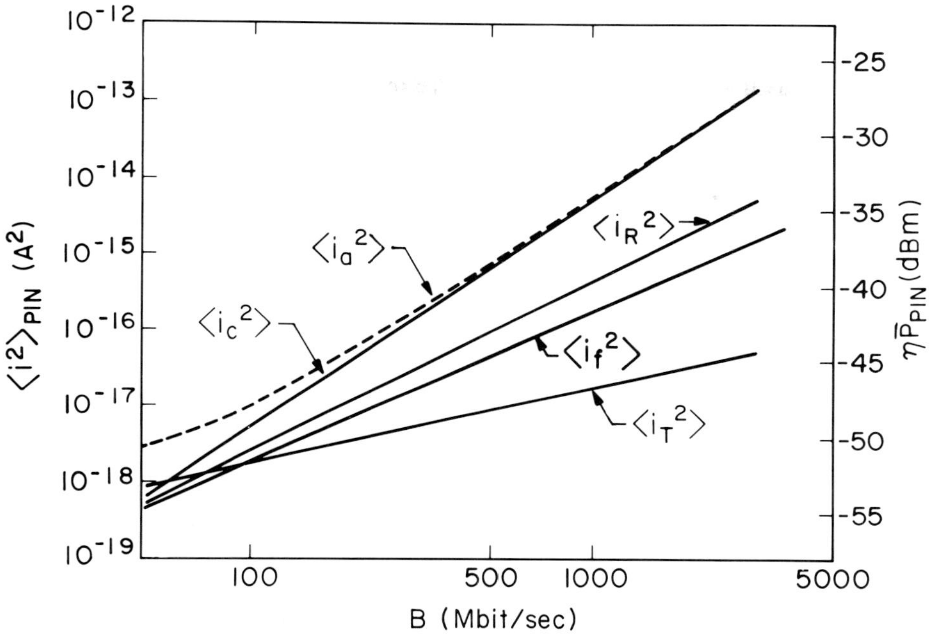

FIG. 4. Noise-current components and $\eta \bar{P}_{p-i-n}$ versus bit rate for low-noise optical receivers with $T = 20°C$.

designed receivers. Finally, the $1/f$ noise term is also significant only at the lowest bit rates considered. In Fig. 4, the bit-rate dependences of the various noise terms are readily apparent. The B^3 dependence of the channel noise quickly dominates the linear dependence of $\langle i_T^2 \rangle$ on B. Note also that $\langle i_R^2 \rangle$ is roughly proportional to B^2, which is in apparent contradiction to the behavior given in Eq. (15), where we have written $\langle i_R^2 \rangle \propto B/R_F$. As indicated earlier, however, R_F must be reduced at high bit rates to allow for the required increase in amplifier bandwidth. In practical designs, $R_F \underset{\sim}{\propto} 1/B$, which leads to the dependence of $\langle i_R^2 \rangle$ on B^2 shown in Fig. 4.

The dashed curve indicates the total noise current $\langle i_a^2 \rangle_{p-i-n}$, as defined in Eq. (22). The receiver sensitivity $\eta \bar{P}_{p-i-n}$ can then be found via Eq. (10). Thus, using the curve for $\langle i_a^2 \rangle$ in conjunction with the scale on the right-hand ordinate of Fig. 4 gives $\eta \bar{P}_{p-i-n}$ at $\lambda = 1.3\ \mu m$ and BER $= 10^{-9}$. The receiver sensitivity is observed to vary from -50 dBm at 50 Mbit/sec to -30 dBm at 2 Gbit/sec using the component values listed in Table I.

In Fig. 5a, we plot $\eta \bar{P}_{p-i-n}$ versus B for several values of the total shunt-leakage current I_T, where other values used in the calculation are taken from Table I. Here, the sensitivity of $\eta \bar{P}_{p-i-n}$ to dark current is only significant at

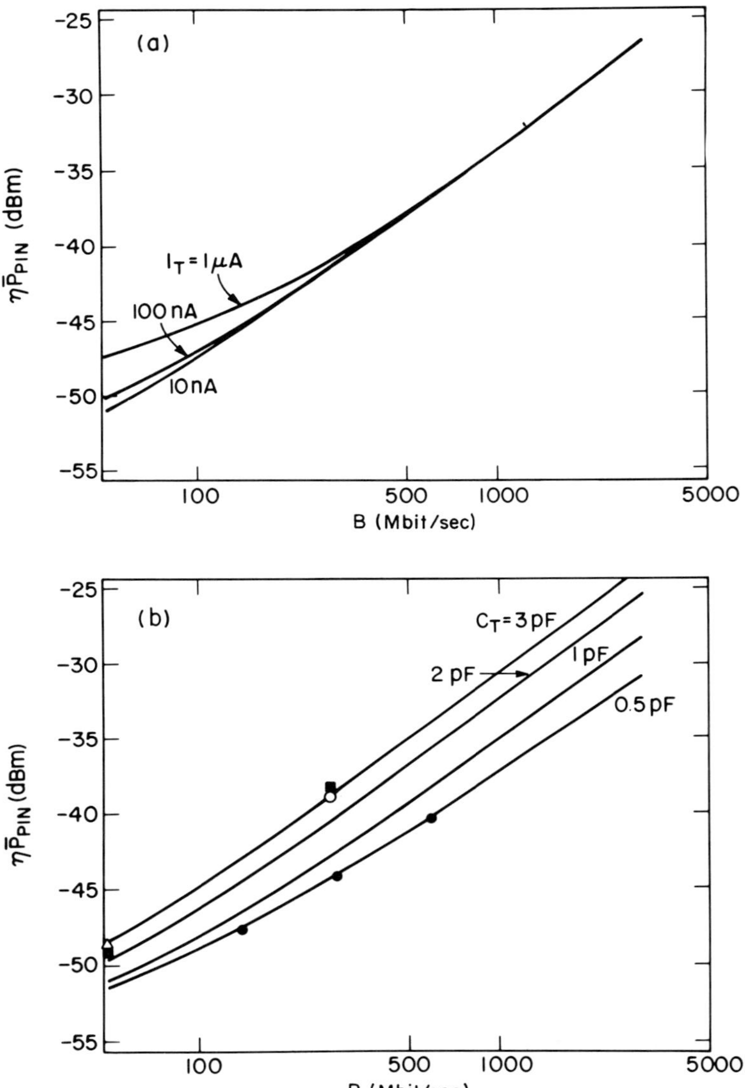

FIG. 5. *p–i–n* Receiver sensitivity versus bit rate and dark current at $T = 20°C$: (a) $C_T = 1.5$ pF; (b) C_T as parameter and $I_T = 100$ nA; experimental results are indicated by data points.

low bit rates (c.f., Fig. 4). For example, an increase in I_T from 100 nA to 1 μA results in a 3-dB degradation in receiver sensitivity at 50 Mbit/sec, whereas this increase in I_T results in a negligible loss in sensitivity at 500 Mbit/sec.

In contrast, Fig. 5b illustrates that the receiver sensitivity depends strongly on front-end capacitance C_T at all bit rates of interest. However, this sensitivity is most pronounced at high bit rates due to the B^3 dependence of the channel-noise term. Thus, at 2 Gbit/sec, roughly 6 dB of sensitivity is lost for $C_T = 3$ pF, compared with $C_T = 0.5$ pF, whereas at 50 Mbit/sec, a similar decrease in C_T results in only an $\sim$ 3-dB sensitivity penalty.

The data points in Fig. 5a indicate experimental results for $In_{0.53}Ga_{0.47}As$ $p-i-n$ photodetector receivers discussed in the literature. The solid circles correspond to data obtained using an HZ GaAs FET front end with a total input capacitance of $C_T = 0.45$ pF (Smith *et al.*, 1982). These data are in excellent agreement with calculation and are among the highest receiver sensitivities that have been achieved to date. Two other measurements made at $B = 45$ Mbit/sec that agree well with the calculations are shown in Fig. 5b. Here, the datum represented by the triangle was obtained using a TZ GaAs FET receiver with $C_T \simeq 2.5$ pF (Forrest *et al.*, 1981). The adjacent solid square indicates data obtained with a receiver similar to that used by Forrest and co-workers, although the input capacitance was slightly lower, at 2 pF (Lee *et al.*, 1980). In the latter reference, results were also obtained at $B = 274$ Mbit/sec with $C_T = 2.5$ pF. At this bit rate, the sensitivity measured was $\sim$ 1 dB worse than the calculated value. The difference may be due to additional noise introduced by the postamplifier (typically, $\sim$0.5 dB) and by nonoptimum equalization of the integrated pulse shape. Also, for comparison with the GaAs FET experiments, a result for a Si bipolar amplifier at 274 Mbit/sec is indicated by the open circle (Boenke *et al.*, 1982).

In all cases, the calculated sensitivities are accurate to within 0.5 to 1 dB when compared with the highest sensitivity amplifiers reported in the literature. The discrepancies are most probably due to additional noise introduced by the postamplifier and by nonoptimum equalization. This latter effect may be expected to become increasingly severe at high bit rates, as is discussed in Part VI.

4. APD Noise Current

The noise current of an optical receiver that employs an APD is calculated in a manner similar to that used in Section 3 for a $p-i-n$-detector receiver. In the case of an APD, however, it is assumed that the shot noise is multiplied by the same avalanche gain as the photocurrent M and is therefore an additional source of excess noise. In this case, Eq. (17) can be replaced with (Smith and Forrest, 1982)

$$\langle i_T^2 \rangle_{APD} = 2q(I_{DU} + M^2 F(M) I_{DM}) I_2 B, \qquad (24)$$

where all the unmultiplied dark-current sources from both the APD and the FET (including I_g) have been included in a single term I_{DU}, and the primary dark current that undergoes multiplication is equal to I_{DM}. Since the first term in Eq. (24) is simply the unmultiplied receiver shunt-current noise, the total noise current for an amplifier using an APD is

$$\langle i_a^2 \rangle_{APD} = \langle i_a^2 \rangle_{p-i-n} + 2qM^2F(M)I_{DM}I_2B. \tag{25}$$

To calculate the APD receiver sensitivity, Eq. (25) is substituted for $\langle i_a^2 \rangle$ in Eq. (12) or (13), which is then minimized to obtain $\eta\bar{P}_{APD}$ at the optimum gain. As I_{DM} increases, M_{opt} decreases from its maximum value obtained at $I_{DM} = 0$, thereby decreasing the ultimate receiver sensitivity achievable (i.e., increasing $\eta\bar{P}_{APD}$). In Section 6, we show that Eq. (25) provides a convenient means whereby the sensitivities of receivers using APDs can be directly compared with those obtained using $p-i-n$s without the necessity of considering the details of the amplifier noise sources. In this manner, we can easily estimate the improvement in receiver sensitivities obtainable using APDs with nonzero multiplied dark currents ($I_{DM} > 0$).

From Eq. (24), we infer that the contribution of the unmultiplied dark currents to the noise are in general negligible in comparison with I_{DM} due to the multiplication of the latter noise source. For example, at $M_{opt} = 20$ and $F(M) \cong M^{1/2}$ (typical of $In_{0.53}Ga_{0.47}As/InP$ APDs), a primary dark current of 1 nA contributes nearly as much to the total receiver noise as $I_{DU} = 2\ \mu A$. Thus, for most practical receivers we can ignore all unmultiplied sources of dark current (e.g., detector surface currents, gate leakage, etc.) and consider only those currents that undergo multiplication.

IV. Sensitivity Calculations

5. Negligible APD Dark Current ($I_{DM} = 0$)

To compare the sensitivity of a receiver employing an APD to one employing a $p-i-n$ diode, it is useful to express $\eta\bar{P}_{APD}$ in terms of $\eta\bar{P}_{p-i-n}$. Thus, combining Eqs. (10) and (13) we obtain (Smith and Forrest, 1982)

$$\eta\bar{P}_{APD} = (\eta\bar{P}_{p-i-n}/M) + (hc/\lambda)Q^2F(M)I_1B. \tag{26}$$

In Eq. (26), $\eta\bar{P}_{p-i-n}$ corresponds to the equivalent noise power introduced by an amplifier used in conjunction with an ideal, noiseless $p-i-n$ detector. If the ideal $p-i-n$ is replaced by an APD with negligible dark current in that same receiver, Eq. (26) would give the receiver sensitivity thus obtained. Recall also that the value chosen for $\eta\bar{P}_{p-i-n}$ depends implicitly on both Q and B [Eq. (9)]. Thus, for Eq. (26) to be a consistent comparison of sensitivities, we require that the bit rate and the error rate for both the APD and $p-i-n$ receivers be the same.

Figure 6 illustrates $\eta\bar{P}_{\mathrm{APD}}$ versus $\eta\bar{P}_{p-i-n}$ at $\lambda = 1.3\ \mu$m using several k values typical of photodiodes fabricated using materials such as Ge, InP, and Si. The numbers in parentheses indicate the optimum gain at which the corresponding receiver sensitivity is obtained. Also, the results are given in terms of decibels per 1 mW, where 0 dBm is equivalent to 1 mW of detected power.

Figure 6a shows the sensitivities obtained at $B = 50$ Mbit/sec. The sensitivity of an APD receiver is observed to increase with a decrease in the ionization coefficient ratio k. The enhanced sensitivity is a result of the reduced excess noise associated with small k values [Eq. (14)], thereby allowing an increase in the optimum gain. Although considerable sensitivity improvement may be obtained by decreasing k, the uniformity and quality of the APD must also be improved to obtain the higher values of gain without the device undergoing premature avalanche breakdown. In practice, high values of uniform gain have been difficult to achieve in InP-based compounds due to microplasmic breakdown frequently observed at crystalline-defect sites (Capasso *et al.*, 1980; Lee and Burrus, 1980; Forrest *et al.*, 1983).

As an example of the sensitivity improvement attainable using an APD, we note that at 50 Mbit/sec, state-of-the-art amplifiers using low dark-current $p-i-n$ detectors have sensitivities of $\eta\bar{P}_{p-i-n} \gtrsim -50$ dBm (cf. Fig. 5b). Thus, using an APD with $k = 1$ in a receiver with $\eta\bar{P}_{p-i-n} = -50$ dBm yields $\eta\bar{P}_{\mathrm{APD}} = -56.3$ dBm with $M_{\mathrm{opt}} = 9$, representing a 6.5-dB sensitivity improvement. However, if k is reduced to $k = 0.025$ typical of Si devices, one obtains $\eta\bar{P}_{\mathrm{APD}} = -62$ dBm with $M_{\mathrm{opt}} = 54$. Thus, a decrease in k by a factor of 40 results in an additional 5.5-dB improvement in sensitivity, leading to a total of 12 dB improvement in sensitivity over that obtained using a $p-i-n$ photodiode receiver.

In Figs. 6b–d, we show $\eta\bar{P}_{\mathrm{APD}}$ at bit rates of 500 Mbit/sec and 1 and 2 Gbit/sec. The latter bit rates represent the highest presently being considered for practical high-capacity lightwave communication systems. As shown later, receivers operating at bit rates in excess of 1 Gbit/sec may be considerably less sensitive than our calculations indicate as a result of several intrinsic component and system speed limitations. In Fig. 6, the range of $p-i-n$ receiver sensitivities plotted on the abscissa has been adjusted at each bit rate to include values typical of state-of-the-art receivers. For zero dark current, the largest improvement is obtained using the least sensitive receivers. Of course, the highest absolute sensitivities are still obtained when an APD is used along with the best receiver available, even though the relative improvement may be less than if the APD were used with an inferior receiver. Indeed, one of the most compelling justifications for employing APDs is to enable one to relax the circuit-design criteria of the front end. In this manner, inexpensive, although inferior, receivers can give acceptable performance when used in conjunction with an avalanche detector.

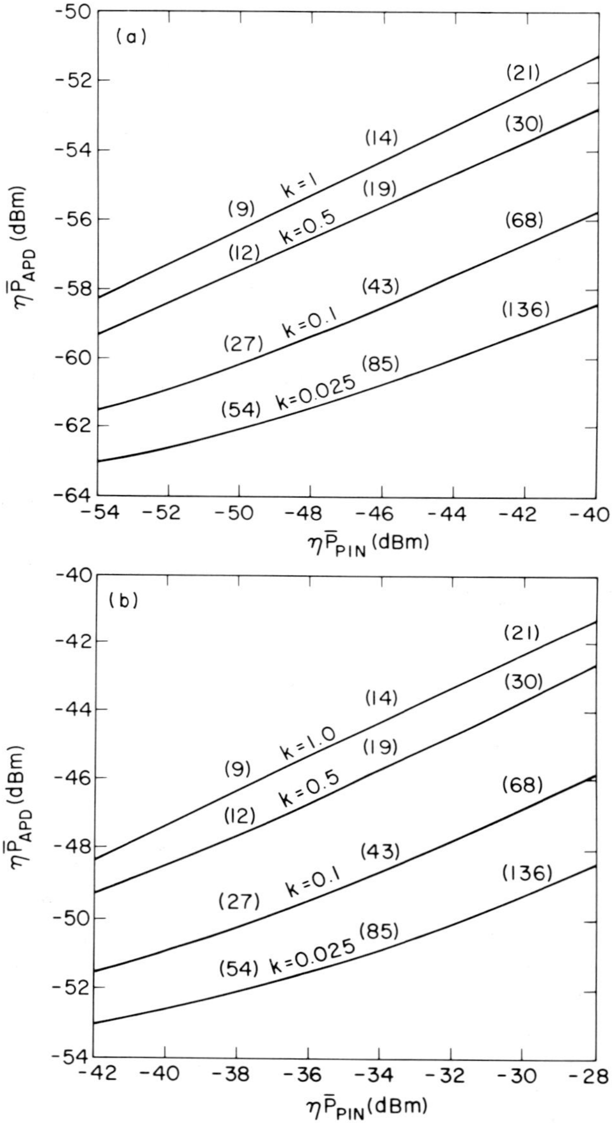

FIG. 6. APD receiver sensitivity versus $p-i-n$ receiver sensitivity for zero APD dark current ($I_{DM} = 0$) with BER $= 10^{-9}$, $\lambda = 1.3$ μm: (a) $B = 50$ Mbit/sec, (b) $B = 500$ Mbit/sec, (c) $B = 1$ Gbit/sec, and (d) $B = 2$ Gbit/sec. Numbers in parentheses indicate optimum gains yielding the corresponding value of APD sensitivity; data are shown for several values of the ionization-coefficient ratio k; dashed line indicates the boundary of the gain–bandwidth-limited region. *(Continued.)*

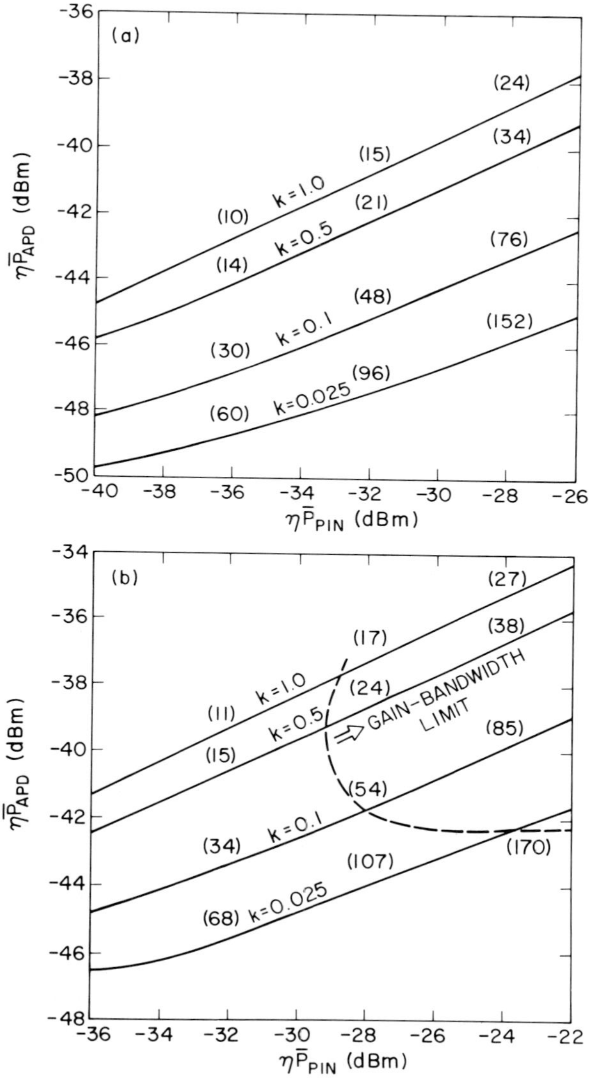

FIG. 6. *(Continued)*

Finally, in Fig. 6d is shown the approximate gain-bandwidth-limit boundary for $In_{0.53}Ga_{0.47}As/InP$ heterostructure APDs—devices that are currently considered to be the most promising avalanche detectors for use in long-wavelength communication systems (Nishida *et al.*, 1979). The gain–bandwidth limit indicates the bit rate at which the ac gain is $1/\sqrt{2}$ of its dc

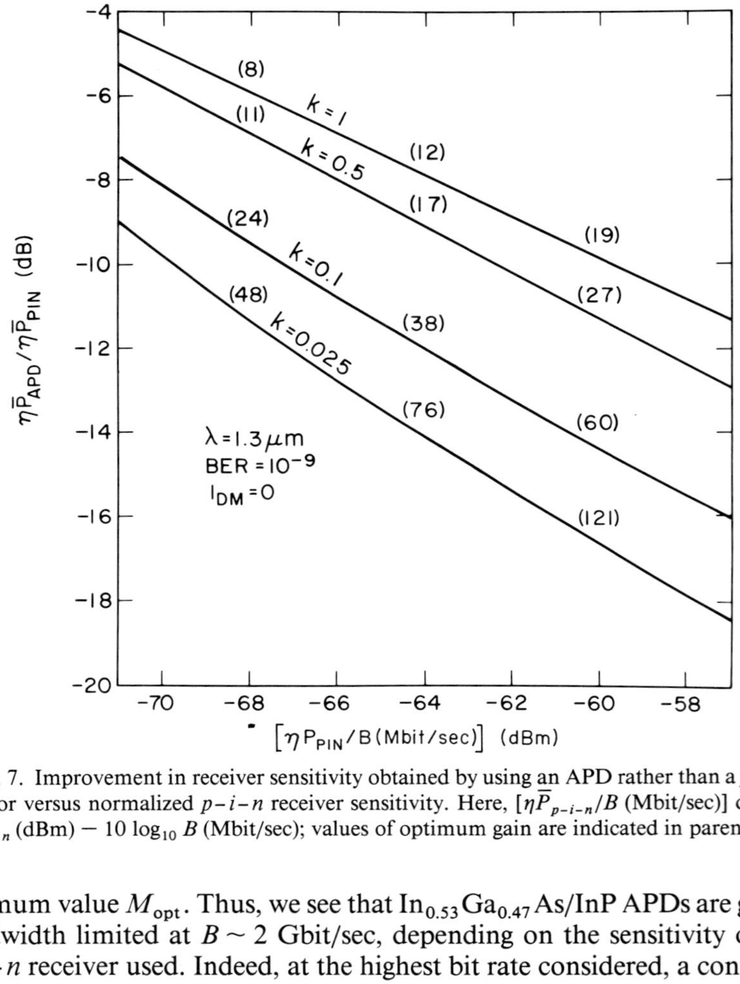

FIG. 7. Improvement in receiver sensitivity obtained by using an APD rather than a $p-i-n$ detector versus normalized $p-i-n$ receiver sensitivity. Here, $[\eta \bar{P}_{p-i-n}/B$ (Mbit/sec)] dBm = $\eta \bar{P}_{p-i-n}$ (dBm) − 10 $\log_{10} B$ (Mbit/sec); values of optimum gain are indicated in parentheses.

optimum value M_{opt}. Thus, we see that $In_{0.53}Ga_{0.47}As$/InP APDs are gain–bandwidth limited at $B \sim 2$ Gbit/sec, depending on the sensitivity of the $p-i-n$ receiver used. Indeed, at the highest bit rate considered, a considerable portion of the plot falls within the gain–bandwidth limit, indicating that the ideal values calculated for $\eta \bar{P}_{APD}$ cannot in practice be achieved. The APD gain–bandwidth limit is discussed in greater detail in Section 11.

The curves in Fig. 6 can be replaced by a single, normalized curve independent of B. Figure 7 illustrates the sensitivity improvement (in decibels) obtained using an APD instead of a $p-i-n$-photodiode versus $p-i-n$-receiver sensitivity normalized to the bit rate. Once again, curves for several k values are given, and the values of M_{opt} are listed in parentheses. From Fig. 7 we see that between 4 and 18 dB of improvement in sensitivity can be obtained using an APD, depending on the various parameters chosen for the normalized sensitivity, $\eta\bar{P}_{p-i-n}/B$, and k. As an example of the use of this plot, we take $B = 100$ Mbit/sec, and assume a typical receiver sensitivity obtained with a $p-i-n$ detector of $\eta\bar{P}_{p-i-n} = -46$ dBm at $T = 20°C$ (c.f. Fig. 5). Thus $[\eta\bar{P}/B(\text{Mbit/sec})]$ (dBm) $= -46 - 10\log_{10} 100 = -66$ dBm. If the $p-i-n$ detector is replaced with an $In_{0.53}Ga_{0.47}As$/InP heterostructure APD where $k = 0.5$ (Umebu *et al.*, 1980; Forrest *et al.*, 1981), we calculate a 7.5-dB improvement at $M_{opt} \simeq 13$. This gives $\eta\bar{P}_{APD} = -53.5$ dBm, which represents a considerable sensitivity improvement for lightwave systems.

6. Nonzero APD Dark Current ($I_{DM} > 0$)

For the case of $I_{DM} > 0$, we substitute the expression for $\langle i_a^2 \rangle_{APD}^{1/2}$ [Eq. (25)] into Eq. (13) to obtain

$$\eta\bar{P}_{APD} = [(\eta\bar{P}_{p-i-n}/M)^2 + 2q(hc/q\lambda)^2 Q^2 F(M) I_{DM} I_2 B]^{1/2} + (hc/\lambda) Q^2 F(M) I_1 B. \tag{27}$$

Here, perfect source extinction ($r = 0$) during the transmission of a space has been assumed. This expression differs from that obtained for $I_{DM} = 0$ by the addition of the second term in brackets. Figure 8 illustrates the dependence of $\eta\bar{P}_{APD}$ on I_{DM} for several k values and bit rates of interest. At each bit rate we have assumed a $p-i-n$ receiver sensitivity that corresponds to state-of-the-art values shown in Fig. 5a, taking $I_T = I_{DU} = 100$ nA. The values obtained for $\eta\bar{P}_{APD}$ are therefore conservative estimates of the advantage gained using an APD, and one can expect even greater sensitivity improvement using less sensitive receivers.

Thus, for example, Fig. 8a is a plot of $\eta\bar{P}_{APD}$ at $B = 50$ Mbit/sec, assuming that $\eta\bar{P}_{p-i-n} = -50$ dBm. The numbers in parentheses correspond to the optimum gains required to obtain a given value of $\eta\bar{P}_{APD}$. Also, the arrows on the ordinate correspond to minimum values of $\eta\bar{P}_{APD}$ achievable in the limit of $I_{DM} = 0$ for each k value considered (c.f., Fig. 6a). From this plot, we see the following:

(i) M_{opt} is a decreasing function of primary dark current I_{DM}. Thus, an increase in I_{DM} results in a decrease in APD receiver sensitivity. For example, assuming that $k = 0.5$, the sensitivity advantage obtained using an APD rather than a $p-i-n$ detector drops by ~3.5 dB for an increase in I_{DM} from 0.1 to 10 nA. Indeed, the advantage completely disappears at $I_{DM} \geq 100$ nA.

(ii) The sensitivity obtained with an APD is greatest for small k values only as $I_{DM} \to 0$. This advantage is rapidly lost as I_{DM} increases above 1 nA at $B = 50$ Mbit/sec.

We emphasize that these conclusions are valid only when APDs are employed in conjunction with very high sensitivity receivers. As pointed out

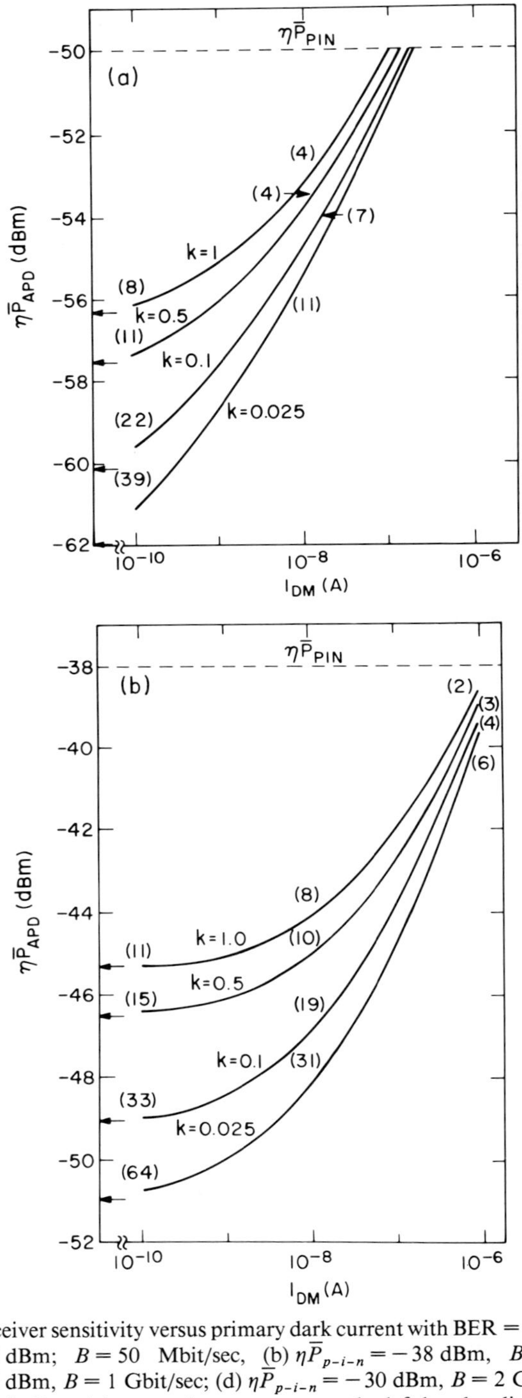

FIG. 8. APD receiver sensitivity versus primary dark current with BER $= 10^{-9}$, $\lambda = 1.3\ \mu$m: (a) $\eta\overline{P}_{p-i-n} = -50$ dBm; $B = 50$ Mbit/sec, (b) $\eta\overline{P}_{p-i-n} = -38$ dBm, $B = 500$ Mbit/sec; (c) $\eta\overline{P}_{p-i-n} = -34$ dBm, $B = 1$ Gbit/sec; (d) $\eta\overline{P}_{p-i-n} = -30$ dBm, $B = 2$ Gbit/sec. Values of optimum gain are indicated in parentheses; arrows on the left-hand ordinate correspond to values of $\eta\overline{P}_{\text{APD}}$ extrapolated to $I_{\text{DM}} = 0$ for the given k value. *(Continued.)*

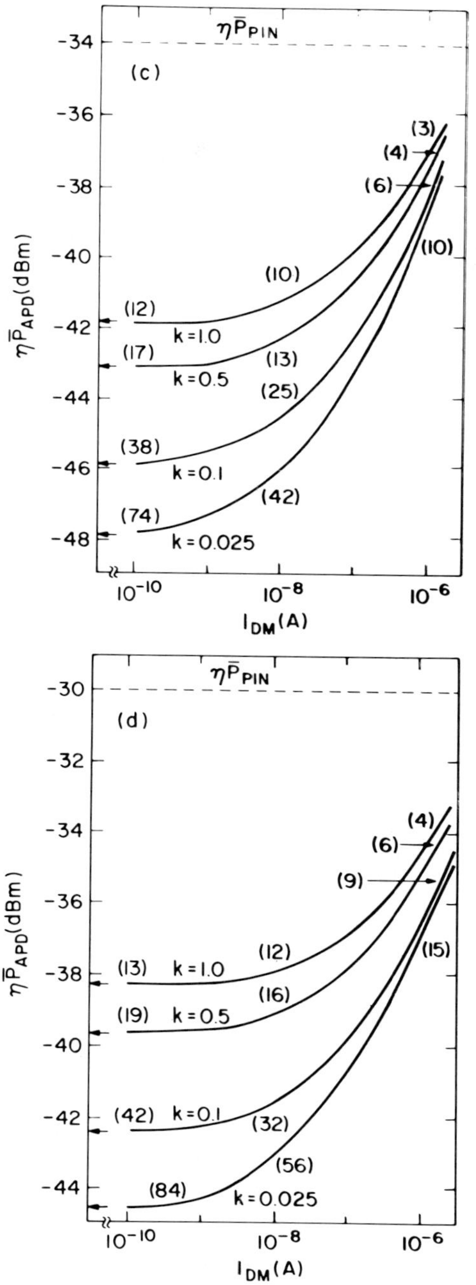

FIG. 8. *(Continued)*

earlier, however, APDs can be advantageously employed with inferior receivers. In this case, detectors with dark currents higher than 10 nA can result in large receiver sensitivity improvements over what can be achieved with a $p-i-n$ photodiode.

In Figs. 8b–d, we plot $\eta\bar{P}_{\mathrm{APD}}$ versus I_{DM} at $B = 500$ Mbit/sec and 1 and 2 Gbit/sec. The effects of I_{DM} on $\eta\bar{P}_{\mathrm{APD}}$ are seen to decrease with increasing bit rate. For example, at 50 Mbit/sec, $\eta\bar{P}_{\mathrm{APD}}$ is strongly dependent on I_{DM} for $I_{\mathrm{DM}} > 0.1$ nA; whereas at 2 Gbit/sec, $\eta\bar{P}_{\mathrm{APD}}$ is relatively unchanged until $I_{\mathrm{DM}} \gtrsim 10$ nA. This is due to the dominance of the signal-current noise term $[(hc/\lambda)Q^2F(M)I_1B$, Eq. (27)] over the multiplied dark-current shot noise at high bit rates.

From Fig. 8 it is also apparent that M_{opt} increases with increasing bit rate for a given value of primary dark current. This leads to APD receivers whose ultimate sensitivity at high bit rates is limited by the speed of response of the detector. However, we note that for the receiver sensitivities considered, the $\mathrm{In_{0.53}Ga_{0.47}As/InP}$ APDs are not gain-bandwidth limited at any bit rate shown. Indeed, since $I_{\mathrm{DM}} > 0$ reduces M_{opt}, we expect that detectors with finite dark currents are less sensitive to gain-bandwidth response-speed limitations than those with $I_{\mathrm{DM}} \to 0$.

In a manner similar to that used to calculate receiver sensitivity at $I_{\mathrm{DM}} = 0$, we can now obtain a single normalized plot for $\eta\bar{P}_{\mathrm{APD}}/\eta\bar{P}_{p-i-n}$ independent of bit rate. This is illustrated in Fig. 9, where $\eta\bar{P}_{\mathrm{APD}}/\eta\bar{P}_{p-i-n}$ is plotted versus normalized dark current (I_{DM} (nA)/B (Mbit/sec)) for several values of the ionization-rate ratio. For this plot we have assumed $\eta\bar{P}_{p-i-n}/B$ (Mbit/sec) $= -67$ dBm, a value typical of the best $p-i-n$ receivers available today. Also shown in parentheses are the values of M_{opt} needed to obtain a given receiver sensitivity. Once again, we observe that significant sensitivity advantage can be obtained at low k values only if the dark current is small. Also, the sensitivity improvement is less strongly dependent on I_{DM} at high rather than at low bit rates, although the APD gain may be bandwidth limited.

To estimate the amount of primary dark current that can be tolerated, we calculate the normalized dark current (I_{DM}/B) at which the sensitivity of an APD receiver is decreased by 1 dB from its value at $I_{\mathrm{DM}} = 0$. The magnitude of this current is calculated by solving Eq. (27) for the value of I_{DM}, which results in an increase in $\eta\bar{P}_{\mathrm{APD}}$ by 26% (1 dB) above the minimum value obtained at $I_{\mathrm{DM}} = 0$. The values of I_{DM} thus obtained can be approximately determined for various values of k and $\eta\bar{P}_{p-i-n}/B$ using the nomogram in Fig. 10. A sample calculation is indicated by the dashed line. In this example, assuming $\eta\bar{P}_{p-i-n} = -47$ dBm at 100 Mbit/sec, we get $[\eta\bar{P}_{p-i-n}/B$ (Mbit/sec)] (dBm) $= -47 - 10\ \log_{10}(100) = -67$ dBm. Using an APD with $k = 0.1$, the normalized dark current yielding a 1-dB degradation in receiver sensitivity is $I_{\mathrm{DM}}/B = 5.5 \times 10^{-3}$ nA/(Mbit/sec), giving $I_{\mathrm{DM}} \simeq 0.55$ nA.

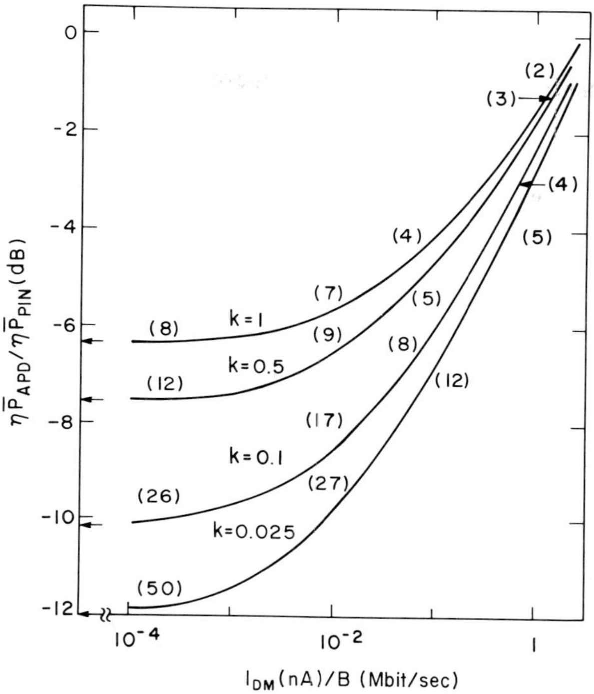

FIG. 9. Improvement in receiver sensitivity with BER $= 10^{-9}$, $\lambda = 1.3\ \mu$m obtained by using an APD rather than a $p-i-n$ detector versus normalized dark current for several k values. Here, $[\eta\bar{P}_{p-i-n}/B$ (Mbit/sec)$] = \eta\bar{P}_{p-i-n}$ (dBm) $- 10\log_{10} B$ (Mbit/sec) $= -67$ dBm; values of M_{opt} are indicated in parentheses, and values of $\eta\bar{P}_{APD}/\eta\bar{P}_{p-i-n}$ extrapolated to $I_{DM}/B = 0$ are indicated by arrows on the left-hand ordinate.

In Table II are listed values of I_{DM} that result in a 1-dB degradation of APD receiver sensitivity for several k values and bit rates of interest. In Table II, $p-i-n$ receiver sensitivities comparable to the best achieved to date are assumed. If the $p-i-n$ receiver sensitivity is decreased, the allowable dark currents listed will increase, relaxing somewhat the stringent conditions placed on the APD leakage current. Note that these values of I_{DM} must be achieved at the highest system operating temperature. However, the dark currents in both $In_{0.53}Ga_{0.47}As$/InP and Ge APDs are thermally activated and are roughly an order of magnitude higher at $T = 70°C$ than at 20°C (Section 9). Typically $T = 70°C$ is the highest acceptable temperature for most transmission-system applications. Thus, the values of I_{DM} that must be realized at room temperature such that $\eta\bar{P}_{APD}$ undergoes only a 1-dB penalty over the entire system operating range are ~ 10 times smaller than those listed in Table II. As is discussed in the Sections 7 and 8, such low values of I_{DM} are extremely difficult to achieve in practical $In_{0.53}Ga_{0.47}As$/InP APDs,

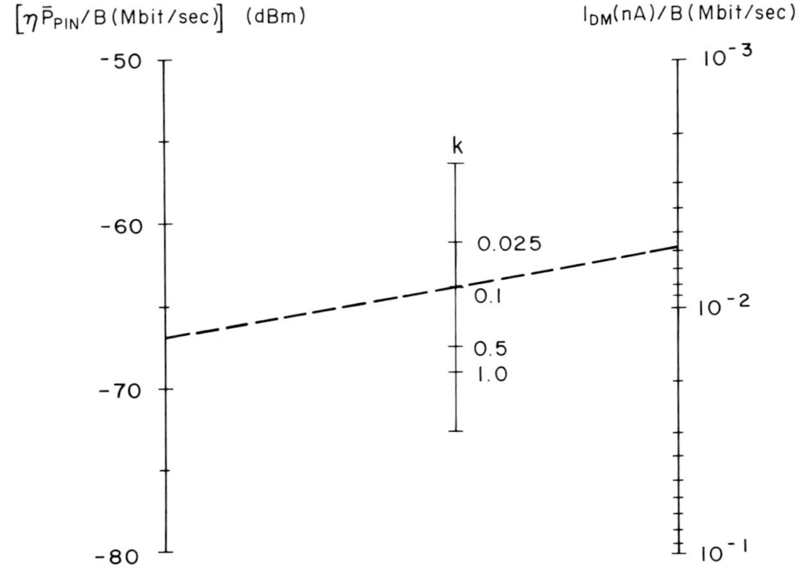

FIG. 10. Nomogram for calculating the normalized bit rate, giving a 1-dB penalty in $\eta \bar{P}_{APD}$ from its ideal value at $I_{DM} = 0$. Here, $[\eta \bar{P}_{p-i-n}/B\,(\text{Mb/s})] = \eta \bar{P}_{p-i-n}\,(\text{dBm}) - 10 \log_{10} B\,(\text{Mbit/sec})$; BER $= 10^{-9}$, $\lambda = 1.3\ \mu\text{m}$.

and are several orders of magnitude smaller than have ever been demonstrated in Ge APDs after nearly 10 years of development. Thus, it is unlikely that high-performance APDs will be obtained using Ge as the detector material, even for use at the highest bit rates presently considered. Nevertheless, Ge APDs can give improvements in the sensitivity of receivers that have

TABLE II

MAXIMUM PRIMARY DARK CURRENT GIVING A 1-dB DEGRADATION IN APD RECEIVER SENSITIVITY

B (Mbit/sec)	$\eta \bar{P}_{p-i-n}$ (dBm)	I_{DM} (nA) $k = 1.0$	0.5	0.1	0.025
50	−50	1.0	0.7	0.3	0.1
100	−47	2.0	1.6	0.6	0.2
500	−38	13	9.5	3.5	1.9
1000	−34	27	19	7.0	3.2
2000	−30	67	45	16	8.1

been designed to achieve high-bit-rate operation at the expense of increased noise. For example, very high-speed receiver operation is achieved by reducing the amplifier load resistor R_L to 50 Ω (see, e.g., Yamada *et al.,* 1980). From Eq. (15) we infer that in this case the receiver sensitivity is limited by Johnson noise generated by R_L and that the receiver is greatly improved, even by the use of inferior APDs with a dark current of several hundred nanoamperes.

Recall that in the previous discussion, $\eta\bar{P}_{p-i-n}$ corresponds to the sensitivity when a noiseless (i.e., zero dark current) $p-i-n$ is used in conjunction with an optical receiver. Thus, $\eta\bar{P}_{\mathrm{APD}}$ is the sensitivity obtained when an APD replaces the $p-i-n$ diode in that same receiver. Since the shot-noise currents of the receiver and the APD are additive [Eq. (16)], the previous comparison is equally valid in the case where the unmultiplied dark current I_{DU} of the APD is equal to the total dark current of the $p-i-n$ detector I_{D}, which it replaces in the optical receiver. This simplified treatment has been shown to be useful in calculating the performance of APD receivers under ideal circumstances. In practice, however, it is often likely that the primary dark current I_{DM}, and not I_{DU}, is comparable to the $p-i-n$ dark current when the $p-i-n$ and the APD are fabricated from the same material system. Here we assume that the dark currents arise predominantly in the bulk rather than on the surface of the detector and therefore undergo multiplication at high electric fields. In Fig. 11, the sensitivity of a receiver using an APD with dark current I_{DM} is compared with one using a nonideal $p-i-n$ detector with dark current $I_{\mathrm{D}} = I_{\mathrm{DM}}$. Parameters other than I_{DM} used to obtain the receiver sensitivities are taken once again from Table I. In Fig. 11a, $\eta\bar{P}_{\mathrm{APD}}/\eta\bar{P}_{p-i-n}$ versus B is plotted for $k = 1.0$ and 0.025. It is apparent that at low bit rates, the sensitivity improvement obtained for detectors with $k = 0.025$ over those with $k = 1$ is dramatically degraded with increasing dark current. Thus, increasing I_{DM} from 0.1 to 100 nA results in a reduction in the difference in sensitivity between APDs with $k = 0.025$ and 1.0 from 4.5 to 1 dB. Further, there is less than 1 dB improvement in sensitivity over that obtained with a $p-i-n$ receiver when using either of the APDs with $I_{\mathrm{DM}} = 100$ nA at 50 Mbit/sec. Note also, that since the room-temperature dark current of a Ge APD (with $k = 1.0$) is > 100 nA, there is little advantage to be gained at $B < 500$ Mbit/sec in using Ge APDs in high-sensitivity amplifiers rather than low-dark-current $p-i-n$ diodes. If the operating temperature is increased, $\eta\bar{P}_{\mathrm{APD}}$ for Ge detectors may be expected to nearly equal $\eta\bar{P}_{p-i-n}$ (see Part VI), even at $B > 500$ Mbit/sec.

In Fig. 11b we plot $\eta\bar{P}_{\mathrm{APD}}/\eta\bar{P}_{p-i-n}$ vs. B for $k = 0.5$ and 0.1. These k values are in the range of what is obtainable using conventional $\mathrm{In_{0.53}Ga_{0.47}As/InP}$ APDs (Susa *et al.,* 1980; Forrest *et al.,* 1981), as well as more sophisticated superlattice structures with enhanced ionization-rate ratios (Capasso *et al.,*

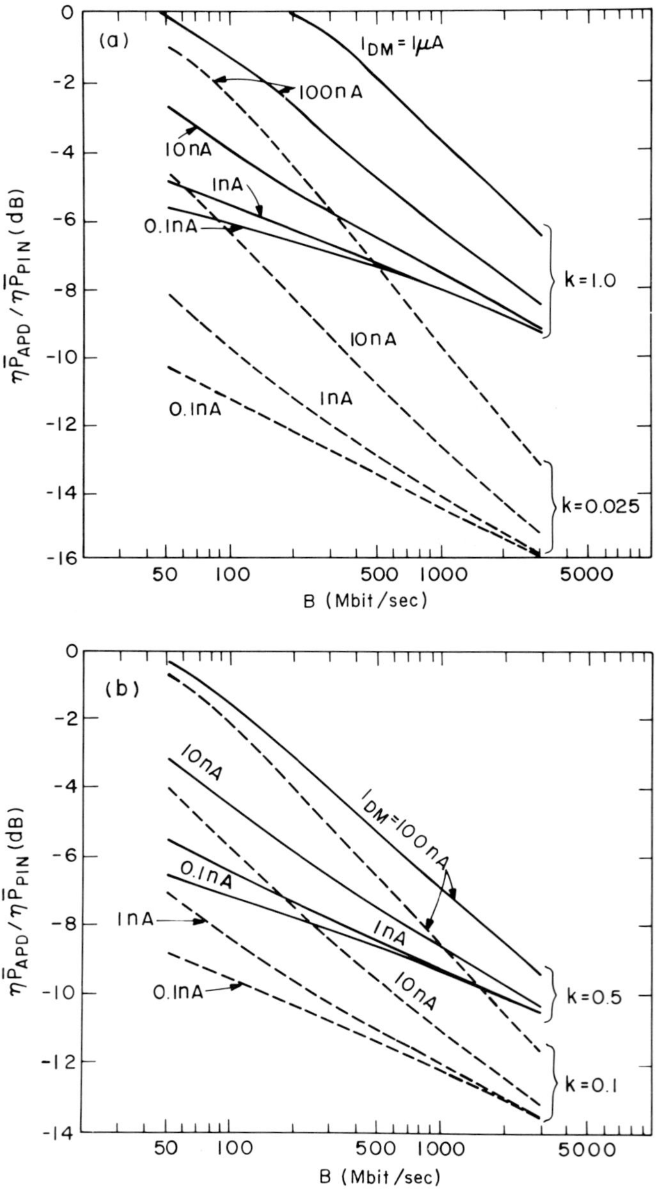

FIG. 11. Improvement in receiver sensitivity obtained using an APD (BER = 10^{-9}, $\lambda = 1.3\ \mu$m, $C_T = 1.5$ pF, $I_g = 15$ nA, $T = 20°$C) rather than a *p–i–n* detector versus bit rate for several values of primary dark current: (a) $k = 1.0$ and 0.025 and (b) $k = 0.5$ and 0.1.

1982). For the former devices, $I_{DM} \sim 1$ nA has been obtained at room temperature (Susa *et al.*, 1980; Kim *et al.*, 1981); values which are indeed comparable to those obtained with $In_{0.53}Ga_{0.47}As$ $p-i-n$ photodiodes (Lee et al., 1980; Leheny *et al.*, 1979). Thus, taking $I_{DM} = 1$ nA and $k = 0.5$, $\eta\bar{P}_{APD}/\eta\bar{P}_{p-i-n}$ decreases from -5.5 dB at 50 Mbit/sec to -10 dB at 2 Gbit/sec. We can therefore expect a significant improvement in sensitivity in systems using $In_{0.53}Ga_{0.47}As$/InP APDs instead of $p-i-n$ diodes made from the same material system, particularly at high bit rates.

For later reference, in Fig. 12 we plot the absolute APD receiver sensitivi-

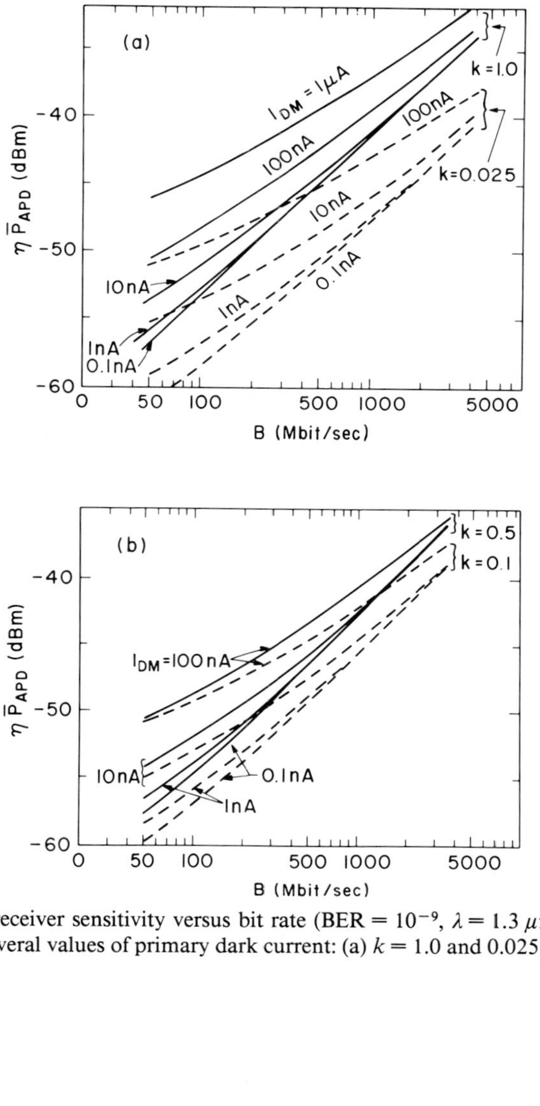

FIG. 12. APD receiver sensitivity versus bit rate (BER = 10^{-9}, $\lambda = 1.3$ μm, $C_T = 1.5$ pF, $I_g = 15$ nA) for several values of primary dark current: (a) $k = 1.0$ and 0.025; (b) $k = 0.5$ and 0.1.

ties as a function of bit rate for several values of I_{DM} and k. As in Fig. 11, parameters other than I_{DM} used in these calculations are listed in Table I.

V. Examples

In Part V we present representative long-wavelength-receiver performance data reported in the literature. Wherever possible these data are compared with the calculated results developed in the previous sections. Direct experimental comparisons made between receivers employing APDs and $p-i-n$ photodiodes provide a convenient means of determining the advantage gained in using an avalanche detector. Unfortunately, there are very few reports of such comparisons, and where they have been attempted, the highest quality $p-i-n$s and APDs have not necessarily been used. Nevertheless, in Section 8, two experiments comparing Ge and $In_{0.53}Ga_{0.47}As/InP$ APDs with high-quality $In_{0.53}Ga_{0.47}As$ $p-i-n$ devices are discussed.

7. Literature Survey of Receiver Performance

Table III is a compilation of a few of the more significant high-bit-rate optical receiver sensitivity measurements made to date at $\lambda = 1.3\ \mu m$. In Table III, the data were obtained at BER $= 10^{-9}$ and at $T \simeq 20°C$. Further, unless otherwise noted, the light source employed is an InGaAsP semiconductor injection laser, and the detector quantum efficiency is assumed to be $\eta = 0.65$. It is apparent that considerable experience has been obtained with high-bit-rate transmission systems using Ge APDs. On the other hand, three experiments using $In_{0.53}Ga_{0.47}As/InP$ heterostructure APDs are listed (experiments 5, 8, and 12 in Table III). In experiment 12, direct comparison indicated an improvement in response over that obtained using an $In_{0.53}Ga_{0.47}As$ $p-i-n$ photodiode.

In some experiments, very long repeaterless transmission distances have been achieved. The longest span between an optical transmitter and receiver reported to date at $\lambda = 1.3\ \mu m$ is 101 km (experiment 9, Table III), where the receiver consisted of an $In_{0.53}Ga_{0.47}As$ $p-i-n$ used in conjunction with a Si bipolar transimpedance front end operated at $B = 274$ Mbit/sec.

At the highest bit rates (≥ 2 Gbit/sec), the transmission distances are reduced due to the inherently lower receiver sensitivities obtainable (see, e.g., Figs. 11 and 12). In addition, system sensitivities are found to be lower for long fiber lengths than for short lengths, particularly at high bit rates. In experiments 1 and 2a, performance results are indicated for spans of 21 and 44.3 km respectively. The receiver sensitivities obtained for very short distance transmission (~ 2 m) in these same experiments are listed in parentheses. This loss of sensitivity at high bit rates with increased distance between regenerators is due to several factors, including transmitter response and fiber dispersion. These effects are considered further in Part VI.

Finally, in experiment 2b, the sensitivity of a receiver using a Si bipolar front end is compared with an FET front end using the same Ge APD. As was suggested in Part II, the receiver employing the FET is ~2 dB more sensitive at the moderate bit rate (100 Mbit/sec) used in the measurement.

A comparison of the data obtained for Ge APDs with the calculated receiver sensitivities in Fig. 12 indicates reasonable agreement at moderate bit rates ($B \lesssim 500$ Mbit/sec) if we assume values of $I_{DM} \lesssim 1\ \mu A$ and $k = 1$ typical of Ge detectors. However, at $B > 500$ Mbit/sec, the calculations predict considerably higher sensitivities (lower $\eta \bar{P}_{APD}$) than are obtained experimentally. One of the reasons for the discrepancy is the low capacitance assumed in the calculations ($C_T = 1.5$ pF). Typically, Ge APDs have a capacitance at breakdown of $C_d > 1$ pF (Kaneda *et al.*, 1979). When C_d is added to the amplifier capacitance of $C_a > 1$ pF, we obtain $C_T > 2$ pF, which can be expected to increase $\eta \bar{P}_{APD}$ significantly from the values given in Fig. 12a. Recall also that the calculations assume high-impedance or transimpedance front ends, whereas for many experiments, R_L as low as 50 Ω is employed to facilitate high-speed performance. These factors, in addition to fiber dispersion and laser noise, will result in degraded system sensitivities.

Table IV is a compilation of several experimental results obtained with optical transmission systems operating at $\lambda \simeq 1.55\ \mu m$. Included among the results is the very long repeaterless span of 161.5 km for an optical transmission system (experiment 6). The extremely low fiber loss of <0.3 dB km^{-1} characteristic of fibers at $\lambda = 1.55\ \mu m$ makes this wavelength particularly attractive for very long repeaterless fiber links. However, some experiments (e.g., experiment 11, Table IV) utilize monomode fibers whose dispersion minimum is at 1.3 μm, whereas the loss is minimized at $\lambda = 1.55\ \mu m$. If the laser has a broad-frequency spectrum, dispersion results in increased intersymbol interference at high bit rates. This leads to a reduction in receiver sensitivity below that expected for fibers whose dispersion and loss minima coincide at $\lambda = 1.55\ \mu m$. Thus, considerable care must be taken to minimize dispersion effects. In experiment 11, this was done by injection locking to decrease the laser spectral width $\Delta\lambda$, thus assuring single longitudinal mode operation. In any case, maximum transmission distances obtained are ~50% longer at $\lambda = 1.55\ \mu m$ than at 1.3 μm at all bit rates. Since $\eta \bar{P} \propto 1/\lambda$ [Eq. (9)], we expect at least this additional increase of ~1 dB in sensitivity for receivers operating at $\lambda = 1.55\ \mu m$ rather than at 1.3 μm. (Here it is assumed that the output power of the laser transmitter is independent of wavelength.)

In experiment 10 (Table IV), a wavelength division-multiplexed bidirec-

TABLE III

OPTICAL RECEIVER AND TRANSMISSION SYSTEM PERFORMANCE AT $\lambda = 1.3\ \mu m$

Experiment No.	B (Mbit/sec)	Detector	Signal format[a]	$\eta\bar{P}$ (dBm)[b]	Transmission Distance (km)	Comments	Reference
1	2240	Ge APD	RZ	−24.9 (−28.2)	21	$I_{DM} = 300$ nA; $\eta = 0.75$	Albrecht *et al.* (1982)
2a	2000	Ge APD	RZ	−31.6 (−34.1)	44.3	Si bipolar front end; $\eta = 0.60$; $R_L = 50\ \Omega$	Yamada and Kimura (1982)
2b	100	Ge APD	RZ	−47.4		FET Front end	
				−45.4		Si bipolar front end	
3	1600	Ge APD	NRZ	−24.5	13.1	$R_L = 50\ \Omega$	Yamada *et al.* (1979)
	1200			−31.7	22.7		
	100			−41.8	30.1	$R_L = 500\ \Omega$	Yamada *et al.* (1978)
4	1200	Ge APD	NRZ	−31.3	11	$\eta = 0.60$; $C_T =$ 1.8 pF; $I_{DM} =$ 100 nA; Si bipolar; $R_L = 50\ \Omega$	
	800			−35.5			
	400			−38.8			
	100			−42.1			
5	1000	InGaAs/InP APD	NRZ	−39.4		$C_T = 1.0$ pF; GaAs FET HZ front end	Campbell (1983)
	420			−42.9			
6	800	Ge APD		−32.2	40	Temperature-stabilized laser; $\eta = 0.60$	Nakagawa *et al.* (1979)
	400			−36.4	40		

7	565	InGaAs *p–i–n*	NRZ	−40.2		GaAs FET front end; $\eta = 0.68$; $C_T =$ 0.45 pF	Smith *et al.* (1982)
	280			−44.7			
	140			−47.7			
8	450	InGaAs/InP APD		−36.5			Taguchi *et al.* (1983)
9a	420	InGaAs *p–i–n*	NRZ	−35.0	84	Si bipolar front end	Boenke *et al.* (1982)
9b	274			−38.9	101		
10	274	InGaAs *p–i–n*	NRZ	−36.3	7.5	GaAs FET front end	Ogawa *et al.* (1981)
	44.7			−48.8	23.3	LED source ($I_g = 1$ nA; $g_m = 50$ msec; $\eta = 0.66$)	
11	274	InGaAs *p–i–n*	NRZ	−37.9		$R_L = 500$ kΩ; $C_T = 1.5–2$ pF	Lee *et al.* (1980)
	44.7			−48.6		$R_L = 30$ kΩ; $C_T = 2.5$ pF; $\eta = 0.65$; $I_T = 135$ nA; GaAs FET front end	
12a	45	InGaAs *p–i–n*	NRZ	−48.4		$\eta = 60\%$	Forrest *et al.* (1981)
12b		InGaAs/InP APD		−53.2		$\eta = 65\%$; $R_L = 500$ kΩ; $C_T = 3$ pF; GaAs FET front end	

[a] RZ: return-to zero format; NRZ: non-return-to-zero format.

[b] Parentheses indicate sensitivity measured over very short fiber span.

TABLE IV

OPTICAL RECEIVER AND TRANSMISSION SYSTEM PERFORMANCE AT $\lambda \simeq 1.55\ \mu m$

Experiment No.	B (Mbit/sec)	Detector	Signal format[a]	$\eta\bar{P}$ (dBm)[b]	Transmission distance (km)	Comments	Reference
1	2000	Ge APD	RZ	−33.6 (−34.2)	51.5	Si bipolar front end; $\eta = 0.60$	Yamada *et al.* (1982a)
2	1200	InGaAs $p-i-n$	NRZ	−33.6 (−36.6)	113.7	$\eta = 0.8$; GaAs FET HZ front end	Chidgey *et al.* (1984)
3	1000	InGaAs/InP APD	NRZ	−40.0		$C_T = 1.0$ pF; GaAs FET HZ front end	Campbell (1983)
	420			−44.9			
4	800	Ge APD	RZ	−34.1	20	$\eta = 0.64$; $R_L = 50\ \Omega$; bipolar front end	Yamada *et al.* (1980)
	400		NRZ	−36.1			
5a	420	InGaAs $p-i-n$	NRZ	−35.6	119	Si bipolar front end; $\eta = 0.72$; $r = 0.29$	Tsang *et al.* (1983)
5b	420	Ge APD	NRZ	−40.4	108	GaAs FET HZ front end	
6	420	InGaAs/InP APD	NRZ	−44.3	161.5	$\eta = 0.68$; $C_T = 1.0$ pF; GaAs FET HZ front end	Kasper *et al.* (1983)
7	400	Ge APD	NRZ	−34.4	18	$\eta = 0.64$; $R_L = 50\ \Omega$	Machida *et al.* (1979)
	100			−42.4		$R_L = 500\ \Omega$	
8	400	Ge APD		−41.3	104	$\eta\bar{P}$ normalized to BER $= 10^{-9}$	Iwashita *et al.* (1982)
9	280	Ge APD	RZ	−38.7	21.7	$\eta = 0.60$	Yamamoto *et al.* (1982)
10	144	InGaAs $p-i-n$	NRZ	−38.0	58	Si bipolar TZ front end; wavelength-multiplexed (1.3 and 1.55 μm) bidirectional link	Spalink *et al.* (1983)
11	140	InGaAs $p-i-n$		−45.0 (−46.5)	102	Injection-locked laser diode GaAs FET HZ front end	Malyon and McDonna (1982)

[a] RZ: return-to-zero format; NRZ: non-return-to-zero format.
[b] Parentheses indicate sensitivity measured over very short fiber span.

tional transmission link was constructed using a single 58-km span of fiber. Two InGaAsP lasers operating at different wavelengths ($\lambda = 1.3$ and $1.55\ \mu m$) were used to inject the optical signals at opposite ends of the fiber. Multiplexing and demultiplexing were accomplished with passive, microoptic, thin-film couplers, and the signals were detected using $In_{0.53}Ga_{0.47}As$ $p-i-n$ diodes.

Finally, in experiment 5, the sensitivity obtained using a Ge APD with a GaAs FET HZ front end was found to be ~4.8 dB higher at 420 Mbit/sec than for a receiver consisting of an $In_{0.53}Ga_{0.47}As$ $p-i-n$ photodetector and a Si bipolar TZ amplifier. Once again, this result compares well with calculations for Ge APDs (Fig. 12a) if we assume values of $k = 1$ and $I_{DM} = 100$ nA typical of high-quality diodes fabricated in this material system. Caution must be used, however, in making a comparison when the various detectors are employed in dissimilar optical receivers. In particular, we expect the Si TZ amplifier to be slightly less sensitive than the GaAs HZ front end at 420 Mbit/sec.

8. Comparison of Ge and $In_{0.53}Ga_{0.47}As$/InP APDs with $p-i-n$ Detector Receivers

In Tables III and IV we have seen that there are only three choices of detectors presently under serious consideration for use at $\lambda = 1.3$ and $1.55\ \mu m$. These are Ge and $In_{0.53}Ga_{0.47}As$/InP APDs, and $In_{0.53}Ga_{0.47}As$ $p-i-n$ photodetectors. The predominant use of Ge APDs in system experiments is due to their commercial availability, whereas the $In_{0.53}Ga_{0.47}As$ $p-i-n$ photodiode is just becoming available. On the other hand, the $In_{0.53}Ga_{0.47}As$/InP heterostructure APD has not been widely tested in system experiments since this detector is still in its initial stages of development in several laboratories. In Section 8 we compare the performance of these three detectors. It will be shown that Ge APDs are superior to $In_{0.53}Ga_{0.47}As$ $p-i-n$ detectors only at the highest bit rates considered and that $In_{0.53}Ga_{0.47}As$/InP APDs have the potential for yielding the highest receiver sensitivities for long-wavelength communications systems in the near future.

Several novel structures (Capasso *et al.,* 1982, 1983), as well as devices fabricated from various material systems, such as AlGaAsSb (Hildebrand *et al.,* 1981), are also the subject of investigation. These more complex structures may provide the basis for fabricating APDs with small k values, and therefore higher receiver performance, in the future. Unfortunately, none of these novel structures has been developed to the extent of being useful for long-wavelength communications experiments and are therefore not considered further in this chapter.

To compare a commercially available Ge APD with a $p-i-n$ detector, we

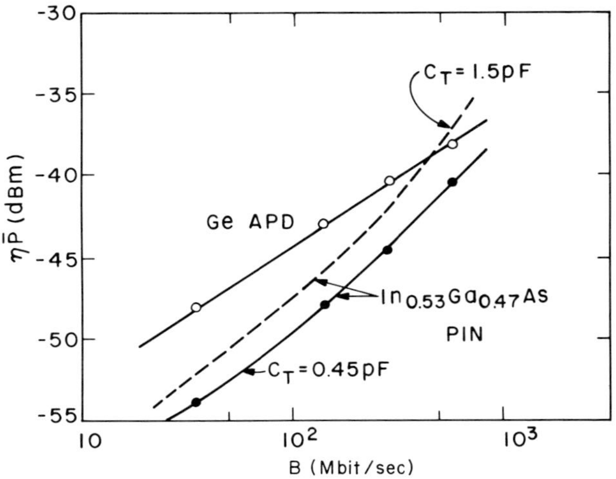

FIG. 13. Experimental comparison of receiver sensitivities obtained using a Ge APD and an $In_{0.53}Ga_{0.47}As$ $p-i-n$ detector [after Smith *et al.* (1982)] ($T = 20°C$, $BER = 10^{-9}$, $\lambda = 1.3\ \mu m$). Plot also shows $\eta\bar{P}_{APD}$ calculated assuming a typical total input capacitance of 1.5 pF.

have plotted in Fig. 13 experimental receiver sensitivities measured by Smith and co-workers (1982). In this experiment, the sensitivity of a very low capacitance ($C_T = 0.45$ pF) receiver consisting of an $In_{0.53}Ga_{0.47}As$ $p-i-n$ detector and a GaAs FET front end was compared with an APD receiver whose noise was dominated by the Ge APD dark current. In the former case, the receiver sensitivities measured over the range 34 Mbit/sec $\leq B \leq$ 565 Mbit/sec are among the best reported. On the other hand, comparison of the results in Fig. 13 with those calculated in Fig. 12a suggests that the Ge APD (with $k = 1$) had a primary dark current at breakdown of $I_{DM} \simeq 1\ \mu A$. The authors found that the sensitivity of the receiver employing the Ge APD was inferior to that using the $In_{0.53}Ga_{0.47}As$ $p-i-n$ detector over the entire bit-rate range considered, although the improvement obtained using the latter receiver was smallest at the highest bit rates, as expected. Further, it was observed that the performance of the Ge APD deteriorated more rapidly with temperature than did the $p-i-n$ detector.

For reference, we have also plotted the $p-i-n$ receiver sensitivity calculated for $C_T = 1.5$ pF (Fig. 5b), a value which is more commonly reported for these circuits than $C_T = 0.45$ pF obtained in the experiment. Comparing this to the measured results for Ge APD receivers, we see that indeed the APD receiver can give a sensitivity improvement over that obtained with a $p-i-n$ detector at $B \gtrsim 500$ Mbit/sec.

Although Ge APDs with somewhat lower dark currents than that used in this experiment have been reported (see, e.g., Kagawa *et al.*, 1981), it is evident that with even the highest quality APDs, only marginal improvement in sensitivity can be obtained over that of an $In_{0.53}Ga_{0.47}As$ $p-i-n$ detector employed in a GaAs FET HZ or TZ front end. In addition, the improvement can be realized only at high bit rates (≥ 500 Mbit/sec) due to the large primary dark currents ($I_{DM} > 100$ nA) and k values ($k \approx 1$) characteristic of Ge detectors. If one takes into consideration the increased complexity of an APD biasing circuit as well as the stringent control necessary in fabricating high-quality APDs, it appears that $In_{0.53}Ga_{0.47}As$ $p-i-n$ detectors may be more attractive for use in long-wavelength optical transmission systems over the entire bit-rate range considered.

One further disadvantage of Ge APDs is that the performance obtained at $\lambda = 1.55$ μm is worse than at 1.3 μm, in contrast to $In_{0.53}Ga_{0.47}As$ $p-i-n$ detectors whose performance is relatively constant over this wavelength range. The loss in Ge APD performance at longer wavelengths arises from a drop in quantum efficiency to $\eta \approx 0.3$. The lower quantum efficiency is due to a reduction in the absorption coefficient of Ge, which in turn results in the absorption of photons in undepleted areas of the diode. Thus, the diode speed of response is limited by slow diffusion of carriers into the depletion region, making some Ge APDs unsuitable for use at high bit rates. However, attempts have been made to improve the speed of response at long wavelengths by employing p^+-n-n^- structures that increase the depletion region width (Yamada *et al.*, 1982b).

In contrast to experiments using Ge APDs, it has been found that significant improvement in receiver performance can be obtained even at moderately low bit rates using $In_{0.53}Ga_{0.47}As$/InP heterostructure APDs in place of $In_{0.53}Ga_{0.47}As$ $p-i-n$ detectors. Figure 14a shows the sensitivity of a receiver employing a heterostructure APD in conjunction with a GaAs FET transimpedance amplifier with $R_F = 500$ kΩ (Forrest *et al.*, 1981). The maximum sensitivity obtained at $B = 45$ Mbit/sec was $\eta\bar{P}_{APD} = -53.2$ dBm at an optimum gain of $M_{opt} = 21$. This represents an improvement of 4.8 dB over the sensitivity achieved using a low dark-current $In_{0.53}Ga_{0.47}As$ $p-i-n$ detector in the same amplifier, where $\eta\bar{P}_{p-i-n} = -48.4$ dBm was obtained (see Fig. 14a). The results in Fig. 11b suggest that $I_{DM} \simeq 3$ nA for the APD. Here we assume that $k = 0.5$, which is consistent with other data obtained in the experiment.

It has been noted that $In_{0.53}Ga_{0.47}As$/InP APDs have a slow speed of response due to the pileup and storage of photogenerated holes at the valence-band discontinuity formed at the heterointerface between the $In_{0.53}Ga_{0.47}As$ absorbing layer and the InP multiplication layer (Forrest *et al.*, 1982a). The effects of charge pile-up are indicated in Fig. 14b, where the response, at 45 Mbit/sec, of the heterostructure APD in a low-noise receiver is indicated in the upper trace. The long tails in the response result in a large degree of intersymbol interference, even at low bit rates. For the sensitivity

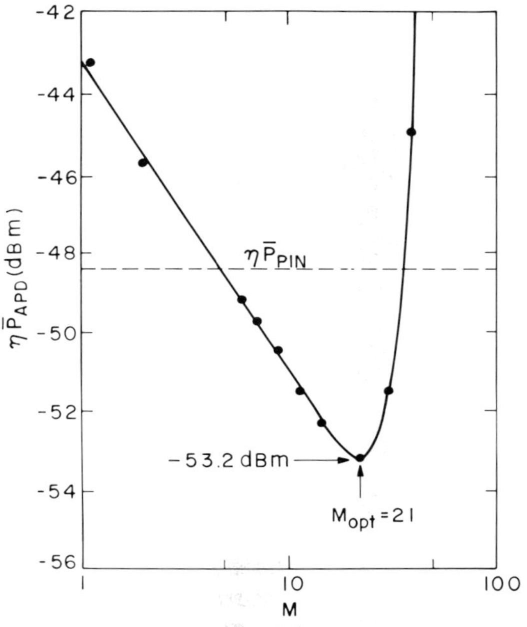

(a)

FIG. 14. (a) Experimental receiver sensitivity versus gain for an $In_{0.53}Ga_{0.47}As$/InP APD; also shown is the sensitivity obtained using an $In_{0.53}Ga_{0.47}As$ $p-i-n$ detector ($\lambda = 1.3\ \mu m$, $T = 24°C$, $B = 45$ Mbit/sec, BER $= 10^{-9}$). (b) Upper trace, response of APD receiver to a 45 Mbit/sec, $\lambda = 1.3\ \mu m$, NRZ data stream; lower trace, response of APD receiver using an equalizer at the preamplifier output. [After Forrest *et al.* (1981). © 1981 IEEE.] *(Continued.)*

measurement shown in Fig. 14a, the slow response was corrected using an equalizer at the preamplifier output. The equalized output is shown in the lower trace in Fig. 14b. Although the intersymbol interference has been reduced, the equalizer has introduced additional noise, evident by the broad oscilloscope traces during the transmission of either a mark or a space. It would not be practical to use such a diode at high bit rates.

It has been suggested, however, that gradually changing the composition at the heterointerface from InP to $In_{0.53}Ga_{0.47}As$ over a distance of several hundred angstroms should result in a reduction in the pile-up and storage of carriers (Forrest *et al.,* 1982a). In this manner, the response time of the APDs could be greatly reduced, thereby making them useful at high bit rates where the sensitivity advantage of using an APD in a receiver is largest. It has been demonstrated that compositional grading does indeed reduce the APD response time (Matsushima *et al.,* 1982; Forrest *et al.,* 1983). Thus, it appears that $In_{0.53}Ga_{0.47}As$/InP heterostructure APDs will offer significant improve-

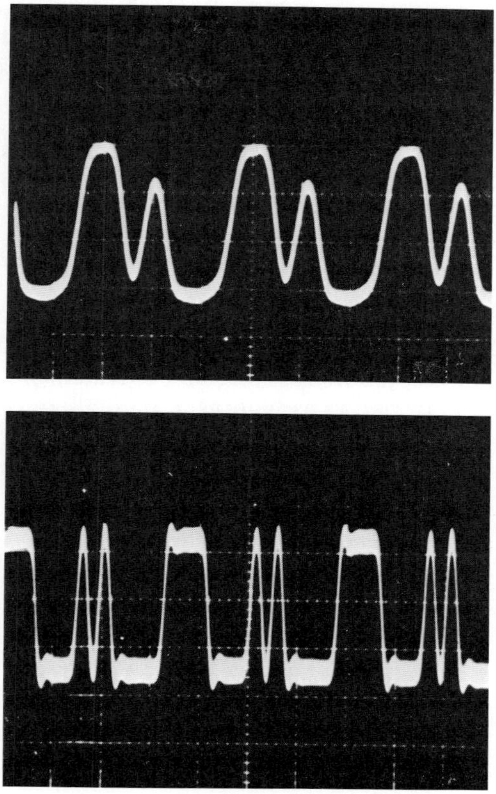

(b)

FIG. 14. *(Continued)*

ment in receiver sensitivity over that obtained using an $In_{0.53}Ga_{0.47}As$ $p-i-n$ detector. Also, the APD primary dark current at breakdown is $I_{DM} \sim 1$ nA, a value $< 10^{-2}$ of that obtainable with Ge APDs. It is likely that the dark currents for heterostructure APDs may be reduced even further; perhaps to as low as $I_{DM} \sim 0.1$ nA in the near future. In addition, the relatively low value of $k = 0.5$ offers improved sensitivity over that obtained using Ge as the detector material.

VI. Sources of Sensitivity Degradation

In Part IV, the sensitivities of APD receivers were calculated assuming nearly ideal circumstances. The only source of sensitivity degradation considered thus far has been detector dark current. In Part VI, we discuss several other performance-limiting mechanisms commonly encountered in optical transmission systems. In particular, we consider the effects of temperature, finite source extinction ($r > 0$), limited APD response time, and fiber dispersion. As pointed out in Parts III and V, the latter three phenomena are particularly important to consider at high bit rates.

9. Temperature Effects

The receiver sensitivity has been shown to be strongly dependent on the primary dark current of the APD. Since most available data on receiver performance are obtained at $T \simeq 20°C$, the discussion thus far has not considered the effects of varying the temperature. However, the dark currents of most high-quality photodiodes are thermally activated. Since system operating temperatures may rise to as high as $T = 70°C$, one thus expects the receiver sensitivity at the highest temperatures to be significantly lower than at $T = 20°C$, particularly at low bit rates. We now compare the sensitivities of receivers employing Ge and $In_{0.53}Ga_{0.47}As$/InP APDs to those employing low dark-current $In_{0.53}Ga_{0.47}As$ $p-i-n$ detectors.

To calculate $\eta\bar{P}_{p-i-n}$ versus T, we note that the Johnson noise current $\langle i_R^2 \rangle$ [Eq. (15)], the channel-noise current $\langle i_c^2 \rangle$ [Eq. (20)], and the $1/f$ noise current $\langle i_f^2 \rangle$ [Eq. (21)] increase linearly with temperature. However, the most rapid dependence on temperature arises from the $p-i-n$ and FET shot noise given by [Eq. (17)]: $\langle i_T^2 \rangle = 2q(I_g(T) + I_D(T))I_2 B$. Here both $I_g(T)$ and $I_D(T)$ are thermally activated, where

$$I_D(T) = I_{D0} \exp(-\Delta\varepsilon_D/kT) \tag{28}$$

with a similar expression applicable for $I_g(T)$, where the "D" subscripts in Eq. (28) would be replaced with "g," referring to FET gate parameters. Here $\Delta\varepsilon_D$ ($\Delta\varepsilon_g$) is the activation energy of the detector (gate) dark current, and I_{D0} (I_{g0}) is the dark current at $T \to \infty$ for the photodiode (FET). It has been shown that $\Delta\varepsilon_D \simeq 0.4$ eV for both $In_{0.53}Ga_{0.47}As$ $p-i-n$ photodiodes and APDs (Forrest *et al.*, 1980, 1983). Typically, the activation energy of the GaAs FET gate leakage is slightly higher at $\Delta\varepsilon_g \simeq 0.5$ eV.

In Fig. 15a, $\eta\bar{P}_{APD}/\eta\bar{P}_{p-i-n}$ at $B = 50$ Mbit/sec is shown as a function of temperature for $In_{0.53}Ga_{0.47}As$ $p-i-n$ and APD receivers employing GaAs FET front ends. For this calculation, we assume I_g (20°C) = 15 nA, and $I_D = I_{DM}$ for the two detectors. The various curves indicate the dependence of sensitivity degradation on both temperature (assuming different values of the room-temperature dark current) and the k value. Here, $k = 0.5$ is typical of standard $In_{0.53}Ga_{0.47}As$/InP APDs (Takanashi and Horikoshi, 1981; Umebu *et al.*, 1980), whereas $k = 0.1$ may be obtainable by tailoring the doping profile (Stillman and Wolfe, 1977) or the structure (Capasso *et al.*,

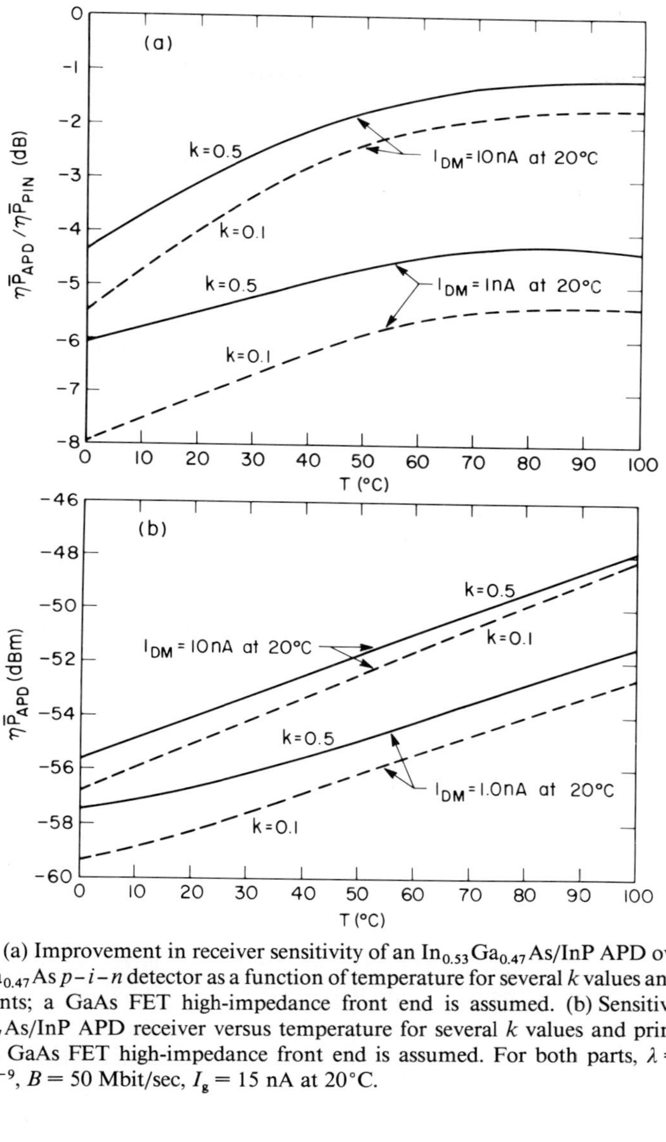

FIG. 15. (a) Improvement in receiver sensitivity of an $In_{0.53}Ga_{0.47}As/InP$ APD over that of an $In_{0.53}Ga_{0.47}As$ $p-i-n$ detector as a function of temperature for several k values and primary dark currents; a GaAs FET high-impedance front end is assumed. (b) Sensitivity of an $In_{0.53}Ga_{0.47}As/InP$ APD receiver versus temperature for several k values and primary dark currents; a GaAs FET high-impedance front end is assumed. For both parts, $\lambda = 1.3\ \mu m$, BER $= 10^{-9}$, $B = 50$ Mbit/sec, $I_g = 15$ nA at 20°C.

1982) of the detector. In this calculation, we assume $\Delta\varepsilon_D = 0.40$ eV and $\Delta\varepsilon_g = 0.50$ eV.

From Fig. 15, we observe that the sensitivity advantage obtained in using the APD decreases with increasing temperature as a result of the increase in $I_{DM}(T)$. The loss in sensitivity is particularly rapid for the curves corresponding to higher dark currents, i.e., I_{DM} (20°C) = 10 nA. Further, note that for $I_{DM} = 1$ nA, the sensitivity advantage $(\eta\bar{P}_{APD}/\eta\bar{P}_{p-i-n})$ is at a minimum at $T \simeq 85°C$, with additional improvement gained at both higher and lower temperatures. This is due to the more rapid increase in I_g with temperature for GaAs FETs compared with I_{DM} for the $In_{0.53}Ga_{0.47}As$ detectors. Thus, at low bit rates and high temperatures, the amplifier noise is greater than the shot noise generated by the APD leakage current. Note, however, that the position of the peak in Fig. 15a depends on the bit rate and activation energies of the leakage currents, I_g and I_{DM}.

In Fig. 15b, we plot $\eta\bar{P}_{APD}$ versus T using the same parameters as those in Fig. 15a. We see that ~4 dB in sensitivity is lost in raising T from 20 to 70°C, for the values of k and I_{DM} (20°C) considered. In comparing these results to those in Fig. 15a, we find that the loss in sensitivity is due to the rise of both I_g and I_{DM} with temperature. Nevertheless, the absolute sensitivities obtained at low k and I_{DM} are considerably better than at higher values of these parameters. From the previous analysis, the dependence of $\eta\bar{P}_{APD}$ on I_{DM} (and therefore T) is expected to be weaker at the higher bit rates.

For Ge, the dark current has two activation energies such that (Hsieh and Card, 1982)

$$I_{DM}(T) = I_{D1}\exp(-\Delta\varepsilon_{Ge}/kT) + I_{D2}\exp(-\Delta\varepsilon_{Ge}/2kT), \qquad (29)$$

where $\Delta\varepsilon_{Ge} = 0.62$ eV is equal to the indirect band-gap energy of Ge. It has already been demonstrated (Section 8) that Ge APDs provide a performance advantage over $In_{0.53}Ga_{0.47}As$ $p-i-n$ detectors only at high bit rates (≥ 500 Mbit/sec) due to the large values of I_{DM} characteristic of Ge diodes at room temperature. In Fig. 16 we plot $\eta\bar{P}_{APD}/\eta\bar{P}_{p-i-n}$ vs. T to compare these detectors over the entire temperature range of interest. Here we assume I_D (20°C) = 10 nA, and take $I_{DM} = 100$ and 500 nA as typical of the best Ge APDs. It is seen that little or no sensitivity advantage in using the APD can be maintained over the range of temperatures plotted, except at the highest bit rates (~2 Gbit/sec) and lowest room-temperature dark currents presently achievable. This is consistent with experimental results discussed in Section 8 (Smith *et al.*, 1982).

Thus, it does not appear to be advantageous to use Ge APDs unless the temperature range of the detector is restricted to values near or below room temperature. Indeed, there is significant sensitivity advantage to be gained in limiting the maximum operating temperature regardless of the structure or material system used for the avalanche detector. One means by which an APD can be conveniently cooled to enhance receiver performance is by the use of thermoelectric coolers.

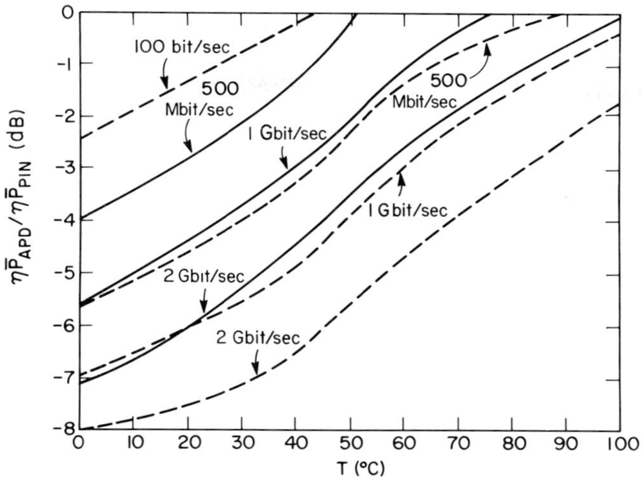

FIG. 16. Improvement in receiver sensitivity of a Ge APD over that of an $In_{0.53}Ga_{0.47}As$ p–i–n detector as a function of temperature, plotted for several bit rates and primary dark currents; a GaAs FET high-impedance front end is assumed. ($\lambda = 1.3\ \mu m$, BER $= 10^{-9}$, $I_D + I_g = 25$ nA at 20°C; dashed line: $I_{DM} = 100$ nA at 20°C, solid line: $I_{DM} = 500$ nA at 20°C.)

10. DEPENDENCE OF $\eta\bar{P}_{APD}$ ON EXTINCTION RATIO

There are several extraneous phenomena introduced by the transmitter at high bit rates (>1 Gbit/sec) that ultimately limit the receiver performance and therefore the distance between regenerators. These include laser modal instabilities and partition noise (Ogawa, 1982); optical pulse delay (Yamada *et al.,* 1978); and spectral broadening (Yamada and Kimura, 1982). It has been found that the latter two effects can be reduced by applying a near-threshold dc level (prebias) to the laser. Superimposed on the dc level is the signal-current pulse. The dc prebias, however, results in light being transmitted in both the logical 0 and 1 states, which in turn leads to a nonzero source extinction ratio r.

An example of the effect of dc prebias current I_{dc} on the longitudinal mode spectrum of the laser is shown in Fig. 17a (Yamada and Kimura, 1982). Here I_{th} is the lasing threshold current. It is evident that the spectral width is narrowed as I_{dc} approaches the threshold current. At very high bit rates, the broad spectral output gives rise to pulse broadening at the receiver resulting from single-mode fiber dispersion (Ogawa, 1982). The effects of dispersion are evident in Fig. 17b, where the bit-error rate is plotted versus I_{dc} for the InGaAsP laser used in Fig. 17a. Using a 44-km span of fiber, when

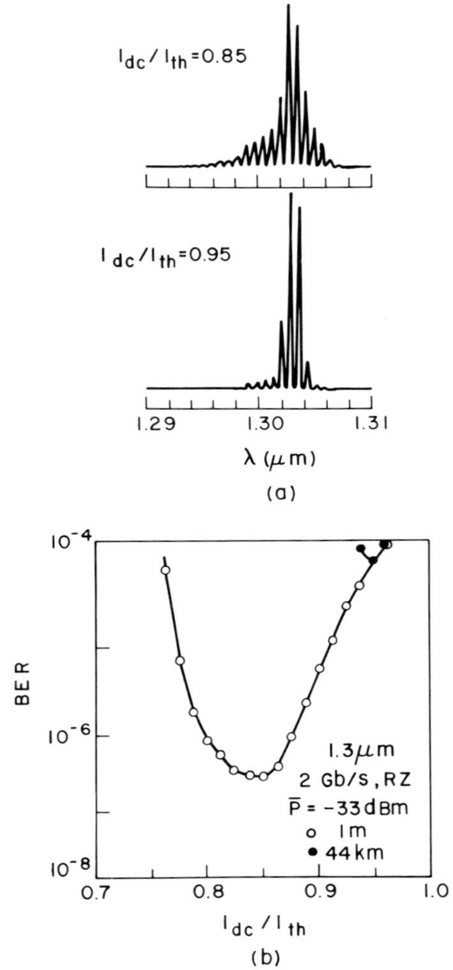

FIG. 17. (a) Laser output spectrum as a function of prebias current I_{dc}; threshold current is I_{th}. (b) Bit-error rate versus pre-bias current for a 2 Gbit/sec RZ transmission experiment. $\lambda = 1.3\ \mu m$ for 2 Gbit, RZ format, $\bar{P} = -33$ dBm; ○, 1 m; ● 44 km. [After Yamada and Kimura (1982). © 1982 IEEE.]

$I_{dc} < 0.95 I_{th}$, the error rate increased due to intersymbol interference arising from fiber dispersion. At $I_{dc} > 0.95 I_{th}$, the error rate was degraded due to the finite extinction ratio. It is noted that alternative means of limiting the width of the output spectrum, and thus the signal dispersion, is via the use of a single frequency laser. A particularly important example of such a structure

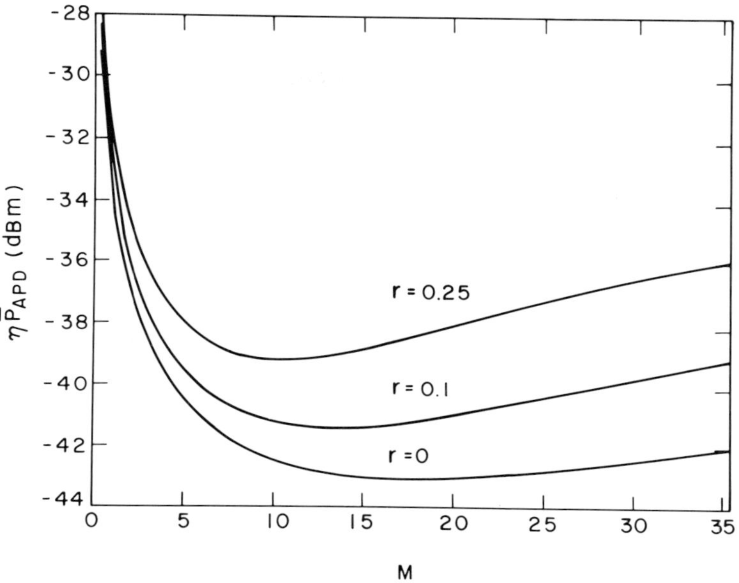

FIG. 18. APD receiver sensitivity versus gain for several values of the extinction ratio r. Here, we have used $k = 0.5$, $\eta\bar{P}_{p-i-n} = -34$ dBm and $B = 1$ Gbit/sec, $\lambda = 1.3\ \mu$m, BER $= 10^{-9}$, $I_{DM} = 0$.

is the distributed feedback (DFB) laser. A full discussion of the means by which laser noise and spectral width can be reduced, however, is beyond the scope of this treatment.

Using Eq. (12), we now calculate the receiver sensitivity for $r > 0$. Figure 18 shows $\eta\bar{P}_{APD}$ versus M at $B = 1$ Gbit/sec for various values of the extinction ratio. In this calculation, we have assumed that $k = 0.5$ along with a $p-i-n$ receiver sensitivity of $\eta\bar{P}_{p-i-n} = -34$ dBm [c.f. Eq's. (10) and (12)]. It is clear that $\eta\bar{P}_{APD}$ increases (sensitivity decreases) with increasing r. The increase in $\eta\bar{P}_{APD}$ results from the decreasing value of M_{opt} under conditions of imperfect source extinction. In Fig. 18, we see that M_{opt} decreases from 17 at $r = 0$ to $M_{opt} = 9$ at $r = 0.25$.

In Fig. 19, the APD receiver sensitivity is plotted for $r > 0$ relative to $\eta\bar{P}_{APD}$ at $r = 0$, assuming conditions similar to those in Fig. 18. Also, the plot shows $\eta\bar{P}_{APD}(r = 0)/\eta\bar{P}_{APD}(r)$ for several values of k. The calculation is relatively insensitive to variations in bit rate. For most high-bit-rate systems, r ranges from a few percent, to reduce the effects of pulse-delay noise (Yamada *et al.*, 1978), to as high as $r = 0.29$ (Tsang *et al.*, 1983). The receiver sensitivity penalty over this range of r can be as high as 4 dB. The loss of

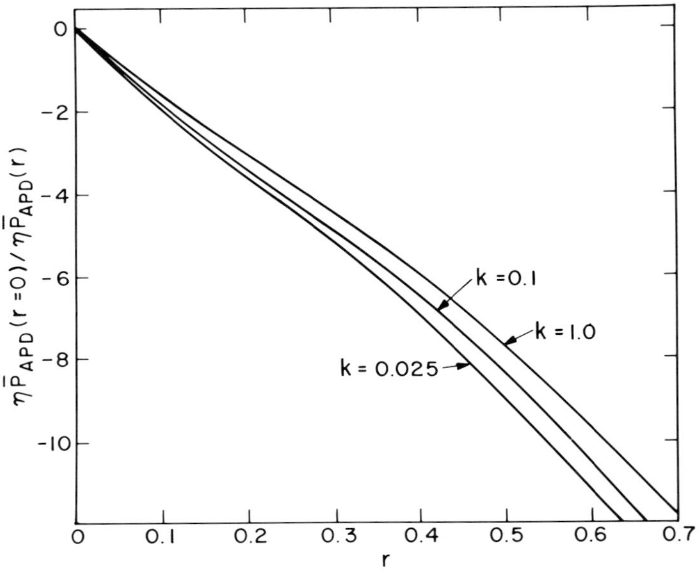

FIG. 19. APD receiver sensitivity penalty versus extinction ratio for several k values using $\eta\bar{P}_{p\text{-}i\text{-}n} = -34$ dBm and $B = 1$ Gbit/sec, $\lambda = 1.3\ \mu$m, BER $= 10^{-9}$, $I_{DM} = 0$.

sensitivity is somewhat greater for lower k values. Recall, however, that the receiver sensitivity at $r = 0$ is higher for APDs with low values of k. The strong dependence on r for devices with low k values is similar to their dependence on dark current. That is, the nonzero extinction ratio increases the noise when a space is transmitted, giving rise to excess shot noise similar to that generated for $I_{DM} > 0$. We therefore expect $\eta\bar{P}_{APD}$ for $r > 0$ to be degraded more rapidly as k decreases, in a manner analogous to the degradation experienced for $I_{DM} > 0$.

11. APD Response Time

In high-bit-rate transmission systems, the response of the photodetector to optical pulses must be sufficiently fast to avoid degrading receiver sensitivity. For example, in Section 8 it was shown that photogenerated charge pileup and storage at the abrupt heterointerface in $In_{0.53}Ga_{0.47}As$/InP APDs resulted in a large receiver sensitivity penalty due to intersymbol interference (ISI). Although the ISI was reduced using an additional stage of equalization, the sensitivity was nevertheless impaired by noise introduced by the nonoptimum equalizer. In addition to the potential increase in noise and complexity involved in having to shape the amplifier output pulse, if the detector response time is considerably longer than the time slot for a mark or space, the maximum photocurrent generated during the transmission of a bit will be smaller than at lower bit rates. This results in a reduction in the overall detection efficiency η and therefore an increase in the mean optical power $\overline{P}$ required to obtain a given signal-to-noise ratio.

Another mechanism limiting the diode speed of response is the diffusion of carriers absorbed in undepleted regions of the semiconductor. In Section 8, we indicated that this phenomenon has frequently been observed in conventional Ge APDs employed at $\lambda = 1.55\ \mu m$, where the absorption coefficient of the semiconductor is relatively small. Indeed, the effect of carrier diffusion in $p^+\text{-}n$ Ge APDs at $B = 2$ Gbit/sec has resulted in a ~1.5-dB penalty in receiver sensitivity compared with $p^+\text{-}n\text{-}n^-$ APDs. These latter devices are relatively free of diffusion tails in the pulse response due to the extension of the depletion region into the lightly doped n^- region (Yamada *et al.*, 1982b).

One further response limitation that can be made very small in properly designed APD structures is the RC_D time constant. In general, one can decrease the cross-sectional area of the diode or increase the depletion-layer width (by decreasing the net carrier concentration in the semiconductor bulk) to reduce C_D to ~0.1 pF. In the range of bit rates considered practical at this time, therefore, RC_D time constants need not limit the APD response.

Thus, we now assume that the response time of well-designed devices is not limited by charge-storage effects, diffusion, or RC_D response. An intrinsic limitation of all APDs, however, is the gain–bandwidth product. That is, the minimum time for an APD with gain $M = 1$ to respond to an optical pulse is determined by the carrier transit time across the depletion region. However, as M increases, the response time is limited by the time it takes for the primary as well as secondary carriers to be swept out of the depletion region. Due to the regenerative nature of the multiplication process, the "effective" transit time across the avalanche region is given by (Emmons, 1967; Kaneda *et al.*, 1976)

$$\tau_1 = Nk\tau_0. \tag{30}$$

Here $k \leq 1$ is the effective ionization-rate ratio for the particular APD material and structure, N a number varying slowly between $\frac{1}{3}$ and 2 with k, and

$$\tau_0 \approx W/v_m \tag{31}$$

the transit time of carriers of mean velocity v_m across the gain region of width W. The avalanche buildup time at a dc gain of M_0 is therefore

$$t = \tau_1 M_0. \tag{32}$$

The ac gain at frequency ω is related to M_0 via

$$M(\omega) = \frac{M_0}{\sqrt{1 + \omega^2 M_0^2 \tau_1^2}}, \tag{33}$$

for $M_0 > 1/k$, which suggests the existence of a gain–bandwidth product, M_0/τ_1.

We define $B_{1/2}$ as the bit rate at which the ac signal power [proportional to $M^2(\omega)$] is equal to one-half its dc value ($\propto M_0^2$). Now the effective bandwidth of a digital receiver is (Smith and Personick, 1979) $B_{\text{eff}} \approx I_2 B$, where I_2 is a Personick integral (see Section 12). To minimize the receiver noise bandwidth, we require that the speed of the photodiode does not limit the receiver response; i.e., the response time of the photodetector must be $t < 1/(2\pi B_{\text{eff}})$. From Eq. (33), $M^2(\omega) = \frac{1}{2} M_0^2$ at frequency $\omega_{1/2} = 2\pi(B_{\text{eff}})_{1/2} = (M_0 \tau_1)^{-1}$. Thus, we have (Forrest, 1984a)

$$I_2 B_{1/2} = (2\pi M_0 \tau_1)^{-1}, \tag{34}$$

for $M_0 > 1/k$. Here, we have let $(B_{\text{eff}})_{1/2} = B_{1/2}$ to simplify notation. The assumption that $t < 1/(2\pi I_2 B_{1/2})$ implies that the APD can practically be employed in an optical receiver at $B < B_{1/2}$ without unduly affecting performance. The receiver will nevertheless require some equalization to compensate for the degraded detector response at $B_{1/2}$. To simplify receiver design and to ensure low-noise performance, it is desirable to avoid equalizing the APD response at high bit rates. In this case, we would require that $t < 1/(2\pi B_{1/2})$, placing even more stringent demands on device speed.

In Fig. 20, we plot the normalized 3-dB bit rate, $B_{1/2}\tau_0$, versus the multiplication factor for several k values of interest (Emmons, 1967), where equal velocities for electrons and holes are assumed. For Fig. 20 we have taken $I_2 = 0.5$; i.e., the bandwidth of the receiver is equal to the Nyquist frequency. The straight lines below the curve corresponding to $M_0 = 1/k$ follow Eq. (34), whereas for $M_0 < 1/k$, the normalized bit rate is roughly independent of gain.

It is of interest to determine the 3-dB bit rate $B_{1/2}$ for $In_{0.53}Ga_{0.47}As/InP$ APDs since, as has been indicated in Sections 8 and 9, these are the most promising avalanche-detector structures for use in long-wavelength optical transmission systems in the near future. In calculating $B_{1/2}$ from Fig. 20, we first need to determine the carrier transit time τ_0 using Eq. (31). It has been shown (Forrest *et al.*, 1982b) that properly designed heterostructure APDs require a depleted $In_{0.53}Ga_{0.47}As$ absorption-layer thickness of ~ 2–$2.5\ \mu m$ to eliminate slow-diffusion tails. Furthermore, the InP multiplication region is typically ~ 2–$4\ \mu m$ thick as well. To obtain τ_0 in Eq. (31), the mean drift velocity of the carriers v_m must also be determined. Although both InP and $In_{0.53}Ga_{0.47}As$ have electron peak velocities in excess of $v_p = 2 \times 10^7$ cm

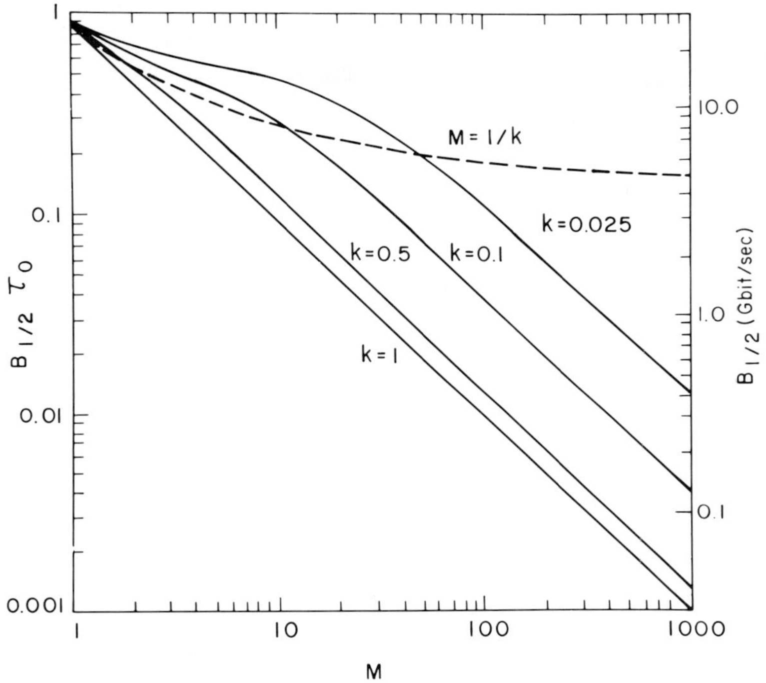

FIG. 20. 3-dB Normalized bit rate versus APD gain for several k values of interest; right-hand ordinate indicates the 3-dB bit rate assuming a transit time of $\tau_0 = 30$ psec typical of $In_{0.53}Ga_{0.47}As$/InP APDs. The transit time limit is $\tau_{tr} = 60$ psec for these devices. [From Forrest (1984). © 1984 IEEE.]

sec^{-1}, the velocity of the carriers decreases rapidly with electric field E once v_p is attained at $E_p \simeq 1 \times 10^4$ V cm^{-1} (Sze, 1981; Hill *et al.*, 1977; Windhorn *et al.*, 1981). Thus, at the high electric fields ($E > 4.5 \times 10^5$ V cm^{-1}) required to obtain significant carrier multiplication in InP, the mean velocity for both electrons and holes is *at most* $v_m = 6-7 \times 10^6$ cm sec^{-1}. Assuming a multiplication region of $\simeq 2$ μm as typical of heterostructure APDs and assuming that the high-field electron and hole velocities are equal, the avalanche-region transit time is approximately $\tau_0 \simeq 30$ psec. Note that this value is consistent with a transit time of 30 psec obtained for fully depleted, transit-time-limited $In_{0.53}Ga_{0.47}As$ $p-i-n$ photodiodes with a depletion-region width of $\simeq 2$ μm (Lee *et al.*, 1981).

Finally, the highest response speed attainable is determined by the carrier transit time across depletion region. From the previous discussion, the total depletion width across both the InP and $In_{0.53}Ga_{0.47}As$ layers is typically

$\gtrsim 4\ \mu$m, giving a transit time of $\tau_{tr} \gtrsim 60$ psec. This value may vary by a factor of 2, depending on the specific device geometry employed.

The scale on the right-hand ordinate in Fig. 20 indicates $B_{1/2}$ (Gbit/sec) assuming $\tau_0 = 30$ psec. We emphasize that the values thus obtained represent what can typically be expected for heterostructure APDs fabricated using InP-based compounds that are not troubled by other, more common sources of speed degradation. For the APDs presently available, $B_{1/2}$ thus corresponds to the $k = 0.5$ curve, although with more sophisticated structures, the lower k values obtained will lead to a higher speed of response $B_{1/2}$ at a fixed value of avalanche gain. From the plot, the avalanche build-up time is thus [Eq. (32)] $t = 7.5 \times 10^{-12} M_0$ sec for $In_{0.53}Ga_{0.47}As/InP$ heterostructure APDs. This is considerably slower than $t = 5 \times 10^{-13} M_0$ sec typical of Si devices (Kaneda and Takanashi, 1975) and is roughly equal to $t = 5 \times 10^{-12} M_0$ sec for Ge (Kaneda and Takanashi, 1973). Note, however, that the avalanche buildup time, and thus the APD speed of response, is linearly proportional to the avalanche-region transit time of $\tau_0 = 30$ psec. As in the case of the total depletion-width transit time, τ_0 is determined by the APD structure, material, and doping profile, and therefore one can expect that it may be reduced from this value in some $In_{0.53}Ga_{0.47}As/InP$ APDs. For example, the length of the avalanche region can be reduced by increasing or tailoring the doping profile (Forrest *et al.*, 1982b) in the high-field region, as is typical of Si APDs. A decrease in the length of the gain region, however, often leads to an increase in the effective k value for the device, thereby resulting in APDs with inferior noise performance. In the case of InP, for example, $k = \alpha/\beta$ is smallest at low values of electric field (Umebu *et al.*, 1980), implying that optimum noise performance is obtained only over long gain regions with only moderate electric fields of $E \gtrsim 4 \times 10^5$ V cm^{-1}.

It is useful to express $B_{1/2}$ as a function of $\eta \bar{P}_{p\text{-}i\text{-}n}$. Thus, for a given receiver sensitivity we can determine if the optimum gain required to achieve a minimum value of $\eta \bar{P}_{APD}$ (see Fig. 2) can be obtained at the desired bit rate. In Section 4, the APD receiver sensitivity for $I_{DM} = 0$ was given by

$$\eta \bar{P}_{APD} = A[(\langle i_a^2 \rangle^{1/2}/M) + qQF(M)I_1 B]. \tag{13}$$

The optimum gain can be obtained by substituting McIntyre's expression for the excess-noise factor $F(M)$ [Eq. (14)] into Eq. (13) and minimizing $\eta \bar{P}_{p\text{-}i\text{-}n}$ with respect to M. In this manner, we obtain

$$M_{opt} = \frac{1}{k^{1/2}} \left[\frac{\langle i_a^2 \rangle^{1/2}}{qQI_1 B} + k - 1 \right]^{1/2}. \tag{35}$$

Using $\eta \bar{P}_{p\text{-}i\text{-}n} = A \langle i_a^2 \rangle^{1/2}$, we express the noise current in terms of $p\text{-}i\text{-}n$

receiver sensitivity, Eq. (35), to yield (Smith and Forrest, 1982)

$$M_{\text{opt}} = \frac{1}{k^{1/2}}\left[\frac{\eta \bar{P}_{p-i-n}}{Q^2(hc/\lambda)I_1 B} + k - 1\right]^{1/2}. \tag{36}$$

Now, the 3-dB bit rate [Eq. (34)] can be expressed in terms of M_{opt}, (Forrest, 1984a) which gives, after some algebra (for $I_{\text{DM}} = 0$),

$$B_{1/2} = \frac{\eta \bar{P}_{p-i-n}}{2Q^2(hc/\lambda)I_1}\frac{1}{1-k}\left\{1 - \left[1 - \frac{(1-k)k}{(\pi\tau_1)^2}\left(\frac{Q^2(hc/\lambda)}{\eta \bar{P}_{p-i-n}}\right)^2\right]^{1/2}\right\}. \tag{37}$$

In writing Eq. (37), we have assumed that $I_2/I_1 \simeq 1$, which is a good approximation for the rectangular family of input pulses (Personick, 1973).

Taking Eq. (37) in the limit of $k = 1$, the 3-dB bit rate is

$$\lim_{k\to 1}(B_{1/2}) = \frac{Q^2(hc/\lambda)}{4(\eta \bar{P}_{p-i-n})I_1}\frac{k}{(\pi\tau_1)^2}. \tag{38}$$

These results imply that as $\eta \bar{P}_{p-i-n}$ increases (i.e., receiver sensitivity decreases), $B_{1/2}$ decreases. Recall, however, that a decrease in receiver sensitivity requires that the optimum gain be increased [Eq. (36) and Fig. 6]. Thus, Eq. (37) yields the expected result that as M_{opt} increases with $\eta \bar{P}_{p-i-n}$, the APD bandwidth as expressed by $B_{1/2}$ must decrease to maintain a constant gain–bandwidth product. We note, however, that $\eta \bar{P}_{p-i-n}$ is dependent on circuit design and is therefore not an independent variable. Indeed, in the preceeding treatment it was shown that $\eta \bar{P}_{p-i-n}$ is a function of Q, B, etc. In practice, one would measure $\eta \bar{P}_{p-i-n}$ at the bit rate of interest and then using Eq. (37) or Eq. (38), calculate the 3-dB bit rate $B_{1/2}$.

Note that Eq. (38) is also valid when

$$\eta \bar{P}_{p-i-n} \gg qQA\sqrt{k(1-k)}(\pi\tau_1). \tag{39}$$

Taking $k = 0.5$ and $\tau_1 = 7.5$ psec characteristic of $In_{0.53}Ga_{0.47}As$/InP APDs, we find that Eq. (38) is a good approximation when $\eta \bar{P}_{p-i-n} \gg -37.8$ dBm. In general, this condition is obtained at $B > 1$ Gbit/sec.

Figure 21 shows the 3-dB bit rate of optimized APD receivers versus $\eta \bar{P}_{p-i-n}$ (assuming $I_1 = 0.5$) for several values of the effective ionization-rate ratio k. The numbers in parentheses indicate the optimum dc gain required at $B_{1/2}$ corresponding to a given receiver sensitivity. To obtain M_{opt}, $B_{1/2}$ has simply been substituted for B in Eq. (36), ignoring the fact that $M(B_{1/2}) = 0.707M_{\text{opt}}$ (Eq. 33). To use Fig. 21, the receiver sensitivity for a noiseless $p-i-n$ detector ($\eta \bar{P}_{p-i-n}$) must be determined at the bit rate B of interest. If the curve corresponding to the k value of the specific APD used gives $B_{1/2} > B$, then the optimized APD receiver is not gain–bandwidth limited.

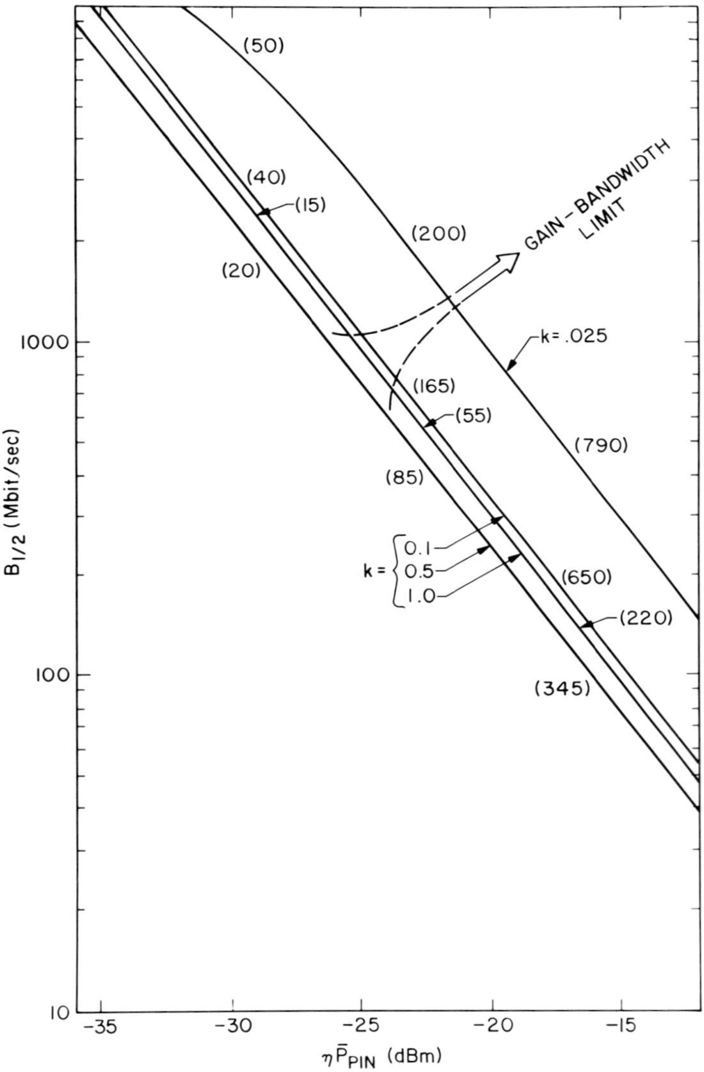

FIG. 21. 3-dB Bit rate versus $p-i-n$ receiver sensitivity for $\tau_0 = 30$ psec typical of optimized receivers employing $In_{0.53}Ga_{0.47}As/InP$ APDs; lines indicate $B_{1/2}$ for several k values, and values of optimum dc gain are in parentheses. If $B \geq B_{1/2}$ for a given $p-i-n$ receiver sensitivity, the APD ac gain is significantly less than its dc gain M_{opt}. $\lambda = 1.3\ \mu m$, BER $= 10^{-9}$, $I_{DM} = 0$. [From Forrest (1984a). © 1984 IEEE.]

However, if $B_{1/2} \leq B$, the value of M_{opt} required to minimize $\eta\bar{P}_{APD}$ exceeds the limit set by the avalanche build-up time, thus giving $M(\omega) < M_0$. Under these conditions, it is not possible to achieve the maximum sensitivity as predicted in Parts II–IV by using an APD. In addition to a reduction in ac gain, the APD photocurrent will be $\geq 90°$ out of phase with the input signal at $B \geq B_{1/2}$. Indeed, measureable phase shifts will be observed at bit rates as low as $\frac{1}{2}B_{1/2}$, which can lead to signal degradation. When the gain–bandwidth limit is exceeded, the sensitivity penalty incurred can be quite large due to the rapid increase in $\eta\bar{P}_{APD}$ for $M < M_{opt}$, especially at low k values (see Fig. 2).

As an example, we take $\eta\bar{P}_{p\text{-}i\text{-}n} = -29$ dBm at $B = 2$ Gbit/sec. Using a conventional $In_{0.53}Ga_{0.47}As/InP$ APD with $k = 0.5$, we obtain $B_{1/2} = 1.8$ Gbit/sec from Fig. 21, a value somewhat lower than the bit rate B. Under these conditions, the APD response time is longer than can be tolerated by the system. Thus, the minimum value calculated of $\eta\bar{P}_{APD} = -39$ dBm (Fig. 6c) cannot be achieved. Indeed, for $\eta\bar{P}_{p\text{-}i\text{-}n} = -29$ dBm and $B = 2$ Gbit/sec, an APD with a transit time of $\tau_0 = 30$ psec must have a k value either above or below 0.5 to operate below the gain–bandwidth limit. Alternatively, use of a lower noise receiver, if available, can result in higher sensitivity using an APD without being limited by device speed. One final approach that can still yield a significant performance advantage is to operate the APD at $M < M_{opt}$. Inspection of Fig. 2, however, indicates that there is a rapid degradation in $\eta\bar{P}_{APD}$ at gains less than optimum for low k value APDs. Thus, lowering M below M_{opt} is useful only with receivers that are marginally gain–bandwidth limited.

Inspection of Fig. 21 reveals the surprising result that optimized receivers using APDs with $k = 0.5$ have a smaller 3-dB bit rate than those with $k = 1.0$ at a given $\eta\bar{P}_{p\text{-}i\text{-}n}$, contrary to the (expected) trend that APD speed increases with decreasing k, as implied by Fig. 20. This apparent "inversion" in Fig. 21 results from two conflicting processes: as k decreases, M_{opt} increases for a given $\eta\bar{P}_{p\text{-}i\text{-}n}$ (Fig. 6); however, an increase in M_{opt} implies that $B_{1/2}$ must decrease such that $B_{1/2}M_{opt} \approx$ constant. For k varying from 1 to 0.5, the optimum gain required for a given value of $\eta\bar{P}_{p\text{-}i\text{-}n}$ increases more rapidly than $B_{1/2}$ (Eq. 34). Thus, conventional $In_{0.53}Ga_{0.47}As/InP$ APD receivers are intrinsically slower than are receivers using devices with k values either higher (Ge) or lower (superlattice APDs) than $k = 0.5$. In addition, the transit time of $\tau_0 = 60$ psec limits the speed of heterostructure APDs using InP-based compounds to values considerably smaller than is typical of either Ge or Si avalanche detectors.

The results shown in Fig. 21 have been used to construct the dashed lines in Fig. 6 that indicate the gain–bandwith-limited region of operation for APDs fabricated using InP-based alloys.

The previous treatment is applicable to APDs with negligible dark current. Recall, however, that M_{opt} decreases with increasing primary dark current (Fig. 8). Thus, values obtained in Fig. 21 represent the lower limit for $B_{1/2}$, assuming ideal detectors with $I_{\mathrm{DM}} = 0$.

We have seen therefore that the gain–bandwidth limit, which is intrinsic to all avalanche detectors, may determine the maximum practical bit rates achieved in long-wavelength communication systems. Indeed, in Section 12, we show that limited response speed that results in intersymbol interference requires an increase in the receiver bandwidth and therefore in the receiver noise. Our analysis also has shown that to achieve the highest speed of response, APDs with low k values are required.

12. Nonrectangular Input Pulses

In the sample calculations of receiver sensitivity, values chosen for the Personick integrals (I_1, I_2, and I_3) assumed a rectangular input-pulse family along with a raised cosine output-pulse shape. There are several effects in practical communication systems, however, that result in nonrectangular input pulses. One such effect discussed in Section 11 was pulse shaping due to finite detector speed of response. Other sources of pulse broadening that are particularly important at high bit rates (>500 Mbit/sec) are finite transmitter response time and fiber dispersion. The broadened pulse at the amplifier input results in an increase in receiver noise and can be calculated using values of the integrals I_1, I_2, and I_3 appropriate to the particular pulse family being received. The source of the increased noise is understood as follows: As the pulse broadens due to the finite response time of the various system components, power transmitted in the pulse tails spills into neighboring time slots, giving rise to intersymbol interference. To decrease the effects of the pulse tails, the equalizer [$E(\omega)$ in Fig. 3] is adjusted to emphasize the high-frequency response of the receiver, which in turn enhances the high-frequency components of the noise. This in turn increases the bandwidth of the receiver and the total noise referred to the amplifier input. In effect, the bandwidth lost by the transmitter and fiber must be compensated for by increasing the receiver bandwidth. Therefore, the input waveform that requires the smallest detection bandwidth is a delta function with a finite mean power $\eta\bar{P}$. This function has the broadest frequency spectrum and therefore permits the receiver to have the narrowest effective bandwidth. Note, however, that to generate increasingly narrow pulses while maintaining a fixed mean power $\eta\bar{P}$ requires that the peak optical power increase in inverse proportion to the pulse width. This is not practical using semiconductor laser transmitters due to the limited peak-coupled power (~0–3 dBm) characteristic of the best devices presently available.

Thus, we find that practical considerations require that the input pulse fill

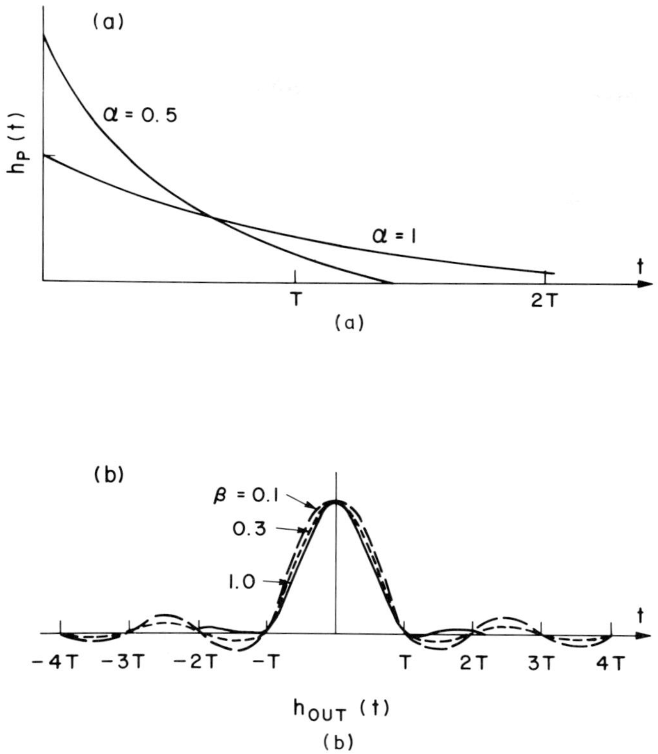

FIG. 22. (a) Exponential input-pulse family; (b) raised cosine output-pulse family. Here, $T = 1/B$ is the bit time slot. [After Personick (1973). Reprinted with permission from The Bell System Technical Journal. © 1973, AT&T.]

a significant fraction of the bit time slot. Pulse broadening due to dispersive effects and slow optical device response can therefore result in increased receiver noise in systems operated at high bit rates or over long distances.

The effects of slow system response on sensitivity can be quantitatively assessed if we assume that these processes can be approximated by an exponentially decaying input-pulse family rather than the rectangular pulses previously assumed. Figure 22a shows an exponential input-pulse shape defined by (Personick, 1973)

$$h_p(t) = (1/\alpha)\exp(-tB/\alpha). \tag{40}$$

Here, α is the pulse-decay constant. To evaluate the integrals I_1, I_2 and I_3, we must also know the output-pulse shape of the linear channel. One such output shape that tends to minimize intersymbol interference is described by the raised cosine function

$$h_{\text{out}}(t) = \frac{\sin(\pi t/T)\cos(\pi\beta t/T)}{(\pi t/T)[1 - (2\beta t/T)^2]}. \tag{41}$$

Using these pulse families, the Personick integrals can be evaluated to determine receiver noise. The results for I_1, I_2, and I_3 are shown in Fig. 23. Also shown is the sum Σ_1, which is due to shot noise arising from the tails of pulses that were incident during time slots other than the decision interval. This term, therefore, exists only in the presence of intersymbol interference. An additional noise-current term must be added to $\langle i_a^2 \rangle$ in Eq. (22) to

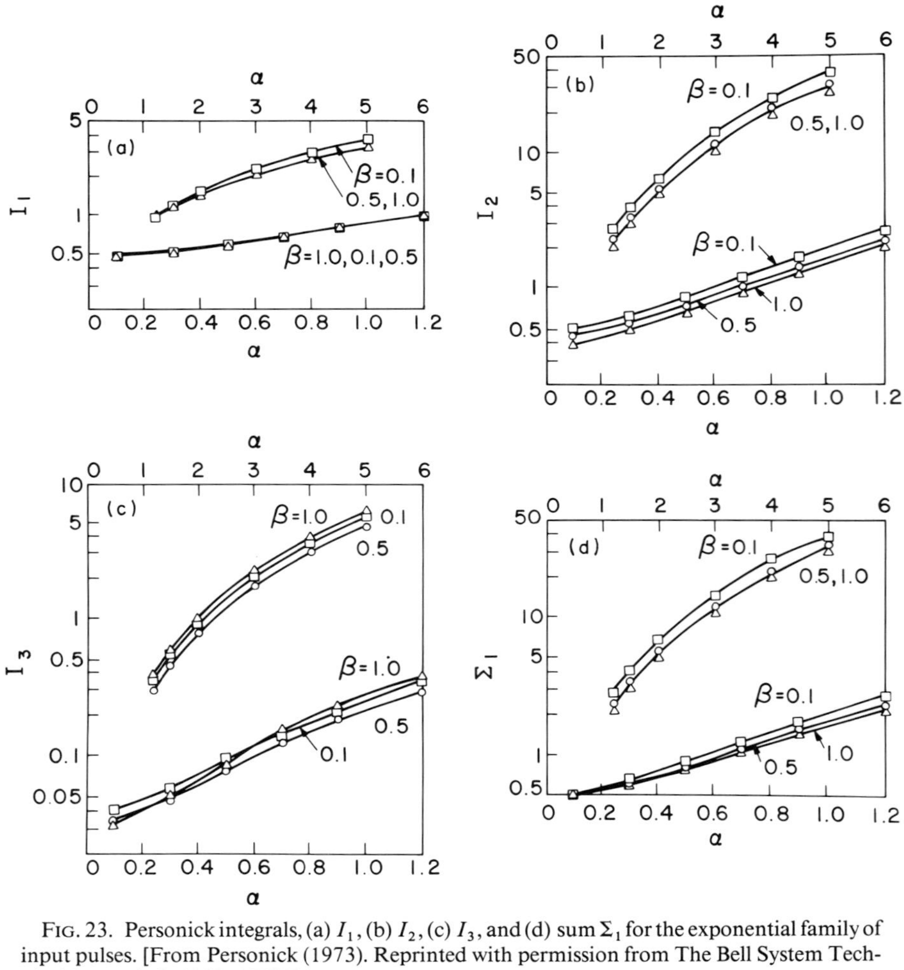

FIG. 23. Personick integrals, (a) I_1, (b) I_2, (c) I_3, and (d) sum Σ_1 for the exponential family of input pulses. [From Personick (1973). Reprinted with permission from The Bell System Technical Journal. © 1973, AT&T.]

account for the noise arising from intersymbol interference. This term can be written (Smith and Personick, 1979)

$$\langle i_x^2 \rangle = 2qI_{sM}M^2F(M)(\Sigma_1 - I_1)B, \tag{42}$$

where I_{sM} is the maximum signal current associated with either a space or a mark. Therefore Eq. (42) represents a worst case for $\langle i_x^2 \rangle$ since we have assumed that all previous time slots were at their maximum signal levels prior to the current time. For rectangular pulse families, $\Sigma_1 \approx I_1 \approx 0.5$, and $\langle i_x^2 \rangle$ vanishes, as expected.

From Fig. 23 we see that all the values of I_1, I_2, and I_3 depend strongly on the pulse width α. If we assume $\alpha = 0.8$ for a pulse transmitted over a long, slightly dispersive fiber, then $I_2 \simeq 1.0$ and $I_3 \simeq 0.18$. These values are roughly twice those obtained for rectangular input pulses, and would reduce the APD receiver sensitivity by a significant amount. Experiments that compare receiver performance using signals transmitted over both short and long fiber spans (>20 km) at high bit rates (≥2 Gbit/sec) indicate that fiber dispersion can typically account for a penalty of between 0.6 (Table IV, experiment 1) and 3.3 dB (Table III, experiment 1) in sensitivity. Moreover, other sources of noise, such as laser mode instabilities resulting in time-dependent dispersion and ISI, may also contribute to the sensitivity degradation.

VII. Conclusions

We have seen that APDs can lead to a considerable improvement in long-wavelength optical receiver sensitivity over that obtained with a $p-i-n$ photodiode, provided that the primary dark current is small. At the time of this writing, $In_{0.53}Ga_{0.47}As/InP$ APDs are the most promising devices for use in long-haul transmission systems, although considerable work still remains to reproducibly reduce the primary dark current at breakdown to ~1 nA or less. Novel APD structures with low k values absorbing at $\lambda \geq 1.3$ μm also have the potential of producing receiver sensitivities comparable to those attainable at shorter wavelengths using Si APDs.

ACKNOWLEDGMENTS

The author thanks P. W. Shumate and R. G. Smith for many helpful discussions.

REFERENCES

Albrecht, W., Elze, G., Enning, B., Walf, G., and Wenke, G. (1982). *Electron. Lett.* **18,** 746.

Barnard, J., Ohno, H., Wood, C. E. C., and Eastman, L. F. (1981). *IEEE Electron Device Lett.* **EDL-2,** 7.

Boenke, M. M., Wagner, R. E., and Will, D. J. (1982). *Electron. Lett.* **18,** 898.

Brain, M. C. (1978). *Electron. Lett.* **14,** 485.
Burrus, C. A., Dentai, A. G., and Lee, T. P. (1979). *Electron. Lett.* **15,** 655.
Campbell, J. C., Burrus, C. A., Dentai, A. G., and Ogawa, K. (1981). *Appl. Phys. Lett.* **39,** 820.
Campbell, J. C., Dentai, A. G., Holden, W. S., and Kasper, B. L. (1983). *Tech. Dig.—Int. Electron Devices Meet.,* 464.
Capasso, F., Petroff, P. M., Bonner, W. A., and Sumski, S. (1980). *IEEE Electron Device Lett.* **EDL-1,** 27.
Capasso, F., Tsang, W. T., Hutchinson, A. L., and Williams, G. F. (1982). *Appl. Phys. Lett.* **40,** 38.
Capasso, F., Tsang, W. T., and Williams, G. F. (1983). *IEEE Trans. Electron Devices* **ED-30,** 381.
Chen, C. Y., Cho, A. Y., Garbinski, P. A., Bethea, C. G., and Levine, B. F. (1981). *Appl. Phys. Lett.* **39,** 340.
Chen, Y.-S. (1978). Personal communication.
Chidgey, P. J., White, B. R., Brain, M. C., Hooper, R. C., Smith, D. R., Smyth, P. P., Fiddyment, P. J., Nelson, A. W., and Westbrook, L. D. (1984). *Electron. Lett.* **20,** 707.
Degani, J., Leheny, R. F., Nahory, R. E., Pollack, M. A., Heritage, J. P., and DeWinter, J. C. (1981). *Appl. Phys. Lett.* **38,** 27.
Diadiuk, V., Groves, S. H., and Hurwitz, C. E. (1980). *Appl. Phys. Lett.* **37,** 807.
Emmons, R. B. (1967). *J. Appl. Phys.* **38,** 3705.
Forrest, S. R. (1984a). *IEEE J. Lightwave Technol.* **LT-2,** 34.
Forrest, S. R. (1984b). *IEEE Electron. Device Lett.* **EDL-5,** 536.
Forrest, S. R. (1985). *IEEE J. Lightwave Technol.* **LT-3,** (in press).
Forrest, S. R., Leheny, R. F., Nahory, R. E., and Pollack, M. A. (1980). *Appl. Phys. Lett.* **37,** 322.
Forrest, S. R., Williams, G. F., Kim, O. K., and Smith, R. G. (1981). *Electron. Lett.* **17,** 917.
Forrest, S. R., Kim, O. K., and Smith, R. G. (1982a). *Appl. Phys. Lett.* **41,** 95.
Forrest, S. R., Smith, R. G., and Kim, O. K. (1982b). *IEEE J. Quantum Electron.* **QE-18,** 2040.
Forrest, S. R., Kim, O. K., and Smith, R. G. (1983). *Solid-State Electron.* **26,** 951.
Fritzche, D., Kuphal, E., and Aulbach, R. (1981). *Electron. Lett.* **17,** 178.
Gloge, D., Albanese, A., Burrus, C. A., Chinnock, E. L., Copeland, J. A., Dentai, A. G., Lee, T. P., Li, T., and Ogawa, K. (1980). *Bell Syst. Tech. J.* **59,** 1365.
Hildebrand, O., Kuebart, W., Benz, K. W., and Pilkuhn, M. H. (1981). *IEEE J. Quantum Electron.* **QE-17,** 284.
Hill, G., Robson, P. N., Majerfeld, A., and Fawcett, W. (1977). *Electron. Lett.* **13,** 235.
Hsieh, Y. P., and Card, H. C. (1982). *IEEE Trans. Electron Devices* **ED-29,** 1414.
Iwashita, K., Nakagawa, K., Matsuoka, T., and Nakahara, M. (1982). *Electron. Lett.* **18,** 938.
Kagawa, S., Kaneda, T., Mikawa, T., Banba, Y., Toyama, Y., and Mikami, O. (1981). *Appl. Phys. Lett.* **38,** 429.
Kaneda, T., and Takanashi, H. (1973). *Jpn. J. Appl. Phys.* **12,** 1091.
Kaneda, T., and Takanashi, H. (1975). *Appl. Phys. Lett.* **26,** 642.
Kaneda, T., Takanashi, H., Matsumoto, H., and Yamaoka, T. (1976). *J. Appl. Phys.* **47,** 4960.
Kaneda, T., Fukada, H., Mikawa, Y., Banba, Y., Toyama, Y., and Ando, H. (1979). *Appl. Phys. Lett.* **34,** 866.
Kasper, B. L., Linke, R. A., Campbell, J. C., Dentai, A. G., Bodhamel, R. S., Henry, P. S., Kaminow, I. P., and Ko, J. S. (1983). *Eur. Conf. Opt. Commun.,* Pap. PD-7.
Kim, O. K., Forrest, S. R., Bonner, W. A., and Smith, R. G. (1981). *Appl. Phys. Lett.* **39,** 402.
Lee, T. P., and Burrus, C. A. (1980). *Appl. Phys. Lett.* **36,** 587.
Lee, T. P., Burrus, C. A., Dentai, A. G., and Ogawa, K. (1980). *Electron. Lett.* **16,** 155.
Lee, T. P., Burrus, C. A., Ogawa, K., and Dentai, A. G. (1981). *Electron. Lett.* **17,** 431.
Leheny, R. F., Nahory, R. E., and Pollack, M. A. (1979). *Electron. Lett.* **15,** 713.

Machida, S., Yamada, J.-I., Mukai, T., Horikoshi, Y., and Tsuchiya, H. (1979). *Electron. Lett.* **15,** 220.

McIntyre, R. J. (1966). *IEEE Trans. Electron Devices* **ED-13,** 164.

Malyon, D. J., and McDonna, A. P. (1982). *Electron. Lett.* **18,** 445.

Matsushima, Y., Akiba, S., Sakai, K., Kushiro, Y., Noda, Y., and Utaka, K. (1982). *Electron. Lett.* **18,** 945.

Nakagawa, K., Hakamada, Y., and Suto, K.-I. (1979). *Electron. Lett.* **15,** 747.

Nishida, K., Taguchi, K., and Matsumoto, Y. (1979). *Appl. Phys. Lett.* **35,** 251.

Ogawa, K. (1981). *Bell Syst. Tech. J.* **60,** 923.

Ogawa, K. (1982). *Bell Syst. Tech. J.* **61,** 1919.

Ogawa, K., and Campbell, J. C. (1982). *Proc. Top. Meet. Opt. Fiber Commun., Phoenix, Ariz.* p. 48.

Ogawa, K., Chinnock, E. L., Gloge, D., Kaiser, P., Nagel, S. R., and Jang, S. J. (1981). *Electron. Lett.* **17,** 71.

Pearsall, T. P., and Papuchon, M. (1978). *Appl. Phys. Lett.* **33,** 640.

Personick, S. D. (1973). *Bell Syst. Tech. J.* **52,** 843.

Shirai, T., Osaka, F., Yamasaki, S., Nakajima, K., and Kaneda, T. (1981). *Electron. Lett.* **17,** 826.

Smith, D. R., Hooper, R. C., Ahmad, K., Jenkins, D., Mabbit, A. W., and Nicklin, R. (1980). *Electron. Lett.* **16,** 69.

Smith, D. R., Hooper, R. C., Smyth, P. P., and Wake, D. (1982). *Electron. Lett.* **18,** 453.

Smith, R. G., and Forrest, S. R. (1982). *Bell Syst. Tech. J.* **61,** 2929.

Smith, R. G., and Personick, S. D. (1979). *Top. Appl. Phys.* **39,** 89.

Spalink, J.-D., Bates, R. J. S., Butterfield, S. J., Lipson, J., Burrus, C. A., Lee, T. P., and Logan, R. A. (1983). *Top. Meet. Opt. Fiber Commun., Tech. Dig., New Orleans, La.* PD-1.

Stillman, G. E., and Wolfe, C. M. (1977). *in* "Infrared Detectors II" (R. K. Willardson and A. C. Beer, eds.), "Semiconductors and Semimetals," Vol. 12, p. 291. Academic Press, New York.

Susa, N., Nakagome, H., Mikami, O., Ando, H., and Kanbe, H. (1980). *IEEE J. Quantum Electron.* **QE-16,** 864.

Sze, S. M. (1981). "Physics of Semiconductor Devices," 2nd Ed. Wiley, New York.

Taguchi, K., Sugimoto, Y., Torikai, T., Makita, K., Minemura, K., and Nishida, K. (1983). *Top. Meet. Opt. Fiber Commun., Tech. Dig., New Orleans, La.* p. 18.

Takanashi, Y., and Horikoshi, Y. (1981). *Jpn. J. Appl. Phys.* **20,** 1907.

Tsang, W. T., Logan, R. A., Olsson, N. A., Temkin, H., Van der Ziel, J. P., Kaminow, I. P., Kasper, B. L., Linke, R. A., Mazurczyk, V. J., Miller, B. I., and Wagner, R. E. (1983). *Top. Meet. Opt. Fiber Commun., Tech. Dig., New Orleans, La.* PD-9.

Umebu, I., Choudhury, A. N. M. M., and Robson, P. N. (1980). *Appl. Phys. Lett.* **36,** 302.

Windhorn, T. H., Cook, L. W., and Stillman, G. E. (1981). *Tech. Dig.–Int. Electron Devices Meet., Washington, D.C.* p. 641.

Yamada, J.-I., and Kimura, T. (1982). *IEEE J. Quantum Electron.* **QE-18,** 718.

Yamada, J.-I., Saruwatari, M., Asatani, K., Tsuchiya, H., Kawana, A., Sugiyama, K., and Kimura, T. (1978). *IEEE J. Quantum Electron.* **QE-14,** 791.

Yamada, J.-I., Machida, S., Kimura, T., and Takata, H. (1979). *Electron. Lett.* **15,** 278.

Yamada, J.-I., Susumu, M., Mukai, T., and Kimura, T. (1980). *Electron. Lett.* **16,** 115.

Yamada, J.-I., Kawana, A., Nagai, H., Kimura, T., and Miya, T. (1982a). *Electron. Lett.* **18,** 98.

Yamada, J.-I., Kawana, A., Miya, T., Nagai, H., and Kimura, T. (1982b). *IEEE J. Quantum Electron.* **QE-18,** 1537.

Yamamoto, S., Utaka, K., Akiba, S., Sakai, K., Matsushima, Y., Sakaguchi, S., and Seki, N. (1982). *Electron. Lett.* **18,** 240.

CHAPTER 5

Phototransistors for Lightwave Communications

J. C. Campbell

AT&T BELL LABORATORIES
CRAWFORD HILL LABORATORY
HOLMDEL, NEW JERSEY

List of Symbols

Subscripts: e, b, c *denote the emitter, base, collector, respectively.* n, p *denote n-type or p-type semiconductors.*

B	Bit rate
$I(J)$	Current (current density)
I_{ph}	Photocurrent
I_{de}	Defect current
I_{dc}	dc Bias current
I_d	Dark current
i_n	Noise current
J_{dpl}	Depletion-layer photocurrent
J_{ir}	Interface–recombination current density
J_{tr}	Tunneling-recombination current density
J_{gr}	Generation-recombination current density
C	Capacitance
D	Minority-carrier diffusion constant
$\Delta E_{g,v,c}$	Band gap, valance band, conduction-band discontinuity
$\bar{f}_c$	Cutoff frequency
F_0	Radiant flux density
G, g	dc, Small-signal optical gain
$h\nu$	Photon energy
$\hbar$	Reduced Planck's constant
kT	Thermal energy

ISBN 0-12-752153-4

L Minority-carrier diffusion length
m^* Effective mass
n, p Density of free electrons, holes
n_0, p_0 Equilibrium minority-carrier density
$\Delta n, \Delta p$ Excess electron-, hole-carrier density
n_i, p_i Interface electron, hole density
N_{IS} Interface state density
P_0 Radiant flux
q Electronic charge
R_e Shockley emitter resistance
R_L Load resistance
S Surface-recombination velocity
V Voltage
v Reduced voltage
v_{th} Thermal velocity
$W_{e,b,c}$ Thickness of emitter, base, collector, including space-charge regions
$w_{e,b,c}$ Thickness of neutral emitter, base, collector
α Common-base current gain
β Common-emitter current gain
ε Semiconductor permittivity
$\sigma_{n,p}$ Capture cross section for electrons, holes
λ Wavelength
η External quantum efficiency
η_e Emitter injection efficiency
η_b Base transport efficiency
$\phi_{e,c}$ Emitter, collector potential
τ Lifetime

I. Introduction

The term "phototransistor" was first coined (Shive, 1949) for a point-contact Ge photoconductive device that bore little resemblence to the bipolar structures to which the term is usually applied today. Shockley *et al.* (1951) first proposed the use of the bipolar configuration ($n-p-n$ or $p-n-p$) as a phototransistor and correctly described the operation of a transistor with an optically generated base current. The first demonstration of this type of photodetector was reported two years later when Shive (1953) described an $n-p-n$ Ge phototransistor that exhibited optical gain in excess of 100. In the late 1950s, the work of Ryvkin and co-workers in the Soviet Union provided a detailed description of the dc gain characteristics and the frequency response of phototransistors (Ryvkin, 1964). Using integrated-circuit technology, monolithic arrays of Si phototransistors were fabricated in the late 1960s for solid-state imaging applications (Schuster and Strull, 1966). The development of charge-coupled devices, however, soon pushed the phototransistor arrays into the background. The most recent interest in phototransistors is due to the development of (1) optical fiber systems, which rely heavily on the performance of photodetectors, and (2) crystal systems from which lattice-matched heterojunction structures can be fabricated. The availability of high-quality heterojunctions has made the wide-band-gap-emitter configuration a practical concept. The idea of using an emitter having a wider band-gap energy than that of the base to improve the emitter injection efficiency of a bipolar transistor was first proposed by Shockley (1951) and later analyzed by Kroemer (1957a,b). Alferov *et al.* (1973) first applied the concept of the wide-band-gap emitter to a phototransistor. They showed that the $Al_xGa_{1-x}As/GaAs$ heterojunction could be used to fabricate a phototransistor with a photoresponse in the wavelength range of interest for initial optical fiber systems; i.e., $0.8\ \mu m \lesssim \lambda \lesssim 0.9\ \mu m$. In this wavelength range the heterojunction phototransistor (HPT) could not compete with the near-perfect Si $p-i-n$ and avalanche photodiodes. However, the attenuation and dispersion characteristics of optical fibers are optimal at longer wavelengths, specifically in the range from 1.0 to 1.6 μm. The inevitable shift of optical-fiber systems to this longer wavelength region, coupled with the poor quantum efficiency of Si photodetectors for wavelengths longer than 1.0 μm, has stimulated research on new "long-wavelength" photodetectors. One approach that has been implemented successfully in numerous long-wavelength systems is a hybrid combination of a $p-i-n$ photodiode followed by a low-noise transistor amplifier. The widespread use of this type of photodetection scheme is due in large part to the inability of present-day long-wavelength avalanche photodiodes to perform as well as their Si counterparts. Since the phototransistor is simply an integrated version of this hybrid combination, it may with further development find application in the front-end circuits of long-wavelength optical receivers. Owing to their high current gain, large electrical output, and demonstrated switching characteristics (Campbell *et al.,* 1983), phototransistors also may prove useful for electrooptic switching and logic applications in the emerging field of photonics.

The emphasis of this chapter is on the use of phototransistors for lightwave systems and photonic circuit applications. First, the basic operating characteristics of a bipolar phototransistor are described with particular attention given to the benefits of the wide-band-gap emitter configuration. In Part I, the optical gain and the common-base and common-emitter current gains are determined from the current transport equations. This leads to a discussion of the defect current and its effect on optical sensitivity, an important consideration for applications where the signal level is small. The bandwidth and transient response characteristics are developed in terms of the charge-control model of Sparkes and Beaufoy (1957). In Part IV, a model for the noise characteristics of the HPT is derived and used to project the sensitivity of an optical receiver with an HPT in the front end. Avalanching in the collector junction of an HPT results in interesting behavior including switching, and this is described in Part V. The discussion on bipolar phototransistors concludes with a review of work on novel structures and the integration of HPTs with other electrooptic components. Finally, a description of the operating characteristics of phototransistors with field-effect transistor-(FET)-like structures is presented in Part VII, along with a review of the results that have been obtained so far.

II. Gain Characteristics

The band-structure diagram in Fig. 1 illustrates the operation of a bipolar HPT. After the incident photosignal passes through the wide-band-gap emitter, which functions as a transparent window, it is absorbed in the base and collector regions, creating electron–hole pairs. The holes, which are generated in the base, in the collector space-charge region, and within a minority-carrier diffusion length of the depletion edge in the bulk collector, accumulate in the base. The resulting change in the base charge alters the emitter-junction potential, causing electrons to be injected from the emitter into the base. If the lifetime of the injected electrons in the base is longer than the transit time across the base, then current gain is achieved by normal transistor action.

Most of the HPTs that have been reported to date have had the floating-base configuration; i.e., no base contact, because the signal base current is generated optically. This has the advantage of eliminating the capacitance of the base contact, but it also removes the possibility of providing a dc base current electrically. The bias voltage is applied between the emitter and collector terminals such that the emitter and collector junctions are forward and reverse biased, respectively. Although a parallel is often drawn to bipolar transistors biased in the common-emitter configuration, since the base-collector junction of a phototransistor functions as a photodiode, the phototransistor is inherently a common-collector device. For most applications this is an insignificant distinction, but as we shall show later, it can be an

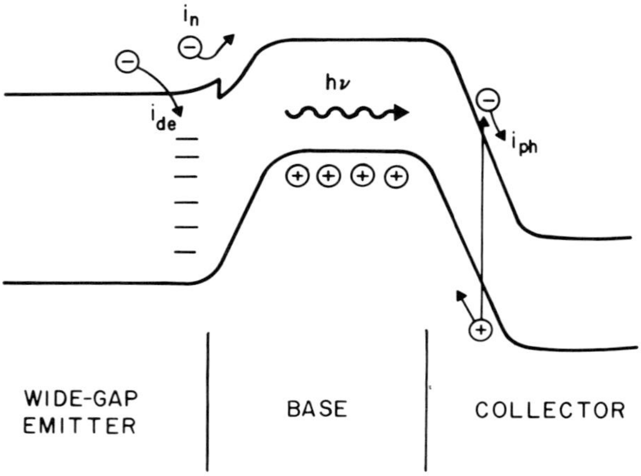

FIG. 1. Energy-band diagram of an $n-p-n$ HPT. The purpose of the wide-band-gap emitter is to suppress reverse injection of holes from the base into the emitter.

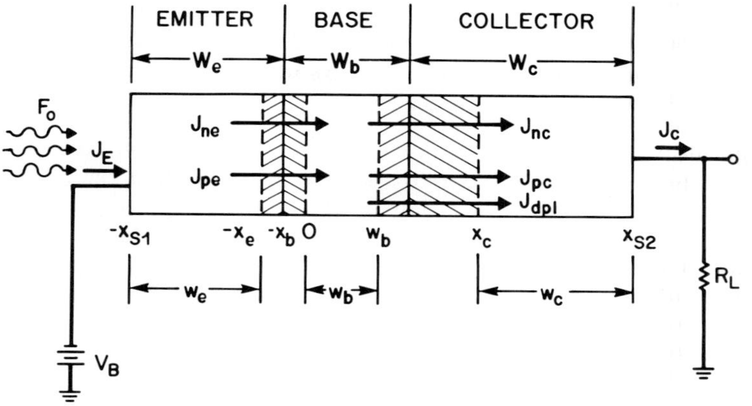

FIG. 2. One-dimensional representation of a phototransistor biased in the floating-base configuration; cross-hatched areas represent space-charge regions in the phototransistor.

important consideration with regard to the noise current of a phototransistor.

1. Transport Equations

The dc optical gain G and the current gain parameters α and β can be derived from the current transport equations. Specifically, the current in each region of the HPT (emitter, base, and collector) is determined from the solutions of the continuity equations for the minority-carrier densities subject to the boundary conditions at the surfaces and at the edges of the space-charge regions of the emitter and collector junctions. The analysis presented here is similar to that presented by Moriizumi and Takahashi (1972) with the important modification, first suggested by Milano (1979), of including the photocurrent generated in the space-charge region of the collector. A planar model (shown in Fig. 2) with spatial variations in only one dimension will be sufficient to illustrate the key features of the gain parameters. The two-dimensional corrections that have been derived for standard bipolar transistors can easily be extended for appropriate geometries to phototransistors. The following list of assumptions helps define the applicability limits of this derivation:

(1) Light is incident through the transparent emitter, and absorption occurs primarily in the base and the space-charge region of the collector.

(2) The space-charge regions in the base and emitter layers are much thinner than the layer thicknesses themselves, and base narrowing is negligible at the operating bias voltage.

(3) The nonequilibrium concentration of minority carriers is much less than the concentration of majority carriers, and hence no electric field exists outside the space-charge regions.

(4) Steady-state conditions are assumed.

We consider an $n-p-n$ structure because in the III – V compounds, which seem better suited for electrooptic applications than other crystal systems, the minority-carrier lifetime in p-type material is approximately 10 times higher than that of comparably doped n-type material. Under steady-state conditions, the constraint of constant current density throughout the device imposes the condition

$$J_{\mathrm{e}} + J_{\mathrm{c}} = 0, \tag{1}$$

where we have adopted the convention that current flowing into the device is positive and that flowing out is negative. The emitter and collector currents consist of the standard injected electron and hole components as well as the primary photocurrent. Written explicitly in terms of the individual current components, Eq. (1) becomes

$$J_{ne} + J_{pe} + J_{nc} + J_{pc} + J_{\mathrm{dpl}} = 0. \tag{2}$$

Explicit expressions for these components are derived in Eqs. (3)–(25).

The emitter hole current J_{pe} can be found from the diffusion equation for the excess hole density Δp_{e} in the emitter,

$$D_{pe}[d^2(\Delta p_{\mathrm{e}})/dx^2] - (\Delta p_{\mathrm{e}}/\tau_{pe}) = 0. \tag{3}$$

Equation (3) is subject to two boundary conditions, one at the emitter surface of the crystal $-x_{\mathrm{s1}}$, and one at the edge of the emitter depletion region $-x_{\mathrm{e}}$:

$$D_{pe}[d(\Delta p_{\mathrm{e}})/dx]|_{x=-x_{\mathrm{s1}}} = S_{\mathrm{e}}\Delta p_{\mathrm{e}}(-x_{\mathrm{s1}}), \tag{4a}$$

$$\Delta p_{\mathrm{e}}|_{x=-x_{\mathrm{e}}} = p_{\mathrm{e0}}[\exp(q\phi_{\mathrm{e}}/kT) - 1]. \tag{4b}$$

Combining Eqs. (3) and (4), we obtain

$$\Delta p_{\mathrm{e}} = \left\{\frac{\cosh[(x + x_{\mathrm{s1}})/L_{pe}] + (S_{\mathrm{e}}L_{pe}/D_{\mathrm{pe}})\sinh[(x + x_{\mathrm{s1}})/L_{pe}]}{\cosh(W_{\mathrm{e}}/L_{pe}) + (S_{\mathrm{e}}L_{pe}/D_{pe})\sinh(W_{\mathrm{e}}/L_{pe})}\right\} x$$
$$\times\, p_{\mathrm{e0}}\left[\exp\left(\frac{q\phi_{\mathrm{e}}}{kT}\right) - 1\right]. \tag{5}$$

The emitter hole current J_{pe} is determined by evaluating the gradient of Δp_{e} at the edge of the emitter depletion region:

$$J_{pe} = -qD_{pe}[d(\Delta p_{\mathrm{e}})/dx]_{x=-x_{\mathrm{e}}}. \tag{6}$$

This yields

$$J_{pe} = -J_{pe0}[\exp(q\phi_e/kT) - 1], \tag{7}$$

where $J_{pe0} = (qD_{pe}/L_{pe})p_{e0}(E_2/E_1)$, and

$$E_2 = \sinh(W_e L_{pe}) + (S_e L_{pe}/D_{pe}) \cosh(W_e/L_{pe}),$$

$$E_1 = \cosh(W_e/L_{pe}) + (S_e L_{pe}/D_{pe}) \sinh(W_e/L_{pe}).$$

If we assume that the emitter junction is located sufficiently far from the surface so that surface recombination can be neglected ($S_e = 0$) and $W_e \gg L_{pe}$, then J_{p0} reduces to

$$J_{pe0} = (qD_{pe}/L_{pe})p_{e0}. \tag{8}$$

The electron-current components J_{nc} and J_{ne} are found from the diffusion equation for the excess electron density in the base Δn_b:

$$D_{nb}[d^2(\Delta n_b)/dx^2] - (\Delta n_b/\tau_{nb}) + \eta\alpha F_0 \exp[-(x - x_b)] = 0. \tag{9}$$

The boundary conditions are

$$\Delta n_b(x = 0) = n_{b0}[\exp(q\phi_e/kT) - 1], \qquad \Delta n_b(x - w_b) = -n_{b0}. \tag{10}$$

The solution is

$$\begin{aligned} \Delta n_b = [\sinh(w_b/L_{nb})]^{-1}\{&[\psi \exp[-(x_b + w_b)\alpha] - n_{b0}] \sinh(x/L_{nb}) \\ &+ [\psi \exp(-\alpha x_b) + \eta_{b0}(\exp(q\phi_e/kT) - 1)] \sinh[(w_b - x)/L_{nb}] \\ &- \psi \exp[-\alpha(x + x_b)] \sinh(w_b/L_{nb})\}, \end{aligned} \tag{11}$$

where

$$\psi = \frac{-\tau_{nb}\eta\alpha F_0}{[(\alpha/L_{nb})^2 - 1]}.$$

The emitter electron current J_{ne} is determined from

$$J_{ne} = qD_{nb}[d(\Delta n_b)/dx]|_{x=0}. \tag{12}$$

Substituting Eq. (11) into Eq. (12) yields

$$\begin{aligned} J_{ne} = -J_{n0}\{&1 + [\exp(q\phi_e/kT) - 1] \cosh(w_b/L_{nb})\} \\ &+ g_b F_0 \exp(-\alpha x_b)[\cosh(w_b/L_{nb}) \\ &- \alpha L \sinh(w_b/L_{nb}) - \exp(-\alpha w_b)], \end{aligned} \tag{13}$$

where

$$J_{n0} = \left(\frac{gD_{nb}n_{b0}}{L_{nb}}\right) \sinh\left(\frac{w_b}{L_{nb}}\right) \quad \text{and} \quad g_b = \frac{qL_{nb}\eta\alpha}{(1 - \alpha^2 L_{nb}^2) \sinh(w_b/L)}.$$

Similarly, J_{nc} can be found from the relation

$$J_{nc} = qD_{nb}[d(\Delta n_b)/dx]|_{x=w_b}. \tag{14}$$

Combining Eq. (11) and (14) yields

$$\begin{aligned} J_{nc} = &-J_{n0}\{\cosh(w_b/L_{nb}) + [\exp(q\phi_e/kT) - 1]\} \\ &+ g_b F_0 \exp(-\alpha x_b)\{1 - \exp(-\alpha w_b) \\ &\times [\cosh(w_b/L_{nb}) + \alpha L \sinh(w_b/L_{nb})]\}. \end{aligned} \tag{15}$$

The diffusion equation for the minority-carrier density in the bulk collector is

$$D_{pc}[d^2(\Delta p_c)/dx^2] - (\Delta p_c/\tau_{pc}) + \eta\alpha F_0 \exp[-\alpha(x + x_b)] = 0, \tag{16}$$

and the boundary conditions are

$$\Delta p_c(x = x_c) = -p_{oc}[\exp(q\phi_c/kT) - 1] \simeq -p_{oc}, \tag{17a}$$

$$D_{pc}[d(\Delta p_c)/dx]|_{x=x_{s2}} = -S_c\, \Delta p_c(x = x_{s2}). \tag{17b}$$

The solution of Eq. (16) subject to the boundary conditions is

$$\begin{aligned} \Delta p_c = (\theta_1)^{-1}\{&[\psi' \exp(-\alpha x_c) - p_{oc}] \\ &\times [\cosh[(x - x_{s2})/L_{pc}] - \xi \sinh[(x - x_{s2})/L_{pc}]] \\ &+ \psi'(\alpha L_{pc} - \xi) \exp(-\alpha x_{s2}) \cosh[(x - x_c)/L_{pc}]\}, \end{aligned} \tag{18}$$

where

$$\xi = \left(\frac{S_c L_{pc}}{D_{pc}}\right), \qquad \psi' = -\frac{\tau_{pc}\eta\alpha F_0 \exp(-\alpha x_b)}{\alpha^2 L_{pc}^2 - 1},$$

$$\theta_1 = \cosh\left(\frac{W_c}{L_{pc}}\right) + \xi \sinh\left(\frac{W_c}{L_{pc}}\right).$$

The hole current in the collector is obtained using

$$J_{pc} = -qD_{pc}(d(\Delta p_c)/dx)_{x=x_c}. \tag{19}$$

The complete expression for J_{pc} is

$$\begin{aligned} J_{pc} = &-\frac{q\eta\alpha L_{pc} F_0}{1 - \alpha^2 L_{pc}^2} \left\{\frac{\theta_2 + (\alpha L_{pc} - \xi)}{\theta_1 - \alpha L_{pc}}\right\} \exp[-\alpha(x_c + x_b)] \\ &+ \left(\frac{qD_{pc}}{L_{pc}}\right) p_{0c} \left(\frac{\theta_2}{\theta_1}\right), \end{aligned} \tag{20}$$

where $\theta_2 = \sinh(W_c/L_{pc}) + \xi\cosh(W_c/L_{pc})$. The coordinate x_{s2} usually denotes the collector–substrate interface. In all practical cases, $S_c = 0$ at this interface, which permits J_{pc} to be written in a more simplified form as

$$J_{pc} = -g_c F_0 + J_{pc0}\tanh(W_c/L_{pc}), \tag{21}$$

where

$$g_c = \left(\frac{q\eta\alpha L_{pc}}{1-\alpha^2 L_{pc}^2}\right)\exp[-\alpha(x_c + x_b)] \times\left[\tanh\left(\frac{W_c}{L_{pc}}\right) + \frac{\alpha L_{pc}[\exp(-\alpha W_c) - 1]}{\cosh(W_c/L_{pc})}\right], \tag{22}$$

$$J_{pc0} = (qD_{pc}/L_{pc})p_{c0}. \tag{23}$$

The photogenerated current in the depletion region J_{dpl} is given by

$$J_{dpl} = +q\int_{w_b}^{x_c} g(x)\,dx, \tag{24}$$

which, after integration, becomes

$$J_{dpl} = (q\eta F_0/h\nu)[\exp(\alpha x_c) - \exp(\alpha w_b)]. \tag{25}$$

2. WIDE-BAND-GAP EMITTER

Many of the standard parameters of bipolar transistors, such as the emitter injection efficiency η_e, the base transport efficiency η_b, the common base current gain α, and the common-emitter (common-collector) current gain β, can be obtained using Eqs. (2), (7), (13), (15), (21), and (25) with $F_0 = 0$.

The base transport efficiency η_b is given by

$$\eta_b = J_{nc}/J_{ne}. \tag{26}$$

Substituting Eqs. (13) and (15) into Eq. (26) and assuming that $\exp(q\phi_e/kT) \gg 1$ yields an approximate expression for η_b:

$$\eta_b \simeq [\cosh(w_b/L_{nb})]^{-1}. \tag{27}$$

The emitter injection efficiency is given by the equation

$$\eta_e = J_{ne}/(J_{ne} + J_{pe}). \tag{28}$$

Equation (28) implies that J_{ne} and J_{pe} account for the total emitter current. In fact, there is another important component to the emitter current, namely, the defect current, which must be considered for actual devices. The effects of the defect current are discussed later, but for the present Eq. (28), which is valid for an ideal junction, suffices to illustrate the benefits of the wide-band-gap emitter. Substituting Eqs. (7) and (13) into (28) and assuming again that $\exp(q\phi_e/kT) \gg 1$, we obtain

$$\eta_e = \frac{\Gamma \cosh(w_b/L_{nb})}{1 + \Gamma \cosh(w_b/L_{nb})}, \tag{29}$$

where

$$\Gamma = \frac{J_{n0}}{J_{p0}} = \frac{\gamma(E_2/E_1)}{\sinh(w_b/L_{nb})}, \tag{30}$$

and E_1 and E_2 are defined in Section 1. Kroemer (1957a,b) has related γ to the material parameters of the crystals that comprise the emitter heterojunction as follows:

$$\gamma = \frac{D_{nb} L_{pe} n_e}{D_{pe} L_{nb} p_b} \left(\frac{m^*_{nb} m^*_{pb}}{m^*_{ne} m^*_{pe}} \right)^{3/2} \exp\left(\frac{\Delta E_g}{kT} \right), \tag{31}$$

where ΔE_g is the difference between the band-gap energies of the emitter and base layers. In fact, ΔE_g should be replaced by ΔE_v the valence-band discontinuity, because it is ΔE_v that effectively prevents the reverse injection of holes from the base into the emitter. For some heterojunctions, such as $Al_xGa_{1-x}As$/GaAs, ΔE_g and ΔE_v are approximately equal (Dingle, 1975). It has been shown, however, for some other heterojunctions, e.g., $In_xGa_{1-x}As_yP_{1-y}$/InP, that $\Delta E_v < \Delta E_g$ (Forrest and Kim, 1981). High emitter-injection efficiency requires that $\gamma \gg 1$. For homojunctions where $\Delta E_g = 0$, this can only be accomplished if the carrier concentration in the emitter n_e is much greater than that in the base p_b. For heterojunctions, on the other hand, the exponential dependence of γ on ΔE_g removes this constraint on the relative doping levels and results, at least in theory, in higher gains, better frequency response, reduced emitter crowding, reduced second-breakdown effects, and improved switching characteristics.† Of these potential benefits the most important for phototransistors are the improvements in current gain and frequency response. The improved frequency response results from the lower emitter capacitance and reduced base resistance, which can be achieved by having a lightly doped emitter and a heavily doped base.

To illustrate the higher current gain afforded by the wide-band-gap emit-

† For more detail on the advantages of the wide-band-gap emitter configuration, with particular emphasis on high-speed integrated circuits, the reader is referred to a review article by Kroemer (1982).

ter, we recall that the common-base current gain α and common-emitter current gain β are given by the relations [see, e.g., Sze (1969)]

$$\alpha = \eta_e \eta_b \tag{32a}$$

and

$$\beta = \alpha/(1 - \alpha). \tag{32b}$$

Substituting Eqs. (27)–(31) into Eq. (32b) yields the expression

$$\beta = \Gamma[1 + \Gamma(\cosh(w_b/L_{nb}) - 1)]^{-1}, \tag{33}$$

which can be used to calculate β as a function of the band-gap discontinuity between the emitter and base. As an example, we consider the case (Fig. 3) of a GaAs base layer and an $Al_xGa_{1-x}As$ emitter. The band-gap discontinuity increases continuously from $\Delta E_g = 0$, which corresponds to the case of a homojunction, to $\Delta E_g = 2.05$ eV as x, the Al content in the emitter, is

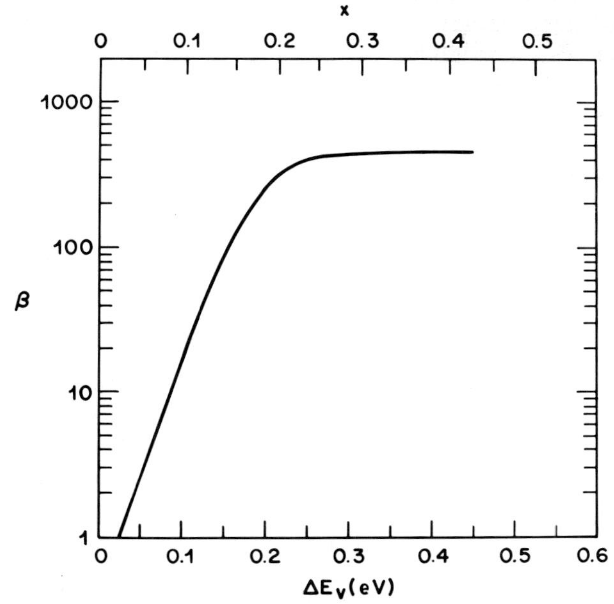

FIG. 3. The dc current gain of an $Al_xGa_{1-x}As$/GaAs HPT versus the valence-band discontinuity between emitter and base.

increased from 0 to 0.5. For this calculation, we have used the following values for the parameters in Eq. (30):

Base	Emitter
$w_b = 0.5\ \mu m$	$\Delta E_v = 0.85\ \Delta E_g = (1.21 + 1.06x)$ eV
$L_{nb} = 6\ \mu m$	$L_{pe} = 1.7\ \mu m$
$D_{nb} = 220\ cm^2\ sec^{-1}$	$D_{pe} = 10\ cm^2\ sec^{-1}$
$p_b = 1 \times 10^{18}\ cm^{-3}$	$n_e = 5 \times 10^{15}\ cm^{-3}$
$m^*_{nb} = 0.067 m_e$	$m^*_{ne} = (0.067 + 0.083x) m_e$
$m^*_{hb} = 0.48 m_e$	$m^*_{he} = (0.48 + 0.31x) m_e$

The plot of gain versus valence-band discontinuity in Fig. 3 can be divided into two regions. For $\Delta E_v \lesssim 0.2$ eV, the current gain β is determined by the emitter-injection efficiency, and is a rapidly increasing function of ΔE_v. It is also evident from Fig. 3 that the wide-band-gap emitter directly results in > 100 increase in the current gain. For $\Delta E_v \gtrsim 0.2$ eV, the gain is constant, independent of ΔE_v. In this region, the gain is limited by the base transport efficiency. A similar calculation to that shown in Fig. 3 for the $InP/In_xGa_{1-x}As_yP_{1-y}$ crystal system has been carried out by Campbell *et al.* (1982). In addition to the high current gain achieved by eliminating reverse injection of holes from the base into the emitter, the wide-band-gap emitter also results directly in improved optical gain because it provides a transparent window allowing absorption to occur away from the surface where the high recombination velocity can significantly reduce the quantum efficiency.

3. Optical Gain

The optical gain G simply relates the number of electrons (or holes) in the collector current to the number of incident photons. Mathematically, this relation is expressed as

$$G = (h\nu/q)[(J_T)_{opt}/F_0], \tag{34}$$

where $(J_T)_{opt}$ is the optical component of the collector current. To derive an expression for J_{opt} it is necessary to evaluate how the emitter potential ϕ_e varies in response to the incident optical signal. An expression for the optical component of $[\exp(q\phi_e/kT) - 1]$ can be obtained from Eq. (2), the constraint of current continuity. Substituting Eqs. (7), (13), (15), (21), and (25) into Eq. (2), and solving for $[\exp(q\phi_e/kT) - 1]$, we obtain

$$\begin{aligned}[\exp(q\phi_e/kT) - 1] = (F_0/h\nu)\{g_c + g_b[&\exp(-\alpha x_b)(1 + \exp(\alpha w_b)) \\ &\times(1 + \alpha L_{nb}\sinh(w_b/L_{nb}) - \cosh(w_b/L_{nb}))] \\ &+ q\eta[\exp(-\alpha w_b) - \exp(-\alpha x_c)]\end{aligned}$$

$$\begin{aligned}&+J_{n0}(1-\cosh(w_b/L_{n0})\\&-J_{p0}\tanh(W_c/L_{pc})\}/\\&[J_{n0}(1-\cosh(w_b/L_{nb}))+J_{pe0}].\end{aligned} \tag{35}$$

The optical component of Eq. (35) is

$$\begin{aligned}[\exp(qQ_e/kT)-1]_{\text{opt}}=(F_0/h\nu)\{&g_c+g_b[\exp(-\alpha x_b)[1+\exp(-\alpha w_b)]\\&\times(1+\alpha L_{nb}\sinh(w_b/L_{nb})-\cosh(w_b/L_{nb})]\\&+\exp(-\alpha w_b)[1-\exp(-\alpha(x_c-w_b))]]\}/\\&[J_{n0}(1-\cosh(w_b/L_{nb}))+J_{pe0}].\end{aligned} \tag{36}$$

The total current density through the phototransistor is

$$J_T=J_{nc}+J_{pc}+J_{\text{dpl}}. \tag{37}$$

If we take only the optical terms from Eqs. (15), (21), and (25), then the current, which is due to the incident optical signal, is given by

$$\begin{aligned}(J_T)_{\text{opt}}=J_{n0}&[\exp(q\phi_e/kT)-1]_{\text{opt}}\\&-(F_0/h\nu)\{g_b\exp[-\alpha(x_b+w_b)]\\&\times(\cosh(w_b/L_{nb})+\alpha L_{nb}\sinh(w_b/L_{nb}))\\&+g_c+q\eta\exp(-\alpha w_b)[1-\exp(-\alpha(x_c-w_b))]\}.\end{aligned} \tag{38}$$

The first term in Eq. (38) is the response of the emitter potential to the accumulation of photogenerated holes in the base layer. The second, third, and fourth terms are the primary photocurrent, which results from absorption in the base, the collector space-charge region, and the neutral region of the collector, respectively. Combining Eqs. (33), (34), (36), and (38) yields a rather complicated expression for the optical gain:

$$\begin{aligned}G=(\beta/q)\{&-g_b\alpha L_{nb}\sinh(w_b/L_{nb})\\&\times[1+(\cosh(w_b/L_{nb})+\sinh(w_b/L_{nb}))\exp(\alpha w_b)]\\&+q\eta\exp(-\alpha w_b)[1-\exp(-\alpha(x_c-w_b))]+g_c\cosh(w_b/L_{nb})\}\\&+(\beta/q\Gamma)\{g_c+q\eta\exp(-\alpha w_b)\\&\times[1-\exp(-\alpha(x_c-w_b))]-g_bf_{nc}\},\end{aligned} \tag{39a}$$

where

$$f_{nc}=1-\exp(-\alpha w_b)[\cosh(w_b/L_{nb})+\alpha L\sinh(w_b/L_{nb})]. \tag{39b}$$

The following three approximations lead to a much simpler representation for G:

(1) The band-gap discontinuity ΔE_g is sufficiently large that $\Gamma \gg 1$. This eliminates the second part of Eq. (39).

(2) The base is thin enough for $w_b/L_{nb} \ll 1$ to be valid.

(3) All of the absorption occurs in the base and space-charge region of the collector so that $g_c = 0$.

Equation (39) then reduces to

$$G = \eta\beta[1 + \exp(-\alpha w_b)(1 - \exp(-\alpha x_c))]. \quad (40)$$

If the base is thin with respect to the absorption length and $x_c > 1/\alpha$, then the second term can be neglected and

$$G \simeq \eta\beta. \quad (41)$$

Thus, for a properly designed phototransistor the optical gain is the product of the quantum efficiency and the current gain. This would imply that many of the design criteria for high-gain bipolar transistors also apply to phototransistors.

4. Defect Current

In actual phototransistors, the gain may be low in spite of the benefit of the wide-band-gap emitter. Also, in experimental devices the optical gain is usually observed to increase with current. These effects are related to the emitter injection efficiency since the base transport factor is usually near unity and essentially independent of current. Recombination centers in the vicinity of the heterojunction emitter are responsible for a "defect" current (labeled i_{de} in Fig. 1), which reduces the emitter injection efficiency. This "defect" component of the emitter current is not injected into the base and hence does not contribute to the gain. At low light levels, it can account for a significant part of the total emitter current, thus limiting the useful gain.

The exact nature of the defect current is probably complex and at the present time not well understood. In transistors with homojunction emitters, generation–recombination current in the space-charge region of the emitter plays a key role in determining the injection efficiency. Hovel and Milnes (1969) have shown that for transistors with heterojunction emitters, recombination through interface states and some form of trapping–tunneling recombination must also be included. They have proposed that a more accurate expression than Eq. (28) for the emitter injection efficiency would be of the form

$$\eta_e = J_{ne}/(J_{ne} + J_{ir} + J_{tr} + J_{gr}), \quad (42)$$

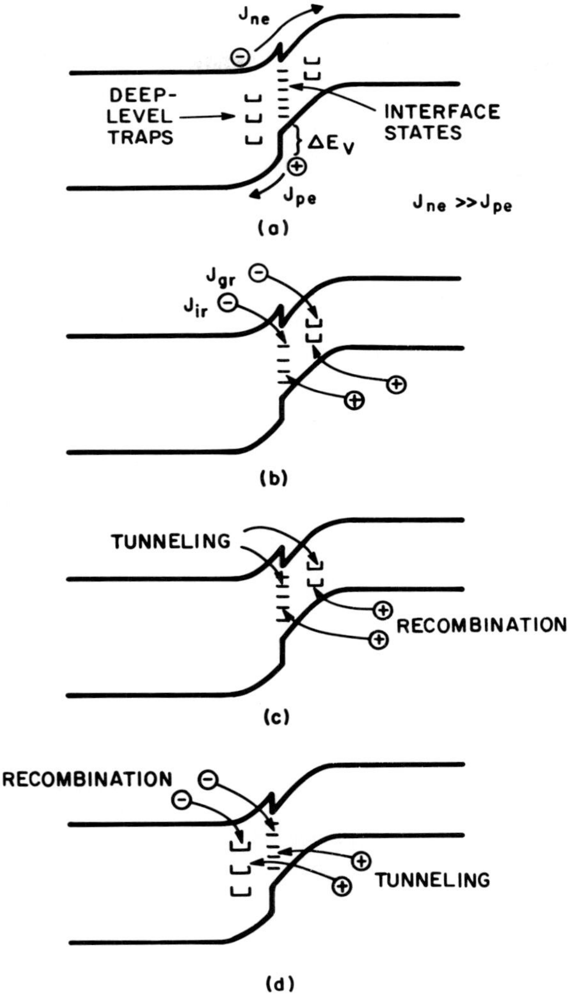

FIG. 4. Possible current-transport mechanisms at the emitter junction of an HPT include (a) normal diffusion current [after Anderson (1962)]; (b) generation–recombination through interface states and deep-level traps; and a combination of electron (c) or hole (d) tunneling and recombination.

where J_{ir} is the current density due to interface recombination, J_{tr} the current density due to trapping–tunneling recombination, and J_{gr} the current density due to generation–recombination. It is assumed in Eq. (42) that the heterobarrier is sufficiently high that J_{pe} can be ignored. These current components are illustrated in Fig. 4.

Hovel and Milnes (1968) have shown that J_{ir} can be determined from

$$J_{\mathrm{ir}} = qn_{\mathrm{i}} \int \frac{N_{\mathrm{IS}}}{(\sigma_n v_{\mathrm{th}})^{-1} + (\sigma_p v_{\mathrm{th}})^{-1}(n_{\mathrm{i}}/p_{\mathrm{i}})}\, dE, \tag{43}$$

where N_{IS} is the density of interface states at an energy E below the conduction band; p_{i} and n_{i} are the hole and electron concentrations at the interface, respectively; v_{th} is the thermal velocity for electrons and holes; and σ_n, σ_p are the capture cross sections. This recombination mechanism is illustrated in Fg. 4b. When $n_{\mathrm{e}} < p_{\mathrm{b}}$, the recombination rate will be limited by the hole-capture rate and $J_{\mathrm{ir}} = gn_{\mathrm{i}}S_{\mathrm{i}}$, where S_{i} is the interface recombination velocity. If J_{ir} is the dominant component of the defect current, then $G \propto J_{\mathrm{ir}}$. The gain will therefore be constant, independent of the emitter current.

The generation–recombination component of the emitter defect current J_{gr} is illustrated in Fig. 4b. This component is comparable to the generation–recombination current in the emitter space-charge region of homojunction transistors. Reddi (1967) has analyzed the homojunction case, both analytically and experimentally, and shown that this type of defect current can be represented by an exponential term as

$$J_{\mathrm{gr}} \propto \exp(q\phi_{\mathrm{e}}/nkT), \tag{44}$$

where $n \neq 1$. The value of n depends on the location of the dominant generation–recombination center in the energy band gap. Using the definitions of G, β, and η_{E}, it is straightforward to show that if J_{gr} is the primary component of the defect current, then the optical gain will vary as a fractional power of the total emitter current:

$$G \propto J^{1-(1/n)}. \tag{45}$$

The most complex and least understood recombination mechanisms that can contribute to the defect current are those where tunneling occurs. Four of these tunneling-recombination mechanisms are illustrated in Fig. 4c and d. In Fig. 4c, electrons from the emitter side of the junction tunnel to interface states or deep-level traps where they recombine with holes. Similarly, Fig. 4d shows holes tunneling to recombination centers from the base region. More complex mechanisms in which electrons are captured by deep-level traps in the emitter depletion region, followed by single- or multistep tunneling to interface recombination states, are also possible (Hovel and Milnes, 1969). For the case of a lightly doped emitter and a heavily doped base, electron tunneling is more probable than hole tunneling. The form of J_{tr} is determined by the balance between the tunneling probability and the electron concentration in the vicinity of the emitter interface. The tunneling probability increases with proximity to the junction, whereas the electron concentration, i.e., the number of candidates for tunneling, decreases as the interface is approached. If the recombination rate is limited by the tunneling probability, which implies that the tunneling probability is low, then the forward current J_{tr} is of the form (Anderson, 1971)

$$J_{tr} = B \exp[-A(\phi_d - \phi_e)], \tag{46}$$

where B is a weak function of voltage and temperature; ϕ_d the built-in voltage; ϕ_e the applied voltage; and A a tunneling constant that depends on the exact shape of the energy barrier, the electron effective mass m_n^*, the equilibrium carrier concentration n_e, and the dielectric constant of the emitter layer ε_e. For the approximation of a linear barrier,

$$A = (4/3\hbar)(m_n^* \varepsilon_e / n_e)^{1/2}. \tag{47}$$

The defect current in Eq. (46) will result in a current-dependent gain of the form

$$G \propto J_{ne}^{(1-\alpha kT/q)}. \tag{48}$$

On the other hand, if the tunneling-recombination rate is limited by the electron concentration, i.e., tunneling very near the interface, then the gain will be independent of the emitter current.

In actual phototransistors, all of the components of the defect current previously described are probably present to some degree, and this combination can lead to complicated gain characteristics. Nevertheless, an accurate characterization of these defect currents and their effect on the optical gain is essential to understanding and improving the performance of phototransistors. A summary of the defect-current recombination mechanisms is presented in Table I.

TABLE I

HETEROJUNCTION DEFECT CURRENT

Recombination mechanism	Defect current	Gain dependence
Generation–recombination	$J_{gr} \propto \exp[qQ_e/mkT]$ $m \neq 1$	$G \propto J_E^{1-1/m}$
Interface–recombination	$J_{ir} \propto S_i$	$G \propto S_i^{-1}$
Tunneling–recombination	$J_{tr} \propto B \exp[-A(V_D - V)]$ (limited by tunneling probability)	$G \propto J_{ne}^{1-\sigma kT/q}$
	$J_{tr} \propto$ constant (limited by electron-capture rate)	$G \propto J_{tr}^{-1}$

5. Junction Displacement

Another factor, in addition to the defect current, that can influence the optical gain in HPTs is the position of the emitter junction relative to the heterojunction interface. Lee and Pearson (1981) have demonstrated that by moving the $p-n$ junction a small distance into the wide-band-gap layer away from the heterojunction interface, the generation–recombination current can be reduced to the extent that the diffusion current J_{ne} dominates even at current densities as low as 10^{-5} A cm^{-2}. Achieving a reduction in the defect current while at the same time maintaining a high injection efficiency with this type of emitter depends critically on the thickness of the p-type wide-band-gap region. This is illustrated in Fig. 5. If this region is too thick (Fig.

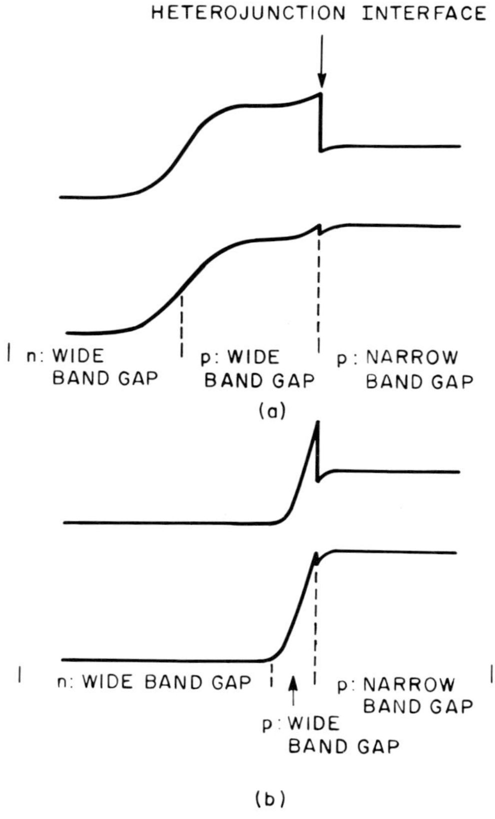

Fig. 5. Energy-band diagram of emitter junction displaced from the heterojunction. In (a) the space-charge regions of the emitter junction and the heterojunction do not overlap; (b) the displacement is small enough for overlap to occur.

5a), the electrons will experience an additional energy barrier, and consequently, there will be a reduction in the injection efficiency. This also increases the effective base width, which also reduces the gain. Campbell *et al.* (1981) have reported that $InP/In_{0.53}Ga_{0.47}As$ HPTs in which the $p-n$ junction is displaced approximately 1 μm from the heterojunction exhibit low gains. On the other hand, the p-type wide-band-gap region must be sufficiently thick to cause a reduction in the defect current. The optimum width is obtained when the space-charge regions of the $p-n$ junction and the heterojunction overlap. This case is shown in Fig. 5b. Sakai *et al.* (1983) have fabricated InGaAsP/InP phototransistors having this type of emitter structure. They report an emitter junction ideality factor n of 1.49, which is an improvement over similar HPTs that they have fabricated that have the $p-n$ junction and heterojunction coincident.

6. Experimental Gain Curves

Working HPTs have been fabricated from several materials, but the first devices to achieve high optical gains used $Al_xGa_{1-x}As/GaAs$ heterojunctions. Most of the $Al_xGa_{1-x}As/GaAs$ HPTs that have been reported have had the structure first described by Alferov *et al.* (1973), an n^--GaAs collector, a p-GaAs base, and an n-$Al_xGa_{1-x}As$ wide-band-gap emitter. The optical gain G of the devices fabricated by Alferov *et al.* was approximately 100 with an estimated current gain β of 300. The wavelength range where these HPTs exhibited high responsivity was determined by the band-gap energies of the $Al_xGa_{1-x}As$ and GaAs layers. The short-wavelength cutoff near 0.65 μm was due to absorption in the emitter. The long-wavelength cutoff occurred near 0.88 μm, the absorption edge of GaAs. Later, two groups, Beneking *et al.* (1976) and Konagai *et al.* (1977), obtained current gains of 2000 and 1600, respectively. All of these HPT structures were grown by liquid-phase epitaxy. Milano *et al.* (1979, 1982) have fabricated HPTs grown by metallo-organic chemical vapor deposition. This is a promising technique for producing the type of structures required for HPTs because thin, uniform layers with defect-free interfaces can readily be obtained. In initial devices, Milano *et al.* have obtained optical gains in excess of 500 and estimated current gains of approximately 700.

The InP/InGaAsP heterojunction has been used to fabricate HPTs with a photoresponse that extends to longer wavelengths than can be achieved with $Al_xGa_{1-x}As/GaAs$. The long-wavelength cutoff for these HPTs is determined by the band-gap energy of the InGaAsP absorbing layer. Optical gains of 600 and 1000 have been reported by Wright *et al.* (1980) and Fritzsche *et al.* (1981), respectively, for HPTs that are photosensitive in the range from $\lambda = 0.9$ to $\lambda = 1.3$ μm. The operating wavelength has been extended to longer wavelengths by Campbell *et al.* (1981) by using the ternary endpoint

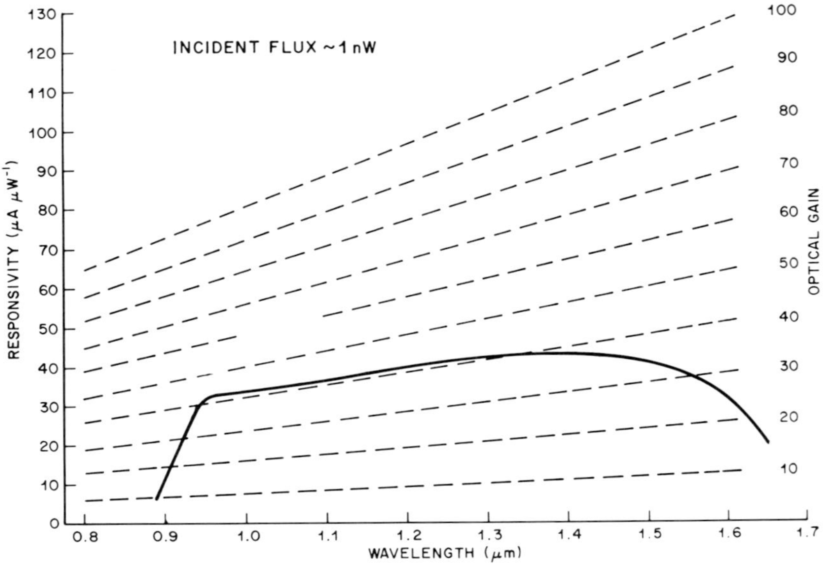

FIG. 6. Spectral response of an InP/InGaAs HPT versus wavelength. The incident optical signal level is 1 nW. [From Campbell *et al.* (1981).]

$In_{0.53}Ga_{0.47}As$ of InGaAsP. The spectral response of this type of HPT is shown in Fig. 6. The long-wavelength cutoff that occurs near 1.65 μm is due to the absorption edge of the $In_{0.53}Ga_{0.47}As$ narrow-band-gap base and collector layers. At shorter wavelengths (<0.95 μm), the photoresponse is limited by absorption in the InP emitter. This curve is relatively flat, from 0.95 to 1.65 μm, and there is appreciable gain ($G \simeq 40$) in spite of the fact that the radiant flux is approximately 1 nW. This indicates that the defect current at the emitter interface is small.

Although very high optical gains have been reported for $Al_xGa_{1-x}As$/GaAs and $In_xGa_{1-x}As_yP_{1-y}$/InP HPTs, for lightwave systems, there is an additional requirement of high sensitivity; i.e., high optical gains must be achieved at very low signal levels. Until recently, phototransistors have not shown much promise in this regard. Appreciable gains could be achieved only at much higher signal levels (usually tens of microwatts) than those encountered in lightwave systems. The most promising result, a gain of 20 at an optical signal level of 1 μW, was reported by Tabatabaie-Alavi *et al.* (1979). Campbell *et al.* (1980), however, have fabricated InP/$In_{0.53}Ga_{0.47}As$ HPTs that exhibited high gains (>100) at low light levels (~ 1 nW). Figure 7a shows a plot of the optical gain of one of these HPTs (diameter, ~ 20 μm) versus the incident optical power level. The same data are plotted in Fig. 7b

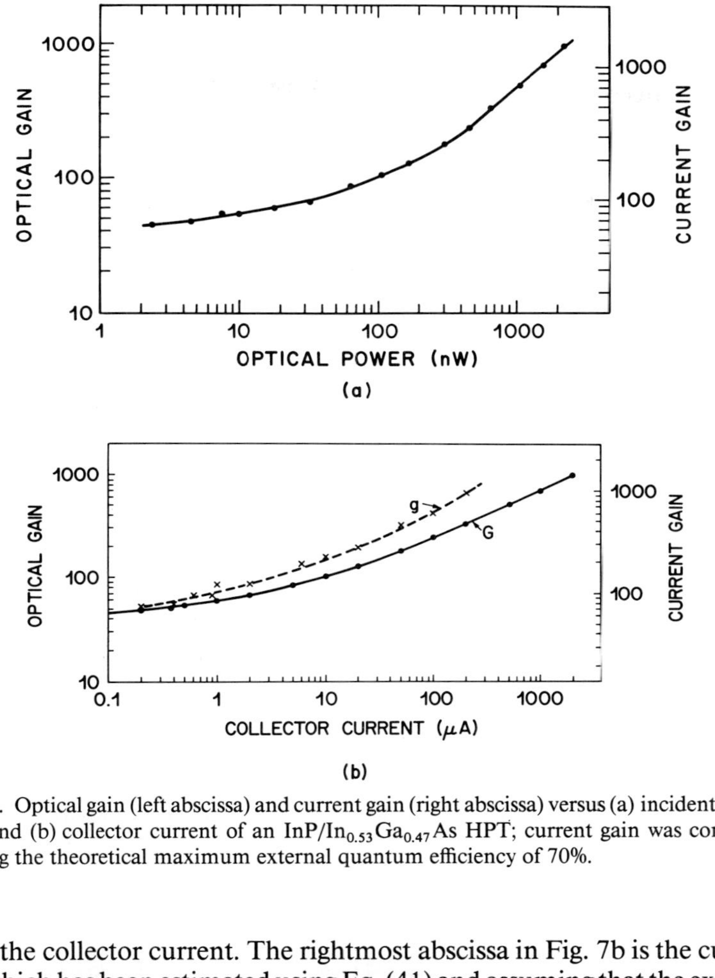

FIG. 7. Optical gain (left abscissa) and current gain (right abscissa) versus (a) incident optical power and (b) collector current of an $InP/In_{0.53}Ga_{0.47}As$ HPT; current gain was computed assuming the theoretical maximum external quantum efficiency of 70%.

versus the collector current. The rightmost abscissa in Fig. 7b is the current gain, which has been estimated using Eq. (41) and assuming that the external quantum efficiency is the theoretical maximum of 70%. This estimate provides a lower limit on the current gain.

The dc optical gain G has been obtained by illuminating the HPT with a radiant flux P_0 from a HeNe laser ($\lambda = 1.15\ \mu m$, $h\nu \cong 1.07$ eV) and by measuring the optical component of the collector current $I_{c\,opt}$:

$$G = (h\nu/q)[(I_c)_{opt}/P_0]. \tag{49}$$

The small-signal gain g is given by

$$g = (h\nu/q)[(\Delta I_c)_{opt}/\Delta P_0], \tag{50}$$

and has been determined by superimposing a small modulated signal on the dc photocurrent.

There appear to be two distinct regions to the gain curves. At the higher values of I_c (Fig. 7b), G increases as a fractional power of the collector current ($G \propto I_c^m$, where m = 0.36). For current gains > 1, $I_c \simeq I_e$ and $G \propto I_E^m$, which is consistent with Eq. (45) when $m = 1 - 1/n$. These results indicate that the dominant component of the defect current for these devices is the generation–recombination term and that the ideality factor of the emitter junction is $n = 1.8$. For these HPTs, ideality factors in the range $1.1 \lesssim n \lesssim 2.0$ have been observed with $n \simeq 1.5$ being typical. In this "high-current" region of the gain curves a maximum gain of 1000 has been achieved at an incident power level of approximately 2 μW. Also in this region, the small-signal gain is larger than the dc gain, a result of the fact that the dc gain is an increasing function of the incident power level.

In the region where $I_c < 10$ μA, the gain curves become relatively independent of the current (or optical signal level P_0). This "flat" portion of the gain curves is due to the leakage current in the reverse-biased collector junction. When this leakage current exceeds the primary photocurrent across the collector junction, the current-dependent emitter injection efficiency will be determined by the leakage current. This is illustrated by the gain curves of another $InP/In_{0.53}Ga_{0.47}As$ HPT shown in Fig. 8. The base-collector leakage current of this particular HPT is approximately 600 nA at the operating voltage (~ 1 V). In the two upper curves, the "leakage" current has been artificially increased to 5 and 20 μA by illuminating the device with an external dc lamp. The effect of increasing the leakage current is to raise the flat portion of the gain curve, causing the intersection with the power-law region to occur at higher current levels. In all three curves, the transition

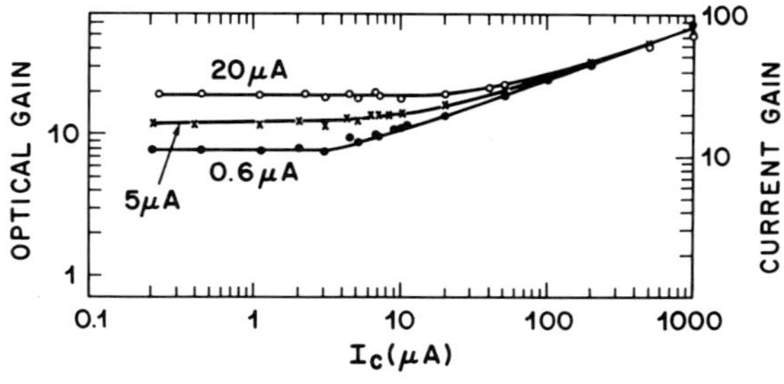

FIG. 8. The dc current gain of a 75-μm-diameter $InP/In_{0.53}Ga_{0.47}As$ HPT for various bias currents. In the lower curve, there is no external bias current, and the collector leakage current due to the dark current of the reverse-biased base–collector junction is 0.6 μA. In the two upper curves, a dc light source has been used to artificially increase the collector leakage current to 5 and 20 μA.

occurs when the signal and leakage components of the collector current are comparable.

In summary, the gain characteristics of HPTs are determined primarily by the properties of the emitter junction. The wide-band-gap emitter permits injection efficiencies very close to unity, and this results, at least in theory, in higher current gains than can be achieved with homojunction transistors. Due to the defect current at the emitter junction, however, the highest gains (>1000) cannot be obtained at very low light levels (<1 μW). On the other hand, useful gains (~ 100) have been demonstrated at incident optical signal levels as low as 1 nW. This is an important development if HPTs are to find application in lightwave systems.

III. Transient Response and Bandwidth

Part II has shown that with proper design and careful attention to materials-related problems, heterojunction phototransistors can achieve high gains at the low optical signal levels that are characteristic of lightwave systems. In addition, the bandwidth of these devices must be large enough to provide gain at the very high bit rates of present lightwave systems; i.e., implementation of optical receivers or photonic circuits utilizing HPTs requires a very high gain–bandwidth product. Since presently available lightwave systems utilize pulse-code-modulation intensity-modulation (PCM IM) and direct detection, the HPT will probably not find application as a small-signal amplifier and thus it is appropriate to deduce the bandwidth from the large-signal transient response.

The kinetics of phototransistors were first analyzed by Ryvkin (1964). More recently, Tsyrlin (1977) has used the standard charge control technique (Sparkes and Beaufoy, 1957) to provide a general treatment of the large-signal transient response of a floating-base HPT. This analysis has shown that the transient response is characterized by a time constant that is determined by the response time of the emitter potential and that depends on the optical signal level. Tsyrlin has also derived approximate expressions for this time constant in terms of materials and device parameters. In their work on HPTs, Milano *et al.* (1982) have extended this analysis to include expressions for the gain–bandwidth product. The following discussion outlines the essential features of this analysis.

It is assumed that the duration of the signal pulse is long with respect to (1) the relaxation of charge in the neutral regions and (2) the relaxation of changes in the minority carrier densities from their steady-state values. The first assumption ensures that any change in the total current is due to a change in the stored charge in the emitter and collector capacitances C_e and C_c. The second assumption is that of quasi-steady state; i.e., the expressions for the minority-carrier distributions and current components derived in Section 1 are valid. The charge in the base is the space charge in the depleted regions of the base near the emitter and collector junctions. This is written as

$$Q_b = Q_{be} + Q_{bc}. \tag{51}$$

The rate of change of Q_b is determined by taking the derivative of Eq. (51) with respect to t:

$$\frac{dQ_b}{dt} = \frac{d(Q_{be} + Q_{bc})}{dV}\frac{dV}{dt}. \tag{52}$$

Using the relation $C = dQ/dV$, Eq. (52) becomes

$$dQ_b/dt = (-kT/q)(C_e + C_c)(dv/dt) \tag{53}$$

where v is the normalized emitter voltage given by $v = (q/kT)(\phi_d - \phi_e)$. The capacitances can be written in terms of v as

$$\begin{aligned} C_e &= \left(\frac{q^2\varepsilon_e\varepsilon_b n_e p_b}{2kT(\varepsilon_e n_e + \varepsilon_b p_b)(Q_d - v)}\right)^{1/2}, \\ C_c &= \left(\frac{q^2\varepsilon_b\varepsilon_c p_b n_c}{2kT(\varepsilon_b p_b + \varepsilon_c n_c)(u - v)}\right)^{1/2}, \end{aligned} \tag{54}$$

where ε is the dielectric constant and $u - v$ the collector potential.

The second charge-control equation is obtained from the charge-conservation condition; i.e., the rate of change of charge in the base is related to the current flowing into and out of the base as given by†

$$dQ_b/dt = I_e + I_c. \tag{55}$$

Using the notation of Section 1, Eq. (55) becomes

$$dQ_b/dt = I_{ne} + I_{pe} + I_{nc} + I_{pc} + I_{dpl}. \tag{56}$$

If ΔE_v is large enough to prevent hole injection from the emitter and the collector depletion region is wide enough that absorption in the neutral portion of the collector can be neglected, then I_{pe} and I_{pc} can be eliminated from Eq. (56). Using Eqs. (13), (15), and (25) for I_{ne}, I_{nc}, and I_{dpl}, respectively, Eq. (56) becomes

$$dQ_b/dt = I_{ph} + 2I_{n0}[\cosh(w_b/L_{nb}) - 1] + I_{n0}[1 - \cosh(w_b/L_{nb})]\exp(v), \tag{57}$$

† At this point it is convenient to switch from current density used in preceeding sections to current. The subscripts of the components are the same, e.g., $J_{ne} \Rightarrow I_{ne}$.

where I_{ph}, the primary photocurrent, is given by

$$\begin{aligned} I_{ph} = g_b P_0 \exp(-\alpha x_b)\{\alpha L_{nb} \sinh(w_b/L_{nb})[1 - \exp(-\alpha w_b)] \\ + [1 - \cosh(w_b/L_{nb})][1 + \exp(-\alpha w_b)]\} \\ + (g\eta P_0/hv) \exp(-\alpha w_b)\{1 - \exp[\alpha(w_b - x_c)]\}. \end{aligned} \tag{58}$$

The steady-state condition is achieved when $dQ_b/dt = 0$, which yields the following expression for the steady-state value of the emitter voltage:

$$\exp(v_f) = \frac{I_{ph} + 2I_{n0}[\cosh(w_b/L_{nb}) - 1]}{I_{n0}[\cosh(w_b/L_{nb}) - 1]}. \tag{59}$$

Using this expression and combining Eqs. (53) and (57), we obtain

$$\frac{dv}{dt} = -\frac{kTI_{n0}[\cosh(w_b/L_{nb}) - 1][\exp(v_f) - \exp(v)]}{q(C_e + C_c)} \tag{60}$$

Integration of Eq. (60) yields the sought-after transient behavior of the emitter voltage

$$\exp(v) = \frac{\exp(v_f)}{1 - [1 + \exp(v_f - v_0)] \exp(-t/\tau_e)}, \tag{61}$$

where v_0 is the initial value of v, and the time constant τ_e is given by

$$\tau_e = \frac{(kT/q)[C_e + C_c]}{I_{ph} + 2I_{n0}[\cosh(w_b/L_{nb}) - 1}. \tag{62}$$

If the gain is $\gg 1$ and $v \gg kT/q$, then Eq. (38) for the optical component of the current is

$$(I_{tot})_{opt} \propto \exp(v),$$

and it follows from Eq. (61) that

$$(I_{tot})_{op} = \frac{I_{ss}}{1 + [(I_{ss}/I_0) - 1] \exp(-t\tau_e)}, \tag{63}$$

where I_{ss} and I_0 are the steady-state and initial values of the current, respectively.

The following three observations provide a clearer understanding of the time constant τ_e:

1. For the case of an effective wide-band-gap emitter, Kroemer's factor $\Gamma \gg 1$, and the expression for the current gain in Eq. (33) becomes $\beta \simeq [\cosh(w_b/L_{nb}) - 1]^{-1}$.

2. The total photocurrent I_{ph} is related to the incident optical signal level as $I_{\text{ph}} = (\eta q/h\nu)P_0$.
3. The collector-junction dark current is $I_{\text{d}} = 2J_{n0}$.

Equation (62) can now be written in the form

$$\tau_{\text{e}} = \frac{(kT/q)[C_{\text{e}} + C_{\text{c}}]}{(\eta q/h\nu)P_0 + I_{\text{d}}/\beta}. \tag{64}$$

The denominator in Eq. (64) is simply I_{c}/β. If we recall that the Shockley emitter resistance R_{e} is given by $R_{\text{e}} = kT/qI_{\text{c}}$, Eq. (64) becomes

$$\tau_{\text{e}} = \beta R_{\text{e}}[C_{\text{e}} + C_{\text{c}}]. \tag{65}$$

The presence of β in this term is due to the Miller feedback effect. From Eqs. (64) and (65) it is evident that the response time of an HPT to a step-function change in the optical signal depends primarily on the charging times of the junction capacitances through the emitter resistance and that this charging is achieved through the primary photocurrent and the dark current. The result is that the time constant is a decreasing function of the signal level. This effect has been observed experimentally by Wright *et al.* (1980) and by Campbell *et al.* (1980). At low signal levels the realization of small time constants, and thus of high operating frequencies, will require that the junction capacitances, particularly the forward-biased emitter capacitance, be minimized.

The time constant τ_{e} can be reduced by introducing an additional dc bias current I_{dc} so that Eq. (65) becomes

$$\tau = \frac{kT}{q}\,\frac{C_{\text{e}} + C_{\text{c}}}{I_{\text{ph}} + I_{\text{d}}/\beta + I_{\text{dc}}}. \tag{66}$$

This bias current could be supplied electrically by adding a base contact or optically with an external dc light source. It should be noted, however, that I_{dc} would also add to the noise current. This is discussed next in Part IV.

It is well known from work on high-frequency microwave transistors that Eq. (66) is only one component of the time constant (Cooke, 1971). The total time constant is

$$\tau = \tau_{\text{e}} + \tau_{\text{b}} + \tau_{\text{d}} + \tau_{\text{c}}, \tag{67}$$

where τ_{b} is the base transit time, τ_{d} the collector depletion-layer transit time, and τ_{c} the *RC* time constant of the collection junction. For lightwave applications, the transit times τ_{b} and τ_{d} will be small relative to τ_{e}. Although τ_{c} is usually much less than τ_{e}, it can in some instances become significant. Hence, a more complete representation of the time constant is

$$\tau = \beta[R_{\text{e}}(C_{\text{e}} + C_{\text{c}}) + R_{\text{L}}C_{\text{c}}]. \tag{68}$$

The average gain–bandwidth product $\bar{f}_c$ is obtained from

$$\bar{f}_c = \beta/(2\pi\tau), \tag{69}$$

which can be combined with Eq. (65) to yield

$$\bar{f}_c = \{2\pi[R_e(C_e + C_c) + R_L C_c]\}^{-1}. \tag{70}$$

The gain–bandwidth product is also referred to in the literature as the cutoff frequency; i.e., the frequency where the gain has dropped to unity.

From the previous discussion, particularly Eq. (64), it can be concluded that a report of the risetime of a phototransistor in response to an optical pulse is of little value unless the optical signal level and the gain are included. Unfortunately, much of the HPT literature is imprecise in this area. The most useful way of characterizing the speed of a phototransistor is to quote the gain–bandwidth product and the optical signal level at which it was measured. For $Al_xGa_{1-x}As/GaAs$ HPTs, Milano *et al.* (1979) obtained $\bar{f}_c \simeq 160$ MHz for $P_0 \simeq 21\ \mu W$, and Beneking *et al.* (1976) reported $\bar{f}_c \simeq 2$ GHz at an unspecified optical signal level. Fritzsche *et al.* (1981) operated an $In_xGa_{1-x}As_yP_{1-y}/InP$ HPT, which had a base contact, at 200 Mbit/sec. For this work they used an external circuit to extract charge from the base and to effect a tradeoff of gain for speed. The gain–bandwidth was determined to be 2 GHz when the optical signal level was 15 μW. Campbell *et al.* (1981) have fabricated a very low capacitance $InP/In_{0.53}Ga_{0.47}As$ HPT, which exhibited a gain–bandwidth product of 2 GHz for a signal level of 2 μW. Figure 9 shows a plot of $\bar{f}_c$ versus P_0 for that HPT. Even at signal levels as low as 100 nW, $\bar{f}_c$ is greater than 100 MHz.

In summary, for signal levels in the range encountered in lightwave systems, the transient response and bandwidth of HPTs is determined primarily by the charging times of the junction capacitances. Furthermore, the time response depends on the signal level. Optimization of the bandwidth of an HPT requires that the junction capacitances be minimized.

IV. Noise Characteristics

The improvements in the optical gain and bandwidth of HPTs, which was described in Parts II and III, has led to several analyses of the noise characteristics of HPTs. The first detailed study of the noise in phototransistors was reported by De La Moneda *et al.* (1971). In that work, the noise current was assumed to consist of only two components: the shot noise of the base current and the shot noise of the collector current. Using a model in which all of the noise generators were referenced to the output, theoretical expressions for the output noise current and the noise equivalent power were derived and compared with experimental data on Si homojunction phototransistors.

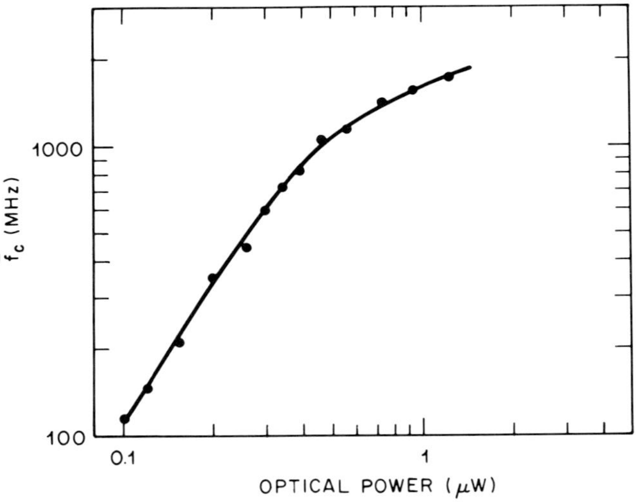

FIG. 9. Cutoff frequency of a small-area (diameter, ~20 μm) $InP/In_{0.53}Ga_{0.47}As$ HPT versus incident optical power; cutoff frequency was determined indirectly from pulse-response measurements.

Milano *et al.* (1982) included the thermal noise in the load resistor and used the modified model to calculate the signal-to-noise ratio of InGaAs/InP HPTs. They demonstrated that the signal-to-noise ratio is very sensitive to the magnitude of the input signal and concluded that (1) optimum performance requires minimization of the junction capacitances, which may be accomplished with an asymmetric structure where the emitter is smaller than the collector; (2) a dc bias current can improve sensitivity by increasing the bandwidth (Part III) only to the point where the shot noise equals the thermal noise; and (3) the overall performance of a well-designed HPT will be comparable to that of a hybrid $p-i-n$/FET combination.

Tabatabaie-Alavi and Fonstad (1981) used what has become the standard analysis of optical receiver design (Smith and Personick, 1980) to derive the sensitivity of an optical receiver utilizing an HPT in the front end. By assuming very low junction capacitances ($C_e + C_c \simeq 0.02-0.05$ pF) and high-current gains ($\beta \simeq 400-1000$), they calculated that the performance of an HPT should be better than that of present-day hybrid $p-i-n$/FET combinations and, in fact, approach that of an avalanche photodiode (APD). Campbell and Ogawa (1982) also used the basic approach of Smith and Personick (1980), but they made modifications to address the fact that a phototransistor is intrinsically a common-collector device. This model is developed later and is used to show the effect of varying device parameters such as gain and leakage current. Using experimental device parameters from InP/$In_{0.53}Ga_{0.47}As$ HPTs, their calculations indicate that the performance of state-of-the-art HPTs (Campbell *et al.*, 1981) should be comparable to that of the best $In_{0.53}Ga_{0.47}As$ $p-i-n$/GaAs FET combination, a somewhat more modest result than that of Tabatabaie-Alavi and Fonstad (1981). At low bit rates, the $p-i-n$/FET would appear to have an advantage, and at higher bit rates, the advantage shifts to the HPT. The bit rate at which both would give the same sensitivity is in the range from 400 to 800 Mbits/sec and is very sensitive to changes in the device parameters.

Brain and Smith (1983) have developed detailed analytic solutions for the sensitivity of long-wavelength optical receivers with HPTs. In their analysis, the receiver sensitivity is expressed explicitly in terms of α, the ratio of the dc bias current to the signal current. The advantage of this approach is that it avoids errors that can result from arbitrary, inconsistent choices of capacitances and load resistance. They point out that in order to keep R_e, the Shockley emitter resistance, from being modulated by the optical signal, α must be greater than unity. They further suggest that the optimistic projections of Tabatabaie-Alavi and Fonstad (1981) are due to an inconsistent treatment of α; i.e., that their derivation of the receiver sensitivity is based on expressions that are valid only if $\alpha > 1$, whereas the parameters used for sensitivity estimates would imply the opposite condition of $\alpha < 1$. In their analysis, Brain and Smith find that the HPT and the hybrid $p-i-n$/FET combination are comparable at bit rates up to approximately 1 Gbit/sec, with the HPT having a slight advantage at the higher bit rates. The crossover of the sensitivity curves occurs between 500 Mbit/sec and 1 Gbit/sec, which is consistent with the calculation of Campbell and Ogawa presented immediately.

Following the theory of Smith and Personick (1980), all noise sources are referenced to the input, and the noise associated with subsequent components of the receiver, such as the post amplifier, equalizer, and filter, are assumed small enough to be neglected. Optical receivers that utilize a *hybrid* $p-i-n$ photodiode/bipolar transistor combination are almost always connected in the common-emitter configuration, a circuit for which the noise-current spectral density is well known. It consists of four principal noise sources: (1) the shot noise of the base current, (2) the shot noise of the collector current, (3) the thermal noise in the load resistor R_L, and (4) the base resistance noise. For the present discussion, the noise associated with the base resistance can be neglected. The base current i_b is the sum of the incident signal photocurrent i_{ph}, the leakage current across the base-collector junction i_d, and an optional dc bias current i_{dc}. This bias current permits the receiver sensitivity to be optimized at the bit rate of interest. Each of these terms contributes full shot noise. However, a correlation effect causes the base-current shot noise to be increased by a factor of 2. This is due to the fact that for each fluctuation in the base current, there is a concomitant fluctuation in that portion of the injected emitter current that compensates the base current. The noise-current spectral density of the base current is thus given by

$$d\langle i^2\rangle_{\text{base}}/df = 4q(i_{\text{ph}} + i_{\text{d}} + i_{\text{dc}}). \tag{71}$$

At the output, the noise current spectral density due to the collector current I_c and the thermal noise of the load resistor R_L are

$$d\langle i^2\rangle_c/df\,|_{\text{output}} = 2qI_c \tag{72}$$

and

$$d\langle i^2\rangle/df\,|_{\text{output}} = 4kT/R_L, \tag{73}$$

respectively. Referenced to the input, these terms become

$$\left.\frac{d\langle i^2\rangle_c}{df}\right|_{\text{input}} + \left.\frac{d\langle i^2\rangle_{\text{th}}}{df}\right|_{\text{input}} = \frac{|Y_{\text{in}}|^2}{|g_m|^2}\left(2qI_c + \frac{4kT}{R_L}\right), \tag{74}$$

where Y_{in} is the input admittance and g_m the transconductance. These parameters are given by

$$Y_{\text{in}} = (1/\beta R_e) + i\omega(C_e + C_c) \tag{75}$$

and

$$g_m = (1/R_e) + i\omega C_c. \tag{76}$$

The base–emitter capacitance is the sum of two terms: a depletion term C_0 and a diffusion term C_d. The diffusion capacitance is given by the relation $C_d = w_b^2/(2R_eD_{nb})$. By combining Eqs. (71)–(76) and using the relations $R_e = kT/qI_c$ and $i_b = I_c/\beta$, the total noise spectral density for the common-emitter circuit can be written as

$$\left.\frac{d\langle i_n^2\rangle}{df}\right|_{\text{CE}} = \frac{4kT}{\beta R_e}\left\{1 + \frac{(0.5 + (R_e/R_L))[1 + (\beta R_e\omega(C_e + C_c))^2]}{\beta[1 + (\omega R_e C_c)^2]}\right\}. \tag{77}$$

It has been pointed out earlier that the phototransistor is inherently a common-collector device. This can be seen clearly in Fig. 10a, which shows a discrete representation of a phototransistor. Figure 10b is the hybrid-π equivalent circuit of an HPT. It can be shown that the signal current of a common-emitter circuit and that of a common-collector circuit that yield the same input voltage are related as

$$i_s(\text{CC}) = (1 + i\omega C_c R_L)i_s(\text{CE}). \tag{78}$$

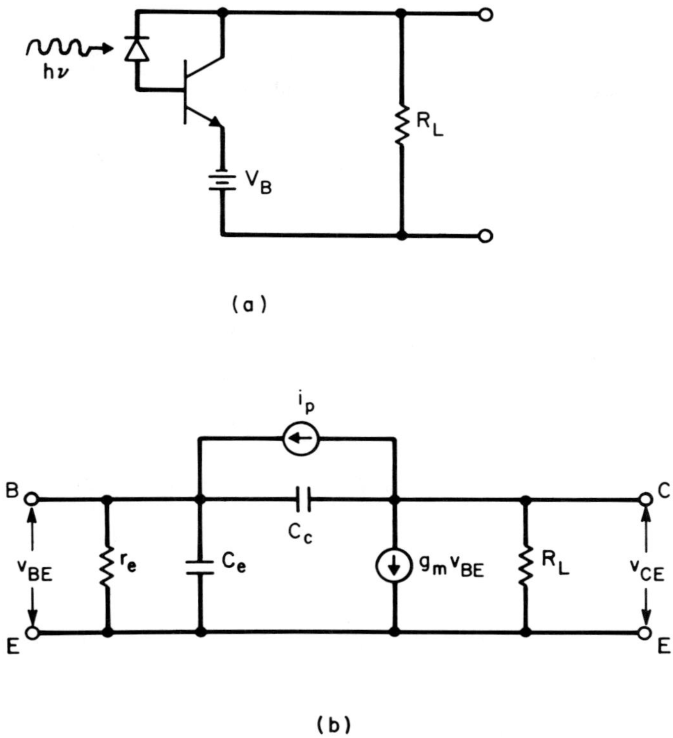

FIG. 10. (a) Circuit model of a phototransistor; note that this circuit is intrinsically common collector; V_B is the bias supply and R_L the load resistor. (b) Equivalent circuit of a phototransistor. [From Campbell and Ogawa (1982).]

If we assume that the noise currents obey the same relation, then the total noise-current spectral density for the HPT becomes

$$\frac{d\langle i_n^2\rangle}{df} = \frac{4kT}{\beta r_e}(1 + (\omega C_c R_L)^2) \times \left\{1 + \frac{(0.5 + (R_e/R_L))[1 + (\beta R_e \omega (C_c + C_e))^2]}{\beta(1 + (R_e \omega C_c)^2)}\right\}. \quad (79)$$

The noise current is obtained by integrating the current spectral density weighted by the frequency transfer function $H(\omega)$:

$$\langle i_n^2\rangle = \int_0^\infty d\langle i_n^2\rangle |H(\omega)|^2. \quad (80)$$

The transfer function is defined as

$$H(\omega) = H_{\text{out}}(\omega)/H_{\text{in}}(\omega), \tag{81}$$

where $H_{\text{in}}(\omega)$ and $H_{\text{out}}(\omega)$ are the Fourier transforms of the input- and output-pulse shapes. If we assume a conventional rectangular input pulse that fills the time slot allotted for each bit, then

$$H_{\text{in}}(\omega) = \frac{\sin(\omega/4B)}{\omega/4B}, \tag{82}$$

where B is the bit rate. An optimum shape for the output pulse is the raised cosine function because this function tends to minimize noise and intersymbol interference. The Fourier transform of the raised cosine function is

$$H_{\text{out}}(\omega) = \cos(\omega/4B). \tag{83}$$

Substituting Eqs. (79), (82), and (83) into Eq. (80) yields the following expression for the total noise current of an optical receiver with an HPT front end:

$$\langle \bar{i}_{\text{n}}^2 \rangle = \frac{32B^2kT}{\pi\beta R_{\text{e}}} \int_0^\infty [1 + (\omega C_{\text{c}} R_{\text{L}})^2] \times \left\{1 + \frac{[0.5 + (R_{\text{e}}/R_{\text{L}})][1 + (\beta R_{\text{e}}\omega(C_{\text{c}} + C_{\text{e}}))^2]}{\beta[1 + (R_{\text{e}}\omega C_{\text{c}})^2]}\right\} \times \tan^2\left(\frac{\omega}{4B}\right) d\omega. \tag{84}$$

Figure 11 shows the noise power if the individual contributions of the base-current shot noise, the collector-current shot noise, and the thermal noise in the load resistor are considered separately. At low bit rates, the base-current shot noise is the significant term, but as B increases, the input admittance rises sharply, causing the other two terms to dominate. The dashed curves in Fig. 11 were computed for a common-emitter circuit. They show that the effect of the common-collector correction term $[1 + (\omega C_{\text{c}} R_{\text{L}})^2]$ can be significant at high bit rates. This term also influences the proper choice of R_{L}. If R_{L} is too high, the effect of the correction term will be large and if R_{L} is too small, the thermal noise ($\propto 1/R_{\text{L}}$) will be excessive. For most cases, the optimum value of R_{L} is approximately $2 \times R_{\text{e}}$, because for this value the thermal noise is equal to the collector-current shot noise. In the following calculations, R_{L} is chosen accordingly.

Figure 12 illustrates the effect of the dc bias current i_{dc}. The solid curves are the same as those in Fig. 11 without a bias current and for the dashed curves $i_{\text{dc}} = 3\ \mu\text{A}$. As the bias current increases, there is a corresponding increase in the base-current shot noise, which causes a degradation of the

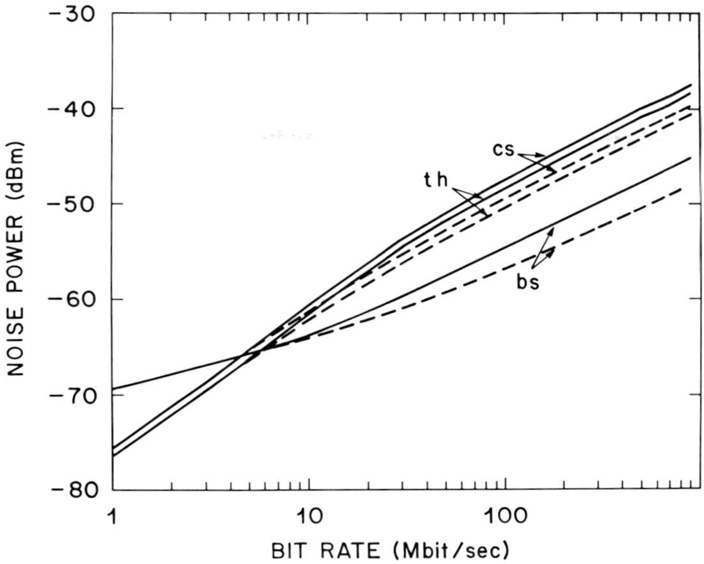

FIG. 11. Base-current shot noise (bs), collector-current shot noise (cs), and thermal noise (th) in the load resistor of an $InP/In_{0.53}Ga_{0.47}As$ HPT; dashed curves show the effect of omitting the common-collector correction term $[1+(\omega C_c R_L)^2]$. Parameters for the plots are $C_e \simeq 1$ pF $+ C_d$, $C_c \simeq 1$ pF, $i_d = 40$ nA, $i_{dc} = 0$, $R_L = 3r_e$, $h_{fe} = 100$. [From Campbell and Ogawa (1982).]

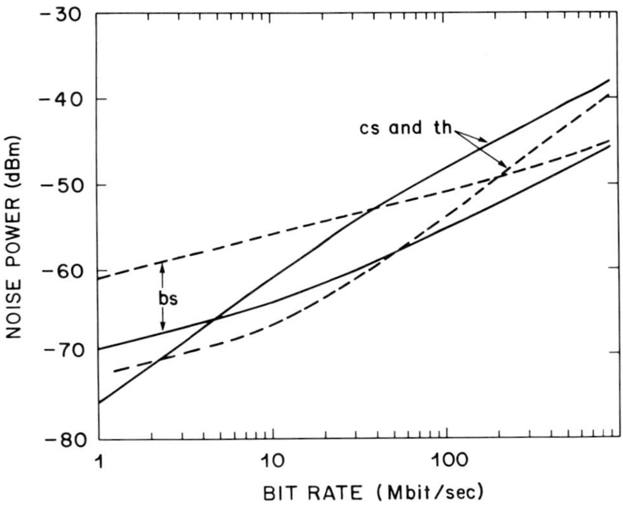

FIG. 12. Noise components of an $InP/In_{0.53}Ga_{0.47}As$ HPT. For the solid and dashed curves, the dc bias current was 0 (solid line) and 3 μA (dashed line), respectively, $C_e \simeq 1$ pF $+ C_d$, $C_c \simeq 1$ pF, $R_L = 2r_e$. The noise components are labeled as in Fig. 11. The load resistor has been selected so that the thermal noise is equal to the collector-current shot noise. [From Campbell and Ogawa (1982).]

receiver sensitivity at low bit rates. On the other hand, at high bit rates where the other two components are largest, the bias current leads to an improvement in receiver sensitivity. This is due primarily to the reduction of the resistance R_e for higher collector current. The net result is that the optimum operating point, i.e., the point where the noise components are comparable, moves to higher bit rates as the bias current is increased. This optimization is a characteristic of bipolar front-end circuits and is an advantage not shared by FET amplifiers.

The minimum signal current required to achieve a 10^{-9} bit-error rate is given by

$$i_s = 6.1\sqrt{\langle i_n^2 \rangle}, \tag{85}$$

and the corresponding average optical power required at the receiver (i.e., the sensitivity) is

$$P = (h\nu/q\eta)i_s. \tag{86}$$

In Fig. 13, the computed sensitivity of InP/InGaAs HPTs (solid curves) is compared with that of a hybrid $In_{0.47}Ga_{0.53}As$ $p-i-n$ photodiode/GaAs–

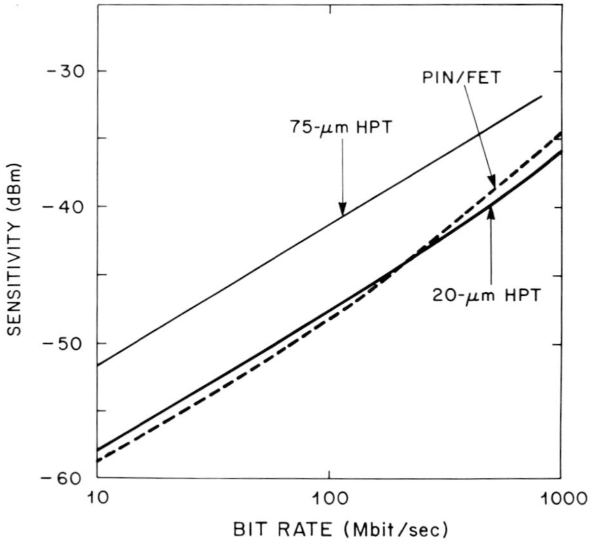

FIG. 13. Light and dark solid curves show the computed sensitivity (using experimental gain curves) of optical receivers with 75- and 20-μm-diameter InP/$In_{0.53}Ga_{0.47}As$ HPTs, respectively; dashed curve is the sensitivity of an $In_{0.53}Ga_{0.47}As$ $p-i-n$ photodiode/GaAs FET combination. [From Campbell and Ogawa (1982).]

FET combination. In these curves $h\nu/\eta q$ has been set to unity. The dashed curve is the sensitivity of the $p-i-n$/FET front end. We have assumed that the $p-i-n$ has a capacitance of 0.3 pF and a leakage current of 10 nA. For the GaAs FET, we have assumed that the gate capacitance is 0.5 pF, the transconductance is 37 msec, the noise factor Γ is 1.78, and the gate-leakage current is 10 nA. The coupling capacitance adds an additional 0.3 pF to the total capacitance. The feedback resistor has been optimized at each bit rate using an open-loop gain of 10. Using experimental gain data and device parameters (Campbell *et al.*, 1981) two HPT curves have been computed. The bias current has been optimized for each bit rate, and the leakage current of the base-collector junction has been estimated as 10 nA. These two curves illustrate the importance of minimizing the junction capacitances. The light solid curve is a 75-μm-diameter HPT having collector and emitter junction capacitances of ~ 1 pF. The sensitivity of this device is significantly less than the hybrid $p-i-n$/FET combination. An improvement of ~ 5 dB is achieved with a small area HPT (diameter, $\lesssim 20\ \mu$m) suitable for use with single-mode fibers. For bit rates less than 400 Mbit/sec, the sensitivity of the $p-i-n$/FET is slightly better (~ 1 dB) than the HPT. At higher bit rates, however, the monolithic HPT structure has a slight advantage. Further improvements in the sensitivity curve of the HPT can be achieved by decreasing the junction capacitances and increasing the gain. Increasing the gain has the effect of lowering the sensitivity curve by approximately $1/\sqrt{\beta}$. It should be pointed out that the sensitivity of the hybrid $p-i-n$/FET combination can also be improved by decreasing the total capacitance and using FETs with higher transconductance. In light of these calculations and the integrated nature of the HPT, along with its ease of fabrication, it would appear that the HPT is an attractive alternative to the hybrid $p-i-n$/FET combination that is used in many present-day optical receivers.

V. Avalanche Effects

The effects of avalanche multiplication on the characteristics of bipolar transistors were first described and analyzed by Miller and Ebers (1955). When biased in the common-emitter (or, equivalently, the common-collector) configuration, avalanche multiplication in the collector junction can lead to (1) an increase in the effective current gain, (2) a softening of the transistor-breakdown characteristic, (3) a lower breakdown voltage than that of the collector junction alone, and (4) switching between a high-voltage low-current "off" state and a lower voltage high-current "on" state. The switching mode of the avalanche transistor exhibits a negative resistance region similar to that of a $p-n-p-n$ switch. Without a base contact, the avalanche transistor switch becomes a bidirectional trigger diode. As one might expect, phototransistors exhibit the same effects due to avalanching as standard bipolar transistors.

The analysis of Miller and Ebers is easily extended to the case of a phototransistor in the floating-base configuration. The equation for the collector current becomes

$$I_c = \frac{(\beta + 1)M}{1 - (M - 1)\beta} [I_{ph} + I_d]. \tag{87}$$

In the derivation of Eq. (87), it has been assumed that the electrons, which are generated in the base, and the holes, which are generated in the collector depletion region, have the same multiplication factors. As in Part II, absorption in the neutral region of the collector is assumed to be negligible. It has been shown (Miller, 1957) that the multiplication factor can be expressed empirically as

$$M = [1 - (V_{BC}/V_a)^m]^{-1}, \tag{88}$$

where V_{BC} is the voltage across the base–collector junction, V_a the breakdown voltage of the collector junction, and m a parameter that is usually in the range 1.5–6.

It is evident from Eq. (87) that avalanche multiplication leads to an effective current gain β', which is expressed as

$$\beta' = M(\beta + 1)/[1 - (M - 1)\beta]. \tag{89}$$

In the limit of no avalanche multiplication ($M = 1$), this expression reduces to the normal current gain β. However, an increase in the bias voltage, and thus the electric field in the space-charge region of the collector, results in an increase in M and leads to an enhancement of β' and I_c. Since β is typically quite large (> 100), M need be only slightly larger than unity for β, and hence I_c, to increase significantly. In fact, there is a singularity when $(M - 1)\beta = 1$, where the apparent increase in I_c and β is limited only by the series resistance in the circuit. This breakdown of the transistor can occur at voltages well below the breakdown of the isolated base–collector junction. Physically, the reason the transistor is so sensitive to such small multiplication factors is due to the regenerative nature of the effect. Initially, the photogenerated holes that are trapped in the base cause electrons to be injected from the emitter. As these pass through the depletion regions of the base–collector junction, secondary electrons and holes are created by impact ionization. The secondary holes are swept back into the base and represent an effective increase in the signal current. This leads to increased injection from the emitter and increased multiplication in the collector; this in turn increases the number of secondary holes swept into the base, calling for more electron injection from the emitter.

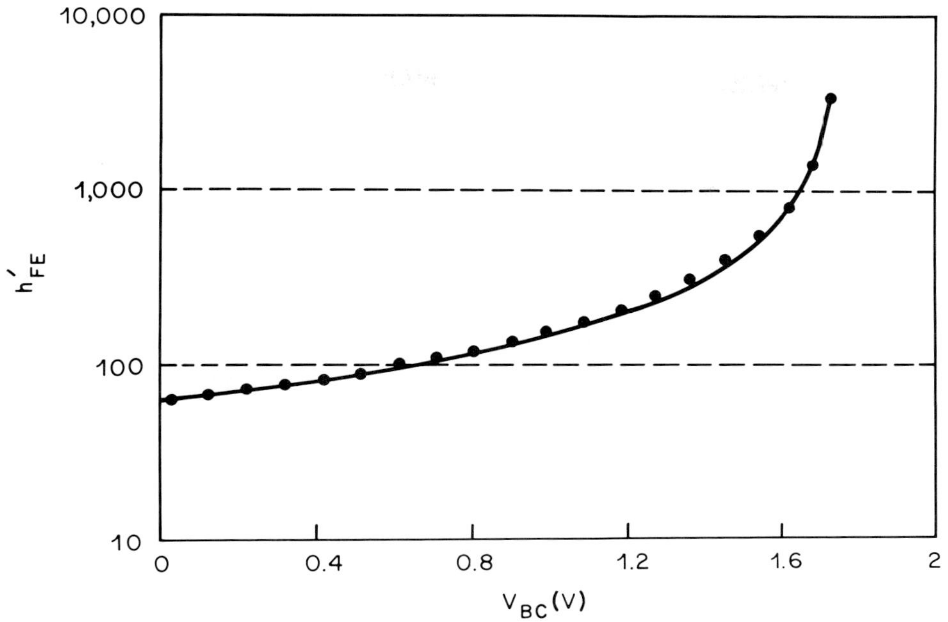

FIG. 14. Effective gain of an avalanche InP/$In_{0.53}Ga_{0.47}As$ HPT versus base-collector bias voltage V_{BC}; (●) experimental points, (—) plot of Eq. (89) with $n = 1$, $V_a = 115$ V. [From Campbell *et al.* (1983). © 1983 IEEE.]

The enhancement of the current gain below the breakdown, or switching point, is illustrated in Fig. 14, which shows the effective gain of an InP/InGaAs HPT as a function of the base–collector voltage V_{BC}. The base–collector voltage was determined by measuring the base–emitter voltage versus current on a separate three-terminal device and subtracting this from the collector voltage of these two-terminal devices. The dots are experimental points obtained at an incident power level of 50 nW. The effective dc current gain β' increases by a factor of 50 from 64 to 3270 as V_{BC} is increased from 0.03 to 1.7 V. The solid line in Fig. 14 is a plot of Eq. (89) using V_a and n as adjustable parameters. A best fit to the data is achieved with $V_a = 115$ V and $n = 1$.

Although it is well known that base narrowing (Early, 1952) can cause bending of the common-emitter characteristics similar to that observed in Fig. 14, this effect is not responsible for the pronounced variation of the collector current with bias shown here. For the range of acceptor concentrations in these HPT's, the effective base width decreases by $<0.02\ \mu m$ (or about 2%) at the maximum applied bias voltage. This is clearly insufficient to account for the observed characteristics.

Changing the optical signal level has a pronounced effect on the shape of the $I-V$ characteristics of avalanche HPTs. For a given signal level, at low bias voltage the normal increase in collector current with signal level is observed. As the bias voltage is increased, however, the curvature in the $I-V$ characteristics increases until switching is observed. Increasing the signal level causes more bending of the $I-V$ characteristics, and the apparent breakdown moves to lower voltages. This is due to the fact that β (and hence G) is an increasing function of the signal level, as shown in Fig. 7. As β increases, the breakdown condition $(M-1)\beta = 1$ is satisfied for lower values of M, which implies lower bias voltage.

Although Fig. 14 shows clearly that avalanching can lead to an enhancement in β, it is important to know whether or not there is an excessive-noise penalty for this additional gain. Related to the output, the noise current in a frequency interval Δf is given by (De La Moneda *et al.*, 1971; Milano *et al.*, 1982)

$$\langle \bar{i}_n^2 \rangle = 4q\beta^2(I_{ph} + I_d)\,\Delta f + 2qI_c\,\Delta f + (4kT\,\Delta f/R_L). \tag{90}$$

As pointed out in Part III, the three noise sources in Eq. (90) correspond to the base-current shot noise, the collector-current shot noise, and the thermal noise in the load resistor, respectively.

The noise current of an $InP/In_{0.53}Ga_{0.47}As$ HPT was measured at 200 kHz, and Eq.(90) was found to provide an excellent fit to the experimental data. At this frequency, the second and third terms in Eq. (90) were approximately 10 times smaller than the first term. In the region of enhanced gain below the switching voltage, as the bias increases the noise current increases in direct proportion to β. Hence, the signal-to-noise ratio should be constant as a function of bias voltage. This is the anticipated result since, in this region, M is so small that there is no appreciable excess noise due to the avalanching process.

It has been shown in Part III that the gain–bandwidth product $\bar{f}_c$ is an important figure of merit for characterizing the large-signal transient response of a phototransistor. Using Eqs. (64) and (69), the effect of avalanching on the gain–bandwidth product has been determined indirectly by measuring the risetime of a small-area (diameter, $\simeq 20\ \mu m$) $InP/In_{0.53}Ga_{0.47}As$ HPT as it responds to a step-function change in illumination. The gain–bandwidth product $\bar{f}_c$, which has been inferred from these pulse-response measurements, is plotted in Fig. 15 versus incident power for various bias voltages. These curves show that even in the regime where the bias voltage and optical signal level are below the switching threshold, the avalanche effect can be used to advantage to achieve a significant increase in the gain–bandwidth product. It must be pointed out, however, that this advantage is achieved at the price of increased complexity in the bias circuitry

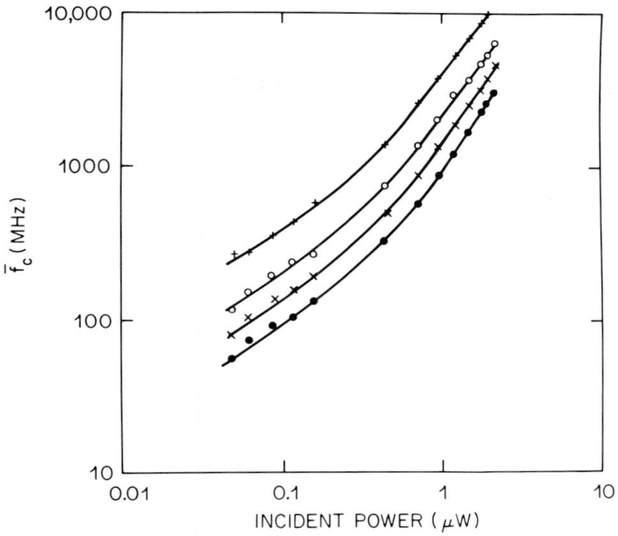

FIG. 15. Gain–bandwidth product $\bar{f}_c$ versus incident optical power for various collector–emitter bias voltages V_{CE}: 0.5 V (●), 1.0 V (×), 1.5 V (○), 2.0 V (+). [From Campbell *et al.* (1983). © 1983 IEEE.]

because the temperature and bias voltage must be controlled more precisely than for a normal HPT.

If the bias voltage is sufficiently close to the breakdown voltage, then the optical signal can initiate switching, and the $I-V$ characteristic will exhibit a region of negative dynamic resistance. This negative resistance is due to the variation of the current gain with collector current. Figure 7 shows that the gain curve of an actual HPT can be a fairly complex function of I_c and hence also of the incident optical power level P_0. Near the breakdown of the avalanche phototransistor, as the collector current begins to increase rapidly, there will also be an increase in h_{FE}, which, if sufficiently large, can result in the condition $(M-1)h_{FE} > 1$. This would represent a state of negative effective gain. As a result, the device will switch to a more stable lower voltage state where $(M-1)h_{FE} \leq 1$.

The $I-V$ curves in Fig. 16 illustrate the switching properties of the avalanche InP/InGaAs HPT. The dashed curve is the dark-current curve. Switching in this case occurs at the voltage, where the leakage current becomes sufficient to trigger the device. The three solid curves illustrate the effect of changing the incident light level. As noted in Fig. 16, these curves correspond to optical power levels of 0.5, 0.95, and 2 μW. All four curves have the same general shape. As the voltage across the device increases, the

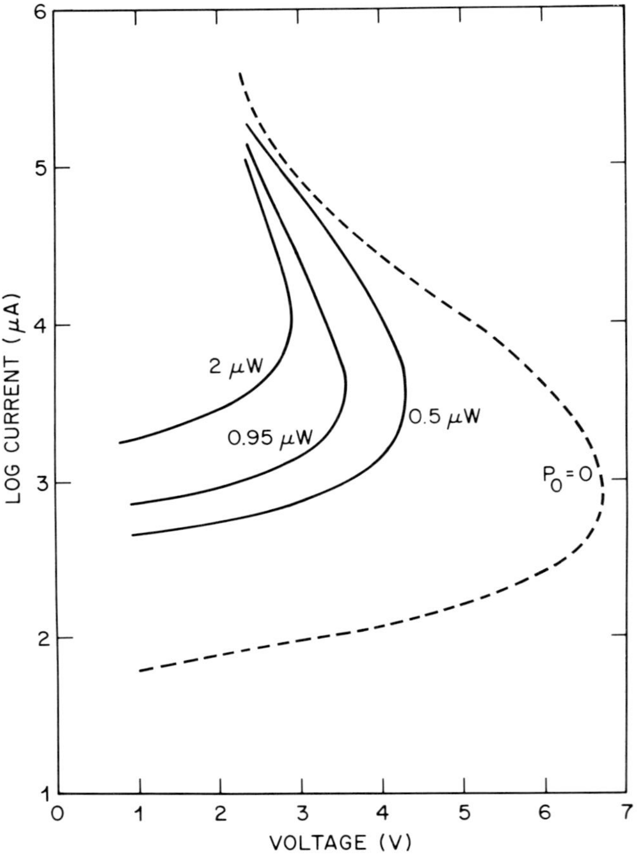

FIG. 16. Current–voltage curves of the avalanche $InP/In_{0.53}Ga_{0.47}As$ heterojunction photodetector for various light levels; dashed curve is the dark-current curve; three solid curves show the response to optical power levels of 0.5, 1, and 2 μW. [From Campbell *et al.* (1983). © 1983 IEEE.]

current increases until a voltage referred to as the turnover voltage V_t is reached. At this point, the slope of the $I-V$ curve becomes infinite. Beyond this point, any increase in the bias voltage results in a voltage drop across the device and an increase in the current; i.e., the dynamic resistance is negative. These curves also illustrate that an increase in the optical signal level causes the turnover voltage to be reduced. Figure 17 is a plot of V_t versus the incident optical power level P_0. On the vertical axis, V_t has been normalized to the dark-current value V_{t0}. For $P_0 = 5$ nW, the device must be biased to within about 1% of V_{t0} for the optical signal to successfully initiate switching.

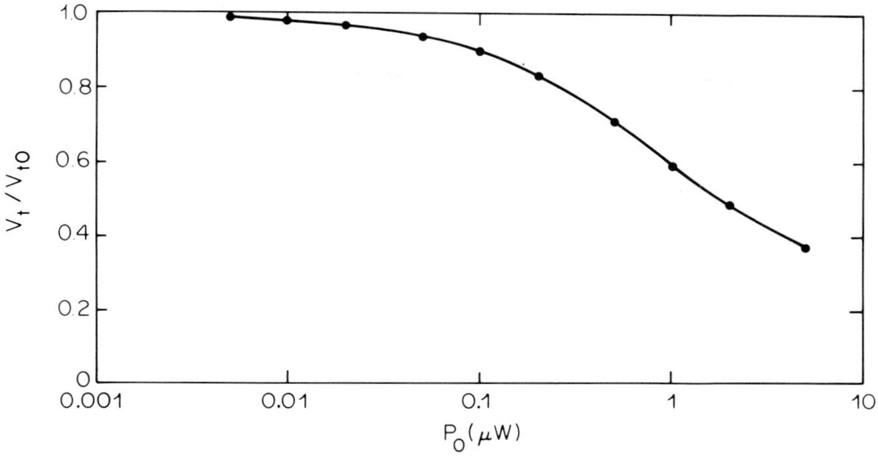

FIG. 17. Turnover voltage V_t versus optical power level P_0; turnover voltage is normalized to its dark-current value V_{t0}. [From Campbell *et al.* (1983). © 1983 IEEE.]

Although this degree of precision for the bias voltage would probably necessitate a relatively complex bias circuit and temperature stabilization, these bias requirements are no more stringent than those commonly encountered for avalanche photodiodes. For higher values of P_0, more latitude is permitted. For example, at $P_0 = 1\ \mu$W, the required bias is only about 50% of V_{t0}.

The dependence of V_t on P_0 described previously suggests that conversely when the bias voltage is set at a specific value below V_{t0}, this device can be used to distinguish between optical pulses having different peak power levels. For example, consider a voltage V_1 that corresponds to V_t for a power level P_1. If an incident optical pulse has a peak power less than P_1, the avalanche HPT will not switch to its high-current state. On the other hand, if the peak power exceeds P_1, a high-current pulse will result. In essence, this device will be in either a low-current "off" state or a high-current "on" state, depending on whether or not the incident pulse exceeds the threshold power required for switching at the set bias voltage. It can be seen in Fig. 17 that the threshold power can be varied by at least three orders of magnitude by changing the bias voltage. Operation as an optical comparator is illustrated in Fig. 18, which shows I_c versus P_0 for three different bias voltages: 5.4, 5.9, and 6.4 V. As expected, higher voltage corresponds to a lower threshold power. The magnitude of I_c in the high-current state is independent of the light intensity, being determined solely by the series resistance, which in this case was 500 Ω. Characteristic of a threshold phenomena, the transition region is very steep, as evidenced by the fact that a 2% change in P_0 is sufficient to cause I_c to increase by two orders of magnitude.

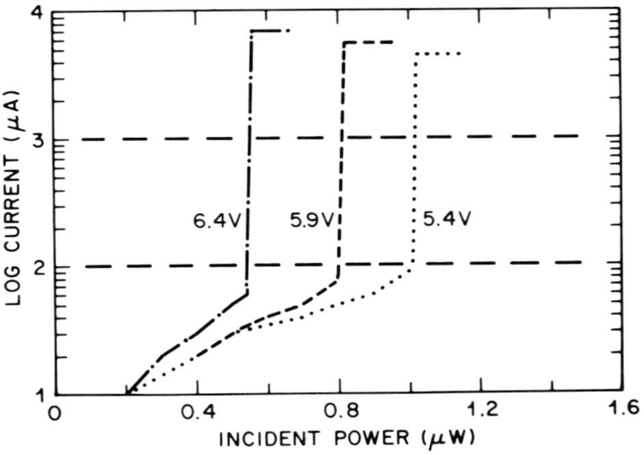

FIG. 18. Collector current versus optical power for various values of the bias voltage; note that a change in this bias voltage causes a shift in the optical threshold level. [From Campbell *et al.* (1983). © 1983 IEEE.]

The pulse response of these devices is interesting and quite different from that of an HPT in its normal mode of operation. Normally, an HPT exhibits minimal time delay after the incident optical pulse and a risetime that is limited at low light levels by the charging time of the base–emitter junction. The risetime is typically in the range 50–500 nsec. In the switching mode, the current gain is very high ($\sim 1 \times 10^5$), the output is a narrow current spike whose width is independent of the width of the optical pulse, the risetime ($\sim$20–30 nsec) is less than in the normal phototransistor mode, and there is an appreciable delay between the leading edge of the optical pulse and the output-current pulse. The risetime represents the actual switching time. However, there is an additional time required for the current to build up to its threshold value. This is the delay time. The delay time will be determined by the charging time of the junction capacitances and is given by the expression

$$t_d = \frac{4.4(kT/q)(C_e + C_c)}{MI_{ph} + [(M+1)/2\beta]I_d} \sum_{m=0}^{N} \frac{1}{[\beta(M-1)]^m}. \tag{91}$$

In the case of no multiplication, Eq. (91) reduces to the standard time constant for the large-signal transient response of an HPT [Eq. (64)]. The number of terms N required in the summation depends on the ratio of the collector current to the initial photocurrent at the turnover point. A typical value for this ratio is on the order of 400, which means that only two or three terms in the summation are required. It is evident from Eq. (91) that the currents available to charge the junction capacitors are the primary photocurrent, the secondary-hole current, and the dark current. Thus, the delay time should decrease as the optical intensity or the bias voltage (and thus M) is increased. This is, in fact, what was observed experimentally. The delay time decreased from 600 to 100 nsec as the incident optical power increased from 1 to 4 μW. The problem of minimizing the delay time in the switching mode is the same as that of optimizing the response times in the normal phototransistor mode; namely, it is necessary to reduce the junction capacitance to the greatest extent possible.

VI. Novel Structures

In Parts II – V, the characteristics of bipolar-type phototransistors with the wide-band-gap emitters have been described in some detail. In the past few years, several variations on the basic transistor structure have been reported. These modified structures have varied a great deal in complexity and, to a lesser extent, also in function. Some of the simpler devices have incorporated a change in the emitter, base, or collector in order to improve one or more of the operating characteristics. In more complex structures, the phototransistor has been integrated with another electrooptic device such as a semiconductor injection laser to form a monolithic photonic circuit. In Part VI, a brief description of some of these novel phototransistor structures is presented, roughly in order of increasing complexity.

In the analysis of Part II, it was assumed that the depleted portions of the base layer were much narrower than the neutral region. It is well known that if the doping level in the base is low, the effective base width decreases with increasing collector voltage (Early, 1952). As a result of this modulation of the base width, an enhancement in the dc gain can be achieved by increasing the bias voltage. Chen *et al.* (1981a,b) have investigated the extreme case of base narrowing; i.e., the condition of total depletion of the base. The depleted-base structure was grown by molecular beam epitaxy (MBE) and is similar to that of previous HPT's in that it has a wide-band-gap $Al_xGa_{1-x}As$ emitter, and light is absorbed in the GaAs base and collector layers. It differs, however, in two important respects. First, although the base is heavily doped ($\sim 1 \times 10^{18}$ cm^{-3}), it is sufficiently thin (~ 60 Å) so as to be completely depleted even at zero bias. The barrier to injection from the emitter is due to the band-gap discontinuity. The gain mechanism is similar to that of a standard phototransistor; the accumulation of photogenerated carriers in the base causes barrier lowering and results in injection from the emitter into the base. The combination of the thin base and high injection efficiency have produced the highest sensitivity reported to date. A dc gain greater than 1000 was achieved with an $n-p-n$ structure at an incident optical signal level of a few nanowatts and the gain of a similar $p-n-p$ structure was approximately

100 at the same signal level. In contrast to the gain curves shown in Fig. 7, the gain of this depleted-base device decreases as the signal level increases. This behavior is due to the saturation of the barrier lowering (Chen, 1981). A second distinction for this device is that the emitter and collector layers are very lightly doped ($\sim 8 \times 10^{13}$ cm^{-3}), and hence the junction capacitances are low. The capacitance measured between the emitter and collector terminals of a device having an area of 1.3×10^{-4} cm^2 was ~ 0.35 pF independent of applied voltage. This is important in light of the analysis of Part III, which demonstrated that optimum frequency response is achieved by minimizing the capacitance. The pulse response of these depleted-base phototransistors was measured using a dye-laser pulse, which had a peak power of 20 mW and a pulse width of 4 psec. For an $n-p-n$ structure, the risetime was 50 psec and the falltime was 600 psec. The complementary $p-n-p$ structure exhibited a full width at half maximum of less than 50 psec. It should be pointed out that at this high signal level, the quantum efficiency of these devices is less than 1% due to the decrease in gain at high signal levels. It is to be expected that at lower peak power levels, gain can be achieved but with longer response times. To date, there is no data available on the pulse response or gain–bandwidth product at low signal levels.

Kroemer (1957a,b) first suggested that the base transit time of a transistor could be reduced by grading the composition (band-gap energy) of the base. This grading creates a quasi-electric field within the base and introduces a drift component to the minority-carrier velocity. The result is a decrease in the transit time, which is usually diffusion limited. Capasso *et al.* (1983) have applied this concept to the phototransistor. A cross-sectional schematic and an energy-band diagram for this MBE-grown $n-p-n$ structure are shown in Fig. 19. The collector is n^--GaAs and the wide-band-gap emitter is n^--$Al_{0.45}Ga_{0.55}As$ ($E_g \simeq 2.0$ eV). The p^+ base layer has been compositionally graded from GaAs at the collector junction to $Al_{0.20}Ga_{0.80}As$ ($E_g \simeq 1.8$ eV) at the emitter junction. The band-gap discontinuity at the emitter junction is $\Delta E_g \simeq 0.2$ eV. Referring to Fig. 3, it is evident that this discontinuity is sufficient to prevent reverse injection into the emitter.

In Part III, it was pointed out that the bandwidth of a phototransistor at low signal levels is limited by the charging time of the emitter and collector capacitors and not by the base transit time. As a result, the benefit of the graded base will be realized only at very high signal levels. In order to observe the effect of the graded base and to determine the ultimate intrinsic speed of this structure, Capasso *et al.* (1983) have examined the pulse response at peak powers in the range 10 mW to 10 W. At zero bias it behaves as an ultrahigh-speed photodetector with a risetime of 20 psec and a full width at half maximum of 40 psec. When biased as a phototransistor, higher responsivity is observed, and the risetime is the same as the zero bias case, but the

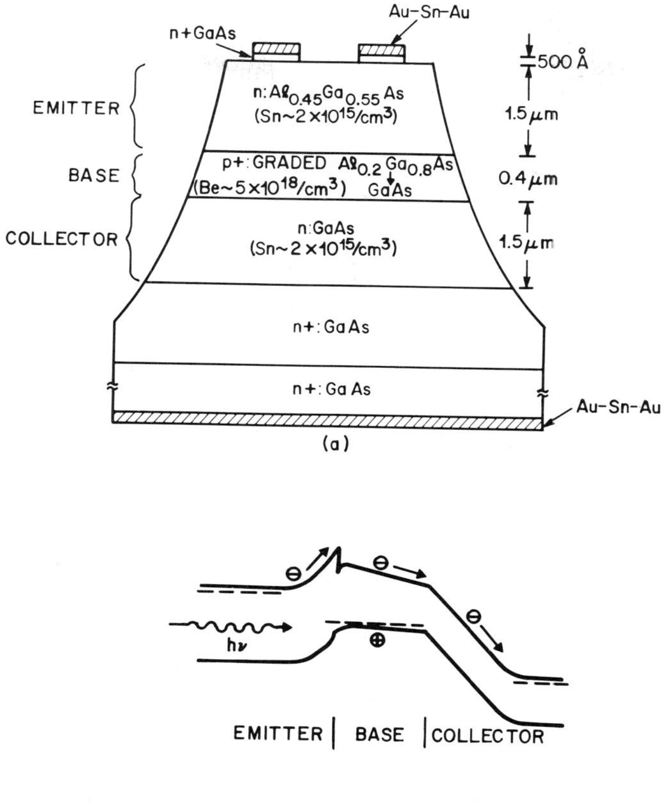

FIG. 19. (a) Schematic cross section and (b) energy-band diagram of a compositionally graded base HPT. [From Capasso *et al.* (1983). © 1983 IEEE.]

decay exhibits a long tail with a time constant of a few hundred picoseconds. For both bias conditions, the gain was less than unity. Analysis of the pulse-response data has shown that the graded-base structure has reduced the base transit time by at least a factor of 4.

A third structure with a modified base layer is the depletion-stop double-base (DSDB) phototransistor (Chen *et al.*, 1982). This device has the standard wide-band-gap-emitter structure, except that the base is composed of two regions with different doping levels. A one-dimensional model and doping profile of this structure is shown in Fig. 20. The region next to the emitter is lightly doped because it has been found, particularly with MBE-grown structures, that reducing the doping levels near the heterojunction

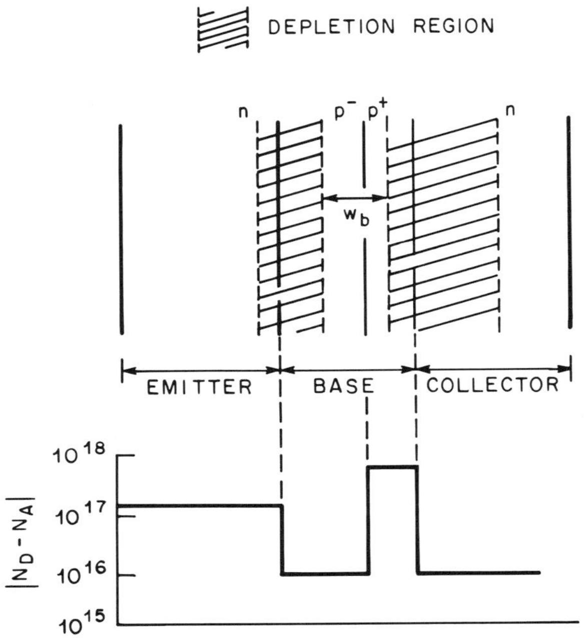

FIG. 20. One-dimensional model and doping profile of the DSDB phototransistor. [After Chen *et al.* (1982a,b).]

interface decreases the defect current. The net result is an improvement in the emitter injection efficiency, especially at low signal levels. On the other hand, low doping of the whole base can lead to excessive base resistance and undesirable modulation of the base width via the Early effect. To circumvent these difficulties a heavily doped base region has been introduced adjacent to the collector. Chen *et al.* (1982) have demonstrated the effectiveness of this double-base approach with an $Al_{0.35}Ga_{0.65}As$/GaAs wide-band-gap-emitter structure. The maximum dc gain obtained was 10 for a 40-μW optical signal, and this decreased to approximately 3 as the signal level was reduced to 2 μW. Although higher phototransistor gains have been reported, this was a significant improvement over similar MBE-grown phototransistors without the double base. In fact, this is the first report of gain in a phototransistor (other than the depleted-base structure) grown by MBE. A consideration for this type of structure is that the various doping levels in the base create an electric field that opposes the normal carrier diffusion through the base. Minimizing the effect of this field on the base transport factor while maintaining high injection efficiency imposes narrow tolerances on the width and doping levels of the two base regions.

The performance of Si bipolar switching transistors is degraded by charge storage in the base region. May (1968) has shown that this effect can be minimized by substituting a Schottky contact for the collector because the metal–semiconductor junction does not inject minority carriers into the base as does a $p-n$-junction collector at saturation. A second advantage of the Schottky collector structure is ease of fabrication and potential for integration. Beneking *et al.* (1980b) were the first to extend the Schottky collector structure to the GaAs material system with a wide-band-gap $Al_xGa_{1-x}As$ emitter. More recently, Naitoh *et al.* (1982) have applied this concept to phototransistors. Their device (Fig. 21) has an n-InP wide-band-gap emitter, a p-$In_{0.69}Ga_{0.31}As_{0.71}P_{0.29}$ base, and on Au Schottky collector. For this structure, absorption occurs only in the base in contrast to the standard structure where much of the absorption occurs in the collector depletion region. If the base is kept thin, a condition for high gain, the absorbing region could be too thin to yield high quantum efficiencies. This can be overcome to some exent, however, if the base is relatively thick ($w_b \gtrsim 3/\alpha$) and lightly doped so that much of the base is depleted and the neutral base region is thin. The maximum current gain achieved with this structure was 1000 at an input signal level of 7 μW, and the low light-level gain was 100 for signal levels of a few nanowatts (Sakai *et al.,* 1983). The leakage current at a bias voltage of 1 V was 20 nA (7.5×10^{-5} A cm^{-2}). One feature of the devices that have been fabricated so far is that the apparent breakdown voltage is strongly depen-

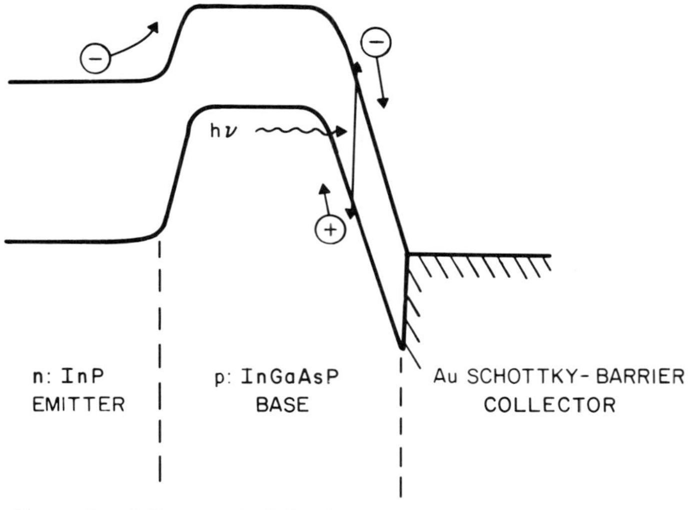

FIG. 21. Energy-band diagram of a Schottky collector $InP/In_xGa_{1-x}As_yP_{1-y}$ HPT. [After Sakai *et al.* (1983). © 1983 IEEE.]

dent on the input optical power level. In fact, the $I-V$ characteristics are very similar to those of the avalanche HPT (Campbell *et al.*, 1983).

The current flow in all of the structures thus far described is perpendicular to the surface. This has the advantage of providing a large optically sensitive area while at the same time maintaining a thin ($<1\ \mu$m) base width. For lightwave applications, this is important because it provides a good match between the area of the phototransistor and the light spot without sacrificing device performance. An alternative is a lateral structure in which the current flows parallel to the surface. The primary advantage to this approach is that it facilitates integration with other devices. Chen and Gustafson (1980) have fabricated an interdigital Si lateral phototransistor. The advantages of this particular structure are that it provides long ($\sim 10\ \mu$m) absorption lengths to extend the wavelength range and a large sensitive area. When operated as a photodiode, this detector exhibited a 40% quantum efficiency and subnanosecond-pulse response. Functioning as a phototransistor a dc current gain of 15 was observed. Ion-implanted III–V lateral $p-n-p$ transistors have been reported (Tabatabaie-Alavi *et al.*, 1982; Krautle *et al.*, 1982). Success in this area may eventually be transferred to phototransistors.

More complex structures in which a phototransistor is integrated with another device have also been reported. One of these is a photo-Darlington (Sakai *et al.*, 1983) circuit. It consists of two InP/InGaAsP wide-band-gap-emitter transistors connected in the Darlington configuration. Figure 22 shows a circuit diagram and a schematic cross section of this structure. One transistor operates as a normal phototransistor and the second serves as an amplifier to provide additional gain. This device has been designed to provide high drive currents so that it can be interfaced directly with devices such as injection lasers or light-emitting diodes (LEDs). To date, the highest gain achieved with this integrated structure was 100 at an incident optical signal level of approximately 1 mW. The total gain was less than the product of gains measured on single phototransistors. This was attributed to crystal nonuniformity and processing difficulties related to etching the emitter mesas. Nevertheless, this structure demonstrated the capacity to deliver up to 100 mA of collector current with no adverse heating effects.

The analysis of the noise characteristics of an HPT in Part IV showed that optimum performance at a given bit rate requires a dc bias current in addition to the signal current. Most of the HPTs that have been reported to date, however, are operated with a floating base, thus eliminating the possibility of providing the dc bias current electrically. Campbell and Dentai (1982) have demonstrated that the required dc bias current can be supplied optically by an LED integrated onto the same chip as the phototransistor. Optimization of the signal-to-noise ratio can then be accomplished by varying the LED drive current. A schematic cross section of these two devices is shown in Fig.

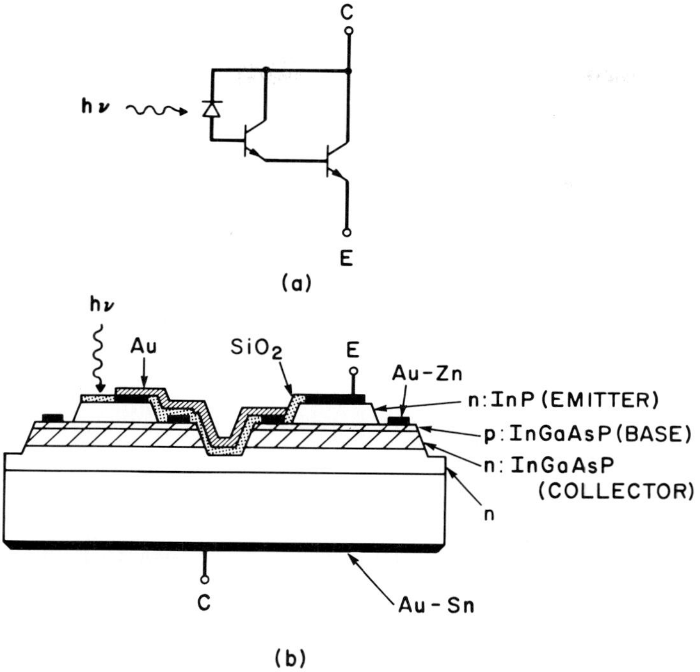

FIG. 22. (a) Circuit model of a photo-Darlington; (b) schematic cross section of a photo-Darlington fabricated from two InP/$In_xGa_{1-x}As_yP_{1-y}$ wide-band-gap-emitter HPTs. [From Sakai *et al.* (1983). © 1983 IEEE.]

23. The HPT is a standard InP/$In_{0.53}Ga_{0.47}As$ wide-band-gap emitter structure, and the LED is a simple InP homojunction that has been created by diffusing Zn into the InP substrate. Electrical isolation between the HPT and the LED has been achieved by etching mesas in a dilute bromine–methanol solution. The shape of the gain curve of the HPT was similar to that shown in Fig. 7; the low light-level gain was approximately 30, and a maximum gain of 170 was achieved at an incident power level of 18 μW. Based on the measured gain of the phototransistor and the power output of the LED, the coupling efficiency was estimated at 1%. This is consistent with the coupling efficiency determined from a purely geometric point of view by evaluating the overlap of the absorbing region of the HPT with the radiation pattern of the LED. Although this seems to be sufficient for the present case, higher coupling efficiencies could be achieved by (1) decreasing the separation between the LED and HPT from its present value of 50 μm, (2) making the HPT concentric to the LED, or (3) using a narrow proton-bombarded region

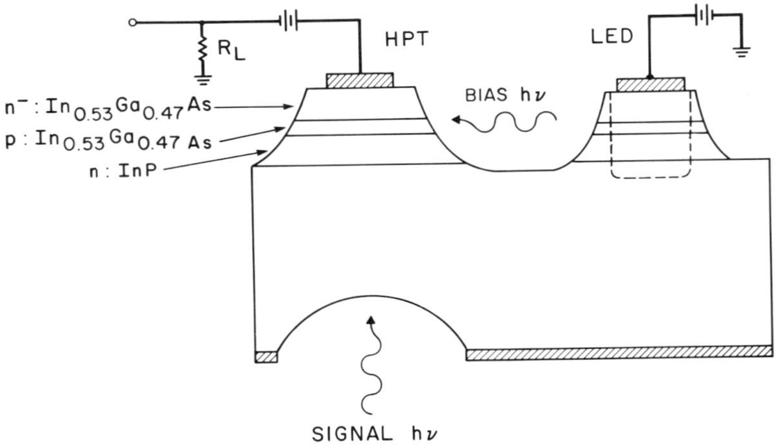

FIG. 23. Schematic cross section of an InP/$In_{0.53}Ga_{0.47}As$ HPT integrated with an LED. [From Campbell and Dentai (1982).]

to provide electrical isolation between the HPT and LED. It has been pointed out earlier that the gain and bandwidth improve with increasing current. Hence it is not surprising that increasing the current through the LED, and thus the dc optical power coupled into the HPT, lead to increased gain and bandwidth. Specifically, a drive current of 6 mA through the LED resulted in a threefold decrease in the risetime (from 210 to 70 nsec) and a twofold increase in the gain (from 30 to 60), compared to the case with no dc optical bias.

A number of applications that utilize a phototransistor to (1) detect an optical signal and (2) supply the drive current for an LED or semiconductor laser have been proposed. These include an optical amplifier (Beneking *et al.*, 1980a; Sasaki and Kuzuhara, 1981); an image converter (Beneking, 1981); a wavelength converter (Beneking *et al.*, 1981); and an optical switch or bistable device (Sasaki *et al.*, 1982). Optical amplification using a hybrid combination of an $Al_xGa_{1-x}As$/GaAs HPT and an $Al_xGa_{1-x}As$/GaAs semiconductor laser was first demonstrated by Beneking *et al.* (1980a). The laser and the HPT were connected in series with the laser biased near its threshold. The collector current resulting from light incident on the HPT provided additional drive current to the laser. In this work, the optical output power from the laser was 27 dB greater than the input to the HPT. Based on the turn-on (~ 16 nsec) and turn-off times (~ 24 nsec), the authors projected that this optical amplifier could be operated up to 30 Mbit/sec. A similar discrete combination using an HPT and an injection laser fabricated from InP/InGaAsP has been reported by Sasaki *et al.* (1982). With an input power of 44 μW, an amplification of 20 dB was achieved.

In the first attempt to integrate this HPT/emitter combination, an $Al_xGa_{1-x}As/GaAs$ HPT structure was grown on one side of a GaAs substrate and an $Al_xGa_{1-x}As/GaAs$ double-heterostructure (DH) LED was grown on the opposite side (Beneking *et al.*, 1981). By adjusting the crystal composition of the active layer so that the LED emission was in the red spectral range (775 nm $\gtrsim \lambda \gtrsim$ 680 nm), this device functioned as an infrared-to-visible wavelength converter as well as an optical amplifier. The amplification that could be achieved with this device was a function of the input optical signal level. A maximum amplification factor of 8 was achieved for approximately 100 μW of input power. A modified version of this device in which both the LED and HPT were grown on the same side of the substrate was integrated into an array and used for image conversion (Beneking, 1981).

A long-wavelength integrated HPT/LED has been fabricated by Sasaki *et al.* (1982). A cross section of this device is shown in Fig. 24. This structure has been realized by the successive growth of a DH LED on an InP/InGaAsP HPT. In principle, light is incident on the HPT through the transparent InP substrate and absorbed in the InGaAsP base and collector layers. The collector current of the HPT then flows into the *n*-type cladding layer of the LED. In practice, the interaction between the LED and the HPT gives rise to a variety of device characteristics. This is due to feedback from the LED to the HPT. The origin of this feedback can be either optical or electrical. Optical feedback occurs when the emission from the LED is absorbed in the HPT, and electrical feedback occurs by the same mechanism as a $p-n-p-n$ switch; i.e., holes are injected back into the base of the HPT. The behavior of this device is determined in large part by the amount of optical and electrical feedback. Electrical feedback can be eliminated by increasing the separation between the HPT and LED. This is accomplished by making the HPT collector and/or the *n*-type cladding layer of the LED wider. Optical feedback can be reduced by decreasing the band-gap energy of the LED active layer so that the base and collector layers are transparent to the LED emission. When both feedback mechanisms are suppressed, this device should operate as a simpe light amplifier. To date, however, Sasaki *et al.* have not observed gain. If either optical or electrical feedback is strong, the characteristics of this device are the same as those of a light-activated electroluminescent switch (Campbell *et al.*, 1982). The $I-V$ curves have regions of negative dynamic resistance, and, depending on the load resistor, bistable operation or switching behavior is obtained. In the switching mode, an incident optical pulse causes the device to switch from a high-voltage low-current state to a low-voltage high-current state. The turnover voltage in the $I-V$ curve decreases with increasing optical input power similar to that shown in Fig. 16. In the high-current state, emission from the LED is observed even after the incident light signal is removed.

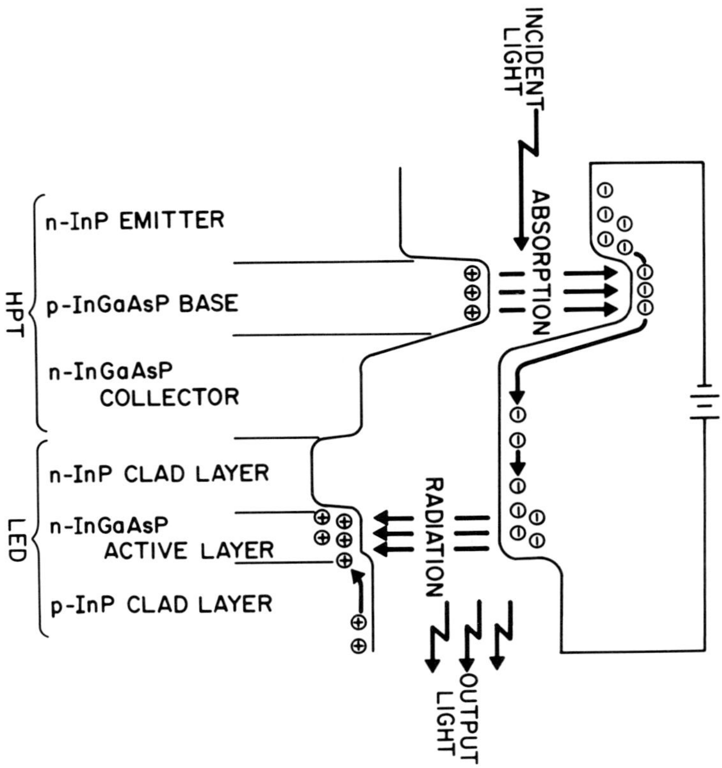

FIG. 24. Energy-band diagram of integrated HPT/LED structure. [After Sasaki *et al.* (1982).]

The number and variety of phototransistor structures that have been reported to date are indicative of the broad range of applications for which the phototransistor is suited. Further progress in discrete phototransistors and monolithic circuits that utilize phototransistors should follow as materials technologies mature and fabrication procedures improve.

VII. Photosensitivity of Field-Effect Transistors

To date, bipolar structures have been the subject of most of the research on phototransistors. Lately, however, the photosensivity of field-effect transistors (FETs) has received a great deal of attention. The advantages of this type of photodetector are very fast response and high optical gain, although these two properties have yet to be demonstrated simultaneously. So far, most of the work on the photoresponse of FETs has been limited to demonstrations of ultrahigh speed and attempts to identify the gain mechanism. Although there is little doubt regarding potential operation of FETs as photodetectors at gigabit data rates, the origin of the gain has been the subject of considerable controversy. Available evidence indicates that the phenomena of photosensitivity and gain in FETs is a complex combination of several mechanisms (Noad *et al.*, 1982), including photoconductivity with gain (Gammel and Ballantyne, 1979, 1980a); changes in the source–drain current resulting from the photovoltaic response of the gate junction (Sugeta and Mizushima, 1980; Wieder *et al.*, 1980) or the substrate-channel junction (Edwards, 1980); and electric-field-aided diffusion of photogenerated carriers from the substrate to the channel (Haythornthwaite, 1980). It is not the purpose of the present discussion to resolve this issue. We shall, however, describe some of the proposed mechanisms and present related experimental results.

Baack *et al.* (1977) first demonstrated the use of an FET as a high-speed photodetector. Using a GaAs metal–semiconductor gate field-effect transistor (MESFET) to detect an optical pulse from a GaAs injection laser, they observed a pulse width of 73 psec, which compared favorably with a pulse width of 178 psec obtained with a high-speed APD. No gain was reported for this work.

Optical gain in a GaAs MESFET photodetector was first reported by Gammel and Ballantyne (1979). In the first stage of this work, a standard GaAs MESFET (gate length, $\sim 1\ \mu m$; gate width, $\sim 200\ \mu m$; channel thickness, $\sim 0.35\ \mu m$; and channel-carrier concentration, $\sim 10^{17}\ cm^{-3}$) was used. Pulse-response measurements and measurements using the 641-MHz beat frequency of a He–Ne laser ($\lambda = 632.8$ nm) and a microwave receiver revealed a gain of 5 and indicated that the most probable explanation for the gain was photoconductivity in the high-field region underneath the gate. This process can be described as follows: Photogenerated electron–hole pairs are created in or near the high-field region. In III–V materials, holes typically have a saturated drift velocity slower than the peak electron velocity. As a result, the photogenerated electrons are quickly swept away leaving the holes. To neutralize the charge of the slowly moving holes, electrons are injected from the source, thus giving rise to an increase in the drain–source current. This process will continue until the photogenerated holes drift out of the high-field region, having been, in effect, collected at that point. A simplified expression for the optical gain in this case is

$$G = \eta(\tau_h/\tau_t), \tag{92}$$

where τ_h is the lifetime of the minority holes and τ_t the transit time for electrons.

Gammel and Ballantyne (1980a,b) confirmed that photoconductivity can be the dominate gain mechanism, at least at high frequencies, with a photodetector that was similar to a GaAs MESFET, except that the Schottky barrier gate was replaced with an etched notch. When a bias is applied between the drain-source contacts of this structure, a high-field region is created on the positive side of the highly resistive notch region. This field has the same effect as the high-field region near the MESFET gate. The photoresponse at 641 MHz was monitored as a focused He–Ne laser beam (diameter, $\sim 1.3\ \mu$m) was scanned between the source and drain. It was found that the peak of the photoresponse, and hence the maximum photoconductive gain, was always on the positive side of the notched region, i.e., the location of the high-field region. Gammel and Ballantyne (1980b) have also fabricated a sputtered 7059 glass/SiO_2 waveguide over the notch region to demonstrate that this type of photodetector may be useful for photonic circuits. The overall coupling efficiency between the waveguide and the detector was approximately 40%.

Sugeta and Mizushima (1980) have reported that their measurements appear to contradict the photoconductivity explanation. They used a 100-psec pulse from an $Al_xGa_{1-x}As$/GaAs laser ($\lambda \simeq 820$ nm) to illuminate the 2-μm space between the gate and drain of a GaAs FET. They suggested that the photoresponse mechanism could be explained as follows: The depletion layer between the drain and gate acts as a simple photodiode to sweep out photogenerated carriers. This gives rise to change in the gate current Δi_g, which is equal to the primary photocurrent i_{ph}. This current and the gate-load resistance R_g combine to change the gate voltage, which in turn modulates the drain current. The change in the drain current Δi_d is given by

$$\Delta i_d = (1 + g_m R_g) i_{ph}, \tag{93}$$

and the optical gain is

$$G = \eta(\Delta i_d \Delta i_g) = (1 + g_m R_g). \tag{94}$$

Sugeta and Mizushima obtained very good agreement between the gain determined by a direct measurement of Δi_d and Δi_g and the term $(1 + g_m R_g)$ calculated from the measured transconductance of the GaAs MESFET.

A similar observation to that of Sugeta and Mizushima has been reported by Wieder *et al.* (1980). In this work, an electron beam was used to create electron–hole pairs in a GaAs MESFET. The energy of incident electrons was varied between 1 and 6 kV, with current densities in the range from 10^{-5} to 10^{-3} A cm^{-2}. In contrast to the relatively low gains ($\lesssim 8$) observed by Gammel and Ballantyne (1980a,b) and by Sugeta and Mizushima (1980) for high frequencies and narrow excitation pulses, Wieder *et al.* (1980) obtained dc-current gains $\gtrsim 10^5$, a value too great to be explained in terms of photoconductivity. Furthermore, this high gain was maintained when the electron beam was pulsed up to 10^4 Hz. They suggested that their observations could be explained in terms of an analogy to the conventional photovoltaic effect observed in Schottky barriers. The electron-beam-generated electron–hole pairs are collected by the Schottky barrier gate. This reduces the gate bias and modulates the drain current by normal FET action. This interpretation is supported by measurements of the gain as a function of the electron-beam current.

Another photoresponse effect has been observed by Edwards (1980). Illumination ($\lambda = 632.8$ nm) of GaAs FET structures with and without gate metallization produced current gains of 2×10^4. The combination of high gain with no gate is inconsistent with previous explanations. Edwards determined that a large fraction of the incident light was being absorbed in the substrate beneath the channel. The field due to the step in the impurity concentration at the channel–substrate interface separates the photogenerated electron–hole pairs and results in a photoinduced voltage that Edwards was able to measure as a function of the incident light power. He demonstrated that as a result of this induced voltage, the substrate acts as a back gate that can effectively modulate the drain current.

Noad *et al.* (1982) have used a combination of optical and electron-beam excitation to clarify the role of the mechanisms just described. They found that the response could be explained in terms of two frequency regimes. For frequencies less than approximately 600 MHz, the photoresponse (gain) is a decreasing function of frequency. In this region, Noad *et al.* stated that although the change in gain with gate bias indicates that a photovoltaic effect in the Schottky barrier may play a role, line scans across the channel and the evidence of high gain in the absence of a gate electrode point to Edward's model of a photoinduced substrate voltage. The combination of device capacitance and high substrate resistance, however, reduces the importance of this effect at high frequencies. At high frequencies, transistor action gives way to the faster lower-gain photoconductive mechanism.

Chen *et al.* (1983) have reported an FET photodetector that has a more promising structure than a standard MESFET. A cross section of this device is shown in Fig. 25. The channel is formed by a lightly doped GaAs layer ($N_A \sim 1 \times 10^{14}$ cm^{-3}) and an n^--$Al_{0.3}Ga_{0.7}As$ layer. It has been shown that band bending at the interface between these two layers creates a triangular potential well that can confine electrons (Stormer *et al.*, 1979). The result is a thin (~ 100 Å) channel of electrons that exhibit high mobility as a result of reduced impurity scattering. The structure in Fig. 25 has the advantage of having a relatively thick GaAs absorbing layer ($\gtrsim 1$ μm) compared to a standard MESFET (~0.2–0.35 μm). A second advantage is the large gate-drain spacing (~8 μm), which permits improved optical coupling, a major limitation for high-speed MESFETs. The transconductance was determined to be 20 msec mm^{-1}, and the source-gate capacitance was 0.3 pF. An optical

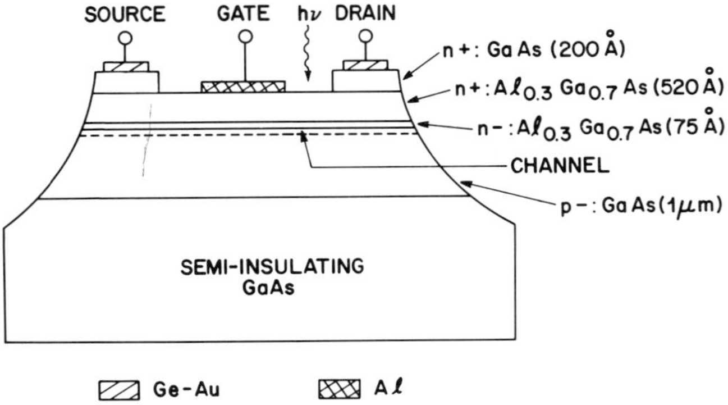

FIG. 25. Schematic cross section of a heterojunction optical FET. [From Chen *et al.* (1983).]

gain of 6.3 was measured with zero gate bias using 50-nsec pulses from an $Al_xGa_{1-x}As/GaAs$ semiconductor laser ($\lambda \sim 8200$ Å). Chen *et al.* reported that their gain data were inconsistent with the model proposed by Sugeta and Mizushima (1980) but were adequately explained by the photoconductivity model. The response time was measured with the channel pinched off in order to eliminate a slow falltime component. Under these conditions, no gain was reported, but a pulse width of 27 psec was obtained. A long-wavelength version of this device has also been fabricated using an $Al_{0.48}Ga_{0.52}As/In_{0.53}Ga_{0.47}As$ heterojunction. Initial results indicate that this type of photodetector has promise for photonic applications; however, additional research on the gain mechanisms and noise characteristics is needed.

Although the work on the photoresponse of FET structures is presently in its infancy, it is already clear that this type of photodetector is capable of ultrafast response with moderate gains. The experience and expertise that has been developed for FET logic circuits should prove useful in the development of optical FET structures. This type of device appears to be well suited for photonic circuits, where a simple structure that can be easily integrated with other devices is required.

VIII. Summary

The fact that a phototransistor is an integrated $p-i-n$ photodetector and transistor amplifier makes it an attractive device for discrete lightwave applications as well as for photonic circuits. It has been shown that useful gains (~ 100) can be achieved at the low light levels encountered in lightwave receivers and that even higher gains can be achieved for higher optical signal levels. As with any $p-i-n$/amplifier combination, the bandwidth is limited by the response of the transistor amplifier; however, phototransistors have been demonstrated with gain–bandwidth products as high as 10 GHz, which should prove adequate for many applications. The fact that a number of modifications of the standard wide-band-gap-emitter bipolar structure have been proposed to improve the device characteristics for specific applications and initial integrated structures designed to perform more complex functions have been reported reflects the utility of the phototransistor in a rapidly developing field of research.

REFERENCES

Alferov, Z. I., Akhmedov, F., Korol'kov, V., and Niketin, V. (1973). *Sov. Phys.—Semicond. (Engl. Transl.)* **7,** 780.

Anderson, R. L. (1962). *Solid-State Electron.* **5,** 341.

Anderson, R. L. (1971). *Proc. Int. Conf. Phys. Chem. Semicond. Heterojunctions, Layer Struct., Budapest, 1970* **2,** 55.

Baack, C., Elze, G., and Walf, G. (1977). *Electron. Lett.* **13,** 193.

Beneking, H. (1981). *IEEE Electron Device Lett.* **EDL-1,** 99.

Beneking, H., Mischel, P., and Schul, G. (1976). *Electron. Lett.* **12,** 395.

Beneking, H., Grote, N., Roth, W., and Svilans, M. N. (1980a). *Electron. Lett.* **16,** 602.

Beneking, H., Grote, N., Roth, W., Su, L. M., and Svilans, M. N. (1980b). *Conf. Ser.—Inst. Phys.* No. 56, p. 385.

Beneking, H., Grote, N., and Svilans, M. N. (1981). *IEEE Trans. Electron. Devices* **ED-28,** 404.

Brain, M. C., and Smith, D. R. (1983). *IEEE J. Quantum Electron.* **QE-19,** 1139.

Campbell, J. C., and Dentai, A. G. (1982). *Appl. Phys. Lett.* **41,** 192.

Campbell, J. C., and Ogawa, K. (1982). *J. Appl. Phys.* **253,** 1203.

Campbell, J. C., Dentai, A. G., Burrus, C. A., and Ferguson, J. F. (1980). *Tech. Dig.—Int. Electron Devices Meet.* p. 526.

Campbell, J. C., Burrus, C. A., Dentai, A. G., and Ogawa, K. (1981). *Appl. Phys. Lett.* **39,** 820.

Campbell, J. C., Qua, G. J., Copeland, J. C., and Dentai, A. G. (1982). *Appl. Phys.* **53,** 5182.

Campbell, J. C., Dentai, A. G., Qua, G. J., and Ferguson, J. F. (1983). *IEEE J. Quantum Electron.* **QE-19,** 1134.

Capasso, F., Tsang, W. T., Bethea, C. G., Hutchinson, A. L., and Levine, B. F. (1983). *Appl. Phys. Lett.* **42,** 93.

Chen, C. W., and Gustafson, T. K. (1980). *Appl. Phys. Lett.* **37,** 1014.

Chen, C. Y. (1981). *Appl. Phys. Lett.* **39,** 979.

Chen, C. Y., Cho, A. Y., Garbinski, P. A., and Levine, B. F. (1981a). *Appl. Phys. Lett.* **39,** 340.

Chen, C. Y., Cho, A. Y., Garbinski, P. A., and Bethea, C. G. (1981b). *IEEE Electron Device Lett.* **EDL-2,** 290.

Chen, C. Y., Cho, A. Y., Garbinski, P. A., and Bethea, C. G. (1982a). *Appl. Phys. Lett.* **40,** 510.

Chen, C. Y., Cho, A. Y., Bethea, C. G., Garbinski, P. A., Pang, Y. M., and Levine, B. F. (1983). *Appl. Phys. Lett.* **42,** 1040.

Cooke, H. F. (1971). *Proc. IEEE* **59,** 1163.

De La Moneda, F. H., Chenette, E. R., and Van der Ziel, A. (1971). *IEEE Trans. Electron Devices* **ED-18,** 340.

Dingle, R. (1975). *Festkoerperprobleme* **15,** 21.

Early, J. M. (1952). *Proc. IRE* **40,** 1401.

Edwards, W. D. (1980). *IEEE Electron Device Lett.* **EDL-1,** 1980.
Forrest, S. R., and Kim, O. K. (1981). *J. Appl. Phys.* **52,** 5838.
Fritzsche, D., Kuphal, E., and Aulbach, R. (1981). *Electron. Lett.* **17,** 178.
Gammel, J. C., and Ballantyne, J. M. (1979). *Tech. Dig.—Int. Electron Devices Meet.* p. 120.
Gammel, J. C., and Ballantyne, J. M. (1980a). *Jpn. J. Appl. Phys.* **19,** L273.
Gammel, J. C., and Ballantyne, J. M. (1980b). *Appl. Phys. Lett.* **36,** 149.
Haythornthwaite, R. F. (1980). Ph.D. Thesis, Carleton Univ., Ottawa, Ontario.
Hovel, H. J., and Milnes, A. G. (1968). *Int. J. Electron.* **25,** 201.
Hovel, H. J., and Milnes, A. G. (1969). *IEEE Trans. Electron Devices* **ED-16,** 766.
Konagai, M., Katsukawa, K., and Takahashi, K. (1977). *J. Appl. Phys.* **48,** 4389.
Krautle, H., Narozny, P., and Beneking, H. (1982). *Electron. Lett.* **18,** 259.
Kroemer, H. (1957a). *RCA Rev.* **18,** 332.
Kroemer, H. (1957b). *Proc. IRE* **45,** 1535.
Kroemer, H. (1982). *Proc. IEEE* **70,** 13.
Lee, S. C., and Pearson, G. L. (1981). *J. Appl. Phys.* **52,** 275.
May, G. A. (1968). *Solid-State Electron.* **11,** 613.
Milano, R. A. (1979). Ph.D. Thesis, Univ. of Illinois, Urbana-Champaign.
Milano, R. A., Windhorn, T. H., Anderson, E. R., Stillman, G. E., Dupuis, R. D., and Dapkus, P. D. (1979). *Appl. Phys. Lett.* **34,** 562.
Milano, R. A., Dapkus, P. D., and Stillman, G. E. (1982). *IEEE Trans. Electron Devices* **ED-29,** 266.
Miller, S. L. (1957). *Phys. Rev.* **105,** 1246.
Miller, S. L., and Ebers, J. J. (1955). *Bell Syst. Tech. J.* **34,** 883.
Moriizumi, T., and Takalashi, K. (1972). *IEEE Trans. Electron Devices* **ED-19,** 152.
Naitoh, M., Sakai, S., and Umeno, M. (1982). *Electron. Lett.* **18,** 656.
Noad, J. P., Hara, E. H., Hum, R. H., and MacDonald, R. I. (1982). *IEEE Trans. Electron Devices* **ED-29,** 1792.
Reddi, V. C. K. (1967). *Solid-State Electron.* **10,** 305.
Ryvkin, S. M. (1964). "Photoelectric Effects in Semiconductors," p. 376. Consultants Bureau, New York.
Sakai, S., Naitoh, M., Kobayashi, M., and Umeno, M. (1983). *IEEE Trans. Electron Devices* **ED-30,** 404.
Sasaki, A., and Kuzuhara, M. (1981). *Jpn. J. Appl. Phys.* **20,** L283.
Sasaki, A., Matsuda, K., Kimura, Y., and Fujita, S. (1982). *IEEE Trans. Electron Devices* **ED-29,** 1382.
Schuster, M. A., and Strull, G. (1966). *IEEE Trans. Electron Devices* **ED-13,** 907.
Shive, J. N. (1949). *Phys. Rev.* **76,** 575.
Shive, J. N. (1953). *J. Opt. Soc. Am.* **43,** 239.
Shockley, W. (1951). U.S. Patent 2, 569, 347.
Shockley, W., Sparks, M., and Teal, G. K. (1951). *Phys. Rev.* **83,** 151.
Smith, R. D., and Personick, S. D. (1980). *In* "Semiconductor Devices for Optical Communications" (H. Kressel, ed.), p. 89. Springer-Verlag, Berlin and New York.
Sparkes, J. J., and Beaufoy, R. (1957). *Proc. IRE* **45,** 1740.
Stormer, H. L., Dingle, R., Gossard, A. C., Wiegmann, W., and Sturge, M. D. (1979). *Solid State Commun.* **29,** 705.
Sugeta, T., and Mizushima, Y. (1980). *Jpn. J. Appl. Phys.* **19,** L27.
Sze, S. M. (1969). "Physics of Semiconductor Devices," p. 69. Wiley, New York.
Tabatabaie-Alavi, K., and Fonstad, C. G. (1981). *IEEE J. Quantum Electron.* **QE-17,** 2259.
Tabatabaie-Alavi, K., Markunus, R. J., and Fonstad, C. G. (1979). *Tech. Dig.—Int. Electron Devices Meet.* p. 643.

Tabatabaie-Alavi, K., Choudhury, A. N. M. M., Alavi, K., Vlcek, J., Slater, N. J., Fonstad, C. G., and Cho, A. Y. (1982). *IEEE Electron Device Lett.* **EDL-3,** 379.
Tsyrlin, L. E. (1977). *Fiz. Tekh. Poluprovodn. (Leningrad)* **11,** 1926 [*Sov. Phys.—Semicond. (Engl. Trans.)* **11,** 1127 (1977)].
Wieder, H. H., Davis, N. M., and Flesner, L. D. (1980). *Appl. Phys. Lett.* **37,** 943.
Wright, P. D., Nelson, R. J., and Cella, T. (1980). *Appl. Phys. Lett.* **37,** 192.

Index

Contents of Volume 22

Part A

Part B

Part C

Part D

Part E